RENAL
EICOSANOIDS

ADVANCES IN EXPERIMENTAL MEDICINE AND BIOLOGY

RENAL EICOSANOIDS

Edited by

Michael J. Dunn

Case Western Reserve University
and University Hospitals of Cleveland
Cleveland, Ohio

Carlo Patrono

Catholic University School of Medicine
Rome, Italy

and

Giulio A. Cinotti

University of Rome "La Sapienza"
Rome, Italy

PLENUM PRESS • NEW YORK AND LONDON

Library of Congress Cataloging in Publication Data

Renal eicosanoids / edited by Michael J.Dunn, Carlo Patrono, and Giulio A. Cinotti.
    p.   cm. — (Advances in experimental medicine and biology; v. 259)
    "Proceedings based on a symposium on renal eicosanoids, held June 9–11, 1987, in
Capri, Italy"—T.p. verso.
    Includes bibliographical references.
    ISBN 0-306-43320-6
    1. Kidneys—Pathophysiology—Congresses. 2. Eicosanoic acid—Derivatives—
Pathophysiology—Congresses. 3. Kidney—metabolism—congresses. I. Dunn, Michael
J., date. II. Patrono, Carlo. III. Cinotti, Giulio A. IV. Series
    [DNLM: 1. Arachidonic Acids—metabolism—congresses. 2. Eicosanoic Acids—
metabolism—congresses. 3. Kidney—drug effects—congresses. 4. Kidney Diseases—
physiopathology—congresses. WD1 AD559 v. 259 / QU 90 R393 1987]
RC903.9.R4722   1989
616.6′107—dc19
DNLM/DLC   89-16264
for Library of Congress   CIP

Based on a symposium on Renal Eicosanoids,
held June 9–11, 1987, in Capri, Italy

© 1989 Plenum Press, New York
A Division of Plenum Publishing Corporation
233 Spring Street, New York, N.Y. 10013

Printed in the United States of America

ACKNOWLEDGEMENTS

We gratefully acknowledge the generous financial support of Merck, Sharp & Dohme who made possible this book as well as the scientific meeting held in June, 1987 in Capri, Italy. This text is not a summary of the meeting but rather represents reviews of the current state of knowledge in the field of renal eicosanoids. Norma Minear provided expert secretarial and editorial assistance in the preparation of the camera-ready copy. Blanche Young was an excellent editorial assistant who reviewed all of the text for accuracy and compiled the index.

v

CONTENTS

THE CELL BIOLOGY OF FIBROBLAST CYCLOOXYGENASE

Amiram Raz, Angela Wyche,
Diana Fagan and Philip Needleman

Department of Pharmacology
Washington University School of Medicine
St. Louis, Missouri 63110

ABSTRACT

We have prepared polyclonal antisera against sheep seminal vesicles cy-clooxygenase (COX) which cross-reacted with human COX. We employed this antisera in studies with human dermal fibroblast cultures to immunoprecipi-tate selectively the COX enzyme. Labeling of the cells with $[^{35}S]$-methio-nine, solubilization of cellular COX followed by its immunoprecipitation, SDS-PAGE electrophoresis and fluorography enabled us to determine directly the synthetic rate of COX protein and its modulation by the monokine inter-leukin-1 (IL-1).

The immunoprecipitated $[^{35}S]$-labeled COX, as judged from SDS-PAGE electrophoresis, has a molecular size of approximately 73,000 daltons, simi-lar to that of native sheep COX and $[^{3}H]$-acetyl COX. IL-1 stimulation of enhanced COX synthesis was time and dose dependent; as little as 0.03 units/ml of IL-1 produced significant stimulation of $[^{35}S]$-labeled COX synthesis. Maximum stimulation was 3-10-fold after preincubation of the cells with IL-1 for 12-16 hours.

IL-1 treatment of cells in serum-free media yielded parallel dose re-sponse curves for stimulation of PGE_2 formation, cellular solubilzed COX activity and synthesis of newly formed COX, suggesting that this IL-1 effect is mediated solely via induction of new COX protein synthesis. In contrast, IL-1 effect on cells incubated in the presence of fetal calf serum is more complex. Serum synergistically augments the IL-1 effect on PGE_2 synthesis

1

in intact cells but concurrently blunts IL-1 induction of COX synthesis, thus suggesting that a factor (or factors) in serum may stimulate PGE_2 production by activating cellular phospholipase(s).

INTRODUCTION

The significance of increased arachidonic acid metabolism in inflammatory disorders has been the subject of numerous studies in recent years. Marked qualitative and quantitative changes in eicosanoid production has been demonstrated in several pathophysiological conditions and tissue injury models including hydronephrosis (1,2), renal vein constriction (3), glomerulonephritis (4), ulcerative colitis (5), rheumatoid arthritis (6), myocardial infarction (7) and pulmonary fibrosis (8). In each of these inflammatory disorders, the enhanced eicosanoid generation is closely associated with the invasion of inflammatory cells into the injured tissue. Our laboratory has been actively involved in studies aimed at understanding the temporal relationships between inflammatory cell influx and exaggerated arachidonate metabolism and correlating biochemical events with the onset and development of tissue injury.

The main renal eicosanoid released in response to a variety of stimuli is PGE_2, a vasodilator and natriuretic agent thought to play a role in the regulation of intrarenal blood flow. Unilateral ureteral obstruction in the rabbit results in increased basal PGE_2 production. When the hydronephrotic kidney (HNK) is perfused *ex vivo*, it exhibits greatly enhanced PGE_2 release in response to stimulatory agents such as bradykinin (BK), angiotensin II, and norepinephrine (1). Furthermore, the HNK, but not the contralateral unobstructed kidney (CLK), has the capacity to synthesize and release the powerful vasoconstrictor thromboxane A_2 (TxA_2) in response to BK stimulation (2). The TxA_2 synthesizing activity is especially pronounced in the HNK cortex (9), the site of preglomerular vasoconstriction.

Histologically, the greatly enhanced arachidonic metabolism in hydronephrosis is accompanied by increased proliferation of fibroblast-like interstitial cells and invasion of inflammatory cells, most notably mononuclear cells, into the renal cortex and outer medulla (10,11). From recent data obtained in our laboratory (12), we propose that a time-dependent sequential invasion of the injury site by various blood cells is taking place. Studies using Indium-111-labeled platelets and leukocytes have shown that following ureter ligation, there is an immediate (within 10 min.) preferential

2

accumulation of platelets in the HNK compared with the CLK. Platelet accumulation maximized 30 min. after ureter obstruction and returned to control level after 2 hours. HNK accumulation of indium-111 polymorphonuclear leukocytes (PMN) was clearly seen 12 hours after ligation, was maximal at 24 hours and returned to control level by 72 hours. The next invading cell, the monocyte, is first detected in the HNK 24 hours after ligation, and its level is maintained for a minimum of 72 hours of obstruction.

Several lines of evidence support a biochemical-cellular relationship between monocyte invasion of the injured tissue and the enhanced eicosanoids production. Ureteral ligation in the cat is not accompanied by inflammatory cell invasion or exaggerated eicosanoid production (13). In addition, in rabbits rendered leukopenic by nitrogen mustard treatment prior to ureteral ligation, agonist-stimulated arachidonate metabolism was not enhanced nor was cortical TxA_2 synthase activity induced (14). Finally, following release of ureteral obstruction, we observed histologically the loss of cortical macrophages and decreased BK stimulation of eicosanoid generation. The possible role of tissue macrophages in upregulating eicosanoid production was verified by the use of the lipopolysaccharide endotoxin (LPS), a previously demonstrated macrophage agonist for eicosanoid production (15,16). LPS administration to the perfused rabbit HNK caused an immediate and prolonged release of TxA_2 and PGE_2. Neither the perfused CLK nor the post-obstructed HNK nor the cat HNK showed similar enhanced release, this being consistent with the absence or greatly reduced number of invading macrophages in these kidney tissue preparations.

The body of evidence outlined above leads us to hypothesize that a sequential release of cell-specific mediators causes the orderly recruitment of platelets, PMN, and macrophages. The initial platelet accumulation and adhesion at the injured site would be accompanied by release of the vasoconstrictor TxA_2 as well as platelet-derived growth factor, a potent chemoattractant for PMN. The PMN is capable of releasing LTB_4, a potent chemotactic agent for monocytes, thereby possibly facilitating their recruitment from the circulating blood into the site of injury. Monocytes, in turn, can release a variety of cell-active products including eicosanoids (17) and monokines (18) which can affect neighboring cells. Indeed, macrophages were previously shown to stimulate proliferation and PGE_2 production of fibroblasts (19). Furthermore, the media from adherent human blood mononuclear cells contain a factor (or factors) that cause a marked stimulation of PGE_2 production by dermal fibroblasts (19), gingival fibroblasts (20), and lung fibroblasts (8).

METHODS AND MATERIALS

Preparation of Renal Cortex Mixed Cell Cultures from Kidney Explants

Cortical cell cultures were prepared and maintained as described previously (21). Thirteen-day-old cultures from HNK and CLK were employed.

Dermal Fibroblast Cultures

Human dermal fibroblasts obtained from the American type cell culture collection (Rockville, MD, USA) were employed in the initial study involving mononuclear cell factors. The cells were maintained, as described previously (22). In subsequent studies with IL-1, fresh normal human skin specimens removed at surgery were used to obtain dermal fibroblasts according to published procedures (23,24). The cells were grown in α-MEM media containing 10% FCS, 1 mM glutamine, 1 mM HEPES pH=7.2, penicillin (100 units/ml) and streptomycin (100 µg/ml). The media was changed twice weekly, and the cells passed when they reached confluency. All cell cultures were grown at 37° in a controlled humidified incubator under 95% air and 5% CO_2.

Preparation of Rabbit Antisera Raised Against Sheep Seminal Vesicle Cyclooxygenase

Cyclooxygenase (COX) from sheep seminal vesicles was purified according to previously described methods (25,26) or purchased from Oxford Biomedical Research (Oxford, MI, USA). Both the purified and the commercial preparation were judged to be at least 95% pure by sodium dodecyl sulfate-polyacrylamide gel electrophoresis (SDS-PAGE) and silver staining.

Rabbits were initially immunized with 100 µg COX dissolved in 0.5 ml of Tris buffer (50 mM, pH=7.4) which was emulsified with equal volume of complete Freund's adjuvant and injected s.c. at multiple sites on the back. Booster injections of 50 µg COX in incomplete Freund's adjuvant were given 7, 14, and 24 days later. A fourth booster was given 60 days after the first immunization, and the rabbits were bled 2 weeks later.

Preparation of ^{125}I-COX and [^{3}H]-acetyl COX.

Iodine-125-labeled sheep seminal vesicles COX was prepared and purified as described previously by Tai and Tai (27). In brief, 25 µg of purified sheep COX were reacted with 1 mCi of carrier-free ^{125}I and chloramine T. The reaction product was purified first on a G-25 (1 x 5 cm) column eluted with 50 mM Tris buffer (pH=7.5) containing 0.1% Tween 20. Material eluted in the void volume (approximately 0.65 mC) was further purified on hydroxy-apetite-cellulose column (0.8 x 6 cm), as described (27). The final protein fraction contained 0.18 mC of ^{125}I-COX. [^{3}H]-acetyl salicylic acid, specific activity 200 mCi/mM, was prepared as described by Roth (28). [^{3}H]-acetic anhydride, 100 mCi of specific activity 400 mCi/mM (NEN, Boston, MA, USA), was reacted with salicylic acid. The [^{3}H]-aspirin obtained was purified as described (28) with final yield of 31% and specific activity of approximately 200 µCi/mM. The acetylated microsomal preparation was solubilized with 1% Tween 20. SDS-PAGE analysis of the solubilized microsomal preparation revealed one major [^{3}H]-labeled band with molecular size identical to that of purified COX.

Culture and ^{35}S-Methionine Labeling of Human Fibroblasts

Human dermal fibroblasts were grown and passed as described above. For experiments, fibroblasts were plated in 6-well, 35 mm culture plates at 4-5 x 10^4 cells/well. The cells were grown in 3 ml of α-MEM media containing 10% fetal calf serum (FCS) (Gibco, Grand Island, NY, USA). Subconfluent cultures (6-8 x 10^4 cells/well, 2-3 days after plating) were used for experiments which were initiated by replacing the α-MEM-10% FCS medium with DMEM medium. Cells were preincubated in 1 ml DMEM for 6-24 hours with 0-2 units/ml of human recombinant interleukin-1 (IL-1) (Cistron, Pinebrook, NJ, USA). Following preincubation, the media were removed, and 1 ml of methionine-poor DMEM (19 parts methionine-free DMEM, 1 part DMEM) was added, followed by addition of the same IL-1 dose and 100 µCi of ^{35}-S-methionine (Amersham, Arlington Heights, IL, USA), respectively. After 6 hours of labeling, the media were removed, the cells quickly washed with 1 ml of phosphate buffered saline (PBS), pH=7.4, and scraped in 1.3 ml of PBS. The cells were washed with 1 ml of cold PBS, and the pellet was kept at -20^0 until further processing. Fibroblast pellets were sonicated in 150 µl solubilization buffer (50 mM Tris, 1 mM diethyldithiocarbamic acid-sodium salt, 10 mM EDTA, 1% Tween 20, pH=8.0, containing 0.2mg/ml α_2-macroglobulin), which effectively solubilized the microsomal COX. The sonicate was centrifuged at 100,000 x g for 10 min. to precipitate residual unsolubilized cellular membranes, and the supernatant was used for immunoprecipitation and activity determination, as described below.

Double Antibody Immunoprecipitation

Titer determination of antisera raised against sheep seminal vesicles COX was done using a double antibody immunoprecipitation. COX (0-16 ng), in 200 µl of Na-phosphate buffer, pH=7.4, containing 1% normal rabbit serum, was incubated for 2 hours at $4°$ with antisera to be titered (final anti-sera dilution 10^{-3}-10^{-4}). Two hundred microliters of goat anti-rabbit IgG (Sigma) diluted 1:4 in Na-phosphate buffer were added and the incubation continued overnight at $4°$, after which the tubes were spun for 5 min. and the supernatant assayed for remaining unprecipitated COX activity, as de-scribed below.

Immunoprecipitation and Quantitation of ^{35}S-Methionine Labeled COX

Purification and quantitation of ^{35}S-methionine labeled COX from fibro-blasts was done by immunoprecipitation, SDS-PAGE electrophoresis, and fluo-rography. The Staph. A immunoprecipitation method used was as described elsewhere (28). In brief, an aliquot (10-30 µl) of tissue homogenate or cell sonicate is diluted with extraction buffer (pH=9.0) to a final volume of 100-150 µl. One microliter of antisera is then added and the mixture in-cubated for 1 hour at $4°$. Staph. A immunoprecipitin (BRL, MD, USA) suspen-sion (30µl) is then added, the tube mixed and left at $4°$ for 30 min. The Staph. A pellet containing bound rabbit IgG (and COX bound to it) is isolat-ed by centrifugation, washed extensively, suspended in nonreducing SDS-PAGE sample buffer (30), and vigorously vortexed. The Staph. A particles are then pelleted down by centrifugation, the supernatant heated to $70°$ for 10 min. and analyzed by 10% SDS-PAGE electrophoresis (30). Following electro-phoresis, the gels are soaked in 40% methanol-10% acetic acid and either stained with Silver staining (Biorad, Richmond, CA, USA) or soaked in 1 M salicylate, dried and exposed to x-ray film (XAR-5, Kodak) under conditions in which band intensity was proportional to radioactivity and exposure time. The positions of the various radioactive bands were calibrated by the use of unlabeled or ^{14}C-labeled proteins (NEN, Boston, MA, USA) which in-cluded ^{14}C-ovalbumin (Mr 46,000 daltons), ^{14}C-bovine serum albumin (Mr 69,000 daltons) and ^{14}C-phosphorylase B (Mr 97,400 daltons). Relative intensities of radioactive COX bands obtained by fluorography were deter-mined by laser scanning of the x-ray films. Preliminary experiments have shown that in x-ray films that were not overexposed, a linear correlation exists between the radioactivity in the band and its intensity, as measured by laser scanning of the film.

6

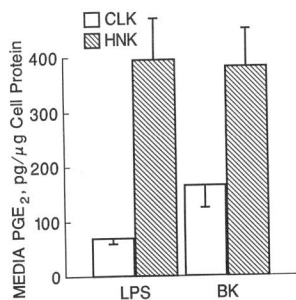

Fig. 1.

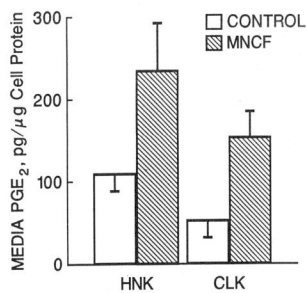

Fig. 2.

Fig. 1. Effect of endotoxin (LPS) and Bradykinin (BK) on PGE_2 biosynthesis by explant cultures from rabbit HNK and CLK (21). Cultures (13 days old) were incubated in the absence or presence of LPS (50 μg/ml) for 24 hours or BK (1 nM) for 1 hour. Aliquots of the media were assayed for PGE_2 by radioimmunoassay.

Fig. 2. Effect of monocytes conditioned media on mixed cultured cells from rabbit HNK and CLK explants (21). HNK and CLK cultures were incubated in 10% control media or in 10% monocytes conditioned media (termed mononuclear cell factor, MNCF). MNCF was generated and prepared as described by us (21). Following incubation for 24 hours at 37°, the media were assayed for PGE_2.

Assay of COX Enzymatic Activity

    Samples of COX, following double antibody immunoprecipitation or of solubilized fibroblasts sonicates, were assayed for COX activity by measuring conversion of added arachidonic acid to PGE_2. Sample aliquots were diluted in SDS-free extraction buffer (29) containing 1 mM epinephrine, 1 MM phenol (pH=9.0). Incubations were initiated by the addition of arachidonic acid (Na-salt solution, pH=9.0) at a final concentration of 100 μM, and allowed to proceed for 30 min. at 37°. Aliquots of the incubation mixture were assayed for synthesis of PGE_2 by radioimmunoassay.

RESULTS

Effect of Mononuclear Cell Factor (MNCF) on Mixed Cell Cultures from HNK and CLK

    Recent studies in our laboratory have examined the effect of monocytes and monocyte-derived media on generation of arachidonate metabolites by the rabbit HNK cortex explants (21,22). Histologically, cell cultures from cortical explants of the HNK contained fibroblast-like cells and macrophages,

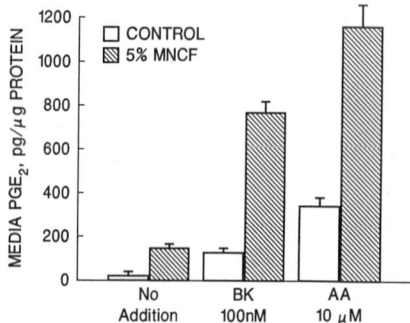

Fig. 3. Effect on mononuclear cell factor (MNCF) on BK-stimulated PGE_2 generation by fibroblasts (22). Fibroblasts preincubated with either 5% control medium or 5% MNCF were stimulated for 1 hour with either BK (100 nM) or arachidonic acid (AA, 10 μM), and the medium assayed for PGE_2. All values obtained in MNCF-treated cells were statistically significant (p $\leq$ 0.01) from control-treated cells.

whereas explants from CLK contained only fibroblasts. Endotoxin-lipolysaccharide (LPS) stimulation of these cultures resulted in increased PGE_2 release from the HNK but not from the CLK-derived cultures (Fig. 1). Significantly, BK stimulation of PGE_2 release was also markedly elevated in the HNK cultures as compared with the CLK cultures. This monocytes-dependent stimulation could also be demonstrated in monocytes-free fibroblasts cultures from CLK if conditioned media from monocytes presumed to contain mononuclear cell factor (MNCF) were added to the CLK-derived fibroblast cultures (Fig. 2). In studies with pure culture of dermal fibroblasts, MNCF strongly augmented the stimulation of PGE_2 release by both BK and exogenous arachidonic acid (Fig. 3). The MNCF effect was blocked by inhibitors of protein synthesis. Furthermore, MNCF increased the Vmax but not the Km of fibroblasts microsomal cyclooxygenase (22). Taken together, these observations strongly suggest that MNCF stimulates synthesis of fibroblasts cyclooxygenase. Further characterization of MNCF showed parallel time and dose-dependent effects of this factor to those of recombinant human interleukin 1 (IL-1) (Fig. 4). MNCF and IL-1 activities comigrated on gel filtration chromatography (Mr 12,000-18,000 daltons) and possessed similar potency for stimulating proliferation of D-10 lymphocytes (24).

The emerging hypothesis which we propose to test and characterize involves the intimate interaction between monocytes-macrophages and fibroblasts or fibroblast-like interstitial cells in the inflamed tissue site. We hypothesize that activated macrophages release IL-1 and possibly other monokines, which in turn stimulate adjacent fibroblasts and fibroblast-like interstitial cells to proliferate and to produce vastly enhanced cyclooxygenase products via direct induction of increased rate of cyclooxygenase

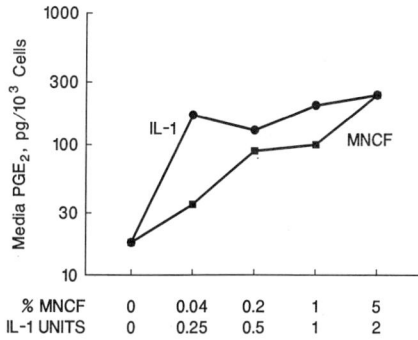

Fig. 4. Comparative biological potency of MNCF and human IL-1 on fibroblasts PGE_2 production (30). Fibroblast cultures were incubated for 72 hours in media containing either 0-5% MNCF or 0-2 units/ml of IL-1 and the medium PGE_2 determined by radioimmunoassay.

synthesis. The studies described below provide mechanistic support of this hypothesis. These studies employed an antisera specific for COX, with which cellular COX can be quantitated and its activity determined. Furthermore, the synthetic rate of cellular COX can be determined by measuring the incorporation of ^{35}S-methionine into the newly formed enzyme.

Characterization of Antisera for COX

Initial studies to determine the titer of the rabbit antisera raised against sheep seminal vesicles COX employed the double antibody immunoprecipitation. Purified COX (0.5-16 ng) was incubated with various dilutions of the antisera as described in Methods. Aliquots of the final supernatant following immunoprecipitation were assayed for remaining COX activity. Results of these assays (Fig. 5) indicate that under the incubation conditions employed, the antisera is effective in quantitatively precipitating up to 15 ng of sheep seminal vesicles COX at a final antisera dilution of 10^{-3}. COX activity could also be quantitatively recovered in the immunoprecipitated pellet, indicating that the antibodies in the antisera we prepared were not directed towards the COX active site.

We next determined the capacity of the antisera to precipitate COX in other tissues and cells extracts. The results (Fig. 5, insert) show that the antisera recognizes sheep and human COX but not mouse or rabbit COX.

Selective Immunoprecipitation of ^{35}S-Methionine-Labeled FB-COX

Primary cultures of dermal fibroblasts were grown as described in Methods. Cells were stimulated with IL-1 (2U/ml) for 16 hours, and then labeled with ^{35}S-methionine for 6 hours. The solubilized cell sonicate

9

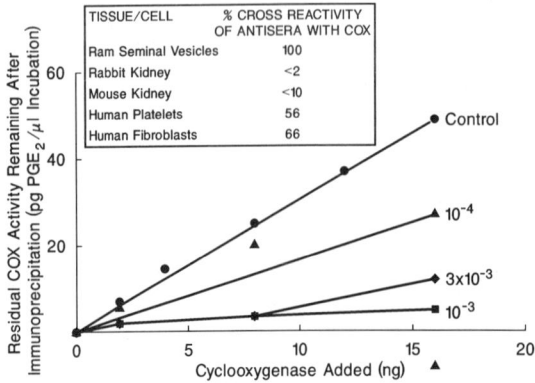

Fig. 5. Determination of COX antisera potency. Purified COX from sheep vesicular glands (0-16) were immunoprecipitated with various dilutions of antisera raised in rabbits. For details, see Methods. Insert: Ability of sheep COX antisera to immunoprecipitate COX from other tissue and cells. Microsomes from sheep seminal vesicles, rabbit kidney medulla, mouse kidney medulla, human platelets and human dermal fibroblasts were solubilized (50 mM tris buffer, pH=8.0 containing 1% Tween 20. Aliquots of the solubilized fractions (containing the bulk of cellular COX activity), with activity equal to that of diluted fractions from sheep seminal vesicles, were immunoprecipitated with the rabbit anti-COX sera, and the residual COX activity determined by incubation with arachidonic acid and assaying for PGE_2 formed. The extent of immunoprecipitation of sheep COX was taken as 100% cross reactivity.

was spun at 100,000 x g and the supernatant subjected to SDS-PAGE electrophoresis before and after immunoprecipitation. The electrophoresis results are shown in Fig. 6. From the numerous radioactive proteins in the cell sonicate, the COX antisera immunoprecipitated a band of molecular size slightly greater than BSA (Mr=69,000 daltons). Furthermore, this band had virtually identical electrophoretic mobility to that of purified COX, [^{3}H]-acetylated COX, and ^{125}I-COX (Fig. 7) with approximate Mr of 75,000 daltons. The antisera also variably precipitated a band with molecular size slightly smaller than BSA (seen in Fig. 7 but more clearly in Fig. 8). This band is a major protein being labeled during incubation with ^{35}S-methionine (Fig. 7) and appears to have high affinity for Staph. A even in the absence of anti-COX since, unlike COX, it is also precipitated by Staph. A when preimmune rabbit serum is used. It is therefore labeled nonspecific band.

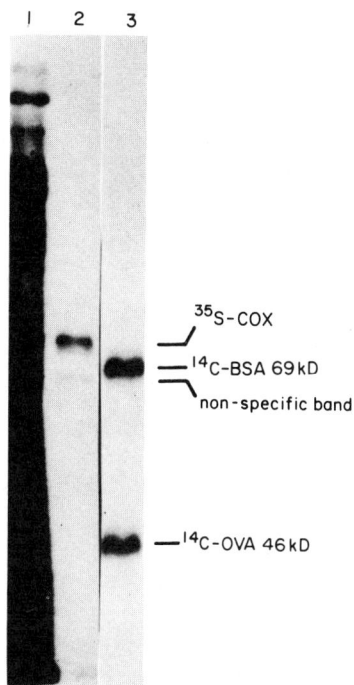

Fig. 6. Characterization of immunoprecipitated ^{35}S-labeled COX
band from fibroblasts. Fibroblasts were preincubated with
IL-1 (2 u/ml) for 18 hrs,
prior to labeling with ^{35}S-
methionine. The resulting
solubilized cell sonicate was
either immunoprecipitated or
directly separated by SDS-PAGE
electrophoresis (for details,
see Methods). Lane 1: cell
sonicate without prior immunoprecipitation. Lane 2: cell
sonicate after immunoprecipitation. Lane 3: ^{14}C-labeled
marker proteins.

The selectivity of the COX antisera for precipitating the ^{35}S-methio-
nine-labeled COX from the solubilized fibroblasts supernatant was further
validated by comparing the displacement of ^{125}I-COX and of ^{35}S-labeled
COX band by added unlabeled purified COX (Fig. 8). The displacement of
^{125}I-COX (Fig. 8) followed the expected sigmoidal shape curve, with 50%
displacement being obtained at 210 ng of unlabeled COX. A similar shape
displacement was obtained for the ^{35}S-labeled FB COX band of fibroblasts
with 50% displacement of the labeled COX at 170 nG (Fig. 8-B). Noteworthy
is the fact the unlabeled COX displaced only the COX band, without affecting
precipitation of the nonspecific band migrating just below BSA (Fig. 8-B,
insert.)

Effect of IL-1 on Fibroblasts Cyclooxygenase

The effect of preincubation time with IL-1 on fibroblast production of
PGE$_2$ and induction of new cyclooxygenase enzyme synthesis was evaluated by
measuring three different parameters: (a) PGE$_2$ generation during the preincubation period with IL-1; (b) cellular cyclooxygenase activity in the solubilized cell sonicate after preincubation and labeling with ^{35}S-methio-
nine; (c) the radioactivity intensity in the COX band following ^{35}S-methio-
nine labeling, immunoprecipitation and SDS-PAGE electrophoresis. The

11

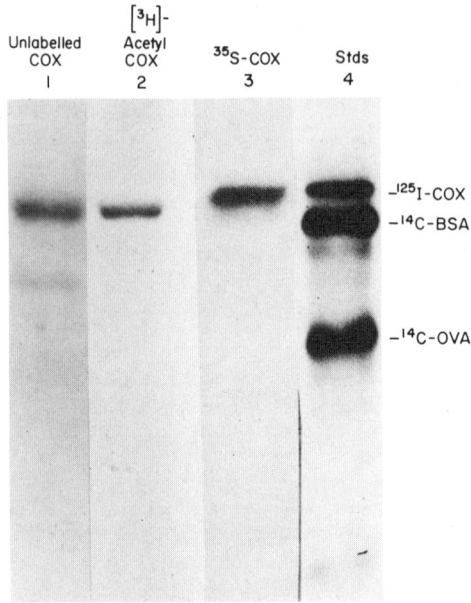

Fig. 7. Comparative migration of purified sheep COX, [^{3}H]-acetylated sheep COX and ^{35}S-labeled fibroblast COX on SDS-PAGE Electrophoresis. Lane 1: purified sheep COX (5 μg) stained with commassie blue. Lane 2: purified sheep COX, acetylated with [^{3}H]-aspirin. Lane 3: ^{35}S-methionine-labeled COX isolated from cell sonicate by immunoprecipitation. FB were preincubated with IL-1 (2 u/ml) for 18 hours prior to labeling with ^{35}S-methionine. For details, see Methods. Lane 4: ^{14}C-labeled protein standards and ^{125}I-COX.

results (Fig. 9) clearly demonstrate that within 6 hours of IL-1 addition, there is a 3-fold increase in the rate of newly synthesized COX, as indicated from the increased ^{35}S-methionine incorporated into the COX band. In agreement with these results, the cellular COX activity was stimulated in parallel. Both of these parameters reach maximal stimulation after 12 hours preincubation with IL-1. As expected, the synthetic rate of the major cyclooxygenase product, PGE_2, is increased dramatically in the first 12 hours of preincubation with IL-1 and then proceeds at a slower rate.

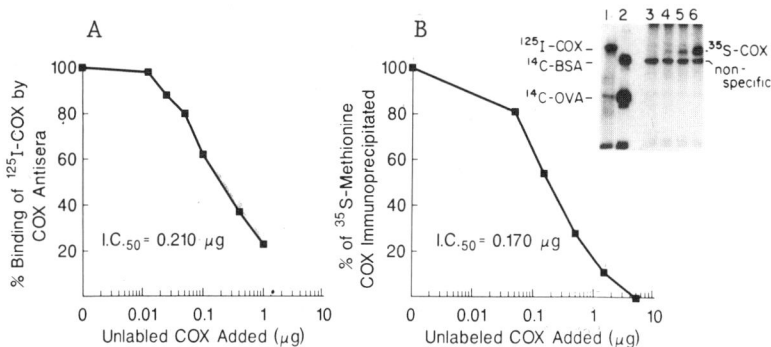

Fig. 8. Binding displacement curve of ^{125}I-COX and ^{35}S-methionine COX by unlabeled sheep COX.

A. Aliquots of ^{125}I-COX solution (approximately 10^5 cpm) were immunoprecipitated as described in Methods. The final pellet was suspended in SDS-PAGE sample buffer and the supernatant counted.

B. Aliquots of fibroblast solubilized cell supernatant from cells labeled with ^{35}S-methionine were mixed with 0-5 ng of sheep COX, immunoprecipitated, and subjected to SDS-PAGE electrophoresis, as described in Methods. Fluorographs obtained were scanned, and the relative intensity of the ^{35}S-COX band was plotted. A value of 100% was arbitrarily assigned to the intensity of the band obtained without added unlabeled COX (Lane 6).

Insert: Fluorograph obtained in a typical experiment. Lane 1: ^{125}I COX; Lane 2: ^{14}C-labeled protein standards; Lanes 3-6: 5 μg, 1.5 g, 0.5 μg, and 0 μg, respectively, of sheep COX added to ^{35}S-methionine-labeled fibroblasts supernatants.

The induction of new fibroblast cyclooxygenase synthesis was found to be exquisitely sensitive to the dose of IL-1 (Fig. 10). As little as 0.03 units/ml of IL-1 caused significant stimulation of new COX synthesis, as evident by increased cellular COX activity, increased ^{35}S-methionine incorporation into the immunoprecipitated COX band and increased production of PGE$_2$. All three parameters increased in parallel with increasing doses of IL-1; maximal stimulation of COX synthesis was attained at 0.3 u/ml.

Effect of Serum on IL-1 Stimulation of Fibroblasts COX

The experiments described above were performed under conditions in which fibroblasts were initially grown in media containing 10% fetal calf serum (FCS) but were subsequently incubated with IL-1 in serum free media. The

13

PGE₂ GENERATED DURING PREINCUBATION (pg/μg protein) (▲)

CELLULAR COX ACTIVITY (pg PGE₂/μg protein) (●)

RADIOACTIVITY in ³⁵S-METHIONINE LABELED COX (■) (Arbitrary Units)

INCUBATION TIME (Hours)

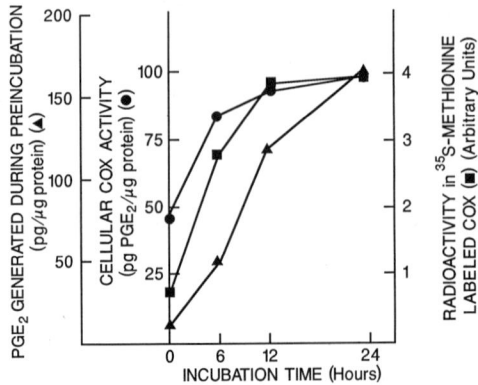

Fig. 9. Time dependent induction of COX synthesis by IL-1. Fibroblasts were preincubated for 0, 6, 12, or 24 hours in DMEM media containing 2 u/ml IL-1. The cells were then labeled with ^{35}S-methionine, and subsequently processed for immunoprecipitation, SDS-PAGE electrophoresis, fluorography, and scanning, as described in Methods. Aliquots of the media after preincubation were analyzed for PGE$_2$, and aliquots of the cell supernatants assayed for COX activity.

reasons for this were two-fold. First, we wanted to determine the effect of IL-1 without a possible modification of this effect by a factor (or factors) present in the serum. Second, in preliminary experiments we have observed dissimilarities between the effect of FCS on the synthesis of PGE$_2$ (as assayed in the cell media) and on the level of cellular cyclooxygenase activity. In the experiments outlined below, the effects of FCS and IL-1 alone and in combination were measured. Fibroblasts were preincubated in serum-free DMEM media or in DMEM containing 10% FCS in the absence or presence of 0.1 or 1.0 u/ml of IL-1 and subsequently labeled with ^{35}S-methionine, as described in Methods. Four parameters of COX activity were followed: (a) PGE$_2$ generation during the 16 hours of preincubation; (b) PGE$_2$ generation during the subsequent 6-hour labeling period; (c) cellular COX activity of the labeled cells; and (d) relative ^{35}S-labeling of the immunoprecipitated COX band. The results clearly indicate a dissociation between the rate of generation and release of PGE$_2$ into the media during preincubation and the measured level of COX activity and rate of synthesis, as indicated by the other three parameters. In the absence of FCS, IL-1

14

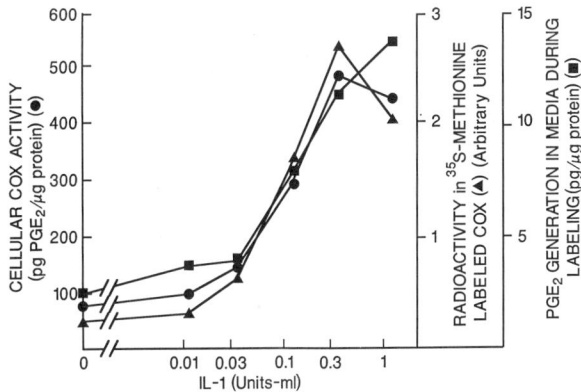

Fig. 10. IL-1 dose-dependent stimulation of fibroblast COX activity, ^{35}S-methionine labeling of COX and PGE$_2$ synthesis. Fibroblasts were preincubated for 16 hours in DMEM media containing 0-1 u/ml of IL-1. The cells were then labeled with ^{35}S-methionine and subsequently processed for immunoprecipitation, SDS-PAGE electrophoresis, fluorography, and scanning, as described in Methods. Aliquots of the labeling media were analyzed for PGE$_2$ and aliquots of the cell supernatants assayed for COX activity.

present during the preincubation stimulated PGE$_2$ generation by only approximately 2.5-fold (from 1.5 to 3.5 pg/μg protein) (Panel A). In contrast, when measured during the subsequent labeling period, IL-1 produced a dose-dependent stimulation of PGE$_2$ synthesis with 30-fold stimulation at 1 u/ml (Panel B). In parallel, cellular COX activity and ^{35}S-methionine labeling of newly formed COX were also markedly stimulated (12-fold and 9-fold, respectively) by IL-1 following preincubation in the absence of serum. Addition of FCS during the preincubation with IL-1 had a profound but differential effect on the four parameters studied. Whereas FCS in the absence of IL-1 did not affect PGE$_2$ generation, it markedly augmented the stimulatory effect of IL-1 on PGE$_2$ release into the media (Panel A), producing 25-fold stimulation, as compared with control, and 10-fold stimulation as compared with 1.0 u/ml of IL-1 alone. However, this serum-dependent augmentation was not due to a stimulatory effect on COX activity or its rate of synthesis. In fact, serum had a paradoxical but consistent inhibitory effect on PGE$_2$ synthesis during the labeling period (Panel B), on the level of cellular COX activity (Panel C), and on the rate of newly synthesized COX (Panel D). This inhibitory effect of serum was not dependent on IL-1, since it was also observed in the absence of this monokine (Panels C and D).

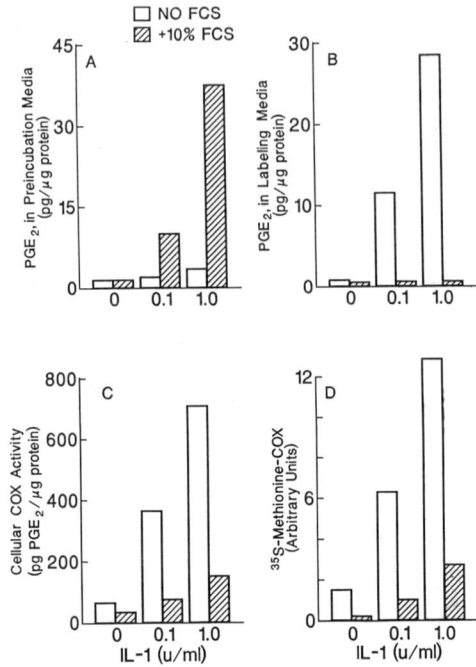

Fig. 11. Effect of FCS on IL-1 modulation of fibroblast COX synthesis and PGE_2 generation. Fibroblasts were preincubated for 16 hours with 0, 0.1, and 1 u/ml of IL-1 in either DMEM (open bars) or DMEM containing 10% FCS (hatched bars) and the media assayed for PGE_2 formed during the preincubation (Panel A). The cells were subsequently labeled with ^{35}S-methionine and aliquots of the labeling media assayed for PGE_2 (Panel B). The cells were processed for immunoprecipitation, SDS-PAGE electrophoresis, fluorography, and scanning (Panel D), as described in Methods. Aliquots of the cell supernatants were assayed for COX activity (Panel C).

DISCUSSION

Previous studies from our laboratory have demonstrated a markedly augmented eicosanoids release in response to agonist stimulation of the hydronephrotic rabbit kidney. Histological and biochemical studies in this inflammatory model, as well as studies by others, led us to propose a working hypothesis which involves intracellular communication via chemical mediators between monocytes/macrophages invading the injured site and the resident tissue fibroblasts. Studies on identification of the molecules mediating the monocyte-fibroblast communication have shown that monocytes-conditioned media contain a factor or (factors), MNCF, which possess molecular and biochemical properties similar to the monokine IL-1. Data presented here, as well as other recent data from our laboratory, clearly document IL-1 stimulation of PGE_2 generation in fibroblasts as well as other cells (30). Similar data have recently been obtained in other laboratories (31,32).

Modulation of prostaglandin production likely occurs at either or both of two key enzymatic steps: (a) rate of release of arachidonic acid from cellular phospholipids by specific phospholipases (PLA_2, PLC); (b) rate of conversion of released arachidonic into prostaglandins, this step being controlled primarily by the activity of cyclooxygenase. Several recent studies with human synovial cells (33) and chonrocytes (34) have indicated that IL-1 stimulation of PGE_2 production is mediated via stimulation of phospholipase(s), causing increased availability of arachidonate for conversion to prostaglandins. Our studies with dermal fibroblasts have demonstrated that MNCF, as well as purified human IL-1, produced increased Vmax of cellular cyclooxygenase which is dependent on new protein synthesis, suggesting that IL-1 directly modulates cyclooxygenase synthesis in fibroblasts. In the experiments described here, we have developed additional experimental tools and techniques which allowed us to determine directly the amount, activity, and rate of synthesis of COX protein in cells and to evaluate the effect of IL-1 on COX activity and synthesis. We have prepared polyclonal antisera against sheep seminal vesicle cyclooxygenase, which cross-reacted with human COX and permitted quantitative immunoprecipitation of the enzyme from cellular extracts. This antisera, together with ^{125}I-labeled COX that we prepared, allowed us to construct a radioassay for immunoreactive COX in tissue and cell extracts (Fig. 8). Finally, labeling of fibroblasts with ^{35}S-methionine, immunoprecipitation and SDS-PAGE electrophoretic separation of the newly synthesized labeled enzyme enabled us to quantitate changes in the synthetic rate of the newly formed COX.

Human recombinant IL-1 was found to stimulate the rate of synthesis of fibroblasts COX in a time- and dose-dependent manner. With cells incubated in serum-free media, IL-1 induction of COX synthesis was clearly evident after 6 hours (sometimes as early as 2 hours) (Fig. 9). This induction is exquisitely sensitive to IL-1 concentration; as little as 0.03 units/ml produced a clear stimulation of COX synthesis and concurrently elevated COX activity (Fig. 10). The parallel dose response curves for IL-1 stimulation of COX synthesis, COX activity and PGE_2 generation during the labeling incubation clearly indicate that (a) the IL-1 effect, under serum-free incubation conditions, appears to be mediated solely by its effect on COX and not on other enzymes involved in PGE_2 formation (e.g., phospholipases); (b) IL-1 effect involves induction of increased synthesis of new COX enzyme and not activation of pre-existing inactive enzyme (this latter possibility would be indicated by a lower rate of ^{35}S-methionine incorporation into COX, as compared with the measured cellular COX activity). The significance of these conclusions is manifested when comparing the effects of IL-1 in the absence and presence of FCS (Fig. 11). Comparison of the results in panel A with those in panels B-D of Fig. 11 clearly shows that (a) the presence of serum during preincubation with IL-1 substantially augments the effect of IL-1 on PGE_2 synthesis and (b) that this serum effect is not mediated via increasing the synthesis of activity of COX; in fact, serum addition during the preincubation markedly blunted IL-1 stimulation of COX synthesis and activity as well as PGE_2 generation during subsequent incubation in serum-free labeling media (Fig. 11, panel B-D). The positive synergistic effect of serum on IL-stimulation of PGE_2 production (Fig. 11, panel A) may be due to stimulation of phospholipase(s), as suggested from recent studies by Chang and coworkers (34) and Godrey and coworkers (33). This possibility and others are currently under investigation in our laboratory.

ACKNOWLEDGEMENT

This work was supported by NIH grants P01-DK3811 and R01-HL20787.

REFERENCES

1. K. Nishikawa, A.R. Morrison, P Needleman, Exaggerated prostaglandin biosynthesis and its influence on renal resistance in the isolated hydronephrotic rabbit kidney, J. Clin. Invest. 59:1143-1150 (1977).
2. A.R. Morrison, K. Nishikawa, P. Needleman, Unmasking of thromboxane A_2 synthesis by ureteral obstruction in the rabbit kidney, Nature 267: 259-260 (1977).

3. R. Zipser, S. Meyers, P. Needleman, Exaggerated prostaglandin and thromboxane synthesis in the rabbit with renal vein constriction, Circ. Res. 47:231-237 (1980).

4. E.A. Lianos, G.A. Giuseppe, M.J. Dunn, Glomerular prostaglandin and thromboxane synthesis in rat nephrotoxic serum nephritis, J. Clin. Invest. 72:1439-1448 (1983).

5. R.D. Zipser, J.B. Patterson, H.W. Kao, C.J. Hauser, R. Locke, Hypersensitive prostaglandin and thromboxane response to hormones in rabbit colitis, Am. J. Physiol. 249:G457-G463 (1985).

6. M.K. McGuire, J.E. Meats, N.M. Ebsworth, L. Harvey, G. Murphy, G.G. Russell, J.J. Reynolds, Properties of rheumatoid and normal synovial tissue in vitro and cells derived from them. Production of prostaglandins and collagenase in response to factors derived from cultured blood mononuclear cells and from synovium, Rheumatol. Int. 2:113-120 (1982).

7. A.S. Evers, S. Murphree, J.E. Saffitz, B.A. Jakschik, P. Needleman, Effects of endogenously produced leukotrienes, thromboxane, and prostaglandins on coronary vascular resistance in rabbit myocardial infarction, J. Clin. Invest. 75:992-999 (1985).

8. J.G. Clark, K.M. Kostal, B.A. Marino, Bleomycin-induced pulmonary fibrosis in hamsters, J. Clin. Invest. 72:2082-2091 (1983).

9. A.R. Morrison, K. Nishikawa, P. Needleman, Thromboxane A_2 synthesis by ureteral obstruction in the rabbit kidney, Nature 267:259-260 (1978).

10. R.B. Nagle, R.E. Bulger, R.E. Cutler, H.R. Jervis, E.P. Benditt, Unilateral obstructive nephropathy in the rabbit: I. Early morphologic, physiologic, and histochemical changes, Lab. Invest. 28:456-467 (1973).

11. T. Okegawa, P.E. Jonas, K. DeSchryver, A. Kawasaki, P. Needleman, Metabolic and cellular alterations underlying the exaggerated renal prostaglandin and thromboxane synthesis in ureter obstruction in rabbits. Inflammatory response involving fibroblasts and mononuclear cells, J. Clin. Invest. 71:81-90 (1983).

12. C.J. Mathia, M.J. Welch, D. Schwartz, S.M. Spaethe, P. Needleman. In-111 labeled cells to differentiate the sequential blood cell invasion of the injured rabbit kidney in vivo, submitted (1987).

13. D.F. Reingold, S. Waters, S. Holmberg, P. Needleman, Differential biosynthesis of prostaglandins by hydronephrotic rabbit and cat kidneys, J. Pharmacol. Exp. Ther. 216:510-515 (1981).

14. J.B. Lefkowith, T. Okegawa, K. DeSchryver-Kecksemeti, P. Needleman, Macrophage-dependent arachidonate metabolite in hydronephrosis, Kidney Int. 26:10-17 (1984).

15. N. Feuerstein, M. Foegh, P.W. Ramwell, Recently reported stimulation of TxB$_2$ and 6-keto PGF$_{1\alpha}$ synthesis by rat peritoneal macrophages incubated with E. coli 055:BS lipopolysaccharide, Br. J. Pharmacol. 72: 389-391 (1981).

16. P.V. Halushka, J.A. Cook, W.C. Wise, Thromboxane A$_2$ synthesis and prostacyclin production by lipopolysaccharide-stimulated peritoneal macrophages, J. Reticuloendothe. Soc. 30:445-450 (1981).

17. S.L. Humes, R.J. Bonney, L. Pelus, M.E. Dahlgren, S.J. Sadowski, F.A. Kuehl, P. Davis, Macrophage synthesize and release prostaglandins in response to inflammatory stimuli, Nature (Lond.) 269:149-151.

18. I. Gery and B.H. Waksman, Potentiation of the T-lymphocyte response to mitogens. II. The cellular source of potentiating mediator(s), J. Exp. Med. 136:143-155 (1971).

19. J.H. Korn, P.V. Halushka, E.C. LeRoy, Mononuclear cell modulation of connective tissue function: Suppression of fibroblast growth by stimulation of endogenous prostaglandin production, J. Clin. Invest. 65:543-554 (1980).

20. S.M. D'Souza, D.J. Englis, A. Clark, R.G. Russell, Stimulation of production of prostaglandin E in gingival cells exposed to products of human blood mononuclear cells, Biochem. J. 198:391-396 (1981).

21. P.E. Jonas, K.M. Leahy, K. DeSchryver-Kecksemeti, P. Needleman, Cellular interactions and exaggerated arachidonic acid metabolism in rabbit renal injury, J. Leukocyte Biol. 35:55-64 (1984).

22. P.E. Jonas and P. Needleman, Mechanism of enhanced fibroblast arachidonic acid metabolism by mononuclear cell factor, J. Clin. Invest. 74: 2249-2253 (1984).

23. P. Hawley-Nelson, J.E. Sullivan, M. Kung, H. Hennings, S.H. Yuspa, Optimized conditions for the growth of human epidermal cells in culture, J. Invest. Dermatol. 75:176-179 (1980).

24. C.R. Albrightson, N.L. Baenziger, P. Needleman, Exaggerated human vascular cell prostaglandin biosynthesis mediated by monocytes: role of monokines and interleukin 1, J. Immunol. 135:1872-1877.

25. F.J. Van Der Ondera, M. Buytonhek, D.K. Nugteren, D.A. Van Dorp. Purification and characterization of prostaglandin endoperoxide synthetase from sheep vesicular glands, Biochem. Biophys. Acta 487:315-331 (1977).

26. A.T. Meukh, G.F. Sudina, N.B. Golub, S.D. Vanfolomeer, Purification of prostaglandin H synthase and a fluorometric assay for its activity, Analytical Biochem. 150:91-96 (1985).

27. C.L. Tai and H.H. Tai, A radioimmunoassay for prostaglandin endoperoxide synthetase, Prostagl. Leuket. Medicine 14:243-254 (1984).

28. G.J. Roth, Preparation of [acetyl-^{3}H] aspirin and use in quantitating PGH synthase, <u>Methods in Enzymology</u> 86:392-400 (1984).

29. M.S. McHardy, S. Schlesinger, J. Lindstrom, J.P. Merlie, The effects of inhibiting oligosaccharide trimming by 1-dexoxynojirimycin on the nicotinic acetylcholine receptor, <u>J. Biol. Chem</u>. 261:14825-14832 (1986).

30. U.K. Laemmli, Cleavage of structural proteins during the assembly of the head of bacteriophage T_4, <u>Nature</u> 227:680-685 (1970).

31. R.J. Zucali, C.A. Dinarello, D.J. Oblon, M.A. Gross, L. Anderson, R.S. Weiner, Interleukin-1 stimulates fibroblasts to produce granulocytes-macrophage colony stimulating activity and prostaglandin E_2, <u>J. Clin. Invest</u>. 77:1857-1863 (1986).

32. J.F. Balaudine, B. Rochemontiex, K. Williamson, P. Seckinger, A. Cruchand, J.M. Dayer, Prostaglandin E_2 and collagenase production by fibroblasts and synovial cells is regulated by urine derived human interleukin-1 and inhibitor(s), <u>J. Clin. Invest</u>. 78:1120-1124 (1986).

33. J. Chang, S.C. Gilman, A.-J. Lewis, Interleukin 1 activates phospholipase A_2 in rabbit chondrocytes: A possible signal for IL 1 action, <u>J. Immunol</u>. 136:1283-1287 (1986).

34. R.W. Godrey, W.T. Johnson, S.T. Hoffstein, <u>Biochem. Biophys. Res. Commun</u>. 142:235-241 (1987).

ARACHIDONIC ACID METABOLISM DURING INTERACTIONS

BETWEEN GLOMERULAR AND BONE MARROW-DERIVED CELLS

Josée Sraer, Marcelle Bens, Jean-Paul Oudinet
and Larent Baud

INSERM 64, Hôpital Tenon, Paris, France

The concept that cell-cell interaction might modify the metabolism of
arachidonic acid (AA) was already suggested nearly ten years ago by several
studies in which platelets and blood vessels were coincubated and AA metabolites were analysed. Over subsequent years most of the researchers in this
field focused their interest on the interaction between either endothelium
and platelets, polymorphonuclear leukocytes (PMNL) and platelets, or PMNL
and endothelium. A brief review of these interactions occurring in nonrenal
tissue will be presented. The hypothesis that the glomerulus, which includes a peculiar endothelium, could be a preferential site for cell-cell
interaction has not been investigated until recently. Yet, it is well documented that activated bone marrow-derived cells may invade the glomerular
capillary in a number of experimental or human glomerulonephritides. Both
cell types--glomerular and bone marrow-derived cells--were recognized to be
the source of various lipidic inflammatory agents such as platelet-activating factor (PAF), prostaglandins (PG), hydroxyeicosatetraenoic acids (HETE)
and leukotrienes (LT). This review focuses upon recent results providing
strong evidence that, during interaction between glomerular and bone marrow-
derived cells, changes occur in arachidonate metabolism. The functional
consequence of these changes will be discussed, but we shall limit this
review to the interactions involving lipidic factors.

I. GENERAL CONSIDERATIONS

Often considered synonyms, "interaction" and "cooperation" are two terms
indiscriminately used to describe the fact that two cells may communicate by
the mean of biochemical signals. In a strict sense, "interaction" should be

reserved to the following event: in response to a product generated by one cell type, a second cell type produces a metabolite acting on the first one. This response usually results in the attenuation of the initial signal and implies, therefore, a local regulatory process which is similar to the well-known general mechanism of feedback. However, in the broad sense, the term "interaction" is often employed even if the response of the second cell on the first one is not proven. In most of these cases, only cooperation can be experimentally demonstrated. Both cell types may produce together an active metabolite which thereby initiates a physiological effect. Alternatively, one cell type may transform an inactive product to an active metabolite produced by the other. Several mechanisms of cooperation in the metabolism of AA have been described: (a) one cell type provides a substrate that is utilized by a specific enzyme of the other; (b) one cell type synthesizes and releases a defined metabolite that either stimulates or inhibits an enzyme involved in the AA cascade of the other.

Whatever the mechanism, such cooperations will result in changes in the overall production or in the nature of locally produced AA intermediates. In vitro, the cooperation between two cell types can be easily studied by incubating the two cell preparations either separately or in combination. AA metabolites are measured, and the amount of metabolites present in the coincubated cells is compared with the predicted theoretical sum (sum of the metabolites produced by each cell preparation incubated separately).

II. CELL-CELL INTERACTION IN NONRENAL TISSUE

A. Sharing of Substrate Between Two Cell Types

Although conflicting results were first reported (1,2), it is now accepted that endothelial cells utilize platelet endoperoxides to produce prostacyclin (PGI_2) (3,4). Such an interaction is often referred to as a substrate steal and probably allows the endothelial cell to respond to the proaggregatory effect of thromboxane (TX) released by stimulated platelets. In this model of cooperation, the cells share a common precursor, endoperoxides, but the transfer of substrate appears to be unidirectional from platelets to endothelial cells (4). The result is an increase in the level of PGI_2 normally synthesized by the endothelium but, in turn, the endothelial cells interact with platelets by decreasing TX production and inhibiting platelet aggregation.

Coincubation of other cells such as PMNL and platelets lead, on the

24

contrary, to the formation of novel AA products. The first report was by Borgeat et al. (5) who demonstrated in PMNL the formation of a novel metabolite, 5S,12S diHETE. From their structural analysis (6), they concluded that this new metabolite had to be the result of the conjugated action of 5- and 12-lipoxygenases. Since purified neutrophils do not possess 12-lipoxygenase, the only possible source of 12-HETE was the contaminating platelets which were present in the PMNL preparation. Thus, in this case 12-HETE of platelet origin substitutes for AA and is utilized by the 5-lipoxygenase in PMNL. Later, Marcus et al. (7), using prelabeled human platelets and unlabeled leukocytes, could demonstrate that upon stimulation with calcium-ionophore, (a) platelet-derived arachidonate could serve as precursor for the neutrophil-derived eicosanoids, LTB_4 and 5-HETE; (b) platelet-derived 12-HETE could be converted to diHETE by neutrophils. Although the role of platelets on PMNL has been well documented (7,8), no results were available concerning the possible subsequent interaction of PMNL on platelets. We, therefore, investigated cooperation between a purified preparation of PMNL (no contaminating platelets) and a suspension of platelets; but in contrast with Marcus's protocol, the combined cells were incubated under basal conditions (no ionophore, no excess of AA). When platelets and pure PMNL were coincubated in the presence of a tracer dose of exogenous [^{3}H] AA, [^{3}H] 12-HETE was greatly diminished (Table 1). This inhibitory effect depended upon PMNL concentration and could be either cancelled or even reversed in the presence of increasing amounts of AA substrate. The mechanism by which 12-HETE was inhibited when P and pure PMNL were coincubated could be attributed to the greater uptake of exogenous AA by PMNL than by platelets (J.P. Oudinet, unpublished results). More recently, the ω-oxidation of platelet 12-HETE by an ω-hydroxylase present in leukocytes was described, and a new

Table 1.   Interaction Between Platelets (P) and Purified Polymorphonuclear Leukocytes (PMNL) on the Conversion of [^{3}H] C20:4 into its Lipoxygenase Products

	12-HETE	12,20 diHETE
	([^{3}H] cpm x 10^3/tube)	
P	296.2 ± 72.3	0
PMNL	3.6 ± 1.0	0
PMNL + P	27.5 ± 7.7**	1.4 ± 0.4**

*Data represent the means $\pm$ s.e.m of 5 to 7 separate experiments. For each [^{3}H] metabolite, the amount synthesized by P (30 x 10^6 cells/ml) plus PMNL (16 x 10^6 cells/ml)(experimental sum) was compared, using Student's t test for paired values, with the sum of the amounts synthesized separately by each of them (predicted sum), ** = p < 0.01*

metabolite was detected in platelet-leukocyte mixtures, namely the 12,20 diHETE (9,10). Under our experimental basal conditions, cooperation between platelets and leukocytes also resulted in 12,20 diHETE formation, although the metabolite was present at very low levels compared with 12-HETE.

Neutrophils and mast cells may also interact as reported by Dahinden et al. (11): LTA_4 generated by neutrophils and stabilized by albumin is efficiently converted by mast cells into LTC_4. Interestingly, it was shown that human erythrocytes, in which the release of membrane-bound AA and its subsequent metabolism by cyclooxygenase and lipoxygenase has never been established, are, however, capable of converting LTA_4 into LTB_4 (12). Similarly, it has been shown recently (13) that platelets possess the necessary machinery to transform LTA_4 to LTC_4 and LTD_4. This suggests that cellular interaction between platelets and LTA_4-forming cells (e.g. PMNL and macrophages) could lead to the formation of these potent peptidolipids in the circulation.

B. AA Metabolite From One Cell Type Interacts with Another Cell Type

Numerous reports clearly indicate that metabolites produced by one cell type may interfere with AA metabolism in another cell type. For example, both 12- and 15-HETE produced by platelets or leukocytes may inhibit PGI_2 generation by endothelial cells (14-16). It was also reported (17,18) that platelet-derived 12-hydroperoxyeicosatetraenoic acid (12-HPETE) promoted activation of the 5-lipoxygenase present in leukocytes, which resulted in LTB_4 and 5-HETE formation. Cellular cooperation between mononuclear blood cells and stimulated platelets led to the formation of 5S,12S diHETE and LTC_4 (19). On the contrary, 15-HETE, possibly from leukocyte origin, was shown to inhibit 12-lipoxygenase in platelets (20) as well as 5-lipoxygenase in leukocytes (21). Vanderhoek et al. (22) also described a stimulating effect of 15-HETE on the 5-lipoxygenase present in mast-cells, the result being the production of LTB_4.

C. An Unknown Factor from One Cell Type Modifies AA Metabolism in Another Cell Type

Recently, two different reports clearly showed that neutrophils activated by either phorbolmyristate acetate (PMA) or the chemotactic peptide f. Met. Leu. Phe (fMLP) were able to stimulate PGI_2 formation by cultured endothelial cells, and this increased PGI_2 was reported to impair the adherence of PMNL on the endothelial cell surface (23-25). The factor that

26

triggered PGI_2 release was not identified. In PMA-stimulated leukocytes, the role of hydrogen peroxide or hydrogen peroxide-derived products was suggested by the decreased release of endothelial PGI_2 observed in the presence of catalase (25). This was also confirmed by the stimulation of PGI_2 release by exogenously generated hydrogen peroxide. On the contrary, the effect of fMLP-stimulated leukocytes on PGI_2 released by endothelial cells was not blocked by catalase (24). The authors excluded the possibility that 15-HPETE could be responsible for the effect observed. However, other organic hydroperoxides have been demonstrated as effective stimulators of PG synthesis (26).

In conclusion, it is clear from all these examples that transcellular metabolism of AA intermediates occurs between vascular and blood cells, which results in changes of cellular function. These changes include increase in permeability of postcapillary venules, adhesion of neutrophils to endothelial cells and leukocyte extravasation. The interaction between these cells leads to either higher levels of a given metabolite or synthesis of new metabolites. Since cell-cell interactions may depend upon the amount of substrate available, caution should be taken in the interpretation of in vitro studies. Interactions between various cell types which release or utilize LTA_4 or possess a cryptic 5-lipoxygenase may represent an important metabolic pathway for the production of leukotrienes, important mediators of inflammation. It is noteworthy that the biological potency of several metabolites formed during cell-cell interaction, such as 5S,12S diHETE or 12,20 diHETE, is not fully demonstrated, although 5S,12S was reported to be chemotactic (27). Thus, cooperation between cells may either result in the synthesis of active compounds such as leukotrienes or represent a detoxification process. A general scheme of all these interactions is proposed in Figure 1.

III. LIPIDS FROM GLOMERULAR ORIGIN MODIFY AA METABOLISM IN BONE MARROW-
    DERIVED CELLS

Glomeruli have been shown to be a major source of lipid mediators of inflammation. Namely, they can release PAF (28), PG (29-33) and lipoxygenase metabolites (34,35) such as 12- and 15-HETE. Species differences exist: rat glomeruli predominantly synthesize PGE_2 and $PGF_{2\alpha}$ (Fig. 2) whereas human glomeruli release mainly PGI_2 detected as 6-keto-$PGF_{1\alpha}$ (32). An unknown cyclooxygenase product, x, was also described in rat glomeruli (33). Whereas 12- and 15-HETE were produced by human glomeruli, 12-HETE was the only peak observed in rat glomeruli (Fig. 2). These HETE

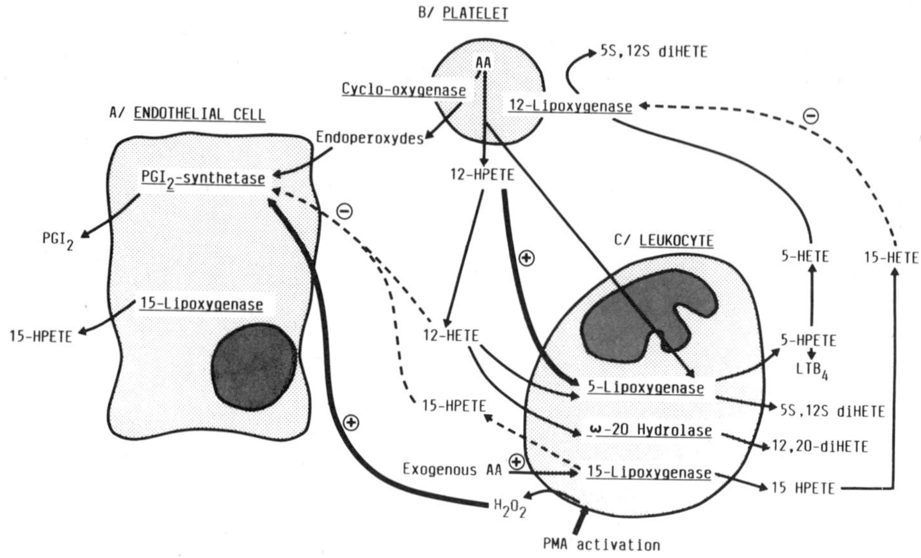

Fig. 1. Transcellular metabolism of arachidonate intermediates between endothelial and blood cells. See text for abbreviations. Solid and dotted lines indicate stimulation and inhibition, respectively.

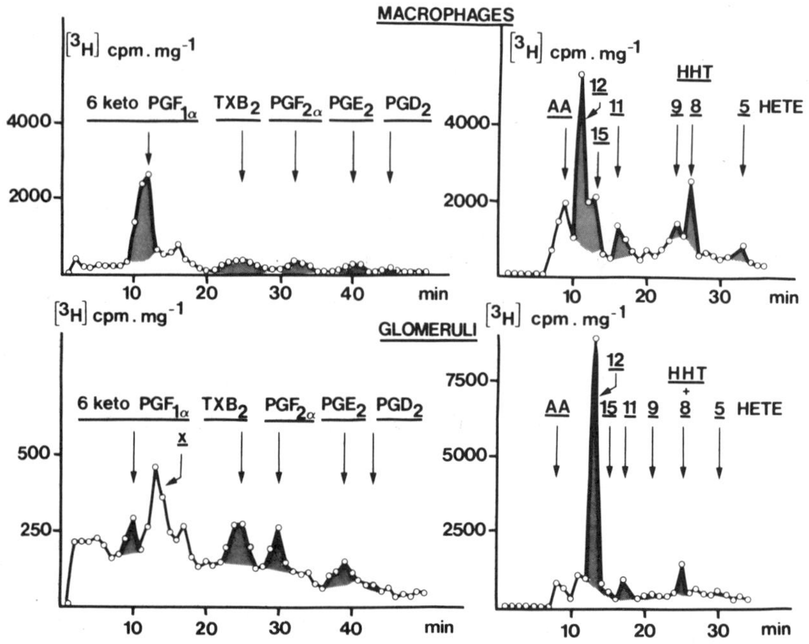

Fig. 2. High performance liquid chromatograms of [^{3}H] AA metabolites (PG, left, or HETE, right) formed by glomeruli (lower part) or macrophages (upper part). The retention time of authentic standards are indicated by arrows. Glomeruli and macrophages were incubated separately for 30 min. at 37°C in the presence of 50 nM [^{3}H] AA.

are, in fact, the end-products detected in vitro but they derive from hydro-
peroxy fatty acids: 12- and 15-HPETE which are reduced into HETE by the
peroxidases present in glomerular cells.

A. Cooperation Between Monocyte-Macrophages

There are two possible sources of monocyte-macrophages within the glomerular tuft: either resident macrophages, normally present as less than 0.1
per tuft (36), or bone marrow-derived macrophages infiltrating the Bowman's
space and the mesangium during experimental nephrotoxic serum nephritis
(37,38) and glomerulonephritis of serum sickness (39). The mechanism of
accumulation of macrophages in diseased glomeruli remains uncertain: immune
adherence via Fc receptors on macrophages (40) or chemotaxis involving complement fractions (41), fibrin deposition (42) or products derived from collagen breakdown (43).

Evidence for binding of rat peritoneal macrophages to rat glomerular
mesangial cells has been reported (44), and endothelial cell-derived chemotactic activity for mouse peritoneal macrophages has also been described
(45). These two studies rendered conceivable the idea that glomerular cells
could release a chemotactic factor responsible for the attachment of macrophages. Indeed, in an in vitro cooperation study between rat peritoneal
macrophages and rat glomeruli (46), we could demonstrate that, during coincubation of these two preparations, the main cyclooxygenase and lipoxygenase
products (47-50) normally produced by macrophages (Fig. 2) were considerably
increased. A previous study of Koyama et al. (51) had already demonstrated
that 6-keto-PGF$_{1\alpha}$ and TxB$_2$ production were associated with adhesion and
spreading of peritoneal macrophages on a glass surface. In accordance with
these results, we reported the following findings: (a) [^{3}H] uridine-
labeled macrophages bound to glomeruli as a function of time and temperature, and this binding was related to glomerular protein and macrophage concentration; (b) nordihydroguaiaretic acid, a lipoxygenase inhibitor, but not
indomethacin, a cycloxygenase blocker, inhibited this attachment. Taken together, these findings suggested that the glomerulus was not a passive supporting structure but released a chemotactic signal acting on the attachment
of macrophages and promoting their activation. In this study (46), glomerular 12-HPETE was proposed to be responsible for macrophage activation and
macrophage binding because a lipid extract of glomeruli as well as the
direct addition of synthetic 12-HPETE stimulated macrophage functions as reflected by an increased production of PG. Thus, lipoxygenase metabolites
produced by the glomerulus may exert chemotactic and stimulatory effects on

monocyte-macrophages. Consequently to this cooperation, the attachment of
these cells on the capillary wall occurs. Because it has been recently
demonstrated (52) that, in the experimental GN induced by anti-basement membrane antibody administration, glomerular 12-HETE--the reduced metabolite of
12-HPETE--was dramatically increased, it may be speculated that the abnormal
production of this chemotactic agent plays a role in the abnormal accumulation of macrophages. Another possible role for 12-HPETE produced in excess
by macrophages has been recently suggested by Lacave et al. (53). These
authors demonstrated that 12- and 5-HPETE--but not their reduced derivatives
--promoted contraction of human epithelial cells in culture. Thus, activation of macrophages by glomerular lipids may be responsible for shape
changes and perhaps functional alteration of epithelial cells already
described in experimental GN (54).

However, if these in vitro studies of cellular cooperation support the
interesting concept that the biochemical disorders observed during GN may be
related to cell-cell interaction, two major flaws exist: (a) The pluricellu-
lar composition of the glomerulus did not allow us to determine which cells
cooperated with macrophages and the exact cellular source of 12-HETE in the
glomerulus. Indeed, epithelial (35) and mesangial (55) glomerular cells
were reported as capable of generating 12-HETE. (b) The possibility that
human macrophages activated by glomeruli may generate leukotrienes, LTB_4
or LTC_4, acting, in turn, on the glomerulus was not investigated. Yet, it
is well documented that, when stimulated by appropriate stimuli, such as
zymosan, ionophore, or IgE, monocyte-macrophages are a source of leukotrienes (50,56-58).

B. Cooperation with Platelets

When a tracer dose of [^{3}H] AA was added to a mixture of human glomeruli and platelets (Fig. 3), striking increase in [^{3}H] TxB_2 was observed
whereas [^{3}H] 12-HETE was unchanged (59). It was verified by aspirin pretreatment that platelets were responsible for the excess of TxB_2 generated
and that the stimulus was from glomerular origin. When TxB_2 generated
from endogenous AA was measured by radioimmunoassay in coincubated cells, a
dramatic increase of this metabolite was detected (Fig. 4). In order to
characterize the glomerular products responsible for this platelet activation, human glomeruli were incubated for 10 min. at 37°C, and the supernate was analysed. Two different biochemical components were present (60):
(a) a procoagulant activity (PCA) resembling thromboplastin, which has already been reported in rat glomeruli (61), was found to be released in the

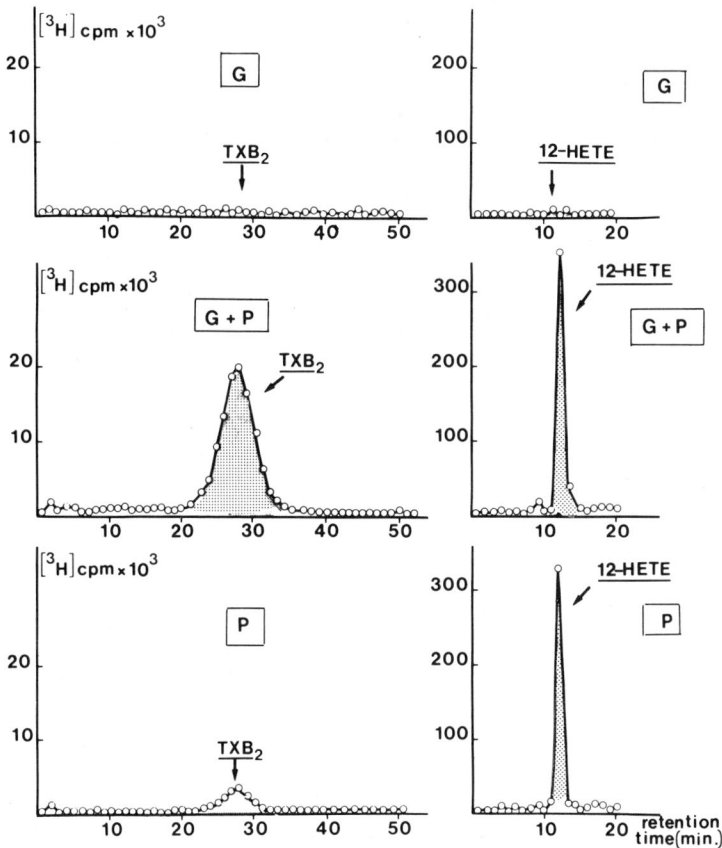

Fig. 3. High performance liquid chromatograms of cyclooxygenase [³H] metabolites (left) and lipoxygenase
[³H] metabolites (right) formed by platelets
(P) and glomeruli (G) incubated either separately
(lower and upper part) or in combination
(middle). Cells were incubated for 30 min. at
37°C in the presence of 50 nM [³H] AA.

glomerular supernate; (b) saturated and monoenoic long-chain fatty acids
were identified by gas chromatography/mass spectrometry. A lipid extract
prepared from the glomerular supernatant media contained up to 80 μM of
myristic, palmitic, stearic and oleic acids; but, surprisingly, no AA was detected. This lipid extract, itself, stimulated generation of TxB$_2$ by
platelets, although at a lesser degree than the whole supernatant. Pretreatment of the lipid extract by charcoal, which is known to avidly bind fatty
acids, completely suppressed the phenomenon. Moreover, an exogenous supply
of synthetic myristic, stearic, palmitic and oleic acids produced the same
stimulatory effect. It was concluded from these experiments that human
glomeruli release fatty acids capable of activating platelets. Since the

31

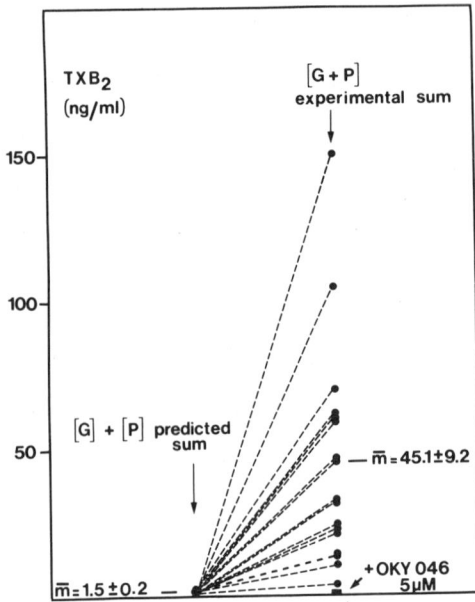

Fig. 4. Radioimmunoassay of thromboxane B_2 performed in human glomeruli and platelets incubated either separately ([G] + [P] predicted sum) or in combination ([G + P] experimental sum). Cells were incubated for 10 min. at 37°C without any addition of exogenous AA. OKY 046 inhibits thromboxane synthase.

glomerular supernate, which contains both PCA and fatty acids, appeared clearly more active than the lipid extract which contained fatty acids alone, it is possible that PCA or another lipoprotein could also participate in platelet activation.

The main question as to whether glomerulus-platelet interaction with stimulation of platelets may also occur in vivo cannot be answered easily. It is likely that, under physiological conditions, such an interaction may not occur since intact glomerular epithelial and mesangial cells were unable to stimulate TxB_2 production in platelets (Table 2). However, the same cells, after having been scraped away from their flasks, were stimulatory. Thus, membranous disturbances or lesions such as those that occur during the mechanical preparation of glomeruli or the dissociation of cultured cells result in the release of the stimulatory long-chain fatty acids in the media. Similar conditions might mimic cellular injury in glomeruli during experimental glomerulonephritis. It is also likely that these membrane alterations liberate PCA, which is considered a dormant factor (62) but has been shown to be released upon stimulation by endotoxin (63) or trypsin (62) (Fig. 5).

The mechanism of the stimulatory effect of fatty acids on TxB_2 production is still unknown. Since long-chain fatty acids are present in all cells, this must be an ubiquitous mechanism possibly related to protein acylation. It has been shown that fatty acids stimulated renin activity

Table 2. Factor of Stimulation of TxB_2 Synthesis Generated by Human Platelets in Contact with Intact or Scraped Cultured Human Glomerular Cells

Experiments (number)	Epithelial cells		Mesangial cells	
	intact	scraped	intact	scraped
1	0.6	2.4	0.3	17.9
2	1	-	0.8	-
3	1.3	2.6	1.2	4.9
4	0.3	8.4	1.3	8.4
5	1.0	2.2		
mean ± s.e.m.	0.84 ± 0.17	3.9 ± 1.5	0.90 ± 0.23	10.4 ± 3.9
Statistical analysis	**		*	

Human platelets [P] and human glomerular epithelial or mesangial cells [C] were incubated for 10 min at 37 °C separately or in combination. The cultured cells were either dissociated from their flasks or maintained in monolayer. The amount of TXB_2 formed under these different conditions ([P], [C] and [P + C]) was measured by radioimmunoassay. The factor of stimulation was calculated as the ratio experimental sum [P + C] over predicted sum ([P] + [C]). The individual results of 5 experiments are shown. Statistical analysis was performed using the rank test of Mann and Whitney. * = p < 0.05 ; ** = p < 0.01

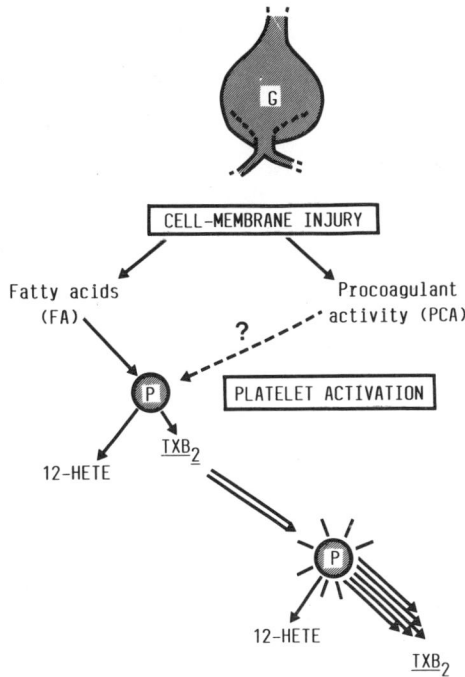

FIG. 5. Hypothetical scheme of platelet activation by glomerular fatty acids and procoagulant activity.

(64) and, more specifically, that myristic acid could be covalently linked to various proteins and thereby modify their activities (65-67). Similarly, myristoylation of the phospholipase protein could occur in our system and could account for the increase in TxB_2 generation. Fatty acids, in the concentration range of 10-100 μM, could also modify phospholipids in plate-let membranes and render them more susceptible to be activated by phospholi-pase C, a mechanism recently advanced by Dawson et al. (68). Whatever the mechanism, the release of fatty acids by injured membranes may represent a novel factor of thrombosis occurring within the glomerulus but also in other tissues.

C. Cooperation with Polymorphonuclear Leukocytes (PMNL)

Because PMNL are constantly present within the glomerulus at the early phase of the nephrotoxic serum nephritis, an in vitro study was undertaken between glomeruli and PMNL. The coincubation of human glomeruli and PMNL was first reported to induce an increase in 12-HETE and in 12,20 diHETE syn-thesis (69). However, it appeared that the preparation of PMNL which was used in this study was contaminated by platelets (P), as shown by the high level of 12-HETE and 12,20 diHETE synthesized by PMNL alone (Table 3). The ratio of P/PMNL could be estimated to be around 3 when using unpurified PMNL suspensions. In order to study specifically the interaction between pure

Table 3. Interaction Between Human Glomeruli (G) and Unpuri-fied Leukocytes (L) on the Conversion of [^{3}H] C20:4 into Lipoxygenase Products

	12-HETE	12,20 diHETE
	([^{3}H] cpm x 10^{-3}/tube)	
G	9.6 ± 1.0	0
L	95.8 ± 18.4	20.5 ± 4.8
(P/L ratio = 3)		
Predicted sum	105.4 ± 18.5	20.5 ± 4.8
Experimental sum	219.2 ± 44.0**	35.7 ± 6.6**

Data represent the mean of 11 experiments. For each [^{3}H] metabolite, the amount synthesized by G + L in combination (experimental sum) was compared, using the Student's t test for paired values, to the sum of the amounts synthesized by each of these preparations incubated separately (predicted sum) ; ** p < 0.01

Table 4. Interaction Between Glomeruli (G) and Purified
Polymorphonuclear Leukocytes (PMNL) on the Conver-
sion of [^{3}H] C20:4 into its Lipoxygenase
Products

	12-HETE	12-20 diHETE
	([^{3}H] cpm x 10^{-3}/tube)	
G	1.8 ± 0.4	0
PMNL	3.2 ± 1.0	0
G + PMNL	5.2 ± 2.1 ns	0.8 ± 0.4 ns

Data represent the means ± s.e.m. of 7 separate experiments. For each [^{3}H] metabolite, the amount synthesized by G plus PMNL (18 x 10^6 cells/ml) (experimental sum) was compared, using Student's t test for paired values, with the sum of the amounts synthesized separately by each of them (predicted sum) ; ns = non significant.

PMNL and glomeruli and not among 3 cellular types, the contaminated preparation of PMNL was further purified, using defibrinated blood, dextran separation and Ficoll-Hypaque gradient, and pure PMNL were obtained (P/PMNL ratio
of 0.05). Under these conditions, no interaction could be detected either
in 12-HETE or in 12,20 diHETE synthesis (Table 4). Thus, in this study, contrary to what was observed in glomerulus-macrophage interaction, glomerular
metabolites did not seem to trigger PMNL. Because PMNL and platelets both
invade the glomerulus at the early phase of GN, in vitro addition of glomeruli to a mixture of PMNL + P was performed. Unexpectedly, this addition
restored a normal level of 12-HETE synthesis by platelets and also stimulated [^{3}H] 12,20 diHETE (Fig. 6). Taken together, these results clearly
indicated that whenever 12-HETE synthesis was increased, this metabolite was
utilized by PMNL, which possess ω-20 hydroxylase, resulting in the increase
of the level of 12,20 diHETE. In the circulating blood, the amount of AA
available for the cells is probably very low, due to the presence of albumin
which binds this fatty acid very avidly (70). It is thus conceivable that,
similar to our in vitro studies, the level of 12-HETE generated by platelets
must be continuously inhibited in the presence of circulating PMNL and thus
remains very low. The possibility exists that glomerulus-derived products
may increase this level although no approach was made in this study to
define which glomerular products were responsible for this interaction. In
conclusion, it is highly probable that glomerular cells influence the metabolism of AA when both platelets and PMNL are present. Such an interaction
could thereby occur when platelets and PMNL infiltrate the diseased glomeruli.

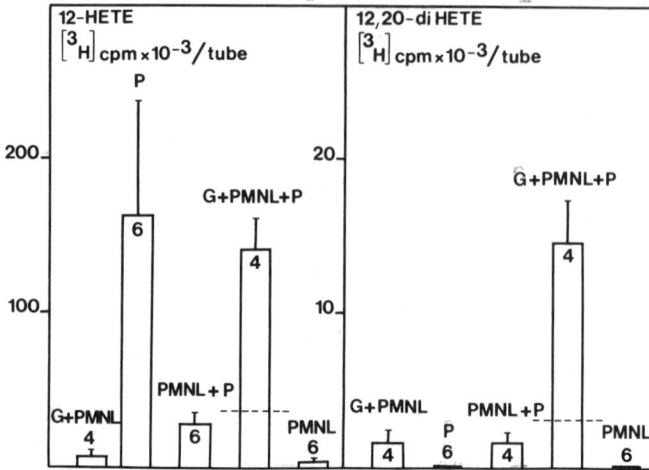

Fig. 6. Conversion of [3H] AA into [3H] 12-HETE and [3H] 12,20 diHETE by various cell preparations incubated separately or in combination: G = human glomeruli; PMNL = purified polymorphonuclear leukocytes; P = platelets. P and PMNL were used at concentrations of 25 and 12 x 10^6 cells/ml respectively. Glomeruli were used at a concentration of 1-2 mg protein/ml. Means ± s.e.m. are given. The number of individual experiments is shown above each bar.

IV. LIPIDS FROM BONE MARROW-DERIVED CELLS MODIFY AA METABOLISM IN GLOMERULAR CELLS

A. Monocyte-macrophages

As already stated, unstimulated monocyte-macrophages produce both PG and HETE. When a lipid extract was prepared from rat peritoneal macrophages and coincubated with rat glomeruli, glomerular production of 12-HETE was completely inhibited (71). Therefore, nonelicited macrophages exert an inhibitory effect on glomerular cells although the lipid responsible for this effect was not identified. A complete scheme of interaction can be proposed (Fig. 7) in which, in response to a glomerular metabolite (12-HPETE), macrophages synthesize a lipid lowering the level of this metabolite. If glomerular 12-HPETE synthesis is dramatically increased as found in GN, the inhibitory response of macrophages may become inefficient and the chemotactic effect of glomerular 12-HPETE may persist.

In pathological circumstances, when invading monocyte-macrophages are triggered, they are presumed to release large amounts of potent inflammatory

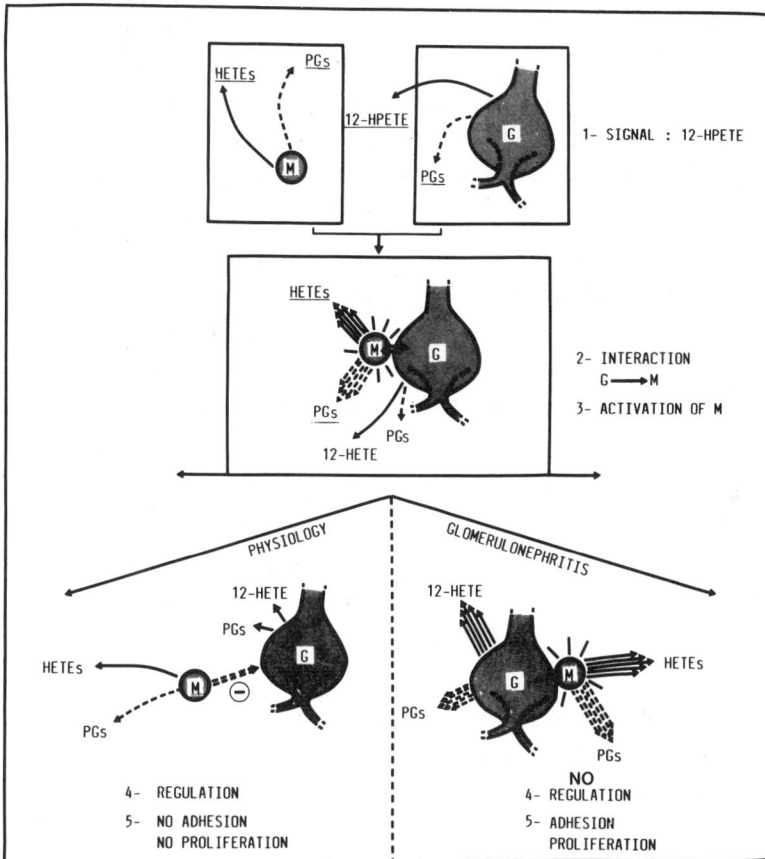

Fig. 7. Hypothetical scheme of activation of macrophages by
glomerular 12-hydroperoxyeicosatetraenoic acid.

agents such as LTB_4 and LTC_4. Rat peritoneal macrophages predominantly
produce LTB_4 whereas human macrophages synthesize mainly LTC_4. If accumulation of LTC_4 occurs within the glomerular tuft, it has been shown (72)
that it can be readily converted by glomeruli via LTD_4 into LTE_4, a less
active metabolite (73). A recent report by Baud et al. (74), working on
HL-60 cells, indicates that LTD_4, but not LTC_4, was able to activate
leukocytes (75). Glomerular cells possess the enzymes necessary to transform inactive LTC_4 into active LTD_4 (γ-glutamyltranspeptidase) and to
limit the excess of active product (dipeptidase which transforms LTD_4 into
LTE_4). Therefore, lipids from monocyte-macrophages may exert their
effects on glomerular cells. Two physiological events may follow this cooperation: (a) cell proliferation: LTC_4 and LTD_4 increased the growth of
cultured human glomerular epithelial cells (76); (b) cell contraction:
Barnett et al. (77) have shown that both LTC_4 and LTD_4 produced a

decrease of 10-14% in the surface area of glomeruli and, more precisely, mesangial cells. Simonson et al. (78) obtained similar results. Interestingly, it has been demonstrated that, unlike other cells, glomerular mesangial cells did not respond to LTC_4 by increasing their production of PGE_2 (77,78). Cramer et al. (79), on the contrary, demonstrated that addition of LTC_4 to endothelial cells promoted PGI_2 synthesis. Similarly, LTD_4 treatment of bovine aortic endothelial cells and murine smooth muscle cells in culture resulted in an increase in prostanoid synthesis (80). Thus, LTC_4 may exert its effect on human mesangial cells without being counteracted by PGE_2 whose role on relaxation is well documented (81). It is interesting to compare this result with that found in cultured human fibroblasts (82). Because of the high level of endogenous PGE_2 production, LTC_4 could not exert its proliferating effect on control cultured fibroblasts, but blockade of PGE_2 by indomethacin unmasked the effect of LTC_4. It may thus be critical for the kidney that, in response to locally produced LTC_4, the glomerular cells do not increase their PGE_2 production.

Stimulated macrophages liberate PAF, a potent mediator of inflammation. This lipid has been shown to interact with AA metabolism in mesangial cells (83): PAF increased PGE_2 synthesis by mesangial cells and also caused their contraction with a maximum effect at 10^{-7}M. This may be of interest since it is known (84) that the increase in vascular permeability observed in response to PAF is enhanced by PG. Therefore, during acute inflammation, PG can be proinflammatory. In turn, the increase in PGE_2, which occurs after PAF addition, limited the production of reactive oxygen radicals (85) known to elicit the release of endogenous AA.

B. Polymorphonuclear Leukocytes (PMNL)

LTB_4 is a major product synthesized by stimulated PMNL. During nephrotoxic serum nephritis (86), active or passive Heymann nephritis (87) and nephropathy induced by cationic gammaglobulins (88), increased production of LTB_4 has been reported, although the exact origin of the eicosanoid may still be questioned. In our experience, it seems that the glomerulus is, unlike PMNL (89), unable to metabolize LTB_4 into the LTB_4 ω-hydroxy and ω-carboxy derivatives. Rat and human glomeruli lack ω-20 hydroxylase. Table 5 shows the comparative results obtained in our lab with PMNL and glomeruli incubated with a tracer dose of [^{3}H] LTB_4. No trace of ω-hydroxy or carboxy derivatives of LTB_4 could be detected in the media. However, in our HPLC system, the 6-trans LTB_4 isomer and authentic LTB_4 were not adequately resolved. Therefore, it cannot be excluded that glomeruli, as demonstrated for a renal homogenate (90), may transform LTB_4 into its 6-trans isomer whose activity is still unknown.

C. Platelets

When platelets were coincubated with human glomeruli, we demonstrated that PGI_2, measured by radioimmunoassay of its stable metabolite 6-keto-$PGF_{1\alpha}$, increased moderately (x 1.5 - 2) (unpublished results). Since platelets are devoid of PGI_2-synthase, this indicates that a platelet metabolite was able to trigger AA metabolism in glomeruli. Here again, an interaction seems to occur between platelets and glomeruli with a glomerular factor triggering platelets that, in turn, stimulate glomeruli to produce PGI_2.

V. CONCLUSION

Interaction between glomerular cells and bone marrow-derived cells may occur either (a) because these cells arrive in contact with glomerular cells or (b) because glomerular cells may release products capable of attracting and/or stimulating bone marrow-derived cells. The results of these interactions will depend on several factors: (a) the nature of the interacting blood cells, (b) the availability of the substrate AA, (c) the ratio between

Table 5. Comparative Results of [^{3}H] LTB_4 Bioconversion by Human Glomeruli and Human Leukocytes

Addition of [^{3}H] LTB_4 to :	LTB_4 + Trans-LTB_4	ω-OH + ω-COOH LTB_4
	Percentage of the sum of cpm recovered after HPLC	
Buffer alone	98.4	1.6
Leukocytes	0.9	99.1
Human glomeruli	98.3	1.7
Rat glomeruli	88.8	11.2

Incubation of the cells with exogenous [^{3}H] LTB_4 was for 15 min at 37 °C. Resolution of LTB_4 and its metabolites was performed by reverse phase HPLC after extraction and purification on SEP-PAK.

39

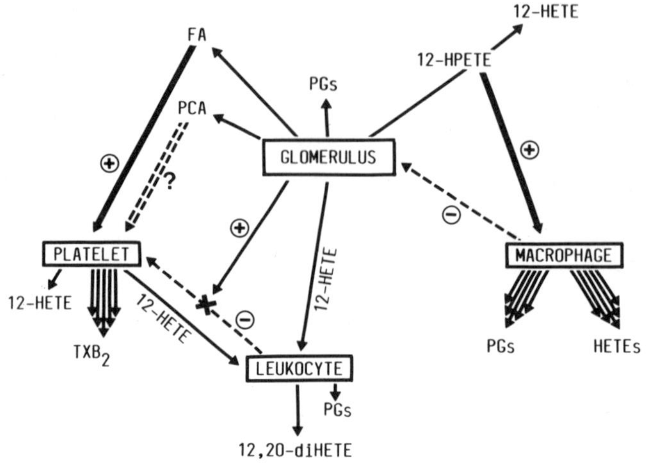

Fig. 8. General scheme of the known interactions
existing among glomeruli and bone marrow-
derived cells.

blood cells and glomerular cells, (d) the degree of activation of the blood
cells. The metabolites formed during the interaction may belong to the
cyclooxygenase and/or the lipoxygenase pathways. Interaction may be reciprocal: it may be initiated by the glomerular or the bone marrow-derived
cells. A general scheme summarizes the main results (Fig. 8) obtained from
these in vitro studies. The in vitro studies that have been performed to
date are probably not representative of an adequate model of the pathological conditions in which bone marrow-derived and glomerular cells are stimulated by various stimuli, such as complement fractions, endotoxins, immune-
complexes, etc. However, they indicate clearly that AA metabolism, even in
unstimulated cells, may be considerably changed by cell-cell interaction.
They also emphasize the role of the glomerulus as a biochemical source of
factors for blood cell activation.

ACKNOWLEDGMENTS

This work was supported by grants of the "Institut National de la
Santé et de la Recherche Médicale" and the "Faculté de Médecine
Saint-Antoine." We thank Mrs. A. Morin for preparing the manuscript.

REFERENCES

1. P. Needleman, A. Wyche, A. Raz, Platelet and blood vessel arachidonate

metabolism and interactions, J. Clin. Invest. 63:345-349 (1979).

2. G. Hornstra, E. Haddeman, J.A. Don, Blood platelets do not provide endo-peroxides for vascular prostacyclin production, Nature (London) 279: 66-68 (1979).

3. A.J. Marcus, B.B. Weskler, E.A. Jaffe, M.J. Broekman, Synthesis of pros-tacyclin from platelet-derived endoperoxides by human endothelial cells, J. Clin. Invest. 66:979-986 (1980).

4. A.I. Schafer, D.D. Crawford, M.A. Gimbrone, Unidirectional transfer of prostacyclin endoperoxides between platelets and endothelial cells, J. Clin. Invest. 73:1105-1112 (1984).

5. P. Borgeat, S. Picard, P. Vallerand, P. Sirois, Transformation of arachi-donic acid in leukocytes. Isolation and structural analysis of a novel dihydroxy derivative, Prostaglandins and Medicine 6:557-570 (1981).

6. P. Borgeat, B. Fruteau de Laclos, S. Picard, J. Drapeau, P. Vallerand, E.J. Corey, Studies on the mechanism of formation of the 5S,12S dihydroxy-6-8-10-14 (E,Z,E,Z)-icosatetraenoic acid in leukocytes, Prostaglandins 23:713-724 (1982).

7. A.J. Marcus, M.J. Broekman, L.B. Safier, H.L. Ullman, N. Islam, Forma-tion of leukotrienes and other hydroxyacids during platelet-neutro-phil interactions in vitro, Biochem. Biophys. Res. Commun. 109:130-137 (1982).

8. B. Fruteau de Laclos, P. Braquet, P. Borgeat, Characteristics of leuko-triene (LT) and hydroxyeicosatetraenoic acid (HETE) synthesis in human leukocytes in vitro: Effect of arachidonic concentration, Prostaglandins, Leukotrienes and Medicine 13:47-52 (1984).

9. P.Y.K. Wong, P. Weslund, M. Hamberg, E. Granström, P.H.W. Chao, B. Samuelsson, ω-hydroxylation of 12-L-hydroxy-5,8,10,14 eicosatetraeno-ic acid in human polymorphonuclear leukocytes, J. Biol. Chem. 259: 2683-2686 (1984).

10. A. Marcus, L.B. Safier, H.L. Ullman, M.J. Broekman, N. Islam, T.D. Oglesby, R. Gorman, 12S,20-dihydroxyicosatetraenoic acid: a new icosa-noid synthesized by neutrophils from 12S-hydroxy icosatetraenoic acid produced by thrombin- or collagen-stimulated platelets, Proc. Natl. Acad. Sci. USA 81:903-907 (1984).

11. C.A. Dahinden, R.M. Clancy, M. Gross, J.M. Chiller, T.E. Hugli, Leukotri-ene C_4 production by murine mast cells: Evidence of a role for extracellular leukotriene A_4, Proc. Natl. Acad. Sci. USA 82:6632-6636 (1985).

12. F. Fitzpatrick, W. Ligget, J. McGee, S. Bunting, D. Morton, B. Samuelsson, Metabolism of leukotriene A_4 by human erythrocytes, J.

Biol. Chem. 259:11403-11407 (1984).

13. C.R. Pace-Asciak, J. Klein, S.P. Spielberg, Metabolism of leukotriene A$_4$ into C$_4$ by human platelets, Biophys. Biochim. Acta 877:68-74 (1986).

14. C. Hadjiagapiou and A. Spector, 12-hydroxyeicosatetraenoic acid reduces prostacyclin production by endothelial cells, Prostaglandins 31:1135-1144 (1986).

15. J. Turk, A. Wyche, P. Needleman, Inactivation of vascular prostacyclin synthetase by platelet lipoxygenase products, Biochem. Biophys. Res. Commun. 95:1628-1632 (1980).

16. Y. Hashimoto, C. Naito, T. Teramoto, H. Kato, M. Kinoshinta, M. Kawamura, H. Hayashi, H. Oka, Time-dependent inhibition of the cyclo-oxygenase pathway by 12-hydroperoxy 5,8,10,14-eicosatetraenoic acid, Biochem. Biophys. Res. Commun. 130:781-785 (1985).

17. J. Maclouf, B. Fruteau de Laclos, P. Borgeat, Stimulation of leukotriene biosynthesis in human blood leukocytes by platelet-derived 12-hydro-peroxy-icosatetraenoic acid, Proc. Natl. Acad. Sci. USA 79:6042-6046 (1982).

18. A. Delmaschio, J. Maclouf, E. Corvazier, M.J. Grange, P. Borgeat, Acti-vated platelets stimulate human neutrophil functions, Nouv. Rev. Fr. Hematol. 27:275-278 (1985).

19. J. Maclouf, A. Delmaschio, M.J. Grange, P. Borgeat, Regulation and manip-ulation of arachidonate cascade in cell-cell interaction, in: "Advances in Prostaglandin, Thromboxane and Leukotriene Research," Raven Press, New York 15:209-211 (1985).

20. J.Y. Vanderhoek, R.W. Bryant, J.M. Bayley, 15-hydroxy-5-8,11,13-eicosate-traenoic acid. A potent and selective inhibitor of platelet lipoxy-genase, J. Biol. Chem. 255:5956-5998 (1980).

21. J.Y. Vanderhoek, R.W. Bryant, J.M. Bayley, Inhibition of leukotriene bio-synthesis by the leukocyte product 15-hydroxy-5,8,11,13-eicosate-traenoic acid, J. Biol. Chem. 255:10064-10066 (1980).

22. J.Y. Vanderhoek, S.N. Tare, J.M. Bayley, A.L. Goldstein, D. Pluznik, New role for 15-hydroxyeicosatetraenoic acid, J. Biol. Chem. 257:12191-12195 (1982).

23. L.A. Boxer, J.M. Allen, M. Schmidt, M. Yoder, R.L. Baehner, Inhibition of polymorphonuclear leukocyte adherence by prostacyclin, J. Lab. Clin. Med. 95:672-678 (1980).

24. D.K. Miller, S. Sadowski, D.D. Soderman, F.A. Kuehl, Endothelial cell prostacyclin production induced by activated neutrophils, J. Biol. Chem. 260:1006-1014 (1985).

25. J.M. Harlan and K.S. Callahan, Role of hydrogen peroxide in the neutro-

phil-mediated release of prostacyclin from cultured endothelial cells, J. Clin. Invest. 74:442-448 (1984).

26. L. Taylor, M.J. Menconi, P. Polgar, The participation of hydroperoxides and oxygen radicals in the control of prostaglandin synthesis, J. Biol. Chem. 258:6855-6857 (1983).

27. R.A. Lewis, E.J. Goetzl, J.M. Drazen, N.A. Soter, K.F. Austen, E.J. Corey, Functional characterization of synthetic leukotriene B and its stereochemical isomers, J. Exp. Med. 154:1243-1248 (1981).

28. E. Pirotsky, E. Ninio, J. Bidault, P. Pfister, J. Benveniste, Biosynthesis of platelet-activating factor. VI. Precursor of platelet-activating factor and acyl transferase activity in isolated rat kidney cells, Lab. Invest. 51:567-572 (1984).

29. A. Hassid, M. Konieczkowski, M.J. Dunn, Prostaglandin synthesis in isolated rat kidney glomeruli, Proc. Natl. Acad. Sci. USA 76:1155-1159 (1979).

30. J. Sraer, J.D. Sraer, D. Chansel, Prostaglandin synthesis by isolated rat renal glomeruli, Mol. Cell. Endocrinol. 16:29-37 (1979).

31. V.W. Folkert and D. Schlondorff, Prostaglandin synthesis in isolated glomeruli, Prostaglandins 17:79-86 (1979).

32. J. Sraer, N. Ardaillou, J.D. Sraer, R. Ardaillou, In vitro prostaglandin synthesis by human glomeruli and papillae, Prostaglandins 23:855-864 (1982).

33. J. Sraer, W. Siess, L. Moulonguet-Doleris, J.P. Oudinet, F. Dray, R. Ardaillou, In vitro prostaglandin synthesis by various rat renal preparations, Biochim. Biophys. Acta 710:45-52 (1982).

34. J. Sraer, M. Rigaud, M. Bens, H. Rabinovitch, R. Ardaillou, Metabolism of arachidonic acid via the lipoxygenase pathway in human and murine glomeruli, J. Biol. Chem. 258:4325-4330 (1983).

35. K. Jim, A. Hassid, F. Sun, M.J. Dunn, Lipoxygenase activity in rat kidney glomeruli, glomerular epithelial cells and cortical tubules, J. Biol. Chem. 257:10294-10299 (1982).

36. R.B. Sterzel, J.H.H. Ehrich, H. Lucia, D. Thomson, M. Kashgarian, Mesangial disposal of glomerular immune deposits in acute malarial glomerulonephritis of rats, Lab. Invest. 46:209-214 (1982).

37. S.R. Holdsworth, N.W. Thomson, E.F. Glasgow, J.P. Dowling, R.C. Atkins, Tissue culture of isolated glomeruli in experimental chronic immune complex glomerulonephritis, Nephron 32:227-233 (1978).

38. H. Shigematsu, Morphological approach on the action and function of monocytes and macrophages in acute experimental glomerulonephritis, in: "Proceedings of the VIIIth International Congress of Nephrology," W. Zurukzogzu, M. Papadimetriou, M. Purpasopoulos, M. Sion, C.

Zamboulis, eds., S. Karger, Basel, pp. 872-878 (1981).

39. G.J. Becker, W.W. Hancock, J.L. Stow, E.F. Glasgow, R.C. Atkins, N.M. Thomson, Involvement of the macrophages in experimental crescentic glomerulonephritis, J. Exp. Med. 147:98-109 (1978).

40. R.C. Atkins, S.R. Holdsworth, W.W. Hancock, N.M. Thomson, E.F. Glasgow, Cellular immune mechanisms in human glomerulonephritis: the role of mononuclear leukocytes, Springer Semin. Immunopathol. 5:269-296 (1982).

41. A.B. Magil, L.O. Wadsworth, M. Loewen, Monocytes and human renal glomerular disease. A quantitative evaluation, Lab. Invest. 44:27-33 (1981).

42. N.M. Thomson, S.R. Holdsworth, E.F. Glasgow, R.C. Atkins, The macrophage in the development of experimental crescentic glomerulonephritis, Am. J. Pathol. 94:223-240 (1979).

43. A.E. Postlethwaite and A. Kang, Collagen and collagen peptide chemotaxis of human blood monocytes, J. Exp. Med. 143:1299-1307 (1976).

44. C.H. Dubois, G. Goffinet, J.B. Foidard, C. Dechenne, J.M. Foidard, P. Mahieu, Evidence for a particular binding capacity of rat peritoneal macrophages to rat glomerular mesangial cells in vitro, Europ. J. Clin. Invest. 12:239-246 (1982).

45. M.T. Quinn, S. Parthasarathy, D. Steinberg, Endothelial cell-derived chemotactic activity for mouse peritoneal macrophages and the effects of modified form of low density lipoprotein, Proc. Natl. Acad. Sci. USA 82:5949-5953 (1985).

46. L. Baud, J. Sraer, F. Delarue, M. Bens, F. Balavoine, D. Schlondorff, R. Ardaillou, J.D. Sraer, Lipoxygenase products mediate the attachment of rat macrophages to glomeruli in vitro, Kidney Int. 27:855-863 (1985).

47. W.A. Scott, N.A. Pawloski, M. Andreach, Z.A. Cohn, Resting macrophages produce distinct metabolites from exogenous arachidonic acid, J. Exp. Med. 155:535-547 (1982).

48. C.A. Rouzer, W.A. Scott, A.L. Hamill, F.T. Liu, D.H. Katz, Z.A. Cohn, Secretion of leukotriene C and other arachidonic acid metabolites by macrophages challenged with immunoglobulin E immune complexes, J. Exp. Med. 156:1077-1086 (1982).

49. H. Rabinovitch, J. Durand, M. Rigaud, F. Mendy, J.C. Breton, Transformation of arachidonic acid into monohydroxy-eicosatetraenoic acids by mouse peritoneal macrophages, Lipids 16:518-524 (1981).

50. N. Doig and A.W. Ford-Hutchinson, The production and characterization of products of the lipoxygenase enzyme system released by rat peritoneal macrophages, Prostaglandins 20:1007-1019 (1980).

51. I. Koyama, H. Yamagami, T. Kuwal, M. Kurata, Release of 6-keto-prosta-

glandin $F_{1\alpha}$ and thromboxane B_2 from mouse peritoneal macrophages during their adhesion and spreading on a glass surface, Prostaglandins 23:777-785 (1982).

52. E.A. Lianos, M.A. Rahman, M.J. Dunn, Glomerular arachidonate lipoxygenation in rat nephrotoxic serum nephritis, J. Clin. Invest. 76:1355-1359 (1985).

53. R. Lacave, F. Delarue, E. Rondeau, J.D. Sraer, 5- and 12-hydroperoxytetraenoic acids promote contraction of cultured glomerular visceral epithelial cells, Kidney Int. 27:339 (Abstr.) (1986).

54. K. Kuhn, G.B. Ryan, S.J. Hein, R.G. Galaske, M.J. Karnovsky, An ultrastructural study of the mechanisms of proteinuria in rat nephrotoxic nephritis, Lab. Invest. 36:375-387 (1977).

55. L. Baud, J. Hagege, J. Sraer, E. Rondeau, J. Perez, R. Ardaillou, Reactive oxygen production by cultured rat glomerular mesangial cells during phagocytosis is associated with stimulation of lipoxygenase activity, J. Exp. Med. 158:1836-1852 (1983).

56. J. MacDermont, C.R. Kelsey, K.A. Maddell, P. Richmond, R.K. Knight, P.J. Cole, C.T. Dollery, D.N. Landon, I.A. Blair, Synthesis of leukotriene B_4 and prostanoids by human alveolar macrophages: analysis by gas chromatography/mass spectrometry, Prostaglandins 27:163-169 (1984).

57. W. Hsueh and F.F. Sun, Leukotriene B_4 biosynthesis by alveolar macrophages, Biochem. Biophys. Res. Commun. 106:1085-1091 (1982).

58. J.A. Rankin, M. Hitchcock, W. Merrill, M.K. Bach, J.R. Braschler, P.W. Askenase, IgE-dependent release of leukotriene C_4 from alveolar macrophages, Nature (London) 297:329-331 (1982).

59. J. Sraer, M. Bens, J.D. Sraer, R. Ardaillou, Specific activation of platelet thromboxane (TX) synthetase by human glomeruli in vitro, Kidney Int. 64:168 (Abstr.) (1984).

60. J. Sraer, C. Wolf, J.P. Oudinet, M. Bens, R. Ardaillou, J.D. Sraer, Human glomeruli release saturated fatty acids which stimulate thromboxane synthesis in platelets, Kidney Int., in press (1987).

61. D. DeProst and A. Kanfer, Quantitative assessment of procoagulant activity in isolated rat glomeruli, Kidney Int. 28:566-568 (1985).

62. J.R. Maynard, C.A. Heckman, F.A. Pitlick, Y. Nemerson, Association of tissue factor activity with the surface of cultured cells, J. Clin. Invest. 52:1427-1434 (1973).

63. M. Colucci, G. Balconi, R. Lorenzet, A. Pietra, D. Locati, M.B. Donati, N. Semeraro, Cultured human endothelial cells generate tissue factor in response to endotoxin, J. Clin. Invest. 71:1893-1896 (1983).

64. M. Poe and J.M. Liesch, Mouse submaxillary gland renin contains a noncovalently attached fatty acid, J. Biol. Chem. 258:9856-9860 (1983).

65. A.A. Aderem, M.M. Keum, E. Pure, Z.A. Cohn, Bacterial lipopolysaccha-rides, phorbolmyristate acetate and zymosan induce the myristoylation of specific macrophage proteins, Proc. Natl. Acad. Sci. USA 83:5817-5821 (1986).

66. D.A. Towler and L. Glaser, Protein fatty acid acylation: enzymatic syn-thesis of an N-myristoyl glycl peptide, Proc. Natl. Acad. Sci. USA 83:2812-2816 (1986).

67. E.N. Olson, D.A. Towler, L. Glaser, Specificity of fatty acid acylation of cellular proteins, J. Biol. Chem. 260:3784-3790 (1985).

68. R.M.C. Dawson, R.F. Irvine, J. Bray, P.J. Quinn, Long-chain unsaturated diacylglycerols cause perturbation in the structure of phospholipid bilayers rendering them susceptible to phospholipid attack, Biochem. Biophys. Res. Commun. 125:836-842 (1984).

69. J. Sraer, M. Bens, J.D. Sraer, L. Baud, E. Podjarny, R. Ardaillou, Changes in arachidonic acid metabolism during interaction between glomerular and bone marrow-derived cells, Advances in Inflammation Research 10:291-293 (1986).

70. E. Dratewka-Kos, D.O. Tinker, B. Kindl, Unsaturated fatty acids inhibit ADP-arachidonate-induced platelet aggregation without affecting throm-boxane synthesis, Biochem. Cell. Biol. 64:906-913 (1985).

71. J. Sraer, L. Baud, M. Bens, E. Podjarny, D. Schlondorff, R. Ardaillou, J.D. Sraer, Glomeruli cooperate with macrophages in converting arachi-donic acid to prostaglandins and hydroxyeicosatetraenoic acids, Prostaglandins, Leukotrienes and Medicine 13:67-74 (1984).

72. J. Sraer, M. Bens, J.P. Oudinet, R. Ardaillou, Bioconversion of leukotri-ene C_4 by rat glomeruli and papilla, Prostaglandins 31:909-921 (1986).

73. K. Bernström and S. Hammarström, Metabolism of leukotriene D by porcine kidney, J. Biol. Chem. 256:9579-9582 (1981).

74. E.J. Goetzl, B.A. Burrall, L. Baud, K.H. Scriven, J.D. Levine, C.H. Koo, Generation and recognition of leukotriene mediators of hypersensitivi-ty and inflammation, Dig. Dis. Sci., in press (1987).

75. E.J. Goetzl, L.L. Brindley, D.W. Goldman, Enhancement of human neutro-phil adherence by synthetic leukotriene constituents of the slow-reacting substance of anaphylaxis, Immunology 50:35-41 (1983).

76. L. Baud, J. Sraer, J. Perez, M.P. Nivez, R. Ardaillou, Leukotriene C_4 binds to human glomerular epithelial cells and promotes their prolif-eration in vitro, J. Clin. Invest. 76:374-377 (1985).

77. R. Barnett, P. Goldwasser, L.A. Scharschmidt, D. Schlondorff, Effects of leukotrienes on isolated rat glomeruli and cultured mesangial cells, Am. J. Physiol. 250:F838-F844 (1986).

78. M. Simonson and M.J. Dunn, Leukotrienes C_4 and D_4 contract rat glomerular mesangial cells, Kidney Int. 30:524-531 (1986).

79. E.B. Cramer, L. Pologe, A. Pawlowski, Z.A. Cohn, W.A. Scott, Leukotriene C promotes prostacyclin synthesis by human endothelial cells, Proc. Natl. Acad. Sci. USA 80:4109-4113 (1983).

80. M.A. Clark, D. Littlejohn, T.P. Conway, S. Mong, S. Steiner, S.T. Crooke, Leukotriene D_4 treatment of bovine aortic endothelial cells and murine smooth muscle cells in culture results in an increase in phospholipase A_2 activity, J. Biol. Chem. 261:10713-10718 (1986).

81. L.A. Scharschmidt, J.G. Douglas, M.J. Dunn, Angiotensin II and eicosanoids in the control of glomerular size in the rat and human, Am. J. Physiol. 250:F348-F356 (1986).

82. L. Baud, J. Perez, M. Denis, R. Ardaillou, Modulation of fibroblast proliferation by sulfidopeptide leukotrienes: effect of indomethacin, J. Immunol. 138:1190-1195 (1987).

83. D. Schlondorff, J.A. Satriano, J. Hagege, J. Perez, L. Baud, Effect of platelet-activating factor and serum-treated zymosan on prostaglandin E_2 synthesis, arachidonic acid release and contraction of cultured rat mesangial cells, J. Clin. Invest. 73:1227-1231 (1984).

84. D. Schlondorff and R. Neuwirth, Platelet-activating factor and the kidney, Am. J. Physiol. 251:F1-F11 (1986).

85. L. Baud, J. Perez, R. Ardaillou, Dexamethasone and hydrogen peroxide production by mesangial cells during phagocytosis, Am. J. Physiol. 250: F596-F604 (1986).

86. E.A. Lianos, Leukotriene synthesis in nephrotoxic serum nephritis: role of complement and polymorphonuclear leukocytes, Clin. Res. 34:972A (Abstr.) (1986).

87. E.A. Lianos and B. Noble, Glomerular arachidonate 5-lipoxygenation in active and passive Heymann nephritis, Kidney Int. 31:277 (Abstr.) (1987).

88. M.A. Rahman, M. Nakasawa, S.N. Emancipator, M.J. Dunn, Increased leukotriene B_4 (LTB_4) synthesis in immune-injured rat glomeruli, Kidney Int. 29:343 (Abstr.) (1986).

89. S. Shak and I.M. Goldstein, ω-oxydation is the major pathway for the catabolism of leukotriene B_4 in human polymorphonuclear leukocytes, J. Biol. Chem. 259:10181-10187 (1984).

90. O. Breuer and S. Hammarström, Enzymatic conversion of leukotriene B_4 to 6-transleukotriene B_4 by rat kidney homogenates, Biochem. Biophys. Res. Commun. 142:667-673 (1987).

LIPOXYGENASE PRODUCTS AND THEIR FUNCTIONS IN GLOMERULI

Raymond Ardaillou, Laurent Baud and Josée Sraer

INSERM 64, Hôpital Tenon, Paris, France

Non-esterified arachidonic acid (C20:4) may be oxygenated by three major lipoxygenase pathways. These include the 5-, 12-, and 15-lipoxygenases which stereospecifically insert one oxygen molecule at the 5, 12, and 15 positions of this fatty acid via mechanisms involving hydrogen abstraction. Each of these enzymes converts C20:4 into a number of metabolites which, for many of them, represent essential mediators of inflammation. C20:4 is also the substrate for cyclooxygenase. Glomerular synthesis of the cyclooxygenase metabolites, prostaglandins (PG) and thromboxane (TX) and the role of these products on glomerular hemodynamics and the inflammatory injury during glomerular diseases have been extensively studied (1-3). This review will discuss the metabolism of C20:4 to its lipoxygenase products in glomeruli and their effects on glomerular functions. Potential pathophysiological action of these metabolites in glomerular inflammation will also be considered.

GENERAL PRINCIPLES OF SYNTHESIS OF THE LIPOXYGENASE PRODUCTS

Apart from the concentrations of the various lipoxygenases in a given tissue, the major factor controlling the synthesis of lipoxygenase products is the availability of free C20:4. C20:4 is esterified in the second position of the phospholipids. Both phospholipases C and A_2 are involved in the release of C20:4. The amount of free intracellular C20:4 serving as substrate for lipoxygenase activity depends on multiple factors which are essentially binding of C20:4 to intracellular proteins, reacylation of C20:4 into membrane phospholipids, and competition for C20:4 with the cyclooxygenase and epoxygenase pathways.

49

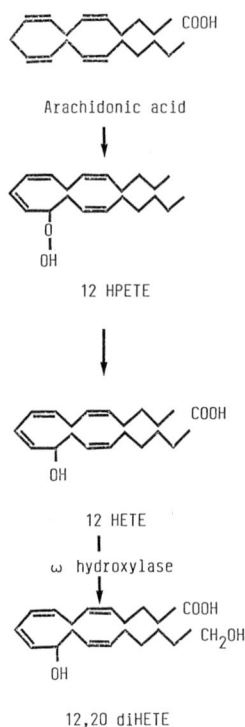

Fig. 1. Schema of the 12-lipoxygenase pathway.

Addition of oxygen to C20:4 in the presence of lipoxygenase results in the formation of hydroperoxyeicosatetraenoic acids (HPETE). HPETEs are unstable species which are converted by reductases into their corresponding stable derivatives, hydroxyeicosatetraenoic acids (HETE). 12-HETE is the only product of transformation of 12-HPETE (Fig. 1). It can be further hydroxylated at position 20 by an ω-hydroxylase (4). A possible biological role for the resulting product, 12-20 diHETE, has not yet been reported.

The 15-lipoxygenase pathway converts arachidonic acid into 15-HPETE which subsequently can be transformed into several products that differ by their degree of hydroxylation (Fig. 2). In addition to the monohydroxylated metabolite, 15-HETE, a novel series of trihydroxy-conjugated tetraenes, the lipoxins, has been described (5). The two main products of this series are lipoxin A (5,6,15 triHETE) and lipoxin B (5,14,15 triHETE). There are also dihydroxylated metabolites, essentially 8, 15 diHETE and 14, 15 diHETE (6). The latter two compounds possess a conjugated triene--e.g., three alternating double bonds--which is specific of the leukotriene (LT) structure. Therefore, they can be considered as LTs of the 15 series.

The 5-lipoxygenase system, which predominates in polymorphonuclear

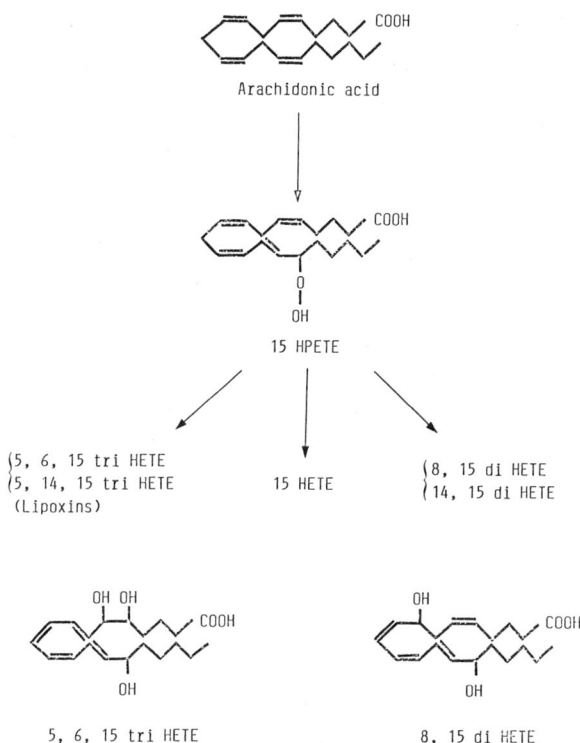

Fig. 2. Schema of the 15-lipoxygenase pathway.

leukocytes (PMNL), macrophages and mast cells (7), gives rise to the most potent of the C20:4-derived mediators of inflammation (Fig. 3). In competition with its reduction into 5-HETE, 5-HPETE is also converted by a dehydrase into LTA_4, an unstable epoxide subsequently transformed into either LTB_4 by an epoxide hydrolase or into LTC_4 by a LTC_4 synthetase. The latter enzyme is present in the microsomes and is distinct from the detoxifying glutathione transferases (8). Non-enzymatic hydrolysis of LTA_4 generates biologically inactive isomers of LTB_4, essentially 6 trans LTB_4 and 6 trans, 12 epi LTB_4 (9). LTB_4 is transformed by ω-hydroxylation or $_\omega$-carboxylation into poorly active or inactive products, 20 OH-LTB_4 and 20 COO-LTB_4 respectively (10). It has been recently proposed that LTB_4 may also be converted by an isomerase present in rat kidney homogenates into 6 trans LTB_4, which could represent the first step in a novel pathway of biological degradation of LTB_4 (11).

LTC_4 is the parent compound of the sulfidopeptide leukotrienes since it includes a molecule of glutathione in position 6. LTC_4 is metabolized into LTD_4 through removal of the γ-glutamyl residue by γ-glutamyl trans-

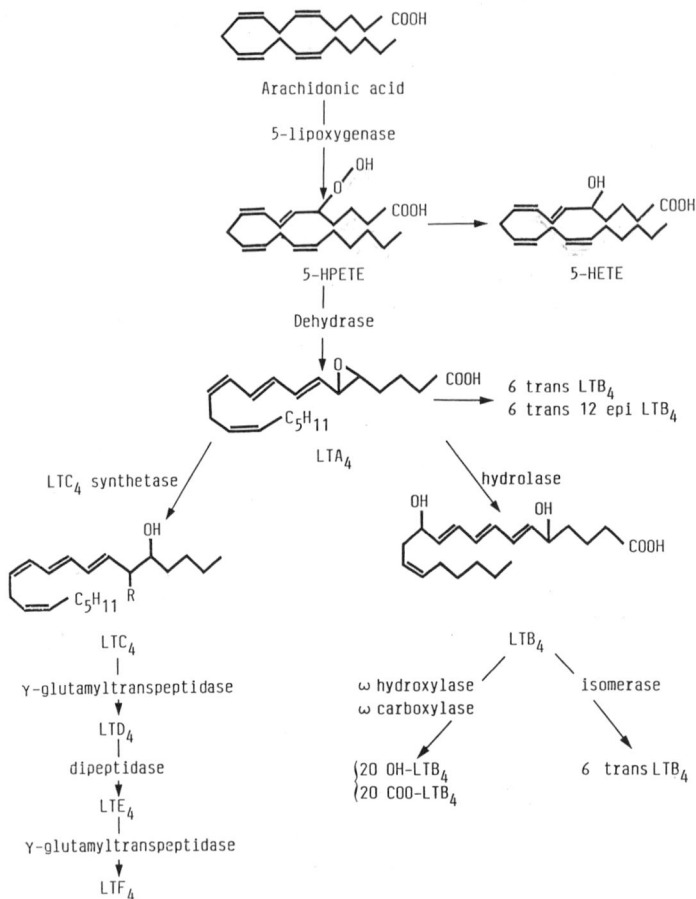

Fig. 3. Schema of the 5-lipoxygenase pathway.

peptidase. If the glycine residue is further removed by a dipeptidase, LTE_4 is formed. Finally, if γ-glutamyl is attached to the cysteine residue on LTE_4, LTF_4 is formed. In contrast to LTB_4, which is degraded by an intracellular process, the sulfidopeptide LTs are inactivated extracellularly by hypochlorous acid. The latter compound results from the interaction of H_2O_2 generated during the oxidative burst, chloride ion and myeloperoxidase released from the leukocytes (12).

METHODOLOGICAL ASPECTS

HETEs differ by the position of their hydroxyl radical, on which depends the polarity of the molecule; but all of them have the same molecular weight. It has been impossible to raise specific antibodies for each of them. In contrast, straight phase high performance liquid chromatography (HPLC), which separates these products with a good resolution according to their polarities, allows an adequate analysis. Because HETEs contain a cis-trans conjugated diene, they absorb UV light with a maximum at 235 nm. However, since the amounts of HETEs produced by most preparations under usual conditions are low, radiometric HPLC is the most frequently utilized. Glomeruli or glomerular cells are incubated in the presence of $[^3H]$- or $[^{14}C]$-C20:4, and the radiolabeled HETEs produced are detected. This technique does not provide quantitative measurements of HETE synthesis since the specific activity of the tracer, which depends on its dilution in its endogenous pool, is unknown.

Analysis of LTs is made difficult by the presence of isomers in addition to the biologically active compounds in the incubation media of the tissues or cells. The antibodies available do not distinguish the LTs from their corresponding isomers. HPLC allows a correct analysis on condition that several systems are used. This is necessary because two products may coelute in one system and be separated in the other. Recently, special columns (chiral columns) for the separation of isomers have been proposed. LTs have UV absorption spectra characteristic of the conjugated triene with a maximum absorption near 270 nm. Like HETEs and due to their low concentrations, only the labeled (^{14}C or ^3H) products are usually detected. Lipoxins are measured at 302 nm, which corresponds to the tetraene conjugates.

Gas chromatography-mass spectrometry (GC/MS) can identify and quantify the lipoxygenase products. However, this is a relatively complex technique which is not currently available. Moreover, although there has been recent technical progress, the detection limit is usually too high and does not allow this technique to be utilized for the tissues or the cells synthesizing low amounts of lipoxygenase products.

Lipoxygenase inhibitors are also frequently used to identify an unknown C20:4 metabolite as a lipoxygenase product or to attribute a biological effect to lipoxygenase activity. Inhibition of synthesis or suppression of effect by nordihydroguaiaretic acid (NDGA), eicosatetraynoic acid (ETYA) or phenidone, in contrast to the inefficiency of cyclooxygenase blockers such as indomethacin or aspirin, represents indirect arguments in supplement of direct identification of the metabolite. However, these various drugs are not totally specific. Indeed, ETYA blocks the cyclooxygenase pathway, and NDGA behaves as a scavenger of reactive oxygen species. Other drugs acting at various steps of the lipoxygenase pathway have also been used. For

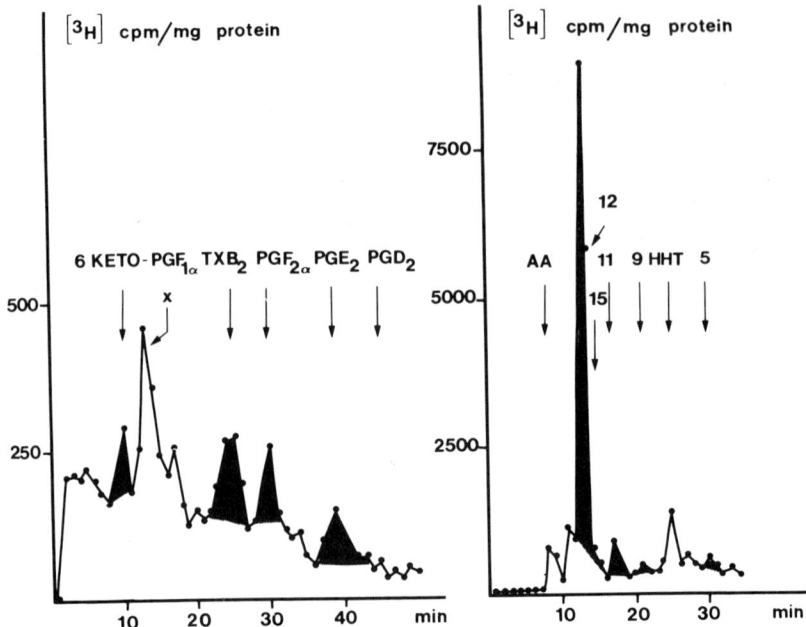

Fig. 4. HPLC profiles of the cyclooxygenase (right) and lipoxygenase (left) products of conversion of ^{3}H-C20:4 (30
ng/ml) after a 30-min. incubation with rat isolated
glomeruli. The retention times of authentic standards
are indicated by arrows.

example, the serine-borate complex inhibits the transformation of LTC_4
into LTD_4, bilirubin and hematin may block glutathione S-transferase activity, and L cysteine inhibits dipeptidase activity. More recently, TMK-688,
a derivative of caffeic acid has been used as an inhibitor of 5-lipoxygenase
(13).

SYNTHESIS OF LIPOXYGENASE PRODUCTS IN GLOMERULI AND ITS CONTROL

Murine and human glomeruli convert C20:4 through both the cyclooxygenase
and the lipoxygenase pathways. This has been demonstrated using freshly isolated glomeruli and cultured glomerular cells. When rat glomeruli are incubated with [^{3}H]- or [^{14}C]-C20:4, they convert this fatty acid into a
number of metabolites. The peaks of lipoxygenase products correspond to
greater amounts of radioactivities than those of cyclooxygenase products, in
the presence of tracer concentrations (30 ng/ml ^{3}H-C20:4) as well as in
the presence of an excess (10 µg/ml [^{14}C]-C20:4) of substrate (Figs. 4 and
5). Approximately 15-40 times more lipoxygenase products than cyclooxygenase products are formed. This can be due to differences in the concentra-

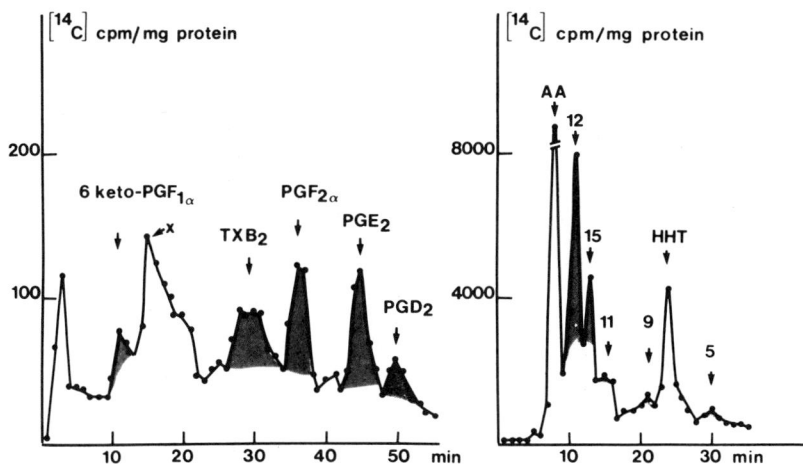

Fig. 5. HPLC profiles of the cyclooxygenase (right) and lipoxygenase (left) products of conversion of ^{14}C-C20:4 (10
g/ml) after a 30-min. incubation with rat isolated glomeruli. The retention times of authentic standards are
indicated by arrows.

tions or the affinities for C20:4 of the various enzymes and perhaps also to
specific pools of C20:4 linked to its different pathways.

We (14) and others (15,16) have identified HETEs in the glomeruli. Rat
glomeruli convert C20:4 into 12-HETE (14-16). They are also capable of synthesizing 15-HETE but only in the presence of high concentrations of C20:4.
No other HETE was detected in our study (14), whereas Jim et al. (15) also
found small amounts of 8- and 5-HETE. Subsequently, we confirmed the synthesis of 5-HETE by rat glomeruli (Figs. 4 and 6) although at a much smaller
rate than for 12-HETE. It is probable that 8-HETE, which coelutes in many
systems with hydroxyheptadecatrienoic acid (HHT), was mistaken for this compound. At low substrate concentrations, human glomeruli synthesize equivalent amounts of 12- and 15-HETE in smaller quantities than in rat (14).
Identification of the HETEs in these studies was achieved by thin layer chromatography (TLC), HPLC and GC/MS.

The blockade of HETE synthesis by lipoxygenase inhibitors such as NDGA,
ETYA and phenidone confirmed that these metabolites were lipoxygenase products. In contrast with PGs, HETEs are essentially stored within the glomeruli and are released only at a small extent into the medium. Our results
(14) demonstrated that the glomerular lipoxygenase activity was distributed
equally in the membranes and the cytosol, was maximum at pH 7.5 (rat) or 9.0

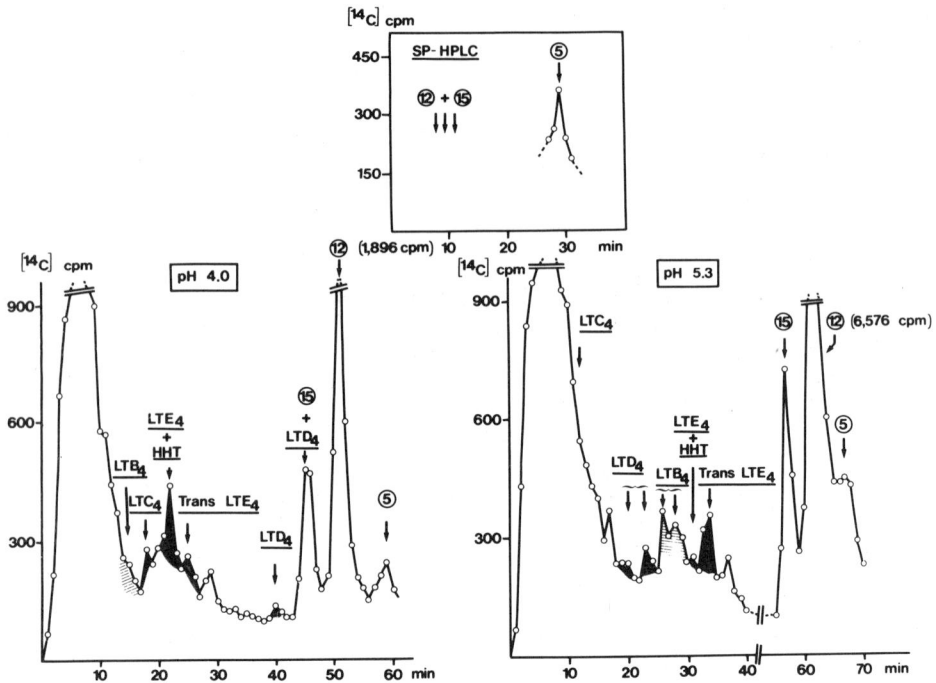

Fig. 6. Reverse phase HPLC of rat glomeruli incubated with ^{14}C-C20:4. Two eluent systems were used: 1-methanol:H$_2$O:acetic acid, 68:32:0.08 (v/v), pH 4 (left); 2-methanol:H$_2$O:acetic acid, 65:35:0.02 (v/v), pH 5.3 (right). The flow rate was 1.2 ml/min. in both systems. The results of straight phase HPLC for better separation of HETEs is given in inset. The retention times of authentic standards are indicated by arrows.

(human) and at 40-42°C (both species). The Km values calculated were 44 μM for 12-HETE in both species, 125 and 667 μM for 15-HETE in the murine and human glomeruli, respectively. These relatively high figures indicate that under normal conditions the velocity of the enzymes is markedly dependent on the concentration of substrate. Cultured epithelial (15) and mesangial (17) glomerular cells also synthesize 12-HETE. No other HETE was detected in epithelial cells. Although 11-HETE was observed in mesangial cells, this HETE is probably a product of the cyclooxygenase pathway. These results support strongly the conclusion that the synthesis of HETEs observed with isolated glomeruli is due to the resident glomerular cells and not to platelets or leukocytes entrapped in the glomerular capillaries.

The problem as to whether glomeruli synthesize LTs has not yet been totally solved. We (18) and others (15) have detected small amounts of radiolabeled 5-HETE after incubation of glomeruli with radiolabeled C20:4. LTB$_4$ synthesis by control rat glomeruli has been demonstrated by Cattell

Table 1. Synthesis of Lipoxygenase Products
          in Glomeruli

Preparation studied	Lipoxygenase product	Reference
Rat glomeruli	12-HETE	Jim et al. (15)
Human glomeruli	12-HETE 15-HETE	Sraer et al. (14)
Rat glomeruli	12-HETE	Sraer et al. (14)
Rat glomerular epithelial cells	12-HETE	Jim et al. (15)
Rat glomerular mesangial cells	12-HETE	Baud et al. (17)
Rat glomeruli	LTB_4 LTC_4	Lianos et al. (20)
Rat glomeruli	LTB_4	Cattell et al. (19)
Rat glomeruli	LTC_4 LTE_4	Sraer et al. (18)

et al. (19) and Lianos (20). Both groups of authors used specific antibodies for LTB_4. It is uncertain whether they could adequately separate
LTB_4 from its different isomers. In contrast, LTB_4 levels in rat glomeruli were below assay sensitivity in the study of Rahman et al. (21). Lianos
(20) and our group (18) also detected small amounts of sulfidopeptide LTs in
rat glomeruli. Using two different systems of HPLC, we observed the presence of LTC_4 and LTE_4 in homogenized rat glomeruli incubated with
$[^{14}C]$-C20:4 in the presence of reduced glutathione. In fact, the amounts
produced were negligible when compared with those of 12- and 15-HETE (Fig.
6). Identification of LTs in these in vitro studies was achieved by HPLC in
different systems, radioimmunoassay and infrared spectroscopy. Taken together, these different reports support the synthesis of LTs by rat glomeruli
under control conditions, although at a limited rate.

This raises the questions of the role of contaminating blood cells in
this synthesis. Indeed, it is difficult to exclude the presence of blood
cells in the renal preparations even after perfusion of the kidney with
saline. The best argument in favor of LT synthesis by renal cells them-

57

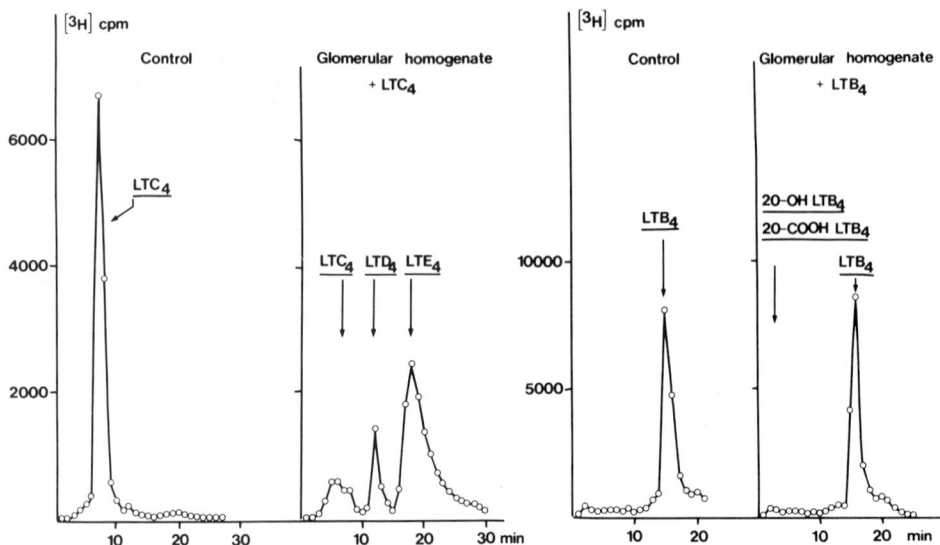

Fig. 7. Reverse phase HPLC of [^{3}H] LTC$_4$ (left) and [^{3}H] LTB$_4$ (right) in buffer or in the presence of a glomerular homogenate. The retention times of authentic standards are indicated by arrows.

selves would be the demonstration of such a synthesis by cultured renal cells, but to our knowledge this demonstration is still lacking. In Table 1, we have summarized the results of the published studies of HETE and LT synthesis by various glomerular preparations under control conditions.

METABOLISM OF LIPOXYGENASE PRODUCTS IN GLOMERULI

We have examined the *in vitro* metabolism of exogenously added [^{3}H]-LTs to rat glomeruli. Both homogenized and intact glomeruli convert LTC$_4$ essentially into LTE$_4$ and, to a smaller extent, into LTD$_4$ (22). Addition of L-cysteine, an inhibitor of dipeptidase, resulted in the accumulation of LTD$_4$. The metabolism of LTC$_4$ by the glomeruli was time- and temperature-dependent. The 10,000 g supernatant and pellet of homogenized glomeruli both retained the ability to metabolize [^{3}H]-LTC$_4$. In contrast with these results, we showed in further studies (Fig. 7) that there was no conversion of LTB$_4$ in its two main metabolites, 20-OH LTB$_4$ and 20-COO LTB$_4$ after incubation of [^{3}H]-LTB$_4$ with homogenized rat glomeruli. These results indicate that rat glomeruli possess the two enzymes necessary to metabolize LTC$_4$ into its further products, γ-glutamyl transpeptidase and dipeptidase, whereas they do not contain ω-hydroxylase. In opposition with the latter findings, Lianos (20) reported that [^{3}H]-LTB$_4$ was degraded to [^{3}H]-ω-LTB$_4$ in normal rat glomeruli.

58

The question as to whether or not glomeruli possess an ω-hydroxylase appears to be essential. In our opinion the metabolism of both classes of LTs by glomeruli is basically different. The processing of LTC_4 and LTD_4 by peptide cleavage can be considered rather as a bioconversion from one active mediator to another, since in many aspects LTD_4 possesses the same potency as LTC_4. In contrast, LTE_4, although exhibiting agonist properties, is much less active than its two precursors (23). The two main metabolites of LTB_4, 20-OH LTB_4 and 20-COOH LTB_4, are chemotactic towards human leukocytes but exhibit a much lower potency than LTB_4 (24). Therefore, the glomerular cells are not well equipped to protect themselves from the noxious effects of LTs. Only the local production of H_2O_2 by stimulated mesangial cells (17) or infiltrating PMNL or macrophages could result, via production of hypochlorous acid in the presence of Cl^- and myeloperoxidase, in inactivation of sulfidopeptide LTs after formation of sulfoxides.

EFFECTS OF LIPOXYGENASE PRODUCTS ON GLOMERULAR FUNCTIONS IN VIVO

Badr et al. (25) reported that LTC_4 infused in anesthetized rats immediately produced a significant elevation of mean arterial pressure and reductions in cardiac output and renal blood flow. LTC_4 administration also resulted in an average loss of 20% in plasma volume which stimulated angiotensin II production. This was responsible for perpetuation of the reduction in renal blood flow and, consequently, a fall in glomerular filtration rate by approximately 50%. The angiotensin II-mediated effects were abolished by saralasin. The results obtained in conscious rats by Filep et al. (26) were somewhat different. Intravenous injection of 8 µg/kg LTC_4 reduced by 33% renal plasma flow but slightly increased glomerular filtration rate. In parallel there was an increase in urine flow and urinary sodium excretion rate and also a rise in mean arterial pressure. Administration of FPL 55712 practically abolished LTC_4-induced changes in renal hemodynamics and water and electrolyte excretion.

Recently, Badr et al. (27), using micropuncture techniques, have studied the glomerular microcirculatory dynamics in Munich-Wistar rats after intravenous injection of LTD_4. Precautions (Saralasin administration, replacement of plasma volume loss, partial aortic constriction to maintain constancy of renal perfusion pressure) were taken to evaluate only the direct renal effects of LTD_4. Under these conditions, LTD_4 caused an increase in efferent but not afferent arteriolar resistance with a consequent fall in glomerular plasma flow rate and rise in glomerular capillary pressure.

Table 2. Glomerular Effects of Lipoxygenase Products In Vivo

Product tested	Technique and preparation	Dose	Effect	Reference
LTD$_4$	rat, micropuncture	1 µg/kg/min	SNGFR GPF	Badr et al. (26)
LTC$_4$	rat, clearance	8 µg/kg	GFR RPF	Filep et al. (25)
LTC$_4$	rat, clearance	2 µg/kg/min	GFR	Badr et al. (24)
Lipoxin A	rat, micropuncture	750 ng/kg/min intra-arterial	GFR SNGFR GPF	Serhan et al. (27)

Single nephron glomerular filtration rate fell, due to combined reductions in glomerular plasma flow and glomerular ultrafiltration coefficient (Kf). These results demonstrate the local glomerular constrictor action of LTD$_4$.

In contrast with the sulfidopeptide LTs, lipoxins are vasodilatory agents. Serhan et al. (28) showed that administration of lipoxin A in the renal artery of anesthetized rats resulted in the parallel increase of single nephron glomerular filtration rate and glomerular plasma flow. There was a selective and marked fall in afferent but not efferent arteriolar resistance and a rise in intraglomerular capillary pressure with decrease of Kf. These authors concluded that lipoxin A produced glomerular hyperfusion, hypertension and hyperfiltration. These effects, which are in opposition with those of sulfidopeptide LTs, suggest a balance between constrictory and dilatory metabolites of C20:4 via the 5-lipoxygenase pathway (Table 2).

EFFECTS OF LIPOXYGENASE PRODUCTS ON GLOMERULAR FUNCTIONS IN VITRO

Lipoxygenase products exhibit many effects in isolated glomeruli or glomerular cultured cells. Sulfidopeptide LTs modify the cell surface area of the glomerular mesangial cells. The PGE$_2$-stimulated release of renin is inhibited by 12-HETE, and 12-HETE and 12-HPETE from glomerular origin may act on the functions of the bone marrow-derived cells present in the glomerular capillaries. Lipoxygenase activity is necessary for production of oxygen reactive species associated with phagocytosis of foreign particles by mesangial cells. Finally, sulfidopeptide LTs promote the proliferation of the glomerular cells (Table 3).

Table 3. Glomerular Effects of Lipoxygenase Products In Vitro

Product tested	Preparation	Effect	Dose (μM)	Effect (percent of control)	Reference
12-HETE	Human mesangial cells	Inhibition of PGE_2-stimulated renin secretion	0.1	0	Ardaillou et al. (33)
5-HPETE 12-HPETE	Human glomerular	Contraction	100	45	Lacave et al. (32)
LTC_4	Rat mesangial cells	Contraction	1	77	Simonson and Dunn (31)
LTC_4	Rat glomeruli	Contraction	1	86	Barnett et al. (30)
	Rat mesangial cells	Contraction	0.1	90	
LTC_4	Human epithelial cells	Proliferation	0.1	248	Baud et al. (29)
LTC_4	Rat mesangial cells	Proliferation	0.01 - 1	160	Simonson and Dunn (31)

The first step in the glomerular effects of lipoxygenase products is, for the sulfidopeptide LTs, their binding to glomerular-specific receptors. LTC_4 receptors have been identified and characterized in isolated rat renal glomeruli (29). The equilibrium dissociation constant for $[^3H]$-LTC_4 binding to glomeruli, calculated from saturation and competitive binding-inhibition studies, was approximately 30 nM; and glomerular LTC_4 receptor density was around 9 pmol/mg protein. The other natural vasoactive sulfidopeptide LTs, LTD_4 and LTE_4, the chemotactic LT, LTB_4, and the sulfidopeptide LT antagonist, FPL 55712, competed for the receptor at concentrations 2-3 orders of magnitude greater than LTC_4. It was concluded that $[^3H]$-LTC_4 bound specifically to glomerular cells and not to retained blood cells since $[^3H]$-LTC_4 binding was identical in glomeruli isolated from saline-perfused and nonperfused kidneys within the same animal. This conclusion was strengthened by the demonstration (30) of the presence of specific receptors for LTC_4 in glomerular cultured human epithelial cells. The apparent dissociation constant was 220 nM, greater than that observed in isolated glomeruli. LTD_4 and LTE_4 also displace $[^3H]$-LTC_4 from its receptors, with ED_{50} values that were 2 orders of magnitude higher than for LTC_4, but FPL 55712 was minimally active. In these two studies, specific binding could be demonstrated only at low temperatures (4 and 8°C respectively) and in the presence of inhibitors of γ-glutamyl transpeptidase activity such as L-serine borate. This confirms that the enzymes metaboliz-ing LTC_4 are very active in the glomeruli.

In vitro studies suggested that the decrease of Kf observed in vivo after administration of sulfidopeptide LTs could be due to contrac-tion of glomerular mesangial cells. Barnett et al. (31) reported that both LTC_4 and LTD_4 at 1 μM produced a fall of 10-14% in the surface area of isolated glomeruli similar to what has been previously observed with other vasoconstrictor agents such as AII, AVP and PAF. The effects of both ago-nists were dose- and time-dependent. It is likely that the shape change represented cell contraction because rapid increases in the wrinkling of the mobile surface of the mesangial cells grown on a flexible silicone rubber support were observed. Simonson and Dunn (32) obtained similar results. Using image analysis microscopy, they measured the number of contracting mesangial cells and the cross-sectional area of these cells in the presence of increasing doses (1 nM - 1 μM) of LTC_4 and reported a dose-dependent decrease (16-23%) in the surface area of approximate 30% of the cells. Con-traction started at 5 min. and was maximum at 10-20 min.

Sulfidopeptide LTs, like AII, AVP and PAF, increase intracellular calci-um concentration, as demonstrated recently by one of us (L. Baud, unpub-lished results) in HL-60 cell-derived myelocytes. This increase in intracel-lular calcium could result in phosphorylation of myosin and contractile ac-tivity via activation of the myosin light chain kinase (Fig. 8). In con-trast to other vasoactive agonists, sulfidopeptide LTs did stimulate PGE_2 production in rat mesangial cells (31). HPETEs have also been shown to induce a dose-dependent reduction of the surface area of human cultured vis-ceral epithelial cells. Both 5- and 12-HPETE were active whereas their hy-droxyderivatives, 5- and 12-HETE, had no effect (33).

Lipoxygenase products may act indirectly on the glomerular microcircula-tion via their effects on the local renin production. At 0.1 μM, 12-HETE did not modify basal renin production, but it inhibited PGE_2-stimulated renin production in human mesangial cells. The mechanism of this effect has not yet been unraveled. Neither the basal nor the PGE_2-stimulated produc-tion of endogenous cyclic AMP was changed by 12-HETE. Like PGE_2, it had no effect on endogenous cyclic GMP (34). Therefore, it may be hypothesized that 12-HETE acts via a change in intracellular calcium which is considered,

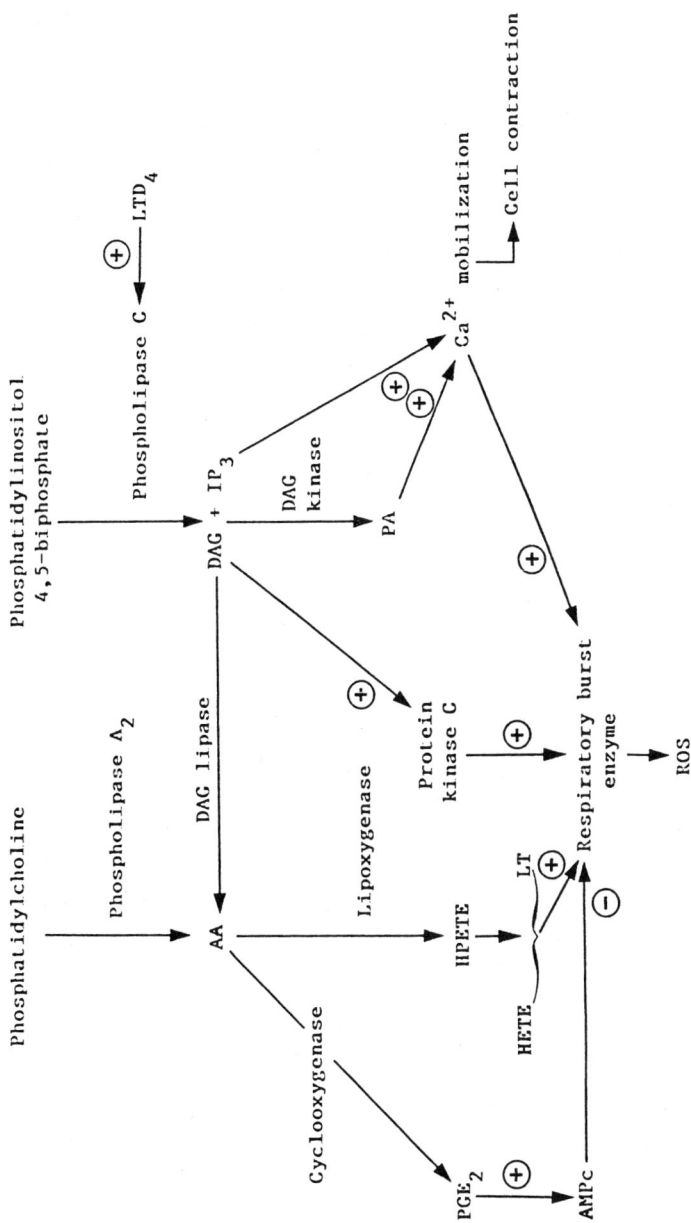

Fig. 8. Effects of lipoxygenase products on oxidative burst and cell contraction.
DAG = diacylglycerol; IP$_3$ = inositoltriphosphate; PA = phosphatidic acid;
ROS = reactive oxygen species; AA = arachidonic acid.

in parallel with cyclic AMP and GMP, as a putative second messenger of renin secretion.

Lipoxygenase products from glomerular origin may act on the functions of monocyte-macrophages present in the glomerular capillaries or in the Bowman's space. Both 12-HETE and its precursor, 12-HPETE, are the main products of the lipoxygenase pathway synthesized in rat glomeruli. Exposure of macrophages to 10 nM 12-HPETE promoted an increase of their PG production. Incubation of macrophages with glomeruli or a lipid extract of glomeruli gave identical results, and this effect was suppressed by pretreatment of glomeruli by NDGA. Inhibition of the lipoxygenase pathway also inhibited the attachment of ^{3}H-uridine-labeled macrophages to isolated glomeruli (35). We also observed that the lipid extract prepared from a macrophage-conditioned medium inhibited the production of 12-HETE by the glomeruli (36).

Taken together, these results suggest the following hypothesis: lipoxygenase products of glomerular origin influence adhesion of monocyte-macrophages to glomeruli probably by increasing PG production in macrophages (37). Macrophages release a lipidic factor which may, in turn, inhibit the synthesis of 12-HPETE by glomeruli and thus limit adherence of macrophages. In addition, 12-HPETE and 12-HETE have been shown to stimulate LTB_4 in PMNL (38) and procoagulant activity in macrophages (39). Therefore, it is possible to hypothesize that glomeruli may influence these functions of bone marrow-derived cells via synthesis of the metabolites of the 12-lipoxygenase pathway.

Lipoxygenase products also play a role in the biochemical events associated with phagocytosis of serum-treated zymosan (STZ) in cultured rat mesangial cells (17). Under these conditions the cells exhibited an oxidative burst with increased synthesis of superoxide anion (O_2-) and hydrogen peroxide (H_2O_2). Incubation with STZ markedly stimulated the release of C20:4 from its phospholipid stores and its transformation into 12-HETE. Two lipoxygenase inhibitors, ETYA and phenidone, inhibited STZ-stimulated H_2O_2 production, whereas they did not modify the phagocytic process, as shown by the absence of any effect on the uptake of ^{125}I-STZ by the mesangial cells. In contrast, indomethacin did not produce any effect (17), and exogenous PGE_2 decreased H_2O_2 production by phagocytozing mesangial cells in a concentration-dependent manner (40). Taken together, these results demonstrate opposite effects of the cyclooxygenase and of the lipoxygenase products on the oxidative burst associated with phagocytosis in the mesangial cells. Lipoxygenase products may affect the activity of the respiratory burst enzyme either directly or via stimulation of protein kinase C (Fig. 8).

Sulfidopeptide LTs promote glomerular cell proliferation. Dose-dependent (1-100 nM) stimulation of [^{3}H] thymidine incorporation, an index of cell proliferation, was observed in cultured human glomerular epithelial cells incubated with LTC_4 and LTD_4 but not with LTB_4. The response was 248% and 172% of control values at 100 nM LTC_4 and LTD_4 respectively, and the effect of sulfidopeptide LTs was abolished by FPL 55712 (30). LTC_4 and LTD_4 increased the cell number in parallel, which confirmed that thymidine incorporation reflected increased cell proliferation. Simonson and Dunn (32) also observed that LTC_4 in the range of 0.01 - 1 μM stimulated the proliferation of rat mesangial cells. Identical results have been obtained in non-renal cells, such as keratinocytes (41). Thus, in addition to being a vasoactive agent for glomerular mesangial cells, LTC_4 is a mitogen, as demonstrated for epidermal growth factor and platelet-derived growth factor which exhibit both properties (42,43).

In further experiments with cultured human fibroblasts, we tried to analyze the mechanism of action of LTC_4 on cell proliferation (44). LTC_4-induced cell growth was observed only when PGE_2 synthesis was blocked by indomethacin. However, LTC_4 modified neither [^{3}H]-C20:4 release from prelabeled fibroblasts nor PGE_2 by fibroblasts. These results suggest that the basal production of PGE_2 by fibroblasts was sufficient to inhibit LTC_4-induced DNA synthesis in these cells. Thus, the initiation and development of the fibrotic process in the different tissues could depend in part on the balance between PG and LT productions.

The biochemical transduction of signals from glomerular receptors for sulfidopeptide LTs leading to cell proliferation is not yet completely understood. Binding may activate the G protein-regulated phospholipase C to induce hydrolysis of phosphatidylinositol, as shown recently by Sarau et al. (45) for LTD_4 in rat basophilic leukemia cells. Thus, the hydrolytic products of phosphatidylinositol, inositoltriphosphate and diacylglycerol could be considered as the intracellular messengers of LTD_4 (Fig. 8). LTD_4 has also been found to increase the intracellular pH of HL-60 cells by enhancing sensitivity of the Na^+/H^+ antiport (46). This would result in cell alkalinization, which has been demonstrated as the prerequisite of cell proliferation with different growth factors (47).

Table 4. Synthesis of Lipoxygenase Products in
Glomerular Diseases

Product studied	Glomerular disease	Glomerular synthesis	Reference
12-HETE	Rat nephrotoxic serum nephritis	+	Lianos et al. (16)
5-HETE LTB$_4$	Active and passive Heymann nephritis	+	Lianos et al. (53)
LTB$_4$	Rat nephrotoxic serum nephritis	+	Lianos (51)
LTB$_4$	Rat nephrotoxic serum nephritis	undetectable	Rahman et al. (52)
LTB$_4$	Rat cationic bovine gamma-globulin-induced nephritis	+	Rahman et al. (52)
LT C$_4$	Rat nephrotoxic serum nephritis	small amounts	Lianos (51)

LIPOXYGENASE PRODUCTS AND GLOMERULAR DISEASES

An increased synthesis of lipoxygenase products has been demonstrated in some forms of experimental glomerulonephritis, and a role for these products in glomerular inflammation or alteration of glomerular hemodynamics has also been proposed (Table 4). Badr et al. (48) have recently shown that FPL 55712, an antagonist at the postreceptor stage of LTC$_4$ and LTD$_4$, ameliorated the decline in renal blood flow and glomerular filtration rate observed in rats with endotoxin-induced acute renal failure. They concluded that sulfidopeptide LTs mediate, in part, the renal functional impairment of experimental endotoxemia.

In this report there was no study of the in vitro production of LTs by the glomeruli. The local glomerular synthesis of HETEs and LTs has been essentially studied in immune glomerulonephritis. Lianos et al. (16) demonstrated an enhanced glomerular synthesis of 12-HETE in rats with nephrotoxic serum nephritis. Increased synthesis commenced 3-5 hours after administration of nephrotoxic serum, was maximum at day 2 with 10-fold enhancement, persisted till day 7, and declined towards control levels by day 14. There was no change in the rate of conversion of C20:4 into 8- and 9-HETE. Even

after alkaline extraction of membrane lipids 5-HETE was not detected showing
it had not been reacylated. It is very likely that 12-HETE was produced by
the glomerular cells and not by the invading blood cells present in the glomeruli since polymorphonuclear leukocytes synthesize esentially 15- and
5-HETE, which were not found. The role of platelets can be also ruled out,
because platelet depletion induced by antiplatelet antiserum administration
did not decrease glomerular 12-HETE synthesis.

Nephrotoxic serum nephritis is the only example of a glomerular disease
with a demonstrated increase in 12-HETE production. Glomeruli from rats with
streptozotocin-induced diabetes converted high concentrations of $[^{14}C]$-
C20:4 into $[^{14}C]$ 15- and 12-HETE at the same rate as glomeruli from control animals (49). In contrast, glomerular cyclooxygenase activity was increased in this model. Similarly, HETE production was not different in glomeruli from normal human kidneys and in those isolated from chronically re-
jected allografts. On the contrary, TxB_2 production was markedly increased in glomeruli from the rejecting kidneys (50).

Increased synthesis of LTs in experimental models of glomerulonephritis
has also been reported recently. Lianos et al. (51) studied rats with nephrotoxic serum glomerulonephritis. In this model LTB_4 but not LTC_4 synthesis was enhanced in the glomeruli within hours after induction of the
heterologous phase of the disease, but it did not persist until the onset of
proteinuria. Increased synthesis of LTB_4 was suppressed after decomplemen-
tation by intravenous administration of cobra venom factor. In contrast to
these results, Rahman et al. (52) were unable to measure detectable levels
of LTB_4 in glomeruli from rats with nephrotoxic serum glomerulonephritis.
LTB_4 production by glomeruli was increased in another model of glomerulonephritis obtained by administration of cationic bovine gammaglobulin. This
form of glomerulonephritis represents a model of membranous nephropathy
devoid of cellular infiltration. In this disease, enhanced LTB_4 synthesis
was sustained for at least a few days after onset of proteinuria. Lianos
and Noble (53) also reported an increased glomerular 5-HETE and LTB_4 synthesis in active and passive forms of Heymann nephritis. This was observed
at the early heterologous phase of the disease, and there was no association
with proteinuria both temporally or in degree.

Although these studies provided conflicting results, they suggest the
role of LTB_4 as a glomerular factor of inflammation and also demonstrate
that the glomerular cells can synthesize LTB_4 upon appropriate stimulation, which could be the in situ formation or the deposition of immune

Table 5.   Cell-cell Interaction Implying Lipoxygenase Products as
Mediators

Producing cell	Mediator	Target cell	Effect	Reference
Macrophage	LTC_4	Mesangial cells	Contraction Proliferation	30-31 31
Macrophage	LTC_4, LTD_4	Epithelial cells	Proliferation	29
Macrophage	12-HPETE 5-HPETE	Epithelial cells	Contraction	32
Macrophage	Unidentified lipid	Glomeruli	Inhibition of 12-HETE synthesis	35
PMNL	LTB_4	Glomeruli	Vascular permeability in synergy with PGE_2	59 *
Glomeruli	12-HPETE	Macrophage	Stimulation of PG and 12-HETE synthesis Adherence	34
Glomeruli	LTB_4	PMNL	Chemotaxis, Secretion of lysosomal enzymes	56, 57

* These authors studied whole kidney rather than glomerular capillary permeability.

complexes as occurs in these different models of glomerulonephritis.   The
source of LTB_4 in the glomerulus is still unknown.   It has been shown that
some resident glomerular cells, namely a fraction of mesangial cells, exhibit in vitro many properties of inflammantory cells and, thus, may derive
from bone-marrow (54).   The number of these cells increase in immune glomerular injury (55) and could be a source of LTB_4 if they metabolize C20:4
through the lipoxygenase pathway similarly to monocytes and macrophages.

CONCLUSIONS

    Lipoxygenase products and LTs are formed in the glomerular capillaries
both by the resident glomerular cells and the invading blood cells when they
are appropriately stimulated.   This occurs mainly in glomerulonephritis.
The local increased synthesis of these products may influence the functions
of the glomerular cells and the bone marrow-derived cells.   In this regard,
lipoxygenase products can be considered as playing a major role in the cell-
cell interaction (Table 5).   Sulfidopeptide LTs released by the monocyte-
macrophage contract the mesangial cells and promote proliferation of the

epithelial and mesangial cells. Both 12-HETE and 12-HPETE from the glomerulus activate the macrophages. LTB_4 synthesized by the glomerular cells or
the PMNL is a potent chemotactic lipid which attracts PMNL and monocytes
(56). It also can produce secretion of lysosomal enzymes from neutrophils
(57). Another possibility is synergism between LTB_4 and PGE_2, which is
synthesized in large amounts in most models of glomerulonephritis (2,3,58)
since it has been shown that LTB_4, although having no effect by itself,
increases vascular permeability in association with PGE_2 (59). Lipoxygenase products may also play a major role in the local formation of reactive
oxygen species (60) which alter the functional properties of the glomerular
basement membrane. In fact, the precise role of lipoxygenase products and
particularly LTs will be estabished only when specific inhibitors are available.

ACKNOWLEDGMENTS

This work was supported by grants of the "Institut National de la
Santé et de la Recherche Médicale" and the "Faculté de Médecine
Saint-Antoine." We thank Mrs. A. Morin for preparing the manuscript.

REFERENCES

1. G. Remuzzi, L. Imberti, M. Rossini, C. Morelli, C. Carminati, G.M.
    Cattaneo, T. Bertani, Increased glomerular thromboxane synthesis as a
    possible cause of proteinuria in experimental nephrosis, J. Clin.
    Invest. 75:94-101 (1985).

2. J.E. Stork and M.J. Dunn, Hemodynamic role of thromboxane A_2 and pros-
    taglandin E_2 in glomerulonephritis, J. Pharmacol. Exp. Ther. 233:
    672-678 (1985).

3. E.A. Lianos, G.A. Andres, M.J. Dunn, Glomerular prostaglandin and throm-
    boxane synthesis in rat nephrotoxic serum nephritis, J. Clin. Invest.
    72:1439-1448 (1983).

4. P.Y.K. Wong, P. Westland, M. Hamberg, E. Granstrom, P.H.N. Chuo, B.
    Samuelsson, ω-hydroxylation of 12-L-hydroxy-5, 8, 10, 14-eicosatetrae-
    noic acid in human polymorphonuclear leukocytes, J. Biol. Chem. 259:
    2683-2686 (1984).

5. B.J. Fitzsimmons, J. Adams, J.F. Evans, Y. Leblanc, J. Rokach, The
    lipoxins-Stereochemical identification and determination of their bio-
    synthesis, J. Biol. Chem. 260:13008-13012 (1985).

6. S. Shak, H.D. Perez, I.M. Goldstein, A novel dioxygenation product of
    arachidonic acid possesses potent chemotactic activity for human

polymorphonuclear leukocytes, <u>J. Biol. Chem</u>. 258:14948-14953 (1983).

7. B. Samuelsson, Leukotrienes: Mediators of immediate hypersensitivity reactions and inflammation, <u>Science</u> 220:568-575 (1983).

8. T. Yoshimoto, R.J. Soberman, R.A. Lewis, K.F. Austen, Isolation and characterization of leukotriene C_4 synthetase of rat basophilic leukemia cells, <u>Proc. Natl. Acad. Sci. USA</u> 82:8399-8403 (1985).

9. P. Borgeat, S. Picard, P. Vallerand, Transformation of arachidonic acid
   in leukocytes. Isolation and structural analysis of a novel hydroxy
   derivative, <u>Prostaglandins and Medicine</u> 6:557-570 (1981).

10. S. Shak and I.M. Goldstein, Leukotriene B_4 ω-hydroxylase in human poly-
    morphonuclear leukocytes. Partial purification and indentification as
    a cytochrome P-450, <u>J. Clin. Invest</u>. 76:1218-1228 (1985).

11. O. Breuer and S. Hammarström, Enzymatic conversion of leukotriene
    B_4 to 6-trans-leukotriene B_4 by rat kidney homogenates, <u>Biochem.
    Biophys. Res. Comm</u>. 142:667-673 (1987).

12. R.A. Lewis and K.F. Austen, The biologically active leukotrienes. Biosyn-
    thesis, metabolism, receptors, functions and pharmacology, <u>J. Clin.
    Invest</u>. 73:889-897 (1984).

13. Y. Iino, K. Tachibana, K. Tomita, N. Yoshiyama, S. Tomura, J. Takeuchi,
    T. Wakabayashi, I. Morita, S. Murota, Effects of 5-lipoxygenase inhib-
    itor (TMK-688) on proteinuria in experimental nephritis, (Abstr.)
    <u>Kidney Int</u>. 31:273 (1987).

14. J. Sraer, M. Rigaud, M. Bens, H. Rabinovitch, R. Ardaillou, Metabolism
    of arachidonic acid via the lipoxygenase pathway in human and murine
    glomeruli, <u>J. Biol. Chem</u>. 258:4325-4330 (1983).

15. K. Jim, A. Hassid, F. Sun, M.J. Dunn, Lipoxygenase activity in rat kid-
    ney glomeruli, glomerular epithelial cells and cortical tubules, <u>J.
    Biol. Chem</u>. 257:10294-10299 (1982).

16. E.A. Lianos, M.A. Rahman, M.J. Dunn, Glomerular arachidonate lipoxygena-
    tion in rat nephrotoxic serum nephritis, <u>J. Clin. Invest</u>. 76:1355-
    1359 (1985).

17. L. Baud, J. Hagege, J. Sraer, E. Rondeau, J. Perez, R. Ardaillou, Reac-
    tive oxygen production by cultured rat glomerular mesangial cells
    during phagocytosis is associated with stimulation of lipoxygenase
    activity, <u>J. Exp. Med</u>. 158:1836-1852 (1983).

18. J. Sraer, M. Bens, R. Ardaillou, J.D. Sraer, Sulfidopeptide leukotriene
    biosynthesis and metabolism by rat glomeruli and papilla, (Abst.)
    <u>Kidney Int</u>. 29:346 (1986).

19. V. Cattell, J. Smith, H.T. Cook, S. Moncada, J.A. Salmon, Leukotriene
    B_4 synthesis in normal rat glomeruli, (Abstr.) <u>Kidney Int</u>. 27:254
    (1985).

20. E.A. Lianos, Glomerular leukotriene biosynthesis and degradation in the
    rat: Effects of immune injury, (Abstr.) Kidney Int. 29:339 (1986).

21. M.A. Rahman, M. Nakazawa, S.N. Emancipator, M.J. Dunn, Increased leuko-
    triene B_4 (LTB$_4$) synthesis in immune injured rat glomeruli,
    (Abstr.) Kidney Int. 29:343 (1986).

22. J. Sraer, M. Bens, J.P. Oudinet, R. Ardaillou, Bioconversion of leukotri-
    ene C_4 by rat glomeruli and papilla, Prostaglandins 31:909-921
    (1986).

23. K. Bernström and S. Hammarström, Metabolism of leukotriene D by
    porcine kidney, J. Biol. Chem. 256:9579-9582 (1981).

24. W. Jubiz, O. Radmark, C. Malmsten, G. Hansson, J.A. Lindgren, J.
    Palmblad, A.M. Uden, B. Samuelsson, A novel leukotriene produced by
    stimulation of leukocytes with formylmethionylleucylphenylalanine,
    J. Biol. Chem. 257:6106-6110 (1982).

25. K.F. Badr, C. Baylis, J.M. Pfeffer, M.A. Pfeffer, R.J. Soberman, R.A.
    Lewis, K.F. Austen, E.J. Corey, B.M. Brenner, Renal and systemic hemo-
    dynamic responses to intravenous infusion of leukotriene C_4 in the
    rat, Circ. Res. 54:492-499 (1984).

26. J. Filep, B. Rigter, J. Fröhlich, Vascular and renal effects of leuko-
    triene C_4 in conscious rats, Am. J. Physiol. 249:F739-F744 (1985).

27. K.F. Badr, B.M. Brenner, M. Wasserman, I. Ichikawa, Evidence for local
    glomerular actions of leukotriene D_4, (Abstr.) Kidney Int. 29:328
    (1986).

28. C.N. Serhan, B. Samuelsson, K.F. Badr, Novel arachidonic acid metabo-
    lites with potent glomerular effects, (Abstr.) Kidney Int. 31:286
    (1987).

29. B.J. Ballerman, R.A. Lewis, E.J. Corey, K.F. Austen, B.M. Brenner, Iden-
    tification and characterization of leukotriene C_4 receptors in iso-
    lated rat renal glomeruli, Circ. Res. 56:324-330 (1985).

30. L. Baud, J. Sraer, J. Perez, M.P. Nivez, R. Ardaillou, Leukotriene C_4
    binds to human glomerular epithelial cells and promotes their prolif-
    eration in vitro, J. Clin. Invest. 76:374-377 (1985).

31. R. Barnett, P. Goldwasser, L.A. Scharschmidt, D. Schlondorff, Effects of
    leukotrienes on isolated rat glomeruli and cultured mesangial cells,
    Am. J. Physiol. 250:F838-F844 (1986).

32. M.S. Simonson and M.J. Dunn, Leukotrienes C_4 and D_4 contract rat glo-
    merular mesangial cells, Kidney Int. 30:524-531 (1986).

33. R. Lacave, F. Delarue, E. Rondeau, J.D. Sraer, 5- and 12-hydroperoxyeico-
    satetraenoic acids promote contraction of cultured glomerular viscer-
    al epithelial cells, (Abstr.) Kidney Int. 27:339 (1986).

34. R. Ardaillou, L. Baud, D. Chansel, J. Sraer, The role of lipoxygenase

products of arachidonate metabolism in the control of glomerular functions, Proceedings of the Xth International Congress of Nephrology, Transmedia Europe Ltd, London, in press (1987).

35. L. Baud, J. Sraer, F. Delarue, M. Bens, F. Balavoine, D. Schlondorff, R. Ardaillou, J.D. Sraer, Lipoxygenase products mediate the attachment of rat macrophages to glomeruli *in vitro*, <u>Kidney Int</u>. 27:855-863 (1985).

36. J. Sraer, L. Baud, M. Bens, E. Podjarny, D. Schlondorff, R. Ardaillou, J.D. Sraer, Glomeruli cooperate with macrophages in converting arachidonic acid to prostaglandins and hydroxyeicosatetraenoic acids, <u>Prostaglandins Leukotrienes Med</u>. 13:67-74 (1984).

37. I. Koyama, H. Yamagami, T. Kuwal, M. Kurata, Release of 6-keto prostaglandin $F_{1\alpha}$ and thromboxane B_2 from mouse peritoneal macrophages during their adhesion and spreading on a glass surface, <u>Prostaglandins</u> 23:777-785 (1982).

38. J. Maclouf, B. Fruteau de Laclos, P. Borgeat, Stimulation of leukotriene biosynthesis in human blood leukocytes by platelet-derived 12-hydroperoxyeicosatetraenoic acid, <u>Proc. Natl. Acad. Sci. (USA)</u> 79:6044-6046 (1982).

39. R. Lorenzet, J. Niemetz, A.J. Marcus, M.J. Broeckman, Enhancement of mononuclear procoagulant activity by platelet 12-hydroxyeicosatetraenoic acid, <u>J. Clin. Invest</u>. 78:418-423 (1986).

40. L. Baud, J. Perez, R. Ardaillou, Dexamethasone and hydrogen peroxide production by mesangial cells during phagocytosis, <u>Am. J. Physiol</u>. 250: F596-F604 (1986).

41. K. Kragballe, L. Desjarlais, J.J. Voorhers, Leukotrienes B_4, C_4, D_4 stimulate DNA synthesis in cultured human epidermal keratinocytes, <u>Brit. J. Dermatol</u>. 113:43-52 (1985).

42. P. Mené, H.E. Abboud, G.R. Dubyak, P. Dicorleto, A. Scarpa, M.J. Dunn, Platelet-derived growth factor modulates contraction and cytosolic free calcium in cultured mesangial cells, (Abstr.) <u>Kidney Int</u>. 31:175 (1987).

43. P. Tsivitse, H.E. Abboud, C. Saunders, T.C. Knauss, Effect of epidermal growth factor on cultured mesangial cells, (Abstr.) <u>Kidney Int</u>. 31: 184 (1987).

44. L. Baud, J. Perez, M. Denis, R. Ardaillou, Modulation of fibroblast proliferation by sulfidopeptide leukotrienes: Effect of indomethacin, <u>J. Immunology</u> 138:1190-1195 (1987).

45. H.M. Sarau, S. Mong, J.J. Foley, H.L. Wu, S.T. Crooke, Identification and characterization of leukotriene D_4 receptors and signal transduction processes in rat basophilic leukemia cells, <u>J. Biol. Chem</u>.

262:4034-4041 (1987).

46. E.J. Goetzl, B.A. Burrall, L. Baud, K.H. Scriven, J.D. Levine, C.H. Koo, Generation and recognition of leukotriene mediators of hypersensitivity and inflammation, <u>Dig. Dis. Sci.</u>, in press (1987).

47. G. L'Allemain, A. Franchi, E. Cragoe, Jr., J. Pouyssegur, Blockade of the Na^+/H^+ antiport abolishes growth factor-induced DNA synthesis in fibroblasts, <u>J. Biol. Chem.</u> 259:4313-4319 (1984).

48. K.F. Badr, V.E. Kelley, H.G. Rennke, B.M. Brenner, Roles for thromboxane A_2 and leukotrienes in endotoxin-induced acute renal failure, <u>Kidney Int.</u> 30:474-480 (1986).

49. M. Schambelan, S. Blake, J. Sraer, M. Bens, M.P. Nivez, F. Wahbe, Increased prostaglandin production by glomeruli isolated from rats with streptozotocin-induced diabetes mellitus, <u>J. Clin. Invest.</u> 75:404-412 (1985).

50. G. Friedlander, L. Moulonguet-Doleris, O. Kourilsky, O. Nussaume, R. Ardaillou, J.D. Sraer, Prostaglandin synthesis by glomeruli isolated from normal and chronically-rejected kidneys, <u>Contr. Nephrol.</u> 41:20-22 (1984).

51. E.A. Lianos, Leukotriene synthesis in nephrotoxic serum nephritis: Role of complement and polymorphonuclear leukocytes, (Abstr.) <u>Clin. Res.</u> 34:972A (1986).

52. M.A. Rahman, M. Nakazawa, S.N. Emancipator, M.J. Dunn, Increased leukotriene B_4 synthesis in immune injured rat glomeruli, (Abstr.) <u>Kidney Int.</u> 29:343 (1986).

53. E.A. Lianos and B. Noble, Glomerular arachidonate 5-lipoxygenation in active and passive Heymann nephritis, (Abstr.) <u>Kidney Int.</u> 31:277 (1987).

54. G.F. Schreiner and E.R. Unanue, Origin of the rat mesangial phagocyte and its expression of the leukocyte common antigen, <u>Lab. Invest.</u> 51: 515-523 (1984).

55. G.F. Schreiner, R.S. Cotran, E.R. Unanue, Modulation of Ia and leukocyte common antigen expression in rat glomeruli during the course of glomerulonephritis and aminonucleoside nephrosis, <u>Lab. Invest.</u> 51:524-533 (1984).

56. M.J.H. Smith, A.W. Ford-Hutchinson, M.A. Bray, Leukotriene B_4: A potent mediator of inflammation, <u>J. Pharm. Pharmacol.</u> 32:517-518 (1980).

57. S.A. Rae and M.J.H. Smith, The stimulation of lysosomal enzyme secretion from human polymorphonuclear leukocytes by leukotriene B_4, <u>J. Pharm. Pharmacol.</u> 33:616-617 (1981).

58. M.A. Rahman, S.N. Emancipator, M.J. Dunn, Immune complex effects on

glomerular eicosanoid production and renal hemodynamics, <u>Kidney Int</u>. 31:1317-1326 (1987).

59. W.F. Stenson, K. Chang, J.R. Williamson, Tissue differences in vascular permeability induced by leukotriene B_4 and prostaglandin E_2 in the rat, <u>Prostaglandins</u> 32:5-17 (1986).

60. L. Baud and R. Ardaillou, Reactive oxygen species: Production and role in the kidney, <u>Am. J. Physiol</u>. 251:F765-F776 (1986).

THE DEVELOPMENT OF NEW ANTILEUKOTRIENE DRUGS:

SPECIFIC LEUKOTRIENE D_4 ANTAGONISTS AND 5-LIPOXYGENASE INHIBITORS

Joshua Rokach and Robert N. Young

Merck Frosst Canada, Inc.
P.O. Box 1005
Pointe Claire-Dorval
Quebec H9R 4P8, Canada

INTRODUCTION

The leukotrienes are a complex set of metabolites derived from oxidative metabolism of arachidonic acid (1). The release of arachidonic acid from phospholipid stores by specific phospholipases, when accompanied by a specific activation of an enzyme known as 5-lipoxygenase, results in the conversion of arachidonic acid to 5-hydroperoxyeicosatetraenoic acid (5-HPETE) (2). Through a specific dehydration step, this same enzyme can convert 5-HPETE to the reactive epoxide known as leukotriene A_4 (LTA_4) (3). LTA_4 can be hydrated by a specific LTA_4 hydrolase enzyme which introduces a hydroxyl group in the 12R position to produce leukotriene B_4 (LTB_4, 5[S], 12[R]-dihydroxy-6, 14[Z]-8, 10[E]-eicosatetraenoic acid) (4). LTB_4 is a potent stimulator of leukocyte migration in vitro (1-3) and in vivo (4) and is comparably potent with chemotactic factors such as C5A and the synthetic peptide f-Met-Leu-Phe. In addition, LTB_4 causes aggregation of polymorphonuclear leukocytes (PMNs) and exudation of plasma and has been shown to stimulate phospholipase A_2. LTB_4 has also been shown to contract human bronchus and guinea pig lung strips. The potent biological effects of LTB_4 and the demonstrated measurable levels of LTB_4 in psoriatic tissue (5) have led to the speculation that LTB_4 may play an important role in mediating inflammatory responses in man.

The active intermediate leukotriene A_4 can also be conjugated by a specific glutathione $\underline{S}$-transferase with glutathione to produce a novel peptido-lipid leukotriene known as leukotriene C_4 (5). Sequential

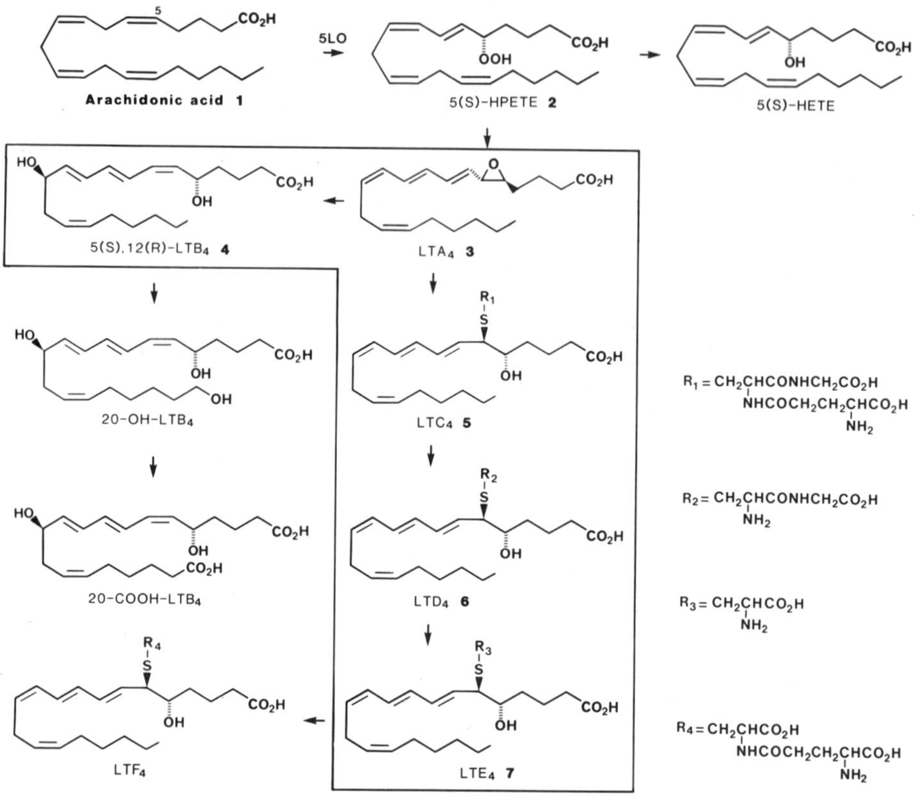

Fig. 1. Metabolism of arachidonic acid via the 5-lipoxygenase pathway.

enzymatic removal of the γ-glutamyl group, followed by removal of the glycine group from the glutathione, results in conversion of leukotriene C_4 to leukotriene D_4 (6) and leukotriene E_4 (7) (Fig. 1).

These leukotrienes are known to account collectively for the biological activity known as slow reacting substance of anaphylaxis (SRS-A) (6). These metabolites have been reported to be produced by human eosinophils (7), macrophages (8) and mast cells (9) and exhibit potent contractile actions on respiratory smooth muscle (10-14), affect mucocillary clearance (15,16), vascular permeability and inflammatory processes in general (17). Thus, the leukotrienes of SRS-A have been implicated as major mediators of bronchial asthma (18-21) as well as bronchial hyperreactivity. The physiological role of leukotrienes in renal tissue is less well defined. Sulfido-peptide leukotrienes have been detected in perfusates of isolated rat and guinea pig kidneys following calcium ionophore and antigen challenge (22,23). Sulfido-peptide leukotrienes (24,25) and leukotriene B_4 (25-27) have been reported to be synthesized by rat glomeruli. Leukotrienes C_4 and D_4 contract rat glomeruli (28) and mesangial cells (28,29) and cause proliferation of human glomerular epithelial cells (30). These leukotrienes are renal vasocon- strictors in the rat (31) and pig (32), causing reductions in renal blood flow (33) and glomerular filtration rate (34,35). The vasoconstrictive effects of LTC_4 and LTD_4 in the rat have been shown to be inhibited by FPL-55712, a putative LTD_4 antagonist (31). Recently, another potent LTD_4 antagonist, SK&F 10453, has been shown to reverse the fall in glomeru- lar ultrafiltration coefficient induced by antiglomerular basement membrane antibody (36). The physiological role of leukotriene B_4 in the kidney is likely to be mainly secondary to its potent chemotactic properties to inflam- matory cells (37). The sum of these observed effects have led to the hypo- thesis that leukotrienes may be important mediators in glomerulonephritis (37).

The biological activities of leukotrienes B_4, C_4, D_4 and E_4 are mediated by specific receptors (38-42). Specific binding sites for leukotri- ene C_4 have been identified in rat glomeruli (43) and human glomerular epi- thelial cells (30). In the case of the SRS-A leukotrienes, it is generally held that in human lung these actions are mediated by a common receptor (44) on which leukotriene D_4 exhibits the most potent activity. In considera- tion of these and other findings, the hypothesis has evolved that drugs that either inhibit the biosynthesis of leukotrienes or act as specific receptor antagonists of leukotrienes could serve as novel and effective therapy for a variety of diseases including human asthma, inflammatory bowel disease, allergic rhinitis, allergic conjunctivitis and psoriasis. The potential of such drugs for the treatment of renal disease remains to be demonstrated, but the evidence of leukotriene synthesis and biological activities in the kidney suggests that they may be useful in the treatment of glomerulo- nephritis. A specific inhibitor of leukotriene biosynthesis would have the considerable advantage of blocking the production of both leukotriene B_4 and the peptidolipid leukotrienes C_4, D_4, and E_4; thus, it should be more effective in diseases such as psoriasis and other inflammatory condi- tions, as well as potentially effective therapy for asthma. Such a drug, however, would run the risk of causing potential side effects if leukotri- enes were found to have some other nonpathological role in the body and the inhibition of their synthesis proved detrimental. On the other hand, a specific leukotriene D_4 antagonist would have a more limited, potential utility but, by blocking only the contractile receptor, would presumably be less prone to causing unrelated side effects. In light of these somewhat conflicting arguments, a number of pharmaceutical companies have set out to produce both 5-lipoxygenase inhibitors and leukotriene D_4 antagonists

with the hopes that one or the other or both would find their ultimate role
in the therapy of human disease. Because of its central role at the branch
point of arachidonic acid metabolism, the 5-lipoxygenase enzyme has proven
to be the favorite target for the development of specific inhibitors.

Many leukotriene D_4 antagonists, of several structural classes, have
been reported in the literature. The majority of these compounds are in a
class related to FPL-55712 and incorporate a 4-acetyl-3-hydroxy-2-propyl-
phenoxy group as a lipophilic component which is critical to potent activity. FPL-55712 has entered limited clinical trials as an aerosol, with equiv-
ocal results (45-48). It is generally felt that the action of FPL-55712 is
very short-lived, which limits its potential as an oral drug. An analog prepared by Lilly, Ly-171883, which incorporated a tetrazole ring in place of
the chromone ring of FPL-55712, has been reported to be orally active (49)
and has shown some efficacy in mild chronic asthma (50). Apparently, this
compound has recently been replaced for toxicological reasons by an analog,
LY-163443 (51). Another leukotriene D_4 antagonist, SK&F-104353, with a
structure more closely related to leukotriene E_4, has been developed by
Smith Kline as a potent (pA_2 = 8.6) aerosol active drug (52). Apparently,
the compound is in Phase II clinical trials in asthma (52). Two 2-quino-
linylmethoxyphenyl derivatives, REV-5901 (53) and WY-48,252 (54), have been
reported to be in development (55). Finally, a highly potent series of
indole and indazole derivatives has been described by workers from Stuart
Pharmaceuticals (56).

Although a large number of inhibitors of leukotriene biosynthesis has
been described in the literature, few have reached advanced stages of development. Lonapalene (RS-43179) from Syntex is a moderately active 5-LO
inhibitor which has shown efficacy in psoriasis as a topical agent (57) and
is currently in Phase III trials. An indole derivative, TZl-41127 (58)
(Teikoku Hormone Manufacturing), is reported to be either in early clinical
trials or about to enter trials. The quinone derivative, AA-861 (Takada),
has been reported to be inactive in early trials in asthmatic patients
(59). REV-5901, the leukotriene D_4 antagonist, is also reported to have
leukotriene biosynthesis inhibitory activity; however, results of early
clinical trials have been disappointing (55).

In order to discover specific inhibitors and antagonists of the leukotrienes, we have set in place a battery of *in vivo* and *in vitro* assays
with which to identify, define and characterize potential drug candidates
and to direct their development and improvement, with the goal of providing

drugs with biological and pharmacokinetic profiles suitable for further
development.

ASSAYS FOR DISCOVERY OF LTD_4 ANTAGONISTS

In Vitro Assays

In the case of the leukotriene D_4 antagonists, the initial screening
assays that were developed are the inhibition of leukotriene D_4-induced
contraction of guinea pig ileum and the inhibition of $[^3H]$-LTD_4 binding
to guinea pig lung membranes. After assay on the guinea pig ileum, an
active drug is subsequently tested for its ability to inhibit a histamine-
induced contraction. Failure to do so indicates a degree of selectivity
required for testing in secondary in vitro assays. Active compounds are
then tested in a $[^3H]$-LTD_4 binding assay to human lung tissue; then
activity is further quantified by the determination of -log K_B or pA_2
values versus LTD_4 on isolated guinea pig tracheal preparations or on
human tracheal tissue.   Specificity can be assessed by the failure to antagonize the contractions induced by a variety of other contractile mediators
such as histamine, serotonin, acetylcholine and prostaglandin $F_{2\alpha}$. Finally, compounds are tested for their ability to inhibit antigen-induced contractions either of tracheal tissue from sensitized guinea pigs (sensitized
to ovalbumin) or of human tissue challenged with goat antihuman IGE-anti-
body.

In Vivo Assays

The compounds identified by in vitro assays as specific LTD_4
antagonists are subsequently tested for ability to inhibit bronchoconstriction induced by intravenous (IV) injection of leukotriene D_4 in anesthetized, artificially ventilated guinea pigs. Compounds can be tested either
by the IV route of administration or by the intraduodenal route of administration (as a measure of oral absorption). By the IV route, repeated admin-
istrations of a fixed dose of leukotriene D_4 are inhibited by increasing
doses of the drug, thereby deriving an ED_{50} estimation. For intraduodenal
(ID) administration, leukotriene challenges are repeated at 10, 30, 50 and
70 minutes post-drug administration to determine the degree and rate of
absorption of the compound. Compounds shown to be active both by the IV and
ID route in the guinea pig are subsequently tested for their ability to
inhibit antigen-induced dyspnea in hyperreactive rats (60) (pretreated with
methysergide). This dyspnea has been shown to be largely leukotriene-

79

dependent. The active drugs are further tested for their ability to inhibit bronchoconstriction induced by aerosol-administered leukotriene D_4 or ascaris antigen in conscious squirrel monkeys (61). In these animals, direct measures of the increase in pulmonary resistance and the decrease in dynamic compliance are available via measurements made with a pulmonary mechanics computer. Drugs can be administered by the aerosol, oral, or IV route. Finally, we have studied the effects of drugs on leukotriene activity in anesthetized pigs as a measure of their ability to block the cardiovascular effects of leukotrienes.

ASSAYS FOR DISCOVERY OF LEUKOTRIENE BIOSYNTHESIS INHIBITORS

In the case of leukotriene D_4 inhibitors (5-lipoxygenase inhibitors), a different battery of *in vitro* assays has been developed. Routine screening of compounds is carried out by measuring the ability of compounds to inhibit the biosynthesis of leukotriene B_4 in rat or human polymorphonuclear leukocytes induced by the calcium ionophore A23187. IC_{50} values are thus readily determinable. Active compounds then are further tested for their ability to inhibit a semipurified 5-lipoxygenase preparation derived from the 100,000 x g supernatant of rat PMNs, in the presence of exogenous arachidonic acid. Finally, potential drug candidates are further characterized for the ability to inhibit the purified human 5-lipoxygenase enzyme, the conversion of added arachidonic acid to 5-HPETE (or 5-HETE) and is measured. In vivo, 5-lipoxygenase inhibitors are tested primarily in the hyperreactive rat assay, as described above, and then in the conscious squirrel monkey in response to ascaris antigen.

LTD_4 Receptor Model

In order to develop specific leukotriene D_4 receptor antagonists, we have sought to gain a better understanding of the nature of the leukotriene D_4 receptor, judged by the study of the interaction of the receptor with its natural ligand and analogs thereof. Thus, a great deal of effort was invested by our group (62) and others (63) in the synthesis of leukotriene analogs in order to define the structural requirements for potent agonism on this receptor. Analysis of the reported biological activities of these compounds and of other restricted analogs suggested the following: (a) The receptor contains at least one hydrophylic site that requires a carboxylic acid functionality or equivalent (the LTD_4 C-1 monoamide retains the activity of the natural product, whereas the LTD_4 glycine, monoamide, has only 10% of this spasmogenic activity). (b) The cysteinyl-glycine chain appears to bind in a polar but non-ionic site. (c) A portion of the receptor binds the lipophilic polyene chain probably in the largely extended conformation but with a definite lipophilic pocket which accepts the C-5 terminus of leukotriene D_4. (d) Some recognition of the triene system is inherent, and the presence of a double bond at C-7 seems to be most important, with the rest of lipophilic unit being somewhat less stringent in its binding requirements. At least in the case of agonists, the spatial disposition of the polar and lipophilic groups is highly important, as suggested by the strong stereochemical requirements at C-5 and C-6 for the maintenance of biological activities of LTD_4 (Figure 2).

At the time we began our efforts to discover LTD_4 receptor antagonists, the only well-known leukotriene D_4 receptor antagonist was the chromone compound, FPL-55712 (64). We attempted to conceptualize how this compound could fit in the LTD_4 receptor and developed the following hypothesis: the carboxylic acid will bind with the carboxylate binding site comparable to the glycine carboxylate of LTD_4; there is apparently no corresponding binding group for the C-1 carboxy group; and the 4-acetyl-3-hydroxy-2-propylphenoxy unit functions as a lipophilic binding unit such that the propyl group binds in the lipophilic pocket that accommodates the terminal three carbons of leukotriene D_4 (C-18, 19 and 20) (Figure 3).

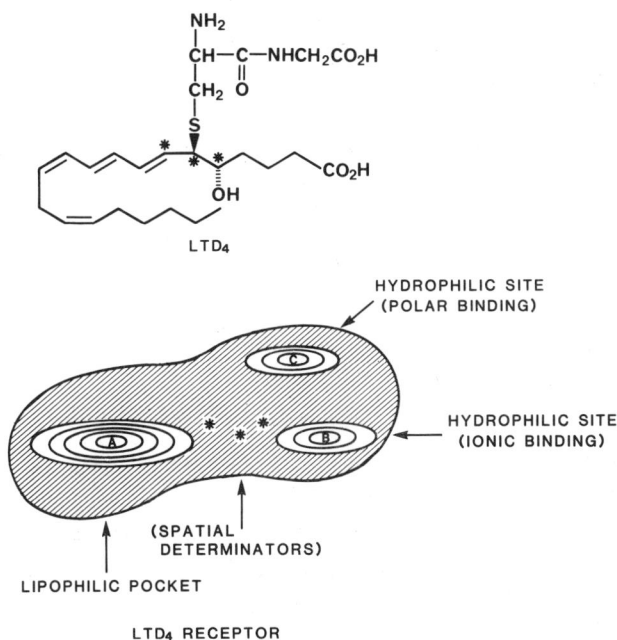

Fig. 2. Conceptualized LTD_4 Receptor.

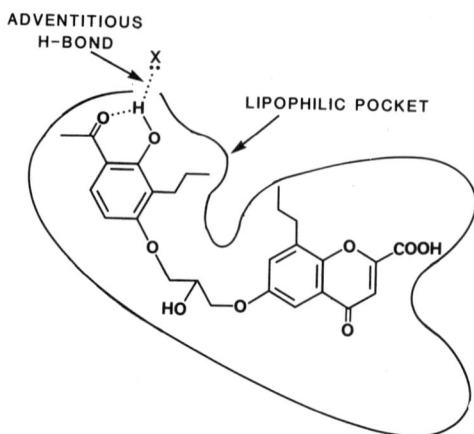

Fig. 3. Binding of FPL-55712 to the
LTD$_4$ receptor

It was noted that the hydroxyacetophenone unit, in spite of its considerable acidity, is a nonpolar unit due to the strong hydrogen bond between the
carbonyl group and the phenol hydroxy group. It is known that this unit is
quite important for activity in this series and, thus, we surmised that this
unit probably interacts through an adventitious hydrogen bond to a component
of the LTD$_4$ receptor which has no binding importance with leukotriene D$_4$
itself. Thus, it seems that the addition of this extra binding energy could
compensate to some degree for the lack of a second carboxy binding group in
FPL-55712.

We have used this working hypothesis as the basis of our attempts to
design novel and potent leukotriene D$_4$ antagonists with improved biological profiles relative to that of the classical antagonist FPL-55712 (which
is particularly hampered by its extremely short half-life *in vivo*). In
this context, we observed, through our screening program, that a 4-alkyl-
phenylketobutenoic acid derivative $\underline{8}$ showed mild but significant activity
against leukotriene D$_4$-induced contractions of the guinea pig ileum.
Noting similarities between $\underline{8}$ and FPL-55712 (9) (i.e., they both are aryl-
ketobutenoic acid derivatives with an appended lipophilic chain of 9 or 10
carbons), we were prompted to prepare a series of compounds $\underline{10}$ incorporating
what we surmised to be the key components of $\underline{8}$ and $\underline{9}$.

The synthesis of these compounds then was realized and led to the observation that these compounds not only exhibited good *in vitro* activity
but also showed significant *in vivo* activity in the guinea pig by the IV

82

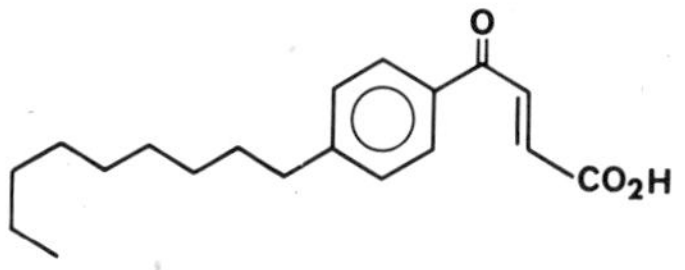

Fig. 4. LTD$_4$ receptor antagonists.

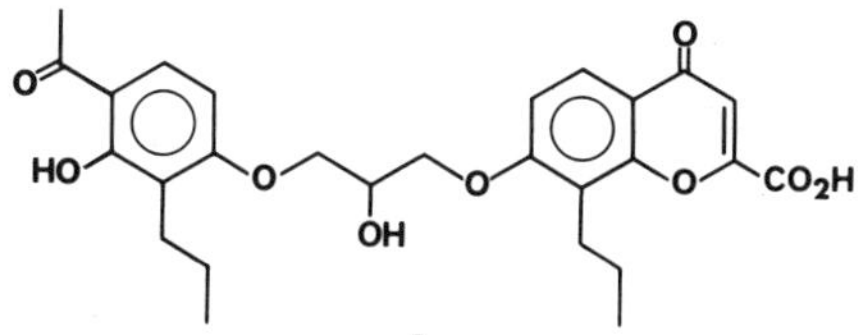

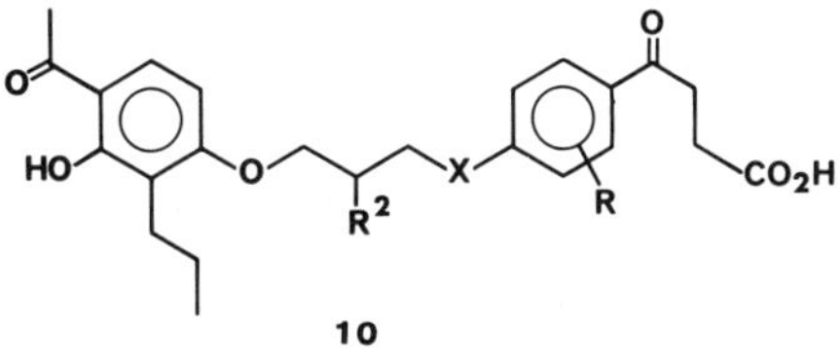

route (when X = S and X = SO$_2$) and by the ID route of administration (when X = S). Generally, the thioether analogues were found to exhibit most consistent oral absorption as indicated by biological activity and blood levels, and the compound 10a was chosen for a more complete characterization (65). Preliminary metabolism studies performed on 10a led to the observation that the ketobutyric acid chain was rapidly metabolized in the dog and the rat via reduction and ß-oxidation to the metabolites 11 and 12 (Figure 5).

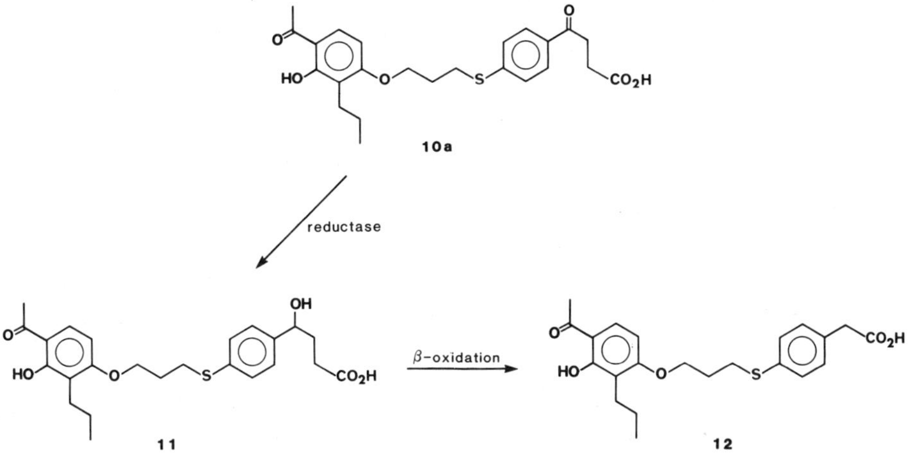

Fig. 5. Metabolism of 10a.

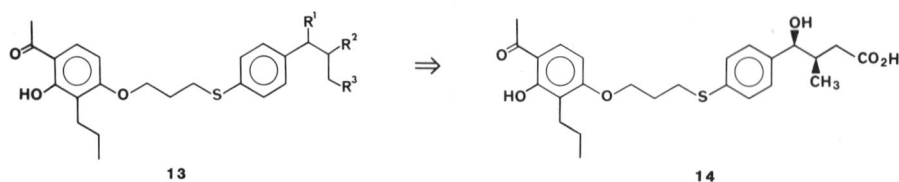

Fig. 6. Structure of L-649,923.

While both metabolites retained intrinsic biological activity, in order to minimize the potential for interspecies differences in metabolism and in order to enhance the biological half-life of this compound, we prepared analogs of 10a having the general formula 13 (Figure 6) with ß-substituents incorporated in order to block the ß-oxidation metabolism. From a series of such analogs, the compound L-649,923 (14) was chosen for further development due to its superior *in vivo* activity by both the IV and ID routes of administration in the guinea pig.

BIOLOGICAL PROFILE OF L-649,923

In Vitro Studies

L-649,923 was developed as a selective receptor antagonist of leukotriene D_4 (66). Receptor-binding studies, using the method of Pong and DeHaven (39), indicated that this compound antagonized the specific binding of [^{3}H]-leukotriene D_4 in guinea pig lung membrane, with a Ki value of 400 nM $\pm$ 213 nM. The compound inhibited [^{3}H]-leukotriene C_4 binding in the same preparation, with a Ki value of 8.6 $\pm$ 0.9 uM. In order to analyze more thoroughly the nature of the interaction of L-649,923 with the LTD_4, we performed saturation experiments and subjected the data to Scatchard analysis. It was shown in these studies that L-649,923 significantly increased the effective K_D without affecting the B_{max}, suggesting that this compound is a competitive antagonist of the leukotriene D_4 binding in guinea pig lung membrane preparations (Figure 7).

The conclusion is supported by Schild analysis (67) which gave a slope of 1.01 $\pm$ .28 and a K_B for L-649,923 of 428 $\pm$ 197 nM. This K_B value agrees well with the Ki value stated above.

Consistent with these findings, L-649,923 also selectively antagonized leukotriene D_4-induced contractions of the guinea-pig ileum and trachea.

84

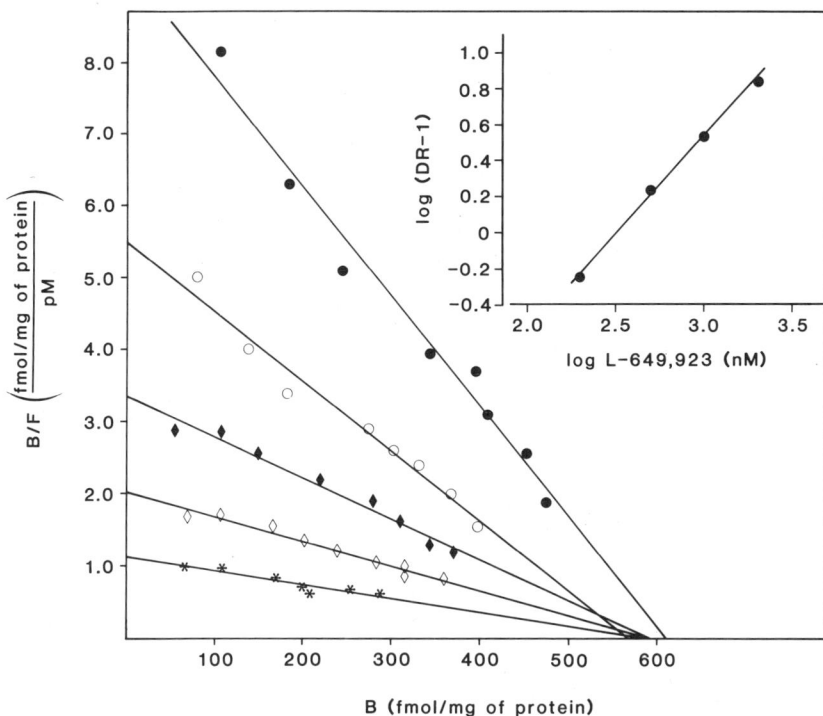

Fig. 7. Scatchard and Schild analysis of the interaction of
L-649,923 with the binding of [^{3}H] leukotriene D_4.
[^{3}H] leukotriene D_4 (28-560 pM) was incubated with
guinea-pig lung membranes in the presence of L-649,923 for
60 min., and specific binding was determined. Data shown
are averages of triplicate determinations from a typical
experiment which was repeated four times and summarized in
Table 1. Control (●); and with L-649,923 at 0.2 (O), 0.5
(◆), 1.0 (◇), and 2.0 μM (*). Inset: Schild plot of
data from Scatchard analysis. The dose ratio (DR) is the
ratio of the K_D value measured in the presence of each
concentration of L-649,923 to the K_D value measured in
the absence of L-649,923.

Quantitative pharmacological analyses indicated that the compound was a
potent competitive antagonist of contractions induced by leukotriene D_4 on
guinea-pig ileum, with a pA_2 value of 8.1 and a Schild plot slope of 0.98
(not significantly different from unity p>0.05). In contrast, the pA_2
value versus LTD_4 on guinea-pig trachea was 7.2, but the interaction was
noncompetitive in nature (slope = 0.7), suggesting differences in receptor
subtypes and/or mechanism between these two tissues. The antagonist activity was selective for leukotriene-induced responses since responses to histamine, acetylcholine, prostaglandin $F_{2\alpha}$, serotonin and U-44069 were minimally affected.

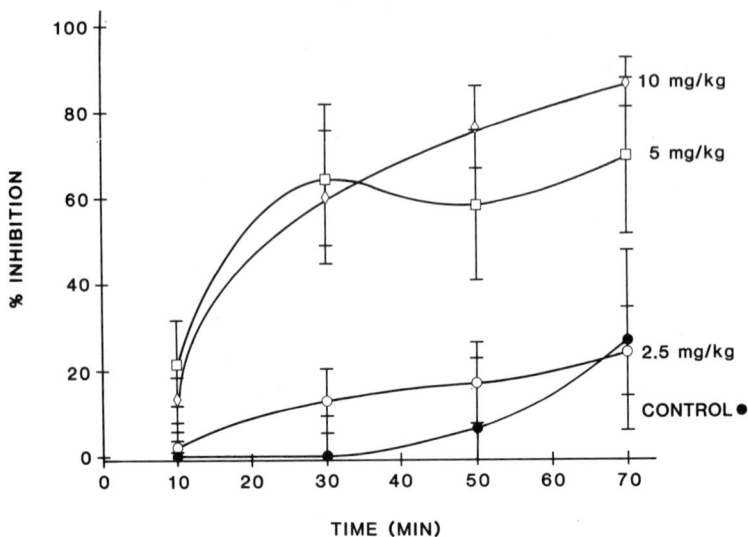

Fig. 8. Anesthetized guinea pigs were artificially ventilated
and insufflation pressures recorded. Increases in
insufflation pressure (bronchoconstriction) were
induced at 20-min. intervals with leukotriene D_4
(0.2 μg/kg IV). Vehicle (1 mL/kg H_2O (O) or 2.5
($\bullet$), 5.0 ($\square$), or 10.0 ($\diamond$) mg/kg L-649,923 was administered intraduodenally (single dose) in solution
as the sodium salt and, 10 min. later, the leukotriene D_4 challenges were continued at 20-min. intervals. The reduction in insufflation pressure (inhibition) was calculated as a percentage of the predrug
(control) response.

In Vivo Studies

L-649,923 has been tested as an antagonist of bronchoconstriction
induced in anesthetized, artificially ventilated guinea pigs. Changes in
insufflation pressure (resistance to inflation) were measured by a modified
Konzett and Rossler method as described before by Jones and Masson (68).
L-649,923 antagonized the bronchoconstriction induced by leukotrienes D_4
and C_4 with mean ED_{50} values of 0.26 $\pm$ 0.1 and 0.38 $\pm$ 0.1 mg/kg, respectively. Following intraduodenal administration, L-649,923 was consistently
active at 5.0 and 10.0 mg/kg against bronchoconstriction induced by leukotriene D_4 (Figure 8).

In order to test the selectivity of L-649,923 *in vivo*, we administered the compound at a dose of 1-10 mg/kg IV to anesthetized guinea pigs in
which bronchoconstriction was elicited by intravenous administration of

either histamine, acetylcholine, serotonin, U-44069 or arachidonic acid. No consistent or significant inhibition of the responses to any of these agonists was observed, confirming the selectivity of L-649,923 observed *in vitro*. Higher doses of L-649,923 (30 mg/kg IV) induced a moderate degree of bronchoconstriction. The reasons for this are unknown, but it appears to be present with other acetophenone compounds tested in this system (68).

The biological activity of L-649,923 was then evaluated in models of antigen-induced bronchoconstriction. In groups of sensitized inbred rats (60) pretreated with methylsergide (3 ug/kg IV, 5 min. prior to challenge), L-649,923 provided significant protection against antigen-induced dyspnea (ED_{50}: 1.5 mg/kg, PO, 1 hr. pretreatment). In parallel studies in the conscious squirrel monkey (63), sensitive to ascaris antigen, L-649,923 (5 mg/kg, PO, 1 hr. pretreatment) significantly inhibited changes in airway resistance and dynamic compliance (Figure 9).

Development of L-648,051

In further developments of the class of compounds 10, a number of sulfone analogs were studied; and it was noted that although these compounds lacked good oral activity, one compound in particular, L-648,051 (15) (Figure 10), exhibited superior properties on isolated tissues, which

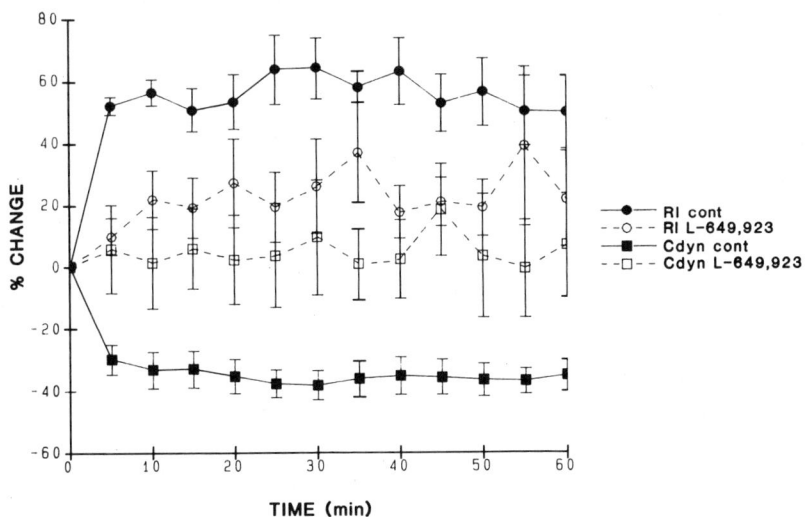

Fig. 9. The actions of L-649,923 (10 mg/kg PO) on antigen-induced bronchoconstriction in five conscious squirrel monkeys. For each animal, the mean of a minimum of three reproducible responses (increase in specific airways resistance [RI] and decrease in dynamic compliance [Cdyn]) to antigen was used for the control curve (63).

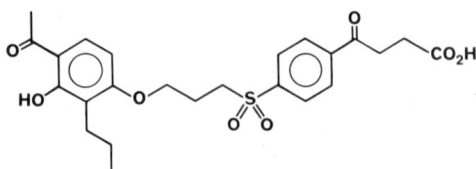

15
L-648,051

Fig. 10. Structure of L-648,051.

suggested that it might warrant further development as an aerosol drug. While there are undoubted advantages inherent in an orally active drug, a drug with the appropriate physical and pharmacokinetic properties to be formulated as an aerosol can have significant advantages over an oral drug by reducing the overall required dose (with concomitant reduction in any systemic side effects) and by delivering relatively higher levels of drug to the site of action. As had been observed in the case of compound 10a previously described, L-648,051 was found to be rapidly reduced to the corresponding hydroxy compound and ß-oxidized in the corresponding acetic acid compound *in vivo*. Whereas these were definite disadvantages for an orally active drug, in the case of an aerosol active compound, it was felt that this metabolism and the rapid elimination of these metabolites from the system should be an advantage for a topical agent by further minimizing systemic exposure. The pharmacological profile that has allowed us to select L-648,051 for further development as an aerosol-active leukotriene D_4 antagonist is described below (70).

PHARMACOLOGICAL PROFILE OF L-648,051

In Vitro

L-648,051 was shown by Scatchard and Schild analysis to interact in a competitive manner with the leukotriene D_4 receptor on guinea pig lung membranes (Figure 11); Ki = 6.2 ± 1.5 uM versus [^{3}H]-leukotriene D_4 and 36.7 ± 6.1 uM versus [^{3}H]-leukotriene C_4. The compound was a competitive antagonist of leukotriene D_4-induced contraction of the guinea pig ileum (pA$_2$ = 7.7, slope 1.0) and the guinea pig trachea (pA$_2$ = 7.3, slope 0.75) at concentrations (6 to 20 X 10^{-6}M) that minimally affected other non-leukotriene spasmogens (e.g., PGF$_{2\alpha}$, U-44069, serotonin, histamine and acetylcholine) (Figure 12).

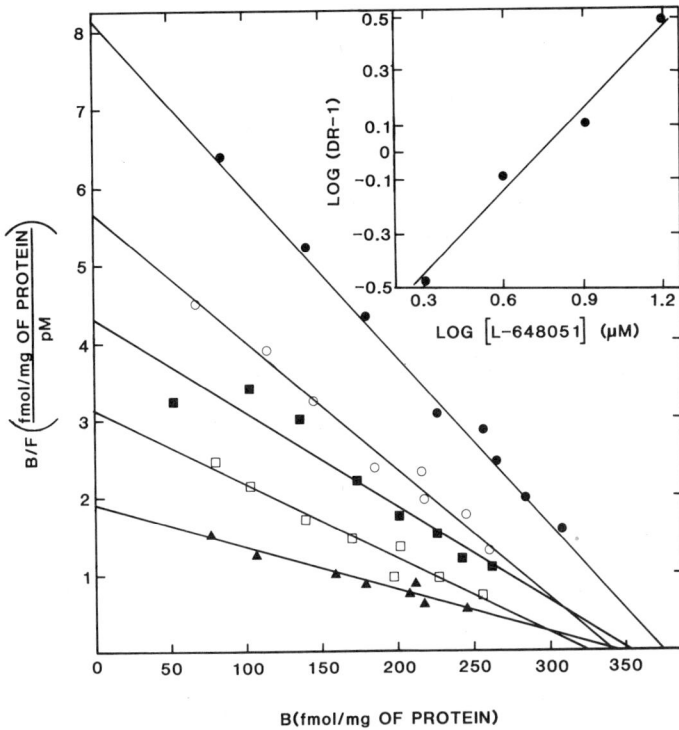

Fig. 11. Scatchard and Schild analysis of [^{3}H] leukotriene D_4 binding in the presence of various
concentrations of L-649,051. A series of concentrations of [^{3}H] leukotriene D_4 was
incubated with guinea pig lung membranes in
the absence or presence of various concentrations of L-648,051 and specific binding determined, as described in Methods. Data shown
are from a typical experiment which was repeated three times. The data from all three
experiments are summarized in Table 1. Control (●); and with L-648,051 at 2.0 (O), 4.0
(■), 8.0 (□), and 16 μM (▲). Inset: Schild
plot of the data from Scatchard analysis. The
slope obtained in this experiment was 1.02.

At higher concentrations (6 X 10^{-5}M), some of this selectivity was
lost. In receptor binding studies on human lung membranes, L-648,051
exhibited an IC$_{50}$ of 6.6 uM (N = 2). The compound inhibited LTD$_4$-
induced contractions of human tracheal preparations with a -log K$_b$ = 6.9.
Thus, while L-648,051 was comparably active on guinea pig trachea relative
to FPL-55712 and L-649,923, it was significantly more active on human
trachea versus LTD$_4$ (see Table 1).

L-648,051 was tested for its ability to block antigen-induced

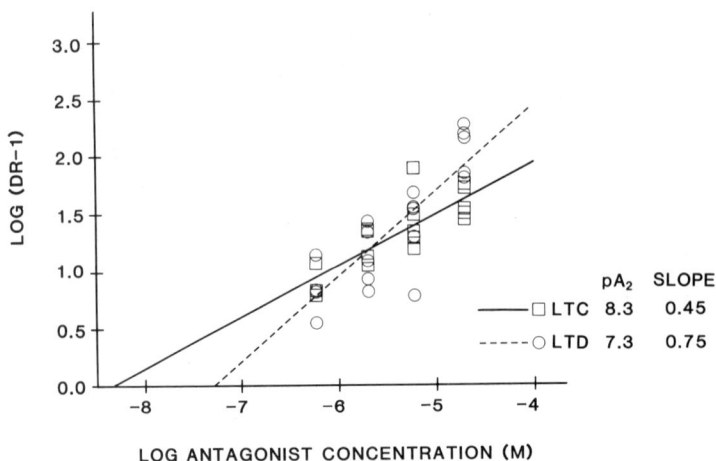

Fig. 12. Schild analysis of L-648,051 (6 x 10^{-7} - 2 x
10^{-5} M) antagonism of leukotriene D_4 (---,O)
and leukotriene C_4 (——,□)-induced contractile responses of isolated guinea pig tracheal
smooth muscle. The slopes of the regression
line for leukotriene C_4 is significantly
different from 1 (p < 0.05). The slope of the
line for leukotriene D_4 is not significantly
different from 1 (p > 0.05).

contractions of guinea pig and human tracheal preparations. Pretreatment of
trachea from guinea pigs previously sensitized to ovalbumin with L-648,051
(2 X 10^{-5}M) gave only partial inhibition of the contraction induced by the
antigen. However, L-648,051 (2 X 10^{-5}M and 2 X 10^{-6}M) effectively
blocked the contractions of isolated human trachealis induced by anti-IGE
antibody (see Figure 13).

Table 1. Comparison of Various Leukotriene D_4 Antagonists (at 2 X
10^{-6}M) on Guinea Pig and Human Tracheal Tissue

GUINEA PIG TRACHEA		HUMAN TRACHEA	
ANTAGONIST	-LOG K_B (VS. LTD$_4$) (N)	ANTAGONIST	-LOG K_B (VS. LTD$_4$) (N)
FPL-55712	6.9 ± 0.04 (5)	FPL-55712[*]	6.5 ± 0.3 (5)
L-649,923	6.9 ± 0.1 (4)	L-649,923	6.0 ± 0.1 (5)
L-648,051	6.9 ± 0.05 (5)	L-648,051	6.9 ± 0.2 (4)

*STUDIES CARRIED OUT IN THE PRESENCE OF 1 x 10^{-7} M ATROPINE AND 7 x 10^{-7}
MEPYRAMINE

90

L-648,051 VS ANTIGEN ON GUINEA PIG TRACHEA
(ATR, MEPY, AND INDO PRESENT)

L-648,051 VS ANTI-IGE ON HUMAN TRACHEALIS
(ATR, MEPY AND INDO PRESENT)

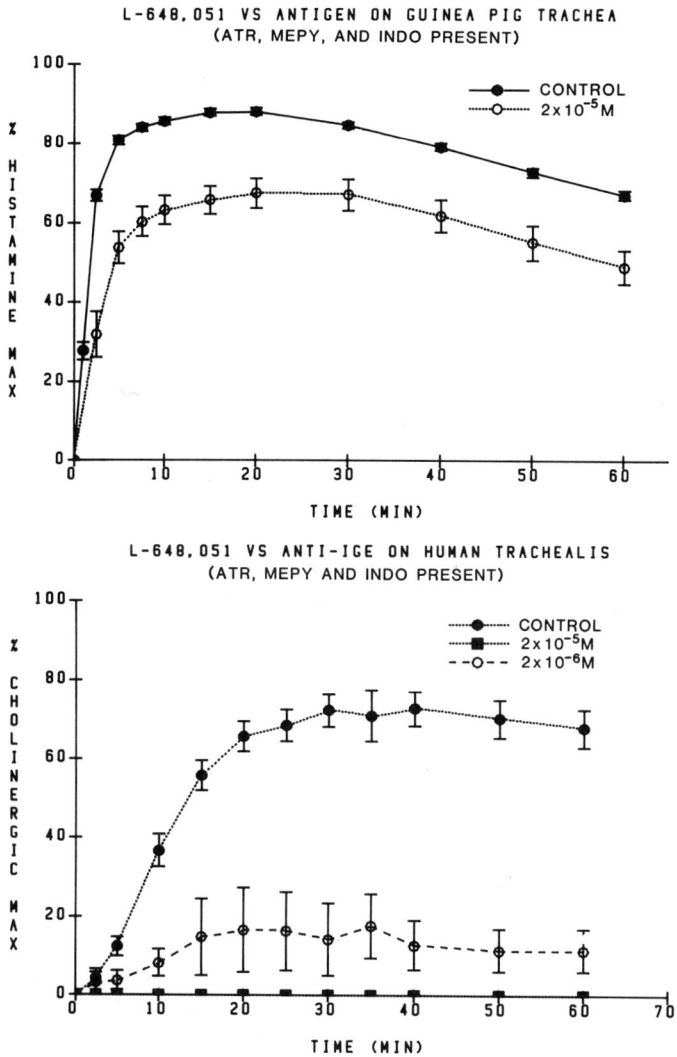

Fig. 13. Upper panel: Effect of L-648,051 (n = 5) on contractions of guinea pig trachea induced by ovalbumin. Lower panel: Effect of 2×10^{-5} M (n =
4) and 2×10^{-6} M (n = 3) L-648,051 on contractions of human trachealis induced by 4 μg/mL
goat antihuman IgE. All studies were carried
out in the presence of 1×10^{-7} M atropine, 8
$\times 10^{-6}$ M mepyramine, and 1.4×10^{-6} M indomethacin. All tissues were pretreated 20 to 30
min. with L-648,051.

Superior properties which indicated the potential of L-648,051 as a topical agent were indicated by experiments in which the compound was compared
with FPL-55712 and L-649,923 for their abilities to reverse an ongoing contraction of guinea pig trachea induced by LTD_4. Using the time for 50%

Table 2. Antagonist Reversal of Ongoing Contraction of Guinea Pig Trachea (0.5 µg/ml Indomethacin Present) Induced By Various Leukotrienes. $t_{1/2}$ (Time (min) for 50% Reversal) (N)

TREATMENT	LTD_4(129 nM)	LTE_4(129 nM)
L-648,051*	5.7 ± .5 (5)	3.5 (2)
FPL-55712*	14.7 ± 1.5 (3)	14.8 ± 5.3 (2)
L-649,923*	45.5 ± 1.5 (3)	45.0 (2)

*ANTAGONIST CONCENTRATION = 6 x 10^{-6} M

reversal of contraction (t 1/2), L-648,051 reversed LTD_4 and LTE_4 contractions significantly faster (3 to 9 times faster). This may be due to the more polar nature of the drug which allows it to partition more rapidly to gain access to the relevant receptor sites in this tissue (Table 2).

In Vivo

Intravenously administered L-648,051 effectively antagonized the bronchoconstriction by LTC_4, LTD_4 and LTE_4 in anesthetized artificially ventilated guinea pigs (Table 3) at 0.2 - 0.3 mg/kg and was highly selective for this blockade relative to other contractile agonists. However, L-648,051 was only weakly active following intraduodenal administration at 10 and 20 mg/kg against bronchoconstriction induced by LTD_4 (30-40% inhibi-

Table 3. Inhibition of Agonist-Induced Bronchoconstriction in the Guinea Pig In Vivo by L-648,051 Administered Intravenously

Agonist	Agonist dose[a] (µg/kg i.v.)	Mean ED_{50} of L-648,051 (mg/kg)[b]	n
Leukotriene D_4	0.2	0.2 ± 0.1	5
Leukotriene C_4	0.4	0.3 ± 0.1	4
Leukotriene E_4	2.0	0.2 ± 0.1	3
Histamine	3.0	<10(63 ± 9%)*	3
U-44069	2.0	>10(15 ± 6%)	3
Arachidonic acid	500	>10(12 ± 8%)	4
Serotonin	10	>10(12 ± 8%)	3
Acetylcholine	50	>10(17 ± 6%)	3

[a]Dose that elicited at least 50% of the maximal bronchoconstrictor response.
[b]Mean ED_{50} values ± SE are doses required to inhibit by 50% the control induced increases in insufflation pressure.
*Significant inhibition at 10 mg/kg compared with control (p<0.05).

tion 70 minutes post-drug administration). This lack of activity appears to be due at least in part to the rapid metabolism and elimination of L-648,051. When administered as a nebulized solution (0.1% for five minutes) to conscious squirrel monkeys, L-648,051 significantly blocked both the increase in airway resistance and decrease in dynamic compliance induced by an aerosol of leukotriene D_4 both 5 and 30 minutes after drug exposure (Figure 14).

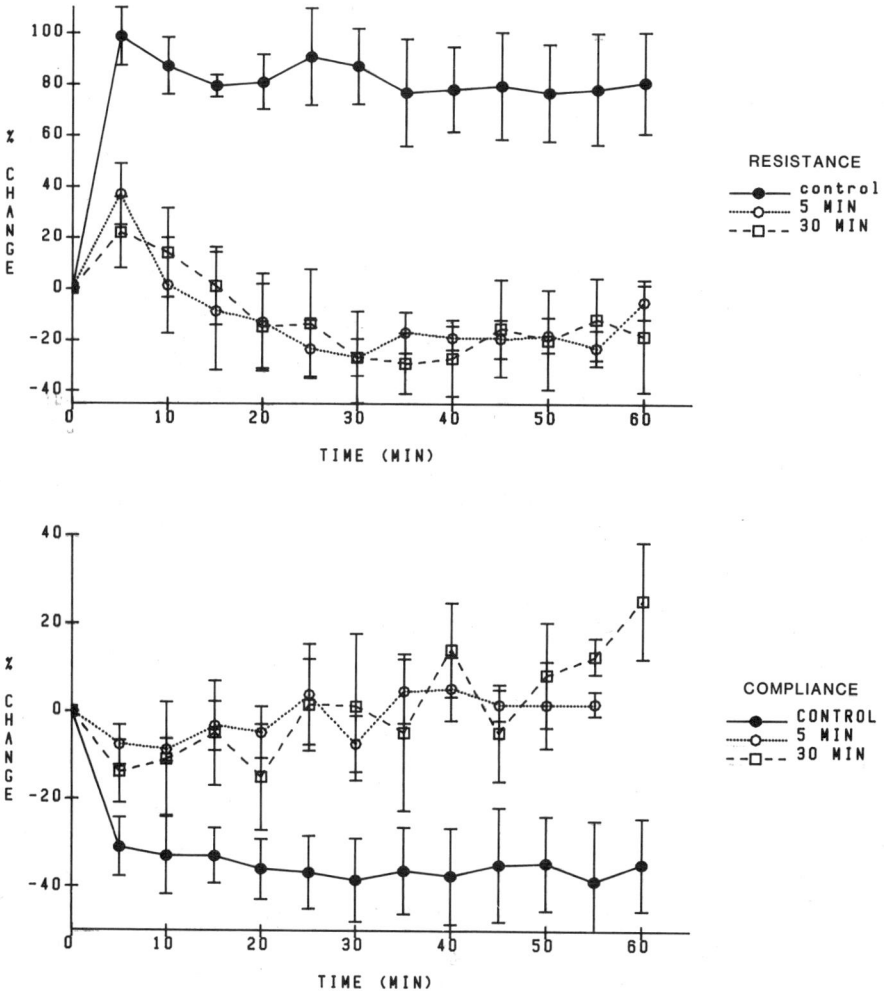

Fig. 14. Effect of 5 (n = 3) and 30 (n = 3) min. pretreatment with L-648,051 aerosol (0.1% for 5 min.) on resistance (upper) and compliance (lower) changes induced in conscious squirrel monkeys by an aerosol of LTD_4 (50 μg/mL, 5 min. aerosol).

SUMMARY AND CONCLUSIONS: LTD$_4$ ANTAGONISTS

L-648,051 (15) is a potent, competitive and selective antagonist of the leukotriene D$_4$ receptor in guinea pig and human lung tissue. Its activity on isolated smooth muscle preparations and its superiority relative to other agents (L-649,923, FPL-55712) in reversing ongoing contraction to LTD$_4$ is somewhat unexpected in light of the results of receptor binding studies (Ki = 6.2 for L-648,051 versus 0.4 for L-649,923 and 2.0 for FPL-55712 on guinea pig lung membrane). The reasons for this observed superiority are unknown but may relate to the rate at which antagonists equilibrate with the leuko-triene receptors on various tissues. The physical properties (polarity) of L-648,051 may contribute to this enhanced rate of equilibration. As has been observed for other acetophenone-containing leukotriene D$_4$ antagonists (e.g. LY-171883 [69]), L-648,051, L-649,923 and FPL-55712 inhibit rat neutro-phil phosphodiesterase with Ki values lower than theophyllene (F.A. Kuehl and H. Bull, unpublished results). This may contribute to the activity of L-648,051, but in that the Ki's for all these compounds are comparable, it does not explain the perceived superiority of L-648,051. The rapid metabo-lism and elimination of L-648,051 in vivo should be an advantage for a topical agent by minimizing systemic exposure. L-648,051 and L-649,923 have been shown to be potent and specific receptor antagonists of leukotriene D$_4$ in vitro and in vivo, but the physical properties of L-648,051 appear to impart to it superior properties as a topical agent. L-648,051 has been formulated as a powder aerosol for delivery via a freon propellant at dosages of from 100 to 750 ug per burst. It appears to be ideally suited for such a formulation. Preliminary studies in man have shown both L-648,051 and L-649,923 to be effective against LTD$_4$ challenge (55,71, 72). Future studies will define the potential role of these drugs as a novel therapy for human asthma.

DEVELOPMENT OF 5-LIPOXYGENASE INHIBITORS

Hypothetical Mechanism of 5-Lipoxygenase Enzyme

In order to develop specific inhibitors of the 5-lipoxygenase enzyme, it was important to gain some understanding of the mode of action of this enzyme, thereby allowing some measure of structural hypothesis to derive potential inhibitors. Although the exact mechanism of the lipoxygenation of arachidonic acid is not well understood, it is generally held that a redox mechanism is involved, in which the primary event is the transfer of a single electron from the Δ5 double bond of arachidonic acid to the enzyme,

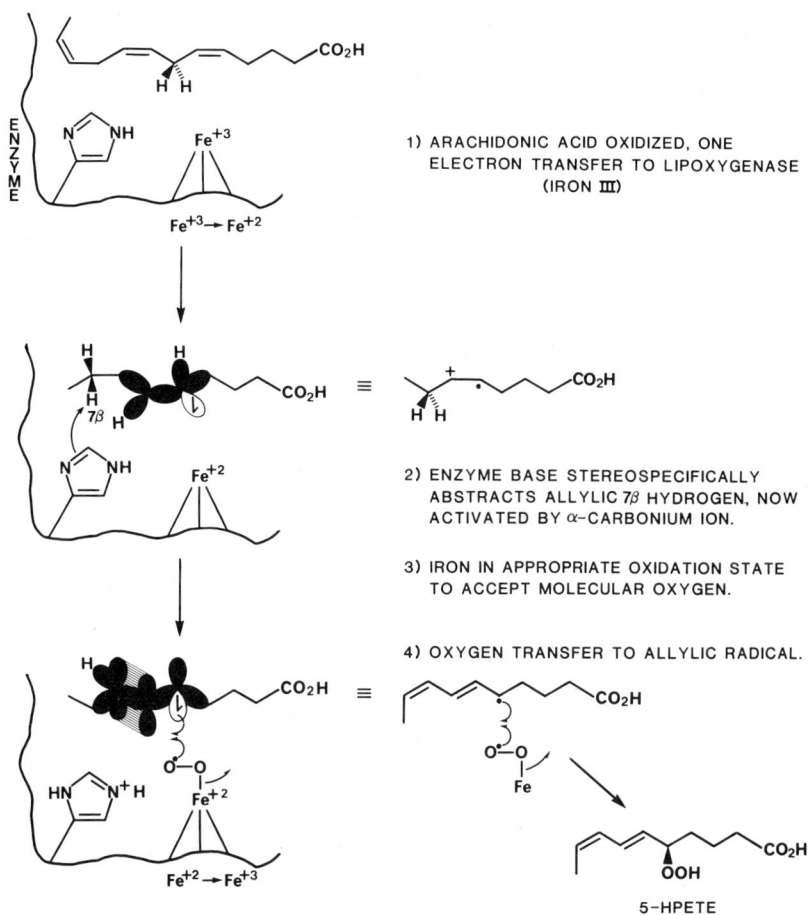

Fig. 15. Hypothetical mechanism of the biosynthesis of 5-HPETE
by the 5-lipoxygenase enzyme.

leaving a radial cation (Figure 15). This is followed by a specific abstraction of a C-6 proton to give a conjugated radical. This radical then traps
a molecule of oxygen either from the air or delivered by the enzyme specifically on the 5S face to give the product 5(S)-hydroperoxeicosatetaenoic acid
(5-HPETE). This same enzyme can then transform 5-HPETE via dehydration to
produce leukotriene A_4. Thus, it is apparent that the enzyme recognizes
specifically the lipophilic character of arachidonic acid and, in particular, the π character of the 5 double bond. In order to hypothesize potential inhibitors, one direction of our thinking involved attempts to rigidify
arachidonic acid as represented in a number of planar analogs (see Figure
16). From this analysis we hypothesized that tricyclic structures such as
Figure 17 shows would serve to fit well in the lipophilic pocket of the
5-lipoxygenase. We further felt that phenothiazinones (phenothiazinols

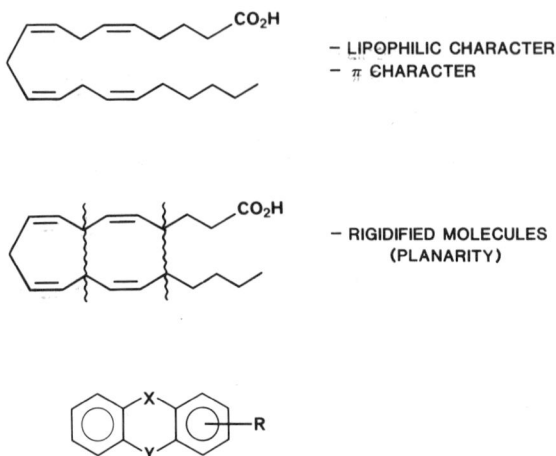

Fig. 16. Working hypothesis.

which could be derived from the reduction of phenothiazinones) could interact with the 5-lipoxygenase enzyme in a manner similar to that of arachidonic acid (i.e., the delivery of a single electron to produce a radial
cation) (Figure 17).

It was generally found, however, that phenothiazinols were highly
unstable and were readily oxidized to the corresponding phenothiazinones.
Furthermore, we found that these compounds, while effective inhibitors of
the 5-lipoxygenase in whole cell preparations (by human or rat PMN), were
ineffective in broken cell preparations or purified enzyme preparations
unless they were accompanied by a reducing agent such as NADH. It thus was
apparent that in whole cells the phenothiazinone was first reduced to the
phenothiazinol which then served as the active inhibitor. A number of such
phenothiazinones were studied and, in their development and optimization, a
number of considerations were considered. The basic unsubstituted phenothiazinone structure, while very active, contains a certain functionality which

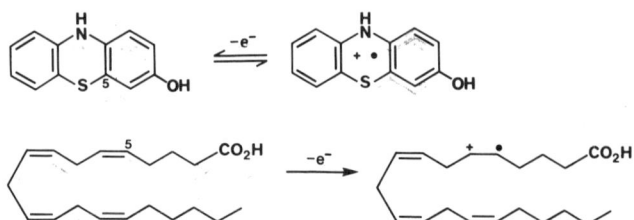

Fig. 17. 5-Lipoxygenase forms radical cations.

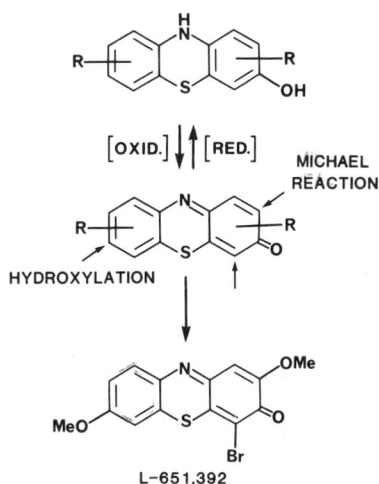

Fig. 18. L-651,392 and the phenothiazinone structure.

suggests the possibility of toxic interactions or metabolic opportunities
within the body.   Thus, the phenothiazinone structure appeared to be prone
to Michael reaction with nucleophiles such as thiols at sites either in con-
jugation with the nitrogen at the 3-position or the 5-position or potentially ß to the carbonyl at the 2-position (see Figure 18).   Also, it seemed
likely that the a-ring of the phenothiazinone would be prone to metabolic
hydroxylation, particularly in the position para to the nitrogen.

    Thus, a series of substitutions were effected and finally evolved to the
structure L-651,392 incorporating a 3-methoxy group, a 5-bromo group and a
7-methoxy group (71).   This compound again required NADH to derive the
active species (the phenothiazinol) which could then take part in electron
transfer reaction to provide a radical cation which, like arachidonic acid,
could further lose a proton to produce a phenolic radial (see Figure 19).
As such, the compound could act as an efficient competitive substrate for
the 5-lipoxygenase enzyme.   The pharmacological profile of L-651,392 is
presented in the following summary.

Pharmacological Profile of L-651,392

    L-651,392 was tested as an inhibitor of the biosynthesis of leukotrienes
in various cell types where lipoxygenase activity is expressed.   Mean concentrations required for 50% inhibition of the production of eicosanoids are
shown in Table 4.

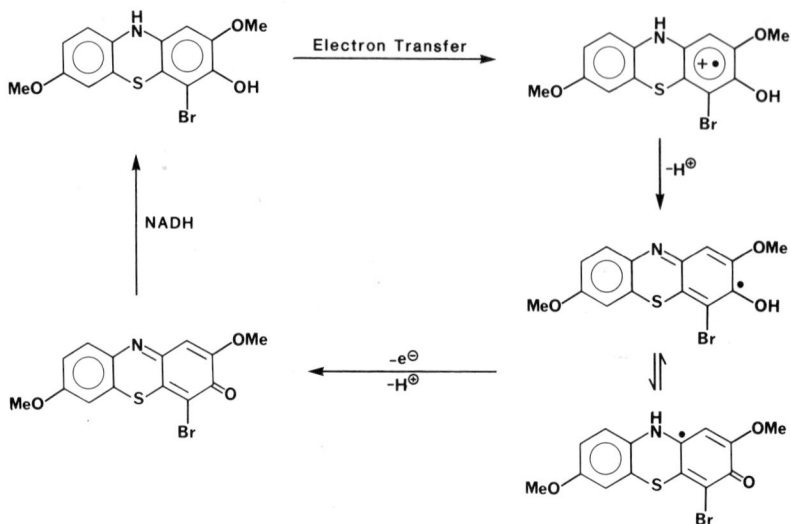

Fig. 19. L-651,392 conversion to a phenolic radical.

Rat peritoneal PMNLs were incubated with L-651,392 and exposed to 10 uM ionophore A23187 for 4 minutes. Leukotriene B_4 levels were determined in the supernatants by HPLC. Human PMN's were purified from human blood and pre-incubated with the drug for 5 minutes, then with cytochalasin B (5 mg/mL) for 5 minutes, followed by f-Met-Leu-phe (10 uM for 5 minutes). As determined by HPLC, production of LTB_4 was significantly inhibited with L-651,392, but no significant inhibition of the release of myloperoxidase or N-acetyl-glucosaminadase was observed. Murine peritoneal macrophages were pre-incubated with [^{3}H]-arachidonic acid, washed, and incubated with zymozan for 3 hours and the production of leukotriene C_4 and thromboxane B_2 assessed by thin layer chromatography and radioimmunoassay. In experiments in which significant inhibition of leukotriene C_4 production was observed, no significant inhibition of prostaglandin E_2, 6-keto-prostaglandin $F_{1\alpha}$

Table 4. Inhibition of Eicosanoids by L-651,392 in Various Cell Types

Cell Type	Product monitored	IC_{50} (± SEM)	n
Rat pertioneal polymorphonuclear leukocytes	LTB_4	$0.6 \pm 0.6 \times 10^{-7}M$	7
Mouse CXBG mastocytoma cells	LTC_4 TBX_2	$2.5 \pm 0.6 \times 10^{-7}M$ $>3 \times 10^{-5}M$	2
Mouse peritoneal macrophages	LTC_4 PGE_2	$2.5 \pm 1.0 \times 10^{-7}M$ $>10^{-6}M$	3
Human polymorphonuclear	LTB_4	$2.6 \pm 0.8 \times 10^7M$	6

or N-acetyl-glucosaminadase was observed. CXBG mastocytomoa cells were pre-incubated for 15 minutes with L-651,392, followed by a 20-minute incubation with ionophore A23187. Eicosanoid levels were determined by radioimmuno-assay.

L-651,392 was also tested on various enzyme preparations. Using a crude 5-lipoxygenase enzyme preparation obtained from rat basophil leukemia (RBL-1) cells (100,000 X g supernatant), significant inhibition of the 5-lipoxygenase enzyme with L-651,392 was observed only in the presence of 100 umol NADH or NADPH. Under these conditions, an IC_{50} value of $0.8 \pm 0.6 \times 10^{-7}$ (n = 5) was obtained. Similar results have been obtained with a highly purified 5-lipoxygenase from porcine leukocytes, indicating that reduction of L-651,392 must occur for inhibition of the 5-lipoxygenase enzyme to be observed. In contrast, L-651,392, in the presence or absence of NADH, failed to inhibit the cyclooxygenase enzyme from ram seminal vesicle microsomes, the 12-lipoxygenase enzyme from either human platelets or porcine leukocytes, the 15-lipoxygenase enzyme obtained from soybean, or the cytochrome P450 from rat liver microsomes. These results indicate that a reduction product of L-651,392 is a potent but selective inhibitor of the 5-lipoxygenase enzyme.

The inhibitory effects of L-651,392 on the 5-lipoxygenase enzyme are manifested *in vivo* through inhibition of antigen-induced bronchoconstriction in inbred rats and squirrel monkeys and antigen-induced changes in vascular permeability in the guinea pig conjunctiva. When L-651,392 was administered PO 1 hr. prior to antigen challenge to inbred hyperreactive rats (pretreated with methysergide [3 ug/kg IV] 5 min. prior to antigen challenge), significant inhibition of the antigen-induced dyspnea was obtained (74) (effective dose of 50% 1.3 mg/kg) (Figure 20).

Similarly, treatment with L-651,392 (5 mg/kg PO) 2 hr. prior to ascaris suum antigen challenge of squirrel monkeys produced significant inhibition of the increases in pulmonary resistance and decreases in dynamic compliance normally observed in these primates (75). Topical application of L-651,392 to the guinea pig eye significantly inhibited both antigen-induced changes in vascular permeability and the 5-lipoxygenase activity in the conjunctiva (determined by *ex vivo* challenge with ionophore A23187 (76). These results suggest that L-651,392 can be used as either an oral or a topical drug for inhibiting leukotriene-mediated allergic responses. Such compounds may prove useful in elucidating the role of 5-lipoxygenase products in physiological and pathological processes both *in vitro* and *in vivo*.

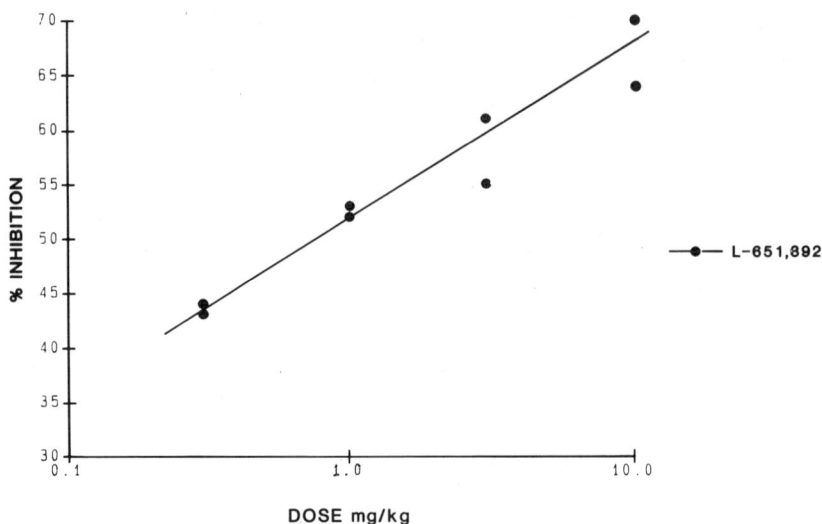

Fig. 20. The percentage inhibition of antigen-induced dyspnea in sensitized and conscious inbred rats by L-651,392 when administered orally 1 hr. prior to antigen exposure. The rats were also treated with methylsergide (3 μg/kg IV) 5 min. prior to antigen. Each point represents the mean of 6 to 10 animals.

Recent studies in our laboratories (44) showed L-651,392 administered topically at 0.05 to 0.5% cream formulations or at 40 ug/20 uL DMSO per ear significantly inhibited, respectively, the A23187-induced increase in incorporation of tritiated thymidine and the increase in immunoreactive LTB_4 in the guinea pig ear. Coggeshal et al. studied the role of leukotrienes in the responses of chronically instrumented sheep to endotoxin (78). They found that L-651,392 (5 mg/kg), administered PO 24 hours before experimentation, blocked effectively the E coli endotoxin-induced increase in leukotriene B_4 concentration in lung lymph and attenuated by over 70% LTB_4 production from stimulated, isolated granulocytes, while not inhibiting thromboxane B_2 production. L-651,392 also increased dynamic compliance in these sheep (9.0 $\pm$ 5.1% for endotoxin + L-651,392 versus 54.6 $\pm$ 5.1% for endotoxin alone). Yared et al. (79) noted that infusion of leukotriene D_4 led to an increase in urinary protein excretion, and L-651,392 (5 mg/kg PO for 2 days) prevented proteinuria induced by antiglomerular basement membrane antibody. In another study, Beck et al. (80) in Kingston, Ontario studied the role of 5-lipoxygenase products in ethanol-induced intestinal plasma protein loss in rabbits. They noted that the introduction of intraluminal ethanol into the jejunum caused an increase in leukotriene B_4 generation. Pretreatment with 20 mg/kg L-651,392 significantly decreased

100

this LTB$_4$ formation in both a control segment of jejunum and the ethanol-
treated segment of jejunum. Ethanol perfusion also caused an increase in
total protein loss, and this increase was significantly inhibited by
L-651,392 treatment.

The stage is now set to rigorously evaluate the role of 5-lipoxygenase
products in human disease. As clinical trials on these selective antagonists and inhibitors go forward over the next few years, we can expect to
learn much about the underlying processes involved with incapacitating
diseases such as asthma, adult respiratory distress syndrome, toxic shock,
psoriasis, and glomerulonephritis.

ACKNOWLEDGEMENTS

The work described herein is the result of the sustained efforts of a
large group of researchers at Merck Frosst Canada Inc. and at Merck and Co.,
Inc., USA., as well as many collaborators in the academic scientific community. Their dedicated and skilled efforts are gratefully acknowledged. The
skilled help and patience of Suzanne Quesnel in preparing this manuscript is
also acknowledged.

REFERENCES

1. A.W. Ford-Hutchinson, M.A. Bray, M.V. Doig, M.E. Shipley, M.J.H. Smith,
   Leukotriene B: a potent chemokinetic and aggregating substance released from polymorphonuclear leucocytes, Nature 286:264-265 (1980).
2. M.H.J. Smith, A.W. Ford-Hutchinson, M.A. Bray, Leukotriene B: a potential mediator of inflammation, J. Pharm. Pharmacol. 32:517-518
   (1980).
3. E.J. Goetzl and W.C. Pickett, The human PMN leukocyte chemotactic activity of complex hydroxy-eicosatetraenoic acids (HETEs), J. Immunol.
   125:1789-1791 (1980).
4. M.A. Bray, F.M. Cunningham, A.W. Ford-Hutchinson, M.J.H. Smith, Leukotriene B$_4$: an inflammatory mediator in vivo, Br. J. Pharmacol.
   22:483-486 (1981).
5. S.D. Brain, R.D.R. Camp, R.M. Dowd, A. Kobza-Black, P.M. Wollard, A.I.
   Mallet, M.W. Greaves, Psoriasis and leukotrienes B$_4$, Lancet ii:762
   (1982).
6. B. Samuelsson, Leukotrienes: mediators of immediate hypersensitivity
   reactions and inflammation, Science (Washington, D.C.) 220:568-575
   (1983).
7. J. Verhagen, P.L.B. Braynzee, J.A. Koedam, G.A. Wassink, M. Boer, G.K.

Terpstra, J. Kreukniet, G.A. Veldink, J.F.G. Vliegenthart, Specific
leukotriene formation by purified human eosinophils and neutrophils,
FEBS Letters, 168:23-28 (1984).

8. M. Damon, C. Chavis, P. Goodard, F.B. Michel, A. Crastes de Paulet,
    Purification and mass spectrometry identification of leukotriene D_4
    synthesized by human alveolar macrophages, Biochem. Biophys. Res.
    Commun. 111:518-524 (1983).

9. D.W. MacGlashan, Jr., R.P. Schleimer, S.P. Peters, E.S. Schulman, G.K.
    Adams III, H.H. Newball, Generation of leukotrienes by purified human
    lung mast cells, J. Clin. Invest. 70:747-751 (1982).

10. S.E. Dahlen, P. Hedqvist, S. Hammarstrom, B. Samuelsson, Leukotrienes
    are potent constrictors of human bronchi, Nature (London) 288:484-
    486 (1980).

11. J.M. Drazen, K.F. Austen, R.A. Lewis, D.A. Clark, G. Goto, A. Marfat,
    E.J. Corey, Comparative airway and vascular activities of leukotri-
    enes C-1 and D in vivo and in vitro, Proc. Nat'l. Acad. Sci.
    USA 77:4354-4358 (1980).

12. C.J. Hanna, M.K. Bach, P.D. Pare, R.R. Schellenberg, Slow-reacting sub-
    stances (leukotrienes) contract human airway and pulmonary vascular
    smooth muscle in vitro, Nature (London) 290:343-344 (1981).

13. P.J. Piper, M.N. Samhoun, J.R. Tippins, T.J. Williams, M.A. Plamer, M.J.
    Peck, Pharmacological studies on pure SRS-A and synthetic leukotri-
    enes C_4 and D_4. In: "SRS-A and leukotrienes," P.J. Piper, ed.,
    Wiley, Chichester, England, pp. 81-99 (1981).

14. T.R. Jones, C. Davis, E.E. Daniel, Pharmacological study of the contrac-
    tile activity of leukotriene C_4 and D_4 on isolated human airway
    smooth muscle, Can. J. Physiol. Pharmacol. 60:638-643 (1982).

15. H. Bisgaard and M. Pederson, SRS-A dampens the activity of human cilia,
    Fed. Proc. 42:1381 (1983).

16. J.D. Lundgren, J.H. Shelhamer, M.A. Kaliner, The role of eicosanoids in
    respiratory mucus hypersecretion, Ann. Allergy 55:5-8 (1985).

17. C.L. Malmsten and I.M. Goldstein, Leukotrienes: mediators of inflamma-
    tory and immediate hypersensitivity reactions, CRC Crit. Rev.
    Immunol. 4:307-333 (1984).

18. J.W. Weiss, J.M. Drazen, N. Coles, E.R. McFadden, Jr., P.F. Weller, E.J.
    Corey, R.A. Lewis, K.F. Austen, Bronchoconstrictor effects of leuko-
    triene C in humans, Science (Washington, D.C.) 216:196-198 (1982).

19. M. Griffin, J.W. Weiss, A.G. Leitch, E.R. McFadden, Jr., E.J. Corey,
    K.F. Austen, J.M. Drazen, Effects of leukotriene D on the airways in
    asthma, N. Engl. J. Med. 308:436-439 (1983).

20. J.C. Delehunt, A.P. Perruchoud, L. Yerger, B. Marchette, J.S. Stevenson,

W.M. Abraham, The role of slow-reacting substance of anaphylaxis in the late bronchial response after antigen challenge in allergic sheep, Am. Rev. Respir. Dis. 130:748-754 (1984).

21. W.M. Abraham, E. Russi, A. Wanner, J.C. Delehunt, L.D. Yerger, G.A. Chapman, Production of early and late pulmonary responses with inhaled leukotriene D_4 in allergic sheep, Prostaglandins 29:715-726 (1985).

22. E. Pirotzky, J. Bidault, C. Burtin, M.C. Gubler, J. Benveniste, Release of platelet activating factor, slow-reacting substance and vasoactive amines from isolated rat kidney, Kidney Int. 25:404-410 (1984).

23. E.S.K. Assem and N. Azizan Abdullah, Release of thromboxane B_2 and leukotriene C_4 and reduction in renal perfusion in experimental anaphylactic reaction of isolated guinea pig kidney, Int. Archs. Allergy Appl. Immun. 82:212-214 (1987).

24. J. Sraer, M. Bens, R. Ardaillou, J.D. Sraer, Sulfidopeptide leukotriene biosynthesis and metabolism by rat glomeruli and papilla (abstr.), Kidney Int. 29:346 (1986).

25. E.A. Lianos, Glomerular leukotriene biosynthesis and degradation in the rat: effects of immune injury (abstr.), Kidney Int. 29:339 (1986).

26. V. Cattell, J. Smith, H.T. Cook, S. Moncada, J.A. Salmon, Leukotriene B_4 synthesis in normal rat glomeruli (abstr.), Kidney Int. 27:254 (1985).

27. M.A. Rahman, M. Nakazawa, S.N. Emancipator, M.J. Dunn, Increased leukotriene B_4 (LTB_4) synthesis in immune-injured rat glomeruli (abstr.), Kidney Int. 29:343 (1986).

28. R. Barnett, P. Goldwasser, L.A. Scharschmidt, D. Schlondorff, Effects of leukotrienes on isolated rat glomeruli and cultured mesangial cells, Am. J. Physiol. 250:F838-F844 (1986).

29. M.Simonson and M.J. Dunn, Leukotriene C_4 and D_4 contract rat glomerular mesangial cells, Kidney Int. 30:524-531 (1986).

30. L. Baud, J. Sraer, J. Perez, M.P. Nivez, R. Ardaillou, Leukotriene C_4 binds to human glomerular epithelial cells and promotes their proliferation in vitro, J. Clin. Invest. 76:374-377 (1985).

31. A. Rosenthal and C.R. Pace-Asciak, Potent vasoconstriction of the isolated perfused rat kidney by leukotrienes C_4 and D_4, Can. J. Physiol. Pharmacol. 61:325-328 (1983).

32. P.J. Piper, A.W.B. Stanton, L.J. McLeod, The actions of leukotrienes C_4 and D_4 in the porcine renal vascular bed, Prostaglandins 29:61-73 (1985).

33. K.F. Badr, C. Baylis, J.M. Pfeffer, M.A. Pfeffer, R.J. Soberman, R.B. Lewis, K.F. Austen, E.J. Corey, Renal and systemic hemodynamic

responses to intravenous infusion of leukotriene C_4 in the rat, Circ. Res. 54:492-499 (1984).

34. K.F. Badr, B.M. Brenner, M.Wasserman, I. Ichikawa, Evidence for local glomerular actions of leukotriene D_4 (abstr.), Kidney Int. 29:328 (1986).

35. K.F. Badr, B.M. Brenner, I. Ichikawa, Effects of leukotriene D_4 on glomerular dynamics in the rat, Am. J. Physiol. 22:F239-F243 (1987).

36. K.F. Badr, A. Gung, G.F. Schreiner, M. Wasserman, I. Ichikawa, Reversal of antiglomerular basement membrane antibody-induced fall in the glomerular ultrafiltration coefficient by the leukotriene D_4 antagonist SK&F 10453 (abstr.), Kidney Int. 31:363, 1987.

37. R. Ardaillou, L. Baud, J. Sraer, Leukotrienes and other lipoxygenase products of arachidonic acid synthesized in the kidney, Am. J. Med. 81(suppl. B):12-22 (1986).

38. D.W. Snyder and R.D. Krell, Pharmacological evidence for a distinct leukotriene C_4 receptor in guinea pig trachea, J. Pharmacol. Exp. Ther. 231:616-622 (1984).

39. S.S. Pong and R.N. DeHaven, Characterization of a leukotriene D_4 receptor in guinea pig lung, Proc. Natl. Acad. Sci. USA 80:7415-7419 (1983).

40. S.S. Pong, R.N. DeHaven, F.A. Keuhl, Jr., R.W. Egan, Leukotriene C_4 binding to rat lung membranes, J. Biol. Chem. 258:9615-9619 (1983).

41. G.K. Hogaboom, M.S. Mong, H. Wu, S.T. Crooke, Peptido-leukotrienes: distinct receptors for leukotriene C_4 and D_4 in the guinea pig lung, Biochem. Biophys. Res. Commun. 116:1136-1143 (1983).

42. S. Mong, H.L. Wu, M.O. Scott, M.A. Lewis, M.A. Clark, B.M. Weichman, C.M. Kinzig, J.G. Gleason, S.T. Crook, Molecular heterogeneity of leukotriene receptors: correlation of smooth muscle contraction and radioligand binding in guinea-pig lung, J. Pharmacol. Exp. Ther. 234:316-325 (1985).

43. B.J. Ballerman, R.A. Lewis, E.J. Corey, K.F. Austen, B.M. Brenner, Identification and characterization of leukotriene C_4 receptors in isolated rat renal glomeruli, Circ. Res. 56:324-330 (1985).

44. C.K. Buckner, R.D. Krell, R.B. Laravusa, D.B. Coursin, P.R. Bernstein, J.N. Will, Pharmacological evidence that human intralobar airways do not contain different receptors that mediate contractions to leukotriene C_4 and leukotriene D_4, J. Pharmac. Exp. Ther. 237:558-562 (1986).

45. T. Ahmed, D.W. Greenblatt, S. Birch, B. Marchette, A. Wanner, Abnormal mucociliary transport in allergic patients with antigen-induced bronchospasm: role of slow-reacting substance of anaphylaxis,

American Review of Respiratory Disease 124:110 (1981).

46. M.C. Holroyde, R.E.C. Altounyan, M. Cole, M. Dixon, E.V. Elliott, Bronchoconstriction produced in man by leukotrienes C and D, _Lancet_ 11:17 (1981).

47. R.E.C. Altounyan and M. Cole, Therapeutic implications: "Proceedings of the Xth International Congress of Allergology and Clinical Immunology," J.W. Kerr, ed., Macmillan, London, p. 271 (1983).

48. T.H. Lee, M.J. Walport, A.H. Wilkinson, M. Turner-Warwick, A.B. Kay, Slow-reacting substance of anaphylaxis antagonist FPL-55712 in chronic asthma, _Lancet_ ii:304 (1981).

49. W.S. Marshall, T. Goodson, G.J. Cullinan, D. Swanson-Bean, K.D. Haish, L.E. Rinkema, J.H. Fleish, Leukotriene receptor antagonists. 1. Synthesis and structure-activity relationships of alkoxyacetophenone derivatives, _J. Med. Chem._ 30:682-689 (1987).

50. M. Cloud, G. Eras, J. Kemp, T. Platts-Mills, L. Altnan, R. Townley, D. Tinkelman, T. King, E. Middleton, A. Sheffer, E. McFadden, Efficacy and safety of LY171883 in patients with mild chronic asthma, _J. Allergy, Clin. Invest._ (abstr.) 79:525 (1987).

51. J.H. Fleish, L.E. Rinkema, K.D. Haisch, D. McCullough, F.P. Carr, R.D. Dillard, Evaluation of LY163443, 1-[2-hydroxy-3-propyl-4-{[4-(1H-tetrazol-5-ylmethyl)phenoxy]methyl}phenyl]ethanone, as a pharmacological antagonist of leukotriene D_4 and E_4, _Naunyn-Schmiedeberg's Arch. Pharmacol._ 333:70-77 (1986).

52. J.G. Gleason, R.F. Hall, C.D. Perchonock, K.F. Erhard, J.S. Frazee, T.W. Ku, K. Kondrad, M.E. McCartney, S. Mong, S.T. Crooke, G. Chi-Rosso, M.A. Wasserman, T.J. Torphy, R.M. Muceitelli, D.W. Hay, S.S. Tucker, L. Vickery-Clark, High-affinity leukotriene receptor antagonists. Synthesis and pharmacological characterization of 2-hydroxy-3[(2-carboxyethyl)thio]-3-[2-(8-phenylocyl)phenyl]propanoic acid, _J. Med. Chem._ 30:959-961 (1987).

53. R.G. Van Inwegen, A. Khandwala, R. Gordon, P. Sonnino, S. Coutts, S. Jolly, REV5901: an orally effective peplidoleukotriene antagonist, detailed biochemical/pharmacological profile, _J. Pharmacol. Exp. Ther._ 241:117-123 (1987).

54. J.M. Hand, L.M. Marshall, J.H. Musser, A.F. Kreft, L. Marinari, S. Schwalm, I. Englebach, A. Auen, J. Chang, The pharmacology of WY-48, 252: a novel selective and orally active antagonist of leukotriene D_4 (abstr.), Conference on the Biology of the Leukotrienes, (N.Y. Acad. Sci.), Philadelphia, June 28-July 1 (1987).

55. N.C. Barnes, The actions of leukotriene agonists, antagonists and synthesis inhibitors in man (abstr.), Xth Int. Congress Pharmacol., Sydney,

Australia, Aug. 23-28 (1987).

56. P.R. Bernstein, Y.K. Yee, D. Snyder, A novel class of peptidoleukotriene antagonists: the bioisosterism of ketosulfone derivatives and acylsulfonamides (abstr.), 194th ACS National Meeting, New Orleans, August 30-September 4, (1987).

57. A. Lassus and S. Forsstrom, A dimethoxynaphthalene derivative RS-43179 gel compared with 0.025 percent fluocinoline acetonide gel in the treatment of psoriasis, Br. J. Dermatol. 113:103-106 (1985).

58. K. Miyasaka, T. Mikami, K. Miyazawa, M. Hagiwara, Inhibitory effect of 2-phenylindole derivative TZl-44127 of 5-lipoxygenase, Jpn. J. Pharmacol. 43(suppl.):118 (1987).

59. M. Fujimura, F. Sasaki, Y. Nakatsumi, Y. Takahashi, S. Hifumi, K. Taga, J. Mifune, T. Tanaka, T. Matsuda, Effects of a thromboxane synthetase inhibitor (OKY-046) and a lipoxygenase inhibitor (AA-861) on bronchial responsiveness to acetylcholine in asthmatic subjects, Thorax 41:955-959 (1986).

60. A.W. Ford-Hutchinson, G. Brunet, R. Hamel, H. Piechuta, G. Holme, Respiratory responses to leukotriene and biogenic amines in normal and hyperreactive rats, J. Immunol. 131:434-438 (1983).

61. C.S. McFarlane, H. Piechuta, R.A. Hall, A.W. Ford-Hutchinson, Effects of a contractile prostaglandin antagonist (L-640,035) upon allergen-induced bronchoconstriction in hyperreactive rats and conscious squirrel monkeys, Prostaglandins 28:173-182 (1984).

62. J.G. Atkinson and J. Rokach, Synthesis of leukotrienes, In: "Handbook of Eicosanoids," Vol. 1, Part B, A.L. Willis, ed., C.R.C. Press, Boca Raton, Florida (1987).

63. E.J. Corey, D.A. Clark, A. Marfat, In: "The Leukotrienes Chemistry and Biology," L.W. Chakrin and D.M. Bailey, eds., pp. 14-103, Academic Press, London (1984).

64. J. Augstein, J.B. Farmer, T.B. Lee, P. Sheard, M.L. Tattersall, Selective inhibitor of slow-reacting substance and anaphylaxis, Nature New Biol. 245:215 (1973).

65. R.N. Young, P. Belanger, E. Champion, R.N. DeHaven, D. Denis, A.W. Ford-Hutchinson, R. Fortin, R. Frenette, J-Y. Gauthier, J. Gillard, Y. Guindon, T.R. Jones, M. Kakushima, P. Masson, A. Maycock, C.S. McFarlane, H. Piechuta, S.S. Pong, J. Rokach, H. Williams, C. Yoakim, R. Zamboni, Design and synthesis of sodium (ß&, γ S*)-4-[[3-(4-acetyl-3-hydroxy-2-propyl]thio]-γ-hydroxy-ß-methylbenzene-butanoate: a novel, selective and orally-active receptor antagonist of leukotriene D_4, J. Med. Chem. 29:1573-1576 (1986).

66. T.R. Jones, R. Young, E. Champion, L. Charette, R.N. DeHaven, D. Denis,

A.W. Ford-Hutchinson, R. Frenette, J-Y. Gautheri, Y. Guindon, M. Kakushima, P. Masson, A. Maycock, S.S. Pong, J. Rokach, R. Zamboni, L-649,923, sodium (ßS*, γR*)-4-(3-(4-acetyl-3-hydroxy-2-propyl-phenoxy)propylthio)- -hydroxy-ß-methylbenzenebutanoate. A selective, orally active leukotriene receptor antagonist, Can. J. Physiol. Pharmacol. 64:1068-1075 (1986).

67. D. Arunlakshana and H.O. Schild, Some quantitative uses of drug antago-nists, Br. J. Pharmacol. 14:48-58 (1959).

68. T.R. Jones and P. Masson, Comparative study of the pulmonary effects of intravenous leukotrienes and other bronchoconstrictors in anesthe-tized guinea-pigs, Prostaglandins 29:799-817 (1985).

69. J.H. Fleisch, L.E. Rinkema, K.D. Haisch, D. Swanson-Bean, T. Goodson, P.P.K. Ho, W.S. Marshall, LY-171883, 1-(2-hydroxy-3-propyl-4-(4-(1H-tetrazol-5-yl)butoxy)-phenyl)-ethanone, an orally active leuko-triene D_4 antagonist, J. Pharmacol. Exp. Ther. 233:148-157 (1985).

70. T.R. Jones, Y. Guindon, R.N. Young, E. Champsion, L. Charette, D. Denis, D. Ethier, R. Hamel, A.W. Ford-Hutchinson, R. Fortin, G. Letts, P. Masson, C. McFarlane, H. Piechuta, J. Rokach, C. Yoakim, R.N. DeHaven, A. Maycock, S.S. Pong, L-648,051, sodium 4-[3-(4-acetyl-3-hydroxy-2-propylphenoxy)-propylsulfonyl]-γ-oxo-benze nebutanoate: a leukotriene D_4 receptor antagonist, Can. J. Physiol. Pharmacol. 64:1535-1542 (1986).

71. N. Barnes, P.J. Piper, J. Costello, The effect of an oral leukotriene antagonist L-649-923 on histamine and leukotriene D_4-induced bronchoconstriction in normal man, J. Allergy Clin. Immunol. 79:816-821 (1987).

72. N.C. Barnes, J. Evans, J.T. Zakrzewski, P.J. Piper, F. Costello, L-648,051 A leukotriene D_4 antagonist is active by the inhaled route in man (abstr.), Xth Int. Congress Pharmacol., Sydney, Australia, August 23-28 (1987).

73. Y. Guindon, Y. Girard, A. Maycock, A.W. Ford-Hutchinson, J.G. Atkinson, P.C. Belanger, A. Dallob, D. DeSousa, H. Dougherty, R. Egan, M.M. Goldenberg, E. Ham, R. Fortin, P. Hamel, R. Hamel, C.K. Lau, Y. Leblanc, C.S. McFarlane, H. Piechuta, M. Thérien, C. Yoakim, and J. Rokach, L-651,392: a novel, potent and selective 5-lipoxygenase inhibitor, In: "Advances in Prostaglandin, Thromboxane and Leukotriene Research," Vol. 17A, B. Samuelsson, R. Paoletti and P.W. Pamwell, eds., Raven Press, New York (1986).

74. H. Piechuta, A.W. Ford-Hutchinson, G.L. Letts, Inhibition of allergen-induced bronchoconstriction in hyperreactive rats as a model for testing 5-lipoxygenase inhibitors and leukotriene D_4 antagonists,

Agents and Actions, in press.

75. C.S. McFarlane, R. Hamel, A.W. Ford-Hutchinson, Effects of a 5-lipoxy-
    genase inhibitor (L-651,392) on primary and late pulmonary responses
    to ascaris antigen in the squirrel monkey, Agents and Actions, in
    press.

76. D. Garceau, A.W. Ford-Hutchinson, S. Charleson, 5-lipoxygenase inhibi-
    tors and allergic conjunctivitis reactions in the guinea pig, Europ.
    J. Pharmacol., in press.

77. C-C. Chan, L. Dubois, V. Young, Effects of two novel inhibitors of
    5-lipoxygenase, L-651,392 and L-651,896, in a guinea-pig model of
    epidermal hyperproliferation, Europ. J. Pharmacol. 139:11-18 (1987).

78. J.W. Coggeshall, W.E. Serofin, M.A. Philips, P.L. Lefferts, J.R.
    Snapper, Role of leukotrienes in an animal model of the adult respira-
    tory distress syndrome, Am. Rev. Respir. Dis., in press.

79. A. Yared, A. Gung, G. Schreiner, I. Ichikawa, K.F. Badr, Leukotriene
    D_4 (LTD_4)-induced protein urea: role in experimental glomerulo-
    nephritis (abstr.), Xth International Congress of Nephrology, London
    (1987).

80. J.T. Beck, A.J. Boyd, P.K. Dinda, Evidence for the involvement of
    5-lipoxygenase products in the ethanol-induced intestinal plasma
    protein loss, submitted for publication.

THE RENAL CYTOCHROME P450 SYSTEM GENERATES NOVEL

ARACHIDONIC ACID METABOLITES

Michal Schwartzman, Mairead A. Carroll, David Sacerdoti**,
Nader G. Abraham* and John C. McGiff

The Departments of Pharmacology and Medicine*
New York Medical College, Valhalla, New York, USA

The Department of Clinical Medicine**
University of Padua, Padua, Italy

The generation of prostaglandins and other oxygenated metabolites of
arachidonic acid (AA) is a complex process initiated by the release of esterified AA from cellular lipids. Once liberated from membrane lipids by
diverse stimuli (peptide hormones, neurotransmitters and mechanical disrup-
tion), the free AA is rapidly metabolized. Metabolism of AA involves three
pathways: (a) cyclooxygenase, leading to the formation of prostaglandins,
thromboxane A_2 (TxA$_2$), and prostacyclin (PGI$_2$); (b) lipoxygenases,
leading to the formation of hydroxy- and dihydroxyeicosatetraenoic acids
(HETEs and diHETEs) and leukotrienes; (c) cytochrome P450-dependent monoxy-
genase system which metabolizes AA by an NADPH-dependent mechanism to a
variety of oxygenated products such as HETEs, epoxyeicosatrienoic acids or
epoxides (EETs) and their hydrolysis products, the dihydroxyeicosatrienoic
acids or diols (DHTs) as well as ω and ω-1 hydroxylated acids (1-4). The
pattern of AA metabolism in the kidney is distinct and AA metabolites participate in integrated renal function. Among the structures that metabolize
AA via cyclooxygenase are collecting tubules (5), glomeruli (6), medullary
interstitium (7) and blood vessels (8). Other structures such as the convoluted tubules have low or negligible cyclooxygenase activity (8). Lipoxygenase activity is mainly associated with leukocytes or platelets (9,10) and
was reported to be present in isolated glomeruli (11). In epithelial cells
isolated from the thick ascending limb of Henle's loop (TALH) of the rabbit
kidney, AA is specifically metabolized by a cytochrome P450-dependent pathway (12). The cells of the TALH participate in the regulation of extracellular fluid volume and composition by establishing the solute gradient in

109

the medulla. The metabolic energy for active transtubular solute reabsorption is provided by Na^+-K^+-ATPase situated on the contraluminal side of the TALH (13). Novel arachidonate metabolites, which have the capacity to regulate segmental renal tubular function and the contiguous medullary micro-circulation, have been identified as major products of a cytochrome P450 system located in the TALH (12,14). The outer medulla, not the cortex, of the rabbit kidney possesses the highest level of the cytochrome P450 enzyme capable of metabolizing AA (15).

RENAL CYTOCHROME P450 MONOOXYGENASES

The cytochrome P450 mixed-function oxidase system has three components: cytochrome P450 as the hemoprotein, a flavoprotein reductase identified as the NADPH-dependent cytochrome C reductase, and phosphatidylcholine, which serves to facilitate electron transfer in microsomal systems. Cytochrome P450 exists in multiple forms that differ in substrate specificity, positional specificity, and stereospecificity (16). The renal content of the components of the cytochrome P450-dependent mixed-function oxidase system has been measured and, in most species studied, it was found to be small when compared with the liver (17). The highest level of cytochrome P450 and the components of this system were found in the renal cortex (18-21). The renal cortical cytochrome P450 system has been shown to metabolize AA to the ω- and ω-1 hydroxylation products, the EETs which can undergo hydrolysis by epoxide hydrolase to form the corresponding DHTs (diols), and the trihydroxy-eicosatrienoic acids (1). The EETs have been shown to stimulate secretion of hormones (22,23) and to inhibit chloride transport in renal tubules (24). Recently, two cytochrome P450-dependent AA metabolites formed by TALH cells have been found to possess biological activity: one inhibits Na^+-K^+-ATPase activity, and the other relaxes blood vessels (25). Increased production of these metabolites by TALH cells has also been described in rabbits made hypertensive by aortic coarctation (26).

THE ASCENDING LIMB AND ARACHIDONATE METABOLITES

Our studies of the past several years have addressed potential regulators of Na^+-K^+-ATPase and local blood flow which are generated within the TALH. These novel arachidonate metabolites, based on characterization of their biological activity, may link tubular Na^+-K^+-ATPase activity to changes in local blood flow and, thereby, couple a metabolic event to the regulation of regional blood flow. They are formed from AA via a cytochrome P450-dependent monooxygenase pathway in response to specific hormonal stimuli, such as arginine vasopressin (AVP) and calcitonin, through an inter-mediate step mediated by cyclic AMP (25). This has been designated the third pathway of AA metabolism, assuming that the several lipoxygenases re-present the second pathway and cyclooxygenase the first. The initial de-scription of biological activities of AA metabolites arising from cytochrome P450 monooxygenases revealed similarities to prostanoids; i.e., they can act as secretagogues and local modulators of circulating hormones, and their activity is usually circumscribed to the microenvironment of the cell of origin (23). The studies of Capdevila, Falck and their associates were the first to suggest the importance of the biological activities of AA metabo-lites generated by cytochrome P450-related monooxygenases (27,28). The epoxides of AA were shown to act as secretagogues of peptide hormones.

We entered the field by misadventure. After isolation of cells from the TALH we, initially and incorrectly, identified a lipoxygenase as the major pathway of AA metabolism (12). Cyclooxygenase activity was negligible and unresponsive to hormonal stimuli in agreement with other studies which did not detect appreciable cyclooxygenase in the TALH, having localized it pri-marily to the collecting tubules of the nephron (29). Thus, segmentation of nephron function with regard to transport activity has corresponding zones that differ with respect to AA metabolism. The cells of the TALH transform AA to newly discovered products which appear to modulate transport function of the TALH as well as modify blood vessel tone; i.e., they function as local hormones. The TALH has one of the highest concentrations in mammalian tissues of the enzyme, Na^+-K^+-ATPase, which drives active sodium trans-port (13). We have found that one of the major products of AA metabolism in TALH is a potent inhibitor of Na^+-K^+-ATPase (ID_{50}AA metabolite 120 nM vs. ID_{50} ouabain 50 nM) (25). The biochemical pathway that generates these hormones from AA was identified, unexpectedly, as a cytochrome P450 species specific for AA.

Incubation of TALH cells with ^{14}C-labeled AA resulted in formation of oxygenated metabolites separated into peaks 1 and 2 by reverse-phase chroma-tography. Peak 1 yielded material that relaxed blood vessels, and peak 2 contained an inhibitor of Na^+-K^+-ATPase. The metabolites accounted for 30% to 40% of the recovered radioactivity, and their formation was not affected by indomethacin. These products of arachidonate metabolism were considered to arise from a cytochrome P450-related monooxygenase based on the following findings:

1. They did not demonstrate UV absorbance at 234 nm and 276 nm, indicat-ing the absence of a conjugated diene or triene structure, respectively.

2. Generation of the metabolites by cell-free homogenates of TALH was dependent on the presence of NADPH.

3. SKF-525A and carbon monoxide inhibited their formation.

4. The metabolites recovered from peaks 1 and 2 had different retention times on reverse-phase HPLC from AA products formed by lipoxygenases.

5. Induction of cytochrome P450 with 3-methylcholanthrene and ß-naphthoflavone increased formation of the products.

IMMUNOCHEMICAL STUDIES ON CYTOCHROME P450-DEPENDENT METABOLISM OF AA

Known inhibitors of mixed-function oxidases such as SKF-525A and metyrapone inhibited, in a variable manner, cytochrome P450-dependent AA metabolism; e.g., SKF-525A diminished formation of these AA metabolites whereas metyrapone was ineffectual. This variability probably signifies involvement of a specific cytochrome P450 isozyme in AA metabolism, and complicates the use of such drugs as definitive proof for the role of cytochrome P450-dependent monooxygenases in AA metabolism. Therefore, we prepared antibodies to the cytochrome c reductase component which is common to all NADPH cytochrome P450 species. We used purified human-liver NADPH cytochrome c reductase to induce antibodies.

Renal cortical microsomes converted ^{14}C-AA to several oxygenated metabolites, identified as ω- and ω-1-hydroxylated AA, epoxides and diols. Involvement of cytochrome P450 in metabolism of AA was assessed by using the aforementioned antibodies to the reductase. Preincubation of cortical microsomes with antibodies against the reductase for 30 min. at room temperature resulted in 80-90% inhibition of arachidonate conversion. Thus, antibodies against NADPH cytochrome c reductase, an integral component of the monooxygenase system, can provide definitive evidence for a cytochrome P450-dependent mechanism in the metabolism of AA.

ALTERATIONS OF RENAL CYTOCHROME P450-RELATED AA METABOLISM

The flux of AA through the TALH can be manipulated by induction or depletion of cytochrome P450 in this zone of the kidney (15). In this study, because of quantitative limitations of cell separation techniques, microsomes rather than isolated cells were used in order to answer the following questions concerning regional zonal differences in cytochrome P450-dependent oxygenation of AA:

1. As the content of cytochrome P450 can be altered in either direction intrarenally, are these changes in cytochrome P450 associated with corresponding changes in monooxygenase activity?
2. If so, are there associated changes in AA metabolism?
3. Are the changes in AA metabolism reflected to a greater degree in a particular region within the kidney?
4. Do changes in AA metabolism within the outer medulla correspond to those in the TALH?

The renal content of cytochrome P450 was increased by treatment with 3-methylcholanthrene and ß-naphthoflavone and decreased by treatment with cobalt. Cobalt induces heme oxygenase resulting in accelerated degradation of heme and reduction of cytochrome P450, a hemoprotein. The outer medulla displayed an exceedingly high rate of heme oxygenase activity. The kidney, therefore, must be considered another site for heme degradation in addition to the liver which is well known for prominent heme degradative activity. Renal microsomes, obtained from the three major zones--cortex, outer medulla and inner medulla--were incubated with AA, indomethacin and an NADPH-generating system. AA product formation was inhibited by SKF-525A. Determination of changes in cytochrome P450 function was assessed by measuring aryl hydrocarbon hydroxylase activity expressed in terms of hydroxylation of benzo-(a)-pyrene. Under control conditions, aryl hydrocarbon hydroxylase activity was detectable in all zones of the kidney, with activity ranging from 3 to 10% of that of the liver. This activity was increased and decreased proportionately to those directional changes in cytochrome P450 content produced by 3-methylcholanthrene and ß-naphthoflavone and by cobalt, respectively. For example, the decrease in hepatic and renal cytochrome P450 produced by cobalt treatment was accompanied by a substantial decrease in aryl hydrocarbon hydrolase activity in the liver and kidney. Thus, mixed-function oxidase activity can be manipulated by changing the tissue content of cytochrome P450 intrarenally. This observation substantiates the finding on the presence and induction of cytochrome P450 in the inner strips of the outer medulla. Perturbations in cytochrome P450 systems were also reflected in AA metabolism by TALH cells. AA conversion by TALH cells via cytochrome P450 systems was increased by 3-methylcholanthrene and ß-naphthoflavone treatment and decreased by cobalt treatment.

Thus, changing the cytochrome P450 content results in corresponding alterations in cytochrome P450-dependent monooxygenase activity and associated changes in AA metabolism expressed either in renal microsomes or in TALH cells. The finding that outer medullary cytochrome P450 isoenzyme(s)-

specific activity directed towards AA metabolism is much higher than that observed in the liver is important in terms of the potential contribution of these AA metabolites to renal function. The cytochrome P450 species that is specific for AA metabolism appears to predominate in the kidney, an interpretation based on the comparison of the capacities of liver and kidney for the metabolism of AA by cytochrome P450-dependent enzymes. Moreover, the highest concentration of this cytochrome P450 species is present in the inner stripe of the outer medulla, the zone from which TALH cells are isolated. The outer medullary microsomes converted AA in the presence of NADPH mainly via ω and ω-1 hydroxylation pathways to form the 20- and 19-hydroxyeicosatetraenoic acids. This is in agreement with the results reported by other investigators (1,4). In addition, renal microsomes possess epoxygenase activity as well as epoxide hydrolase activity and, thereby, can convert AA into epoxides and DHTs. As noted, the epoxides of AA have been shown to stimulate secretion of hormones and to inhibit chloride transport (22,24).

The following conclusions regarding metabolism of AA by renal cytochrome P450-linked oxygenase can be made:

1. The highest concentration of the cytochrome P450 species that metabolized AA is present in the outer medulla, i.e., the zone from which TALH cells are obtained.

2. The flux of AA can be controlled by changing the levels of cytochrome P450.

3. Arachidonate metabolites of this pathway probably contribute to the regulation of renal transport function and to local renal circulatory changes.

These studies have revealed discontinuous regional/zonal stratification of AA metabolism not only within the kidney but also within the nephron, the functional and anatomical unit of the kidney. Furthermore, we have shown that the cytochrome P450-dependent metabolic pathway in the TALH is highly selective in responding to circulating hormones (25). AVP and SCT stimulated AA metabolism in TALH cells (Fig. 1). The concentrations of AVP and SCT that stimulated AA metabolism via cytochrome P450-dependent monooxygenase in TALH corresponded to those concentrations that stimulated adenylate cyclase activity in this nephron segment (30). Moreover, the phosphodiesterase inhibitor, 1-isobutyl 3-methylxanthine (IBMX) (10 uM), increased formation of peaks 1 and 2 twofold, as did dibutyryl cyclic AMP (1 mM). Bradykinin and angiotensin II, only at concentrations two to three orders of magnitude greater than AVP and SCT, increased formation of the AA metabolites and then only minimally. Furthermore, parathormone and isoproterenol, which do not

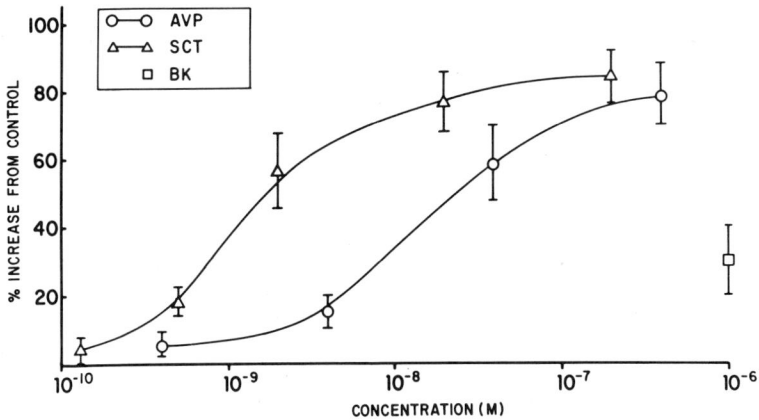

Fig. 1. Hormonal stimulation of arachidonic acid
(AA) metabolism by medullary thick ascending loop of Henle cells prelabeled with
^{14}C-AA (0.4 uCi per 3×10^6 cells for
90 min. at 37°C). Following incubation
with AVP (o), SCT (Δ) and bradykinin
(□), the AA metabolites were separated
by thin-layer chromatography. Radioactive zones corresponding to peaks 1 and
2 were cut and counted using a liquid
scintillation counter. The percentage
increase in the formation of peaks 1 and
2 above the unstimulated control value
was plotted. The means ± SEM for four
experiments are shown.

affect adenylate cyclase in this nephron segment (30), did not increase AA

metabolism by TALH. This study indicates that adenylate cyclase in TALH

transduces the hormonal signal elicited by either AVP or SCT. After forma-

tion of a hormone-receptor complex, cyclic AMP acts as a second messenger to

initiate the events leading to cytochrome P450-dependent AA product genera-

tion.

These findings may provide answers and future directions to major ques-

tions that have been raised by physiologists regarding regulation of the

activity of Na^+-K^+-ATPase in the short term; i.e., Na^+-K^+-ATPase can

participate in rapid adjustments of transport processes. In addition, abnor-

malities of Na^+-K^+-ATPase activity have been described in essential

hypertension (31) as well as in experimental hypertension; i.e., a reduction

in the reabsorptive capacity of the proximal nephron has been proposed to be

an early lesion in hypertension (32).

We regard the products formed from AA metabolism resulting from interactions of hormones with the cytochrome P450-dependent monooxygenases as
having broad implications; i.e., this might serve as a paradigm for a regulatory system that is distributed widely. We had an opportunity to test this
hypothesis. Endemic diseases of the rabbit colonies and technical problems
in TALH cell isolation resulted periodically in insufficient recovery of
TALH cells. We, therefore, turned to the corneal epithelium because of similarities in transport properties of this tissue with those of the TALH,
i.e., active chloride transport coupled to Na-K-ATPase and inhibitable by
furosemide (33). Microsomes prepared from bovine epithelium metabolized AA,
as did TALH, by a cytochrome P450 monooxygenase system (34,35). Based on
the studies in the TALH and cornea, we conclude that these tissues have a
cytochrome P450-dependent monooxygenase that prefers AA as substrate.

We have characterized the kinetics of the enzymic reactions involving
the cytochrome P450 monooxygenase in epithelial cells of the cornea and have
shown that for this tissue, unlike other tissues of the eye, a highly efficient metabolic machinery exists for the transformation of AA to novel products (34). Two novel corneal cytochrome P450-AA metabolites have been identified. One metabolite is 12(R) HETE which was found to be a potent inhibitor of Na^+-K^+-ATPase in the corneal epithelium (36). The other metabolite is a potent vasodilator; at concentrations of 1-10 ng, this compound
relaxed the rat-tail artery and dilated the conjunctival blood vessels upon
topical application to the rabbit eye. The structure of the vasodilator is
tentatively identified as 12-monounsaturated HETE, but the stereochemistry
has yet to be determined.

TALH CELL AA METABOLISM IN HYPERTENSIVE RABBITS

We have studied changes in AA metabolism in TALH cells obtained from
rabbits made hypertensive by suprarenal aortic coarctation (37) as the activity of Na^+-K^+-ATPase is decreased in hypertensive states (31,38). The
arterial blood pressure proximal to the aortic coarctation progressively
rose and reached hypertensive levels, usually by the sixth post-operative
day, and stabilized between 7 and 10 days. The arterial pressure distal to
the constriction fell transiently and then rose to normotensive levels by
the time that blood pressure stabilized at hypertensive levels. Since hypertension did not develop when the clip was applied to the aorta below the
renal arteries, high systemic blood pressure appeared to be a means of

reestablishing perfusion pressure to the ischemic kidneys. Indeed, blood
pressure below the coarctation and renal blood flow measured on the eighth
postoperative day had returned to normal levels associated with normal renal
venous oxygen tension. The renal vascular adjustments set in motion by
aortic coarctation were, therefore, complete by the eighth postcoarctation
day. We selected this hypertensive model because of the need to obtain both
kidneys in order to have sufficient TALH cells to conduct these studies.
Comparisons of TALH cell-AA metabolism were made to that of the outer medulla, depleted of TALH cells during the separation procedure. We also measured the biological activity of these cytochrome P450-AA metabolites to
assess whether their biological profile had been altered during hypertension, with respect to effects on Na^+-K^+-ATPase or vascular activity.

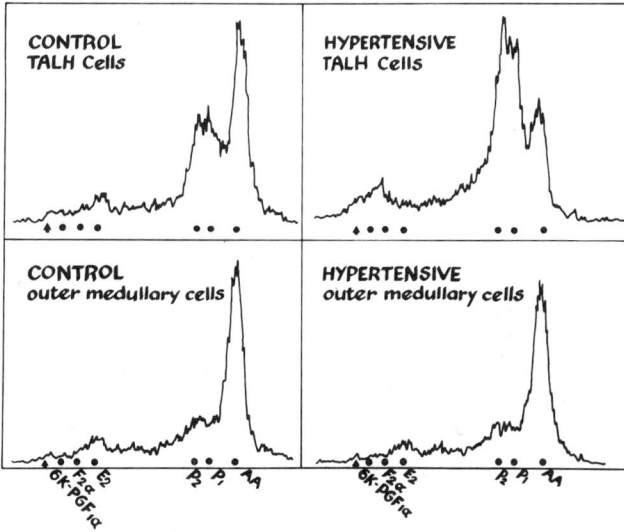

Fig. 2. Arachidonic acid (AA) metabolism by thick ascending loop of Henle (TALH) cells and outer
medullary cells depleted of TALH cells from
sham-operated control and hypertensive rabbits. The TALH and outer medullary cells (3
x 10^6/ml) were incubated with ^{14}C-AA (0.4
uCi) for 30 min. Cells were removed, and released radioactive metabolites were extracted
and separated by thin-layer chromatography
(TLC). Separated radioactive AA metabolites
were visualized by radiochromatographic
scan. Cytochrome P450-AA metabolites, peaks
1 and 2 (P_1 and P_2), not distinguishable
on this TLC scan were later visualized by
autoradiography. 6-K-$PGF_{1\alpha}$, 6-keto-prosta-
glandin $F_{1\alpha}$; $PGF_{2\alpha}$, prostaglandin $F_{2\alpha}$;
PGE_2, prostaglandin E_2.

Hypertension induced by aortic coarctation was associated with a twofold increase in cytochrome P450-AA metabolites, identified as peaks 1 and 2, generated by TALH cells when compared with AA products generated by TALH cells obtained from sham-operated normotensive rabbits (Fig. 2). These AA metabolites were not increased in two experiments in which hypertension did not develop despite coarctation. A selective response of TALH cells in this model of hypertension is suggested as AA metabolism was not elevated in other renomedullary cells, that is, those remaining after separation of TALH cells and obtained from the same hypertensive rabbits. The biological profile of AA metabolites was unchanged in these hypertensive rabbits; i.e., vasodilator material and an inhibitor of Na^+-K^+-ATPase activity was recovered from peaks 1 and 2, respectively. Increased formation of peak 2 may account for the reduced Na^+-K^+-ATPase activity reported in the renal outer medulla of hypertensive animals (39). Preliminary GC-MS identification indicated that peak 2 contains a structure similar to that of 11,12-DHT (25) which presumably arises from transformation of 11,12-EET via an epoxide hydrolase. The 11,12-DHT has been reported recently to be the most potent DHT in inhibiting osmotic water flow in the toad urinary bladder (40).

Peak 1 is a more potent vasorelaxant than peak 2, as noted in an earlier report (25). We have reported that the mass spectrum of a major component in peak 1 possesses many of the structural features of the 5,6-EET, which is a vasodilator (25). Authentic 5,6-EET has been reported to be the most potent epoxide in inducing dose-dependent relaxation of aortic and tracheal strips (41,42) and in effecting dilatation of the intestinal microcirculation (43). In addition, we have shown that the 5,6-EET, unlike 8,9; 11,12 and 14,15-EETs, dilates the perfused rat-tail artery (44) (Fig. 3). The 5,6-EET was the only cytochrome P450-dependent AA metabolite to decrease vascular resistance of the rat caudal artery. Its activity was equipotent to that of acetylcholine. The 5,6-DHT, as well as the γ-lactone of 5,6-EET, were both inactive on the caudal artery. The presence of the nonenzymatically formed 5,6-DHT and γ-lactone may lead to underestimation of the calculated vascular potency of the 5,6-epoxide as these products were without biological activity on the rat-tail artery. The other epoxides of AA, 8,9-,11,12- and 14,15-EETs, were also inactive and failed to relax this blood vessel.

The finding that the 5,6-EET is a potent vasodilator eicosanoid is in accord with a recent report that the 5,6 epoxide of AA is more potent than other epoxides of AA in increasing intestinal blood flow (43). We conclude that the 5,6 epoxide, in contrast to its diol and the other AA epoxides, may

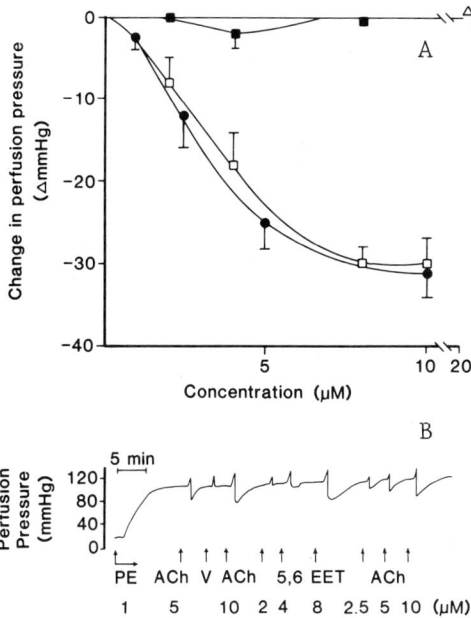

Fig. 3. The effect of intra-arterial injections of epoxyeicosatrienoic acids (EETs) and acetylcholine (ACh) on perfusion pressure (mmHg) of the isolated, perfused rat caudal artery. (A) Concentration-response curves to (□) 5,6-, (■) 8,9-, (△) 11,12-, (○) 14,15-EETs; (◆) 5,6-DHT (dihydroxyeicosatrienoic acid), (◇) 5,6-δ-lactone compared with (●) ACh. Values are shown as mean ± SEM (n=4). (B) Tracing of changes in perfusion pressure of isolated, perfused rat caudal artery after precontraction with phenylephrine (PE). Vehicle control injection (V) was without effect on perfusion pressure. 5,6-EET and ACh induced dose-dependent vasodilatation.

contribute to the regulation of regional vascular tone. This is particularly relevant in the cells of the TALH which produce a substance with similar structural features and biological activity to 5,6 epoxide. The release of this substance into the renomedullary circulation has been suggested to link metabolic activity in this segment of the nephron to changes in local blood flow (45).

The TALH region of the nephron is extremely vulnerable to hypoxic injury, a susceptibility linked to limited blood supply to this region and the high rate of metabolism related to active reabsorption of sodium, potassium and chloride (46). The AA metabolites that we have described may be released from TALH cells, presumably, both in response to hypoxia and as an adaptation to hypoperfusion, in order to reduce the energy-dependent Na^+-K^+-ATPase activity and to induce local vasodilatation so as to maintain blood supply to the medulla, effects that could limit the degree of tissue injury.

ABNORMAL AA METABOLISM IN THE SHR

Our most recent studies have been directed towards possible abnormalities of cytochrome P450-related AA metabolism in the spontaneously hypertensive rat (SHR), the most extensively studied animal model of human essential

119

hypertension (47,48). Alterations of renal AA metabolism have been described in the SHR. Renal prostaglandin catabolism is reduced (49), and medullary prostaglandin biosynthesis and urinary prostaglandin excretion are increased (50). The production of TxA_2 is also increased in isolated glomeruli from the young SHR, indicating that vasoconstrictor metabolites of AA are also produced in excess (51). Furthermore, PGI_2 production by aortic strips taken from the SHR is enhanced (52). No data are available on renal AA metabolism via cytochrome P450 enzymes in the SHR although fragmentary reports suggest that cytochrome P450-related mechanisms may be altered in the SHR (53,54). However, it should be recalled that thromboxane and prostacyclin synthases are specific cytochrome P450 proteins (55,56). Renal thromboxane synthesis and excretion were found to be increased in the young SHR (57). In the 6-week-old SHR, urinary TxB_2 was enhanced, and administration of a thromboxane receptor antagonist increased renal blood flow and GFR, whereas, in the 18-week-old SHR, the antagonist was without effect (57). These findings are in accord with ours that interventions designed to reduce cytochrome P450-related AA metabolism in order to reduce blood pressure in the SHR are effective only in the developmental phase of hypertension. Pursuant to the latter point, it is becoming clear that the resetting of renal function occurs early in the SHR and may be necessary for the development of hypertension (58,59). For example, fractional sodium and water excretion were significantly less in the SHR when compared with age-matched WKY, a difference that disappeared by 8 weeks of age (60). These findings are in agreement with the observed alterations in the relationship between sodium excretion and renal perfusion pressure which occurs as early as the third week in the SHR (59). Rapp and Dahl reported in 1976 that cytochrome P450-related hydroxylation of mineralocorticoids was partially responsible for elevated blood pressure in Dahl-salt-sensitive rats (61). Parenthetically, the SHR also demonstrates salt-sensitivity; i.e., the hypertension is worsened and accelerated by increased salt intake (60). Finally, we have obtained evidence that the cytochrome P450 system is present in the vasculature (62) where it has been described to metabolize AA to products that relax blood vessels (63).

We have demonstrated for the first time a correlation between the level of renal AA metabolism via cytochrome P450 oxygenation and blood pressure elevation in the SHR (64). We also have obtained evidence for formation of a cytochrome P450-AA metabolite in the young SHR that stimulates Na^+-K^+-ATPase activity. This study, described below, endorses the concept that the cytochrome P450-dependent pathway of AA consists of one or more specific isozymes that are tissue specific and can be stimulated and/or regulated by different factors, some of which operate in hypertension.

120

Blood pressure increased in the SHR from 112 to 202 mmHg between the ages of 5 to 20 weeks; the major increase in blood pressure occurred between 5 and 13 weeks. The increase in blood pressure in the WKY for the same period was moderate, reaching 136 mmHg at 20 weeks. Metabolism of AA was determined in microsomes from cortex and outer medulla of kidneys from 5-, 7-, 9-, 11-, 13-, and 20-week-old SHR and WKY. In the absence of an NADPH-generating system, AA was converted primarily to cyclooxygenase products. The percentage of conversion of $[^{14}C]$-AA to prostaglandins by cortical and outer medullary microsomes was low and not significantly different in the SHR and the WKY, in contrast to cytochrome P450-dependent AA metabolism. The latter AA metabolites were defined as those metabolites whose formation was absolutely dependent on NADPH addition, inhibited by SKF-525A, an inhibitor of cytochrome P450-dependent enzymes via type I binding, and unaffected by indomethacin. Addition of an NADPH-generating system to the incubation medium increased the conversion of $[^{14}C]$-AA by several-fold in both cortical and outer medullary microsomes from the SHR and the WKY kidneys, yielding several radioactive peaks (Fig. 4, third lane) that were inhibited by more than 50% with SKF-525A (200 uM) (Fig. 4, fourth lane). The conversion of AA to the cytochrome P450-dependent metabolites peaked at 7 weeks in the SHR (Fig. 5). The cytochrome P450-dependent metabolites were separated by reverse-phase HPLC into three radioactive peaks: peak I had a retention time of 17.5 min. and comigrated with 11,12 DHT standard, peak II of 19 min. comigrated with ω-hydroxylation compounds, and peak III of 27 min. comigrated with 11,12 EET (Fig. 6). Product formation was predictably dependent on the presence of NADPH.

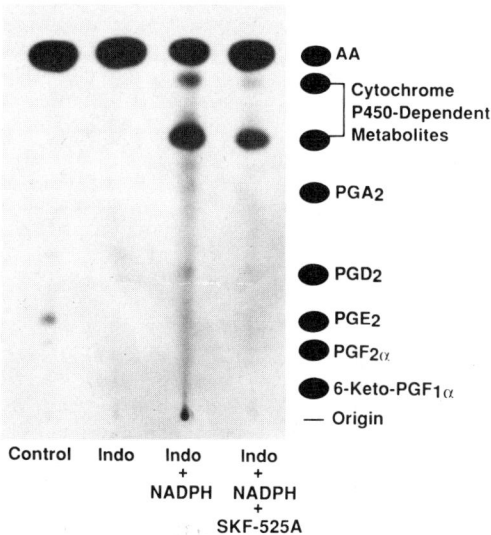

Fig. 4. Autoradiography of thin-layer chromatograph separation of arachidonic acid (AA) metabolites. Microsomes (0.3 mg protein) from the renal cortex of 7-week-old SHR were incubated with $[^{14}C]$-AA (0.2 uCi) for 30 min. with or without indomethacin, SKF-525A and NADPH.

121

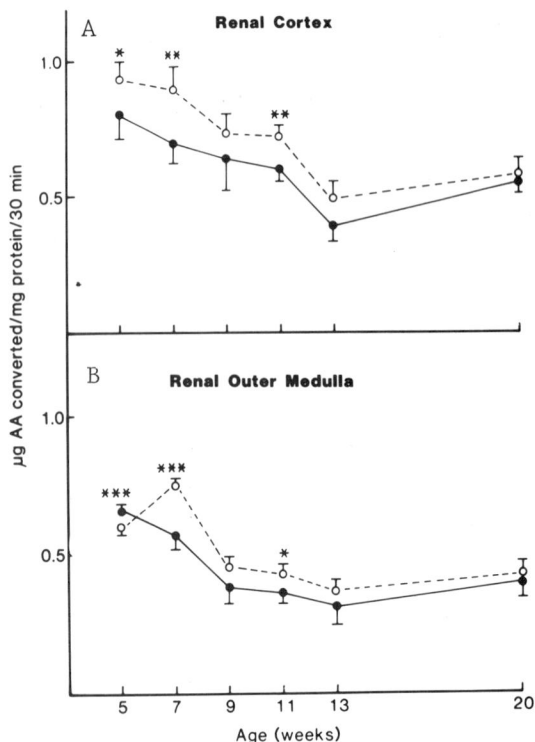

Fig. 5. Renal cortical and outer medullary cytochrome P450-dependent arachidonic acid (AA) metabolites in SHR and WKY. Microsomes from renal cortex (A) and outer medulla (B) of SHR (o) and WKY (●) of 5 (n=6), 7 (n=11), 9 (n=9), 11 (n=11), 13 (n=3) and 20 (n=6) weeks were incubated with [^{14}C]-AA (0.2 uCi), indomethacin and NADPH for 30 min. Metabolites were separated by thin-layer chromatography. Radioactive zones were visualized by autoradiography and counted in a liquid scintillation counter. *p<0.05; **p<0.02; ***p<0.005.

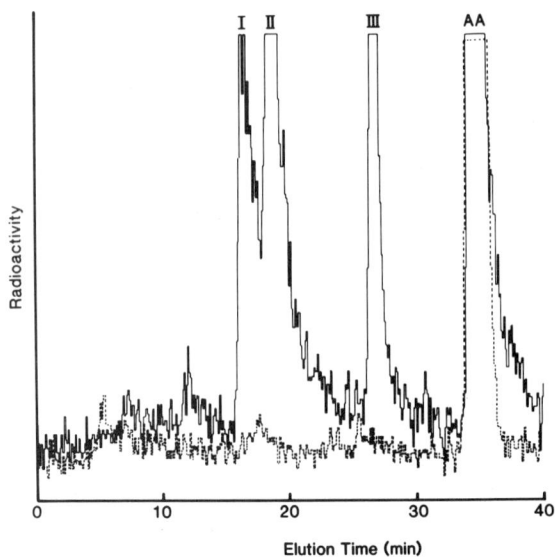

Fig. 6. Separation by reverse-phase HPLC of cytochrome P450-dependent arachidonic acid (AA) metabolites. Microsomes (0.3 mg protein) from the renal cortex of 7-week-old SHR were incubated with [^{14}C]-AA (0.4 uCi), indomethacin and NADPH. Cytochrome P450-dependent metabolites of AA (——) were not formed in the absence of NADPH (---). Peak I comigrated with 11,12-DHT standard, peak II with ω-hydroxylation compounds, and peak III with 11,12-EET.

Once the blood pressure was established by the 20th week of age, there were no differences between the SHR and the WKY in cytochrome P450-dependent metabolism of AA. It should be re-emphasized that the renal functional disturbances described in the SHR, such as decreased renal blood flow and glomerular filtration rate (59) and salt and water retention (58), are evident only in the developmental phase of hypertension and disappear with the established phase. These temporal relationships, while not providing direct evidence, are consistent with an involvement of renal cytochrome P450-dependent AA metabolites in elevation of blood pressure, particularly as some of these metabolites have been found to affect Na^+-K^+-ATPase activity and vascular tone (25). More direct evidence has recently been obtained by depleting cytochrome P450 with stannous chloride, a selective inducer of renal heme oxygenase that degrades cytochrome P450 and other hemoproteins (65). Treatment of the young SHR with stannous chloride prevented elevation of blood pressure but did not affect blood pressure when given at 20 weeks of age (66). We suggest that renal cytochrome P450-dependent metabolites of AA may participate in the renal functional changes of the SHR, particularly during the developmental stage.

SUMMARY AND GENERAL CONCLUSIONS

We have identified AA metabolites in bovine corneal epithelium and rabbit TALH cells having similarities of biological properties and chemical features. Apart from the well-known similarities of the TALH and the corneal epithelium--i.e., highly active Na^+-K^+-ATPase-driven transport and furosemide-inhibitable chloride secretion--we have identified important common features at the biochemical and at the immunochemical level. Thus, the major biological activities of two of the principal products of the corneal epithelium were inhibition of Na^+-K^+-ATPase and vasodilatation, respectively. This recalls the earlier observations on two of the principal products obtained from TALH and designated peaks 1 and 2, based on their retention times on reverse-phase HPLC. Peak 1 was shown to be a potent vasodilator, while peak 2 was a potent inhibitor of Na^+-K^+-ATPase. One of the major products associated with peak 1, as we reported (25), was identified as a 5,6 epoxide of AA. This eicosanoid is a potent dilator of several vascular beds (43,44) which distinguishes it from the other EETs. A parallel development in terms of studies on novel eicosanoids affecting salt and water movement was anticipated by our initial studies on TALH cells (12, 25). Thus, peak 2 produced by TALH inhibited Na^+-K^+-ATPase and yielded a material that, on GC-MS analysis, had the features of an 11,12 diol derivative. A possible derivative of the 11,12 EET is 12(R) HETE (the stereoisomer of 12-hydroxy 5,8,10,14-eicosatetraenoic acid) and is produced by the cornea. It has been shown to inhibit Na^+-K^+-ATPase activity (36). Together, these studies endorse the critical role that novel AA metabolites or closely related substances play in regulating blood flow and extracellular fluid volume.

The studies with TALH cells and the corneal epithelium have provided the impetus to search for similar products in the outer medulla of the SHR. Three principal products were generated from AA by microsomes obtained from the outer medulla. One of these AA products also affects Na^+-K^+-ATPase activity. However, in contrast to the inhibitory activity of the major AA metabolites arising from the cytochrome P450 system in the rabbit renal medulla and bovine corneal epithelium, this principal cytochrome P450-dependent AA metabolite recovered from the SHR kidney stimulated Na+-K+-ATPase. This AA metabolite may contribute to salt retention and the elevation of blood pressure in the developmental phase of hypertension in the SHR. Indeed, the suggested link between the kidney and the eye in terms of biological mediators, particularly AA metabolites, that are responsible for abnormalities of transport function in both organs, is supported by a recent study (67). Increased cataract formation in the Dahl salt-sensitive, hypertensive rat was attributed to a defect in Na^+-K^+-ATPase activity in the lens.

ACKNOWLEDGEMENTS

We wish to thank Gail Price for her secretarial help and Sallie McGiff for editorial assistance. This work was supported by Program Project Grant P01-HL-34300, HL-25394, AHA 86-112, AM-29742, AM-00781 and G. Harold and Leila Y. Mathers Charitable Foundation. D.S. is a Fogarty International Fellow.

REFERENCES

1. E.H. Oliw and J.A. Oates, Rabbit renal cortical microsomes metabolize arachidonic acid to trihydroxyeicosatrienoic acids, Prostaglandin 22:863-871 (1981).
2. E.H. Oliw, F.P. Guengerich, J.A. Oates, Oxygenation of arachidonic acid by hepatic monooxygenases, J. Biol. Chem. 257:3771-3781 (1982).
3. J. Capdevila, L. Parkhill, N. Chacos, R. Okita, B.S.S. Masters, R.W. Esterbrook, The oxidative metabolism of arachidonic acid by purified cytochrome P450, Biochem. Biophys. Acta 101:1357-1363 (1981).

4. A. Morrison and N. Pascoe, Metabolism of arachidonate through NADPH-dependent oxygenase of renal cortex, <u>Proc. Natl. Acad. Sci. U.S.A.</u> 78:7375-7378 (1981).

5. F.C. Grenier, T.E. Rollins, W.L. Smith, Kinin-induced prostaglandin synthesis by renal papillary collecting tubule cells in culture, <u>Am. J. Physiol</u>. 241:F94-F104 (1981).

6. A. Hassid, M. Konieczkowski, M.J. Dunn, Prostaglandin synthesis in rat kidney glomeruli, <u>Proc. Natl. Acad. Sci. U.S.A.</u> 76:1155-1159 (1979).

7. R.M. Zusman and H.R. Keiser, Prostaglandin E_2 biosynthesis by rabbit renomedullary interstitial cells in culture, <u>J. Biol. Chem.</u> 252: 2069-2071 (1977).

8. N.A. Terragno, A. Terragno, J.A. Early, M.A. Roberts, J.C. McGiff, Endogenous prostaglandin synthesis inhibitor in renal cortex. Effects on production of prostacyclin by renal blood vessels, <u>Clin. Sci. Mol. Med.</u> 55:199s-202s (1978).

9. P. Borgeat and B. Samuelsson, Arachidonic acid metabolism in polymorphonuclear leukocytes; Effects of ionophore A23187, <u>Proc. Natl. Acad. Sci. U.S.A.</u> 76:2148-2152 (1979).

10. P. Falardeau, M. Hamberg, B. Samuelsson, Metabolism of 8,11,14-eicosatrienoic acid in human platelets, <u>Biochim. Biophys. Acta</u> 441:193-200 (1976)..

11. K. Jim, A. Hassid, F. Sun, M.J. Dunn, Lipoxygenase activity in rat kidney glomeruli, glomerular epithelial cells, and cortical tubules, <u>J. Biol. Chem.</u> 257:10294-10299 (1982).

12. N.R. Ferreri, M. Schwartzman, N.G. Ibraham, P.N. Chander, J.C. McGiff, Arachidonic acid metabolism in a cell suspension isolated from rabbit renal outer medulla, <u>J. Pharmacol. Exp. Ther.</u> 231:441-448 (1984).

13. P.L. Jorgensen, Sodium and potassium ion pump in kidney tubules, <u>Physiol. Rev.</u> 60:864-917 (1980).

14. N.R. Ferreri, M. Schwartzman, N.G. Ibraham, P.N. Chander, J.C. McGiff, Arachidonic acid metabolism in the rabbit thick ascending limb of Henle's loop, in: "Prostaglandins and Membrane Ion Transport," P. Braquet, R.P. Garay, J.C. Frolich and S. Nicosia, eds., Raven Press, New York, pp. 303-310 (1984).

15. M.L. Schwartzman, N.G. Abraham, M.A. Carroll, R.D. Levere, J.C. McGiff, Regulation of arachidonic acid metabolism by cytochrome P450 in rabbit kidney, <u>Biochem. J.</u> 238:283-290 (1986).

16. R.E. White and M.J. Coon, Oxygen activation by cytochrome P450, <u>Ann. Rev. Biochem.</u> 49:315-356 (1980).

17. M.W. Anders, Metabolism of drugs by the kidney, <u>Kidney Int.</u> 18:636-647 (1980).

18. T.V. Zenser, M.B. Mattamal, B.B. Davis, Differential distribution of the mixed-function oxidase activities in rabbit kidney, J. Pharmacol. Exp. Ther. 207:719-725 (1978).

19. H. Endou, Cytochrome P450 monooxygenase system in the rabbit kidney: its internephron localization and its induction, Jpn. J. Pharmacol. 33:423-433 (1983).

20. J.H. Dees, L.D. Coe, Y. Yasukochi, B.S.S. Masters, Immunofluorescence of NADPH-cytochrome (c) (P450) reductase in rat and mini-pig tissues injected with phenobarbital, Science 208:1473-1475 (1980).

21. J.H. Dees, B.S.S. Masters, U. Muller-Eberhard, F.F. Johnson, Effect of 2, 3, 7, 8 tetrachlorodibenzo-p-dioxin and phenobarbital on the occurrence and distribution of four cytochrome P450 isozymes in rabbit kidney, lung and liver, Cancer Res. 42:1423-1432 (1982).

22. J. Capdevila, N. Chacos, J.R. Falck, S. Manna, A. Negro-Vilar, S.R. Ojeda, Novel hypothalamic arachidonate products stimulate somatostatin release from the median eminence, Endocrinology 113:421-423 (1983).

23. G.D. Snyder, J. Capdevila, N. Chacos, S. Manna, J.R. Falck, Action of luteinizing hormone-releasing hormone: involvement of novel arachidonic acid metabolites, Proc. Natl. Acad. Sci. 80:3504-3507 (1983).

24. H.R. Jacobson, S. Corona, J. Capdevila, N. Chacos, S. Manna, A. Womack, J.R. Falck, Effects of epoxyicosatrienoic acids on ion transport in rabbit cortical collecting tubules. in: "Prostaglandins and Membrane Ion Transport," P. Braquet, R.P. Garay, J.C. Frolich and S. Nicosia, eds., Raven Press, New York, pp. 311-318 (1984).

25. M. Schwartzman, N.R. Ferreri, M.A. Carroll, E. Songu-Mize, J.C. McGiff, Renal cytochrome P450-related arachidonate metabolite inhibits Na^+-K^+-ATPase, Nature (London) 314:620-622 (1985).

26. M. Schwartzman, M.A. Carroll, N.G. Ibraham, N.R. Ferreri, E. Songu-Mize, J.C. McGiff, Renal arachidonic acid metabolism; the third pathway, Hypertension 7(Suppl. I):I-136-I-144 (1985).

27. J. Capdevila, L.J. Marnett, N. Chacos, R.A. Prough, R.W. Estabrook, Cytochrome P450-dependent oxygenation of arachidonic acid to hydroxyeicosatetraenoic acids, Proc. Natl. Acad. Sci. U.S.A. 79:767-770 (1982).

28. J.R. Falck, S. Manna, J. Moltz, N. Chacos, J. Capdevila, Epoxyeicosatrienoic acids stimulate glucogon and insulin release from isolated rat pancreatic islets, Biochem. Biophys. Res. Commun. 114:743-749 (1983).

29. W.L. Smith and A. Garcia-Perez, A two-receptor model for the mechanism of action of prostaglandins in the renal collecting tubule, in: "Prostaglandins, Leukotrienes and Lipoxins," J.M. Bailey, ed., Plenum Press, New York, pp. 35-45 (1980).

30. F. Morel, Sites of hormone action in the mammalian nephron, Am. J. Physiol. 240:F159-F164 (1981).

31. H.E. deWardener and E.M. Clark, The natriuretic hormone: new developments, Clin. Sci. 63:415-420 (1982).

32. D.S. Baldwin, E.A. Gombos, H. Chasis, Urinary concentrating mechanism in essential hypertension, Am. J. Med. 38:864-872 (1965).

33. O.A. Candia, Ouabain and sodium effects on chloride fluxes across the isolated bullfrog cornea, Am. J. Physiol. 223:1053-1057 (1972).

34. M.L. Schwartzman, N.G. Abraham, J. Masferrer, M.W. Dunn, J.C. McGiff, Cytochrome P450-dependent metabolism of arachidonic acid in bovine corneal epithelium, Biochem. Biophys. Res. Commun. 132:343-351 (1985).

35. M.A. Carroll, M. Schwartzman, M. Baba, N.G. Abraham, J.C. McGiff, Formation of biologically active cytochrome P450-arachidonate metabolites in renomedullary cells, Adv. Prost. Leuk. Thromb. Res. 17B:714-718 (1987).

36. M.L. Schwartzman, M. Balazy, J. Masferrer, N.G. Abraham, J.C. McGiff, R.C. Murphy, 12(R)HETE - A cytochrome P450-dependent arachidonate metabolite that inhibits Na^+-K^+-ATPase in the cornea, Proc. Natl. Acad. Sci. U.S.A., in press (1987).

37. M.A. Carroll, M. Schwartzman, N.G. Abraham, A. Pinto, J.C. McGiff, Cytochrome P450-dependent arachidonate metabolism in renomedullary cells: Formation of Na^+-K^+-ATPase inhibitor, J. Hypertension 4(Suppl 4): S33-S42 (1986).

38. G.A. MacGregor, S. Fenton, J. Alaghband-Zadeh, N.D. MarKandu, J.E. Roulston, H.E. deWardener, An increase in a circulating inhibitor of Na^+-K^+-ATPase: a possible link between salt intake and the development of essential hypertension, Clin. Sci. 61:17S-20S (1981).

39. Y.U. Postnov, M. Reznikova, G. Boriskina, Na-K-adenosine triphosphatase in the kidney of rats with renal hypertension and spontaneously hypertensive rats, Pflugers Arch. 362:95-99 (1976).

40. D. Schlondorff, E. Petty, J. Oates, M. Jacoby, S.D. Levine, Epoxygenase metabolites of arachidonic acid inhibit osmotic water flow in response to vasopressin in the toad urinary bladder. Am. J. Physiol., in press (1987).

41. M.J. Finnen, R.J. Flower, A. Lashenko, K.I. Williams, Cytochrome P450-dependent monooxygenase activity and endothelium-dependent relaxation of vascular tissue, Brit. J. Pharmacol., Abstract, p. 85 (1986).

42. M.J. Finnen, R.J. Flower, A. Lashenko, K.I. Williams, Airway epithelium influences responsiveness of guinea pig tracheal strips, Brit. J. Pharmacol., Abstract, p. 86 (1986).

43. K.G. Protor, J.R. Falck, J. Capdevila, Intestinal vasodilation by epoxy-eicosatrienoic acids: arachidonic acid metabolites produced by a cyto-chrome P450 monooxygenase, Circ. Res., 60:50-59 (1987).

44. M.A. Carroll, M. Schwartzman, J. Capdevila, J.R. Falck, J.C. McGiff, Vasoactivity of arachidonic acid epoxides, Eur. J. Pharm. 138:281-283 (1987).

45. J.C. McGiff and M.A. Carroll, Cytochrome P450-related arachidonic acid metabolites, Am. Rev. Respir. Dis., in press (1987).

46. M. Brezis, S. Rosen, P. Silva, F.H. Epstein, Selective vulnerability of the medullary thick ascending limb to anoxia in the isolated perfused rat kidney, J. Clin. Invest. 73:182-190 (1984).

47. J.C. McGiff and C.P. Quilley, Controversies in cardiovascular research. The rat with spontaneous genetic hypertension is not a suitable model of human essential hypertension, Circ. Res. 48:455-463 (1981).

48. E.D. Frolich, Response to "the rat with spontaneous genetic hypertension is not a suitable model of human essential hypertension", Circ. Res. 48:464 (1981).

49. J.M. Armstrong, G.J. Blackwell, R.J. Flower, J.C. McGiff, K.M. Mullane, J.R. Vane, Genetic hypertension in rats is accompanied by a defect in renal prostaglandin catabolism, Nature 260:582-586 (1976).

50. M.J. Dunn, Renal prostaglandin synthesis in the spontaneously hyperten-sive rat, J. Clin. Invest. 58:862-870 (1976).

51. Y. Shibouta, Y. Inada, Z. Terashita, K. Nishikawa, S. Kikuchi, K. Shimamoto, Angiotensin-II-stimulated release of thromboxane A_2 and prostacyclin (PGI_2) in isolated, perfused kidneys of spontaneously hypertensive rats, Biochem. Pharmacol. 28:3601 (1979).

52. C.R. Pace-Asciak and M.C. Carrara, Age-dependent increase in the forma-tion of prostaglandin I_2 by intact and homogenized aortae from the developing spontaneously hypertensive rat, Biochim. Biophys. Acta 574:177 (1979).

53. B.A. Merrick, M.H. Davies, D.E. Cook, T.L. Holcshaw, R.S. Schnell, Alter-ations in hepatic microsomal drug metabolism and cytochrome P450 pro-teins in spontaneously hypertensive rats, Pharmacology 30:129-135 (1985).

54. P. Greenspan and J. Braon, Hepatic microsomal oxidative drug metabolism in the spontaneously hypertensive rat, Biochem. Pharmacol. 30:687-691 (1981).

55. M. Haurand and V. Ullrich, Isolation and characterization of thromboxane synthase from human platelets as a cytochrome P450 enzyme, J. Biol. Chem. 260:15059-15067 (1985).

56. D.L. Dewitt and W.L. Smith, Purification of prostacyclin synthase from

bovine aorta by immunoaffinity chromatography, J. Biol. Chem. 258: 3258-3293 (1983).

57. Y. Shibouta, Z. Terashita, Y. Inada, K. Nishikawa, S. Kikuchi, Enhanced thromboxane A_2 biosynthesis in the kidney of spontaneously hyperten- sive rats during development of hypertension, Eur. J. Pharmacol. 70: 247 (1981).

58. J.R. Dilley, C.T. Stier, Jr., W.J. Arendshorst, Abnormalities in glomeru- lar function in rats developing spontaneous hypertension, Am. J. Physiol. 246:F12-F20 (1984).

59. R.J. Roman, Altered pressure-natriuresis relationship in young spontane- ously hypertensive rats, Hypertension, in press (1987).

60. W.H. Beierwaltes, W.J. Arendshorst, P.J. Klemmer, Electrolyte and water balance in young spontaneously hypertensive rats, Hypertension 4:908- 915 (1982).

61. J.P. Rapp and L.K. Dahl, Mutant forms of cytochrome P450 controlling both 18- and 11-B-steroid hydroxylation in the rat, Biochem. 15:1235- 1241 (1976).

62. N.G. Abraham, A. Pinto, K.M. Mullane, R.D. Levere, E. Spokas, Presence of cytochrome P450-dependent monooxygenase in intimal cells of the hog aorta, Hypertension 7:899-904 (1985).

63. A. Pinto, N.G. Abraham, K.M. Mullane, Arachidonic acid-induced endothe- lial-dependent relaxations of canine coronary arteries: contribution of a cytochrome P450-dependent pathway, J. Pharmacol. Exp. Ther. 240: 856-863 (1986).

64. D. Sacerdoti, N.G. Abraham, J.C. McGiff, M.L. Schwartzman, Renal cyto- chrome P450-dependent metabolism of arachidonic acid in spontaneously hypertensive rats, Biochem. Pharmacol., in press (1987).

65. N.G. Ibraham, M.L. Friedland, R.D. Levere, Heme metabolism in erythroid and hepatic cells, in: "Progress in Hematology, Vol. VIII," E. Brown, ed., Grune and Stratton, New York, pp. 75-130 (1983).

66. D. Sacerdoti, T.V. Mazzilli, N.G. Abraham, J.C. McGiff, M.L. Schwartzman, The role of cytochrome P450-dependent arachidonic acid metabolites in the development of hypertension in SHR, Clin. Res. 35, Abs, p. 449 (1987).

67. C. Rodriguez-Sargent, J.L. Cangiano, J.L. Cahan, E. Marrero, M. Martinez-Maldonado, Cataracts and hypertension in salt-sensitive rats: a possible ion transport defect, Hypertension 9:304-308 (1987).

THE BIOSYNTHESIS AND ACTIONS OF PROSTAGLANDINS

IN THE RENAL COLLECTING TUBULE AND THICK ASCENDING LIMB

William L. Smith, William K. Sonnenburg, Margaret L. Allen,
Tsuyoshi Watanabe, Jianhua Zhu and E.A. El-Harith

Department of Biochemistry
Michigan State University
East Lansing, Michigan, USA 48824

INTRODUCTION

The intent of this review is to describe our current perceptions of the
biosynthesis and function of prostaglandins in the renal collecting tubule
and thick ascending limb of Henle's loop. Although the presentation will
have a biochemical bias, we have attempted to couch our discussion in the
context of the regulation of water reabsorption by the kidney. Specifically, our goal will be to summarize the data that support the model depicted
in Fig. 1. The major features of the model are as follows: (a) the collecting tubule is the principal site of prostaglandin formation in the renal
tubule; (b) PGE_2 is the major prostaglandin product formed by the collecting tubule; (c) PGE_2 synthesis is stimulated by bradykinin (BK) and arginine vasopressin (AVP); (d) PGE_2 acts physiologically as an inhibitory
regulator of AVP-induced water reabsorption by the collecting tubule and
AVP-induced NaCl reabsorption in the medullary thick ascending limb; (e) the
inhibitory actions of PGE_2 are mediated through a PGE receptor coupled to
the inhibitory guanine nucleotide regulatory protein G_i which attenuates
AVP-induced cAMP formation; (f) there may also be a PGE receptor coupled to
G_x (designated PGE-G_x) which, in turn, is coupled to the activation of
cAMP phosphodiesterase activity, perhaps via Ca^{++}-calmodulin; (g) at high
concentrations, PGE_2 acts through a PGE-G_s receptor to stimulate adenylate cyclase; and (h) the PGE-G_s receptor may be involved in feedback inhibition of PGE_2 formation in the collecting tubule.

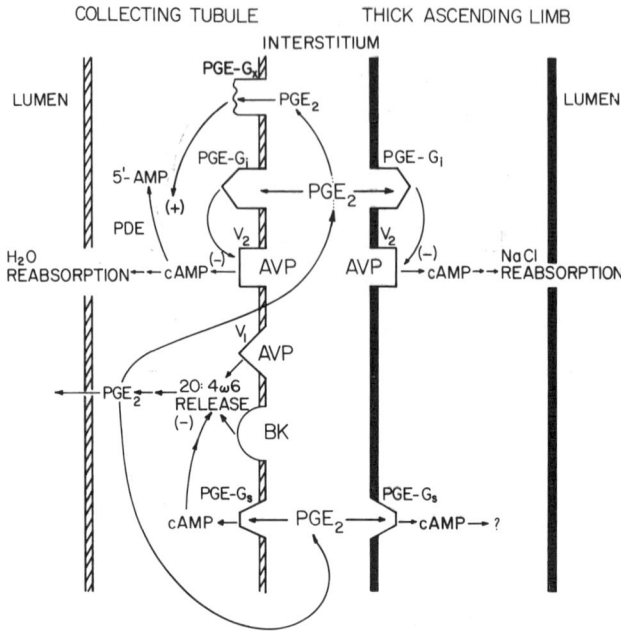

Fig. 1. PGE₂ actions in renal tubule. This
model for hormonal interactions in the collecting tubule and thick ascending limb of
Henle's loop shows proposed cellular loca-
tions for different receptors for prostaglandin E (PGE), arginine vasopressin
(AVP) and bradykinin (BK). Receptors are
arbitrarily assigned to the basolateral
surface for simplicity, although, in some
instances, there is an asymmetry to the
location of receptors (see text). Abbreviations include: PDE, phosphodiesterase;
AC, adenylate cyclase; G, guanine nucleotide binding regulatory protein; V_1 and
V_2, vasopressin receptor types.

Compartmentation of Renal Prostaglandin Biosynthesis

Work reported by Dunn et al. in 1977 indicated that the concentrations
of PGE_2 and $PGF_{2\alpha}$ entering the kidney via the renal artery were too low
to account for the renal effects of these prostaglandins (1). This observation (a) was consistent with the concept that prostaglandins are not circulating hormones (2,3) and (b) suggested that only prostaglandins formed within the kidney have significant renal actions. The study of Dunn et al. (1)
also served to emphasize the need to define the cellular sites of prostaglandin synthesis in the kidney.

Lee et al. (4) and Daniels et al. (5) were the first to show that prostaglandins could be extracted from renal tissue. Shortly thereafter, Hamberg
reported the synthesis of prostaglandins from tritiated arachidonic acid by
homogenates of rabbit renal medulla (6) and identified PGE_2 as the major
product. In the first examination of the localization of prostaglandin formation in kidney, Crowshaw (7) showed that PGE_2 biosynthesis took place
primarily, if not exclusively, in the renal medulla; no synthesis, as measured by PGE_2 production, could be detected in the cortex. These observations were perplexing at the time, since the known renal effects of prostaglandins were to modulate blood flow (8-10)--a phenomenon associated with
control of the tone of cortical arterioles--and there was no adequate explanation of how prostaglandins synthesized in the medulla could reach the
cortex (9). The dilemma was later resolved as it became apparent that synthesis of both PGI_2 and PGE_2 can take place in the renal cortex (11-13)
but that there is a factor present in broken cell preparations of renal
cortex that inhibits prostaglandin formation (13).

The first study of the cellular location of prostaglandin formation was
that of Janszen and Nugteren (14). They used an ingenious histochemical
technique to detect the hydroperoxidase activity associated with PGH synthase in the medullary collecting tubules of rabbit, rats, guinea pigs and
goldhamsters and in the medullary interstitial cells of rats; the histochemical response was eliminated in controls treated in the absence of the fatty
acid substrate (8,11,14-eicosatetraenoic acid) or in the presence of indomethacin. Several years later, we published a series of reports (15-18) in
which we determined the location of the PGH synthase protein using monospe-
cific rabbit antisera prepared against the purified enzyme (19). It is now
clear from this work and related studies in which prostaglandin synthesis by
microdissected tubule (20-23) and vascular segments (12) and isolated cells
(24-27) and glomeruli (28) has been examined that prostaglandin formation is
compartmentalized in the kidney. At least seven different renal cell types
have an appreciable capacity to synthesize cyclooxygenase products (Fig.
2). These include renal medullary interstitial cells, vascular endothelial
and smooth muscle cells, glomerular epithelial and mesangial cells and thin
limb and collecting tubular epithelia. In addition, cells of the thick ascending limb (25,27), and perhaps other cells (21), have a modest capacity
for prostaglandin synthesis. Based on intensities of histochemical (14) and
immunohistochemical staining (15-18) for PGH synthase and the relative abilities of different tubule segments to synthesize PGE_2 from arachidonic acid
(20,21), it is clear that the renal cells with the greatest capacity for
prostaglandin synthesis are the epithelial cells of the collecting tubule
and, in particular, the medullary collecting tubule (17,21).

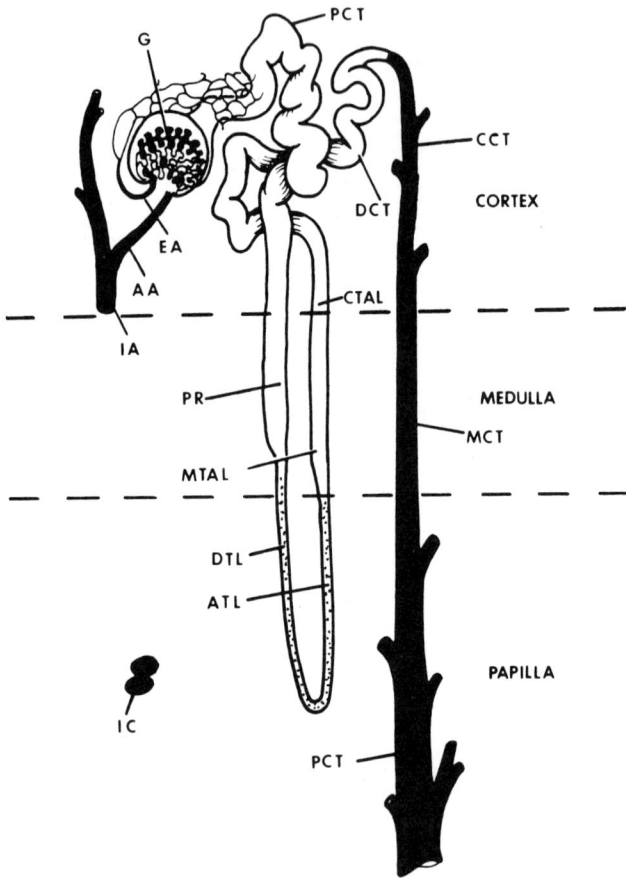

Fig. 2. Localization of prostaglandin biosynthesis in the kidney. Darkened areas indicate sites of prostaglandin formation which include vascular endothelial and smooth muscle cells, glomerular epithelial and mesangial cells, medullary interstitial cells, thin limb epithelia, and collecting tubule epithelia. Abbreviations are: IA, interlobular artery; AA and EA, afferent and efferent arteriole; G, glomerulus; PCT and DCT, proximal and distal convoluted tubule; PR, pars recta; DTL and ATL, descending and ascending thin limb; CCT, MCT and PCT, cortical, medullary and papillary collecting tubule; IC, interstitial cells.

Prostaglandin Products of the Collecting Tubule

There have been two studies aimed at determining the nature of the prostaglandin products formed by freshly isolated collecting tubule cells.
Kirschenbaum et al. (22) found that PGE_2 was the major prostaglandin
formed by incubates of dissected rabbit cortical collecting tubules which
had been prelabeled with tritiated arachidonic acid; however, they also
found substantial amounts of 6-keto-$PGF_{1\alpha}$, $PGF_{2\alpha}$, and even TxB_2. Similarly, Grenier and Smith (29) analyzed the products formed by homogenates of
freshly isolated rabbit papillary collecting tubule cells and determined
that PGE_2 was the major product, but that some 6-keto-$PGF_{1\alpha}$ was also
produced and was actually the major product at low concentrations (≤ 2 μM)
of arachidonate. Later, Grenier et al. determined that PGE_2 was formed
enzymically by homogenates of cultured rabbit papillary collecting tubule
cells and that PGE_2 formation required, as expected, reduced glutathione
(30). However, we have recently prepared several monoclonal antibodies reactive with different PGH-PGE isomerase activities in sheep seminal vesicles,
and none of these antibodies cause immunohistochemical staining of collecting tubule epithelia (31). Presumably, the enzyme is in low abundance in
the collecting tubule, or possibly there are tissue-specific PGH-PGE isomerase isozymes.

Although 6-keto-$PGF_{1\alpha}$ has been reported to be a product of rabbit collecting tubule cells (22,29), we have been unable to detect the PGI_2 synthase antigen in these cells under immunocytochemical conditions where the
antigen is clearly visible in capillary endothelial and vascular smooth
muscle cells. We speculate that small amounts of PGI_2 are formed by collecting tubules, that PGE_2 and PGI_2 are synthesized concurrently, and
that PGI_2 signals capillary endothelial cells that prostaglandins are
being produced by the collecting tubule. Alternatively, capillaries may
"steal" excess PGH_2 produced by the collecting tubule and convert it to
PGI_2 (32). The concept of intercellular crosstalk involving prostaglandins and medullary tubule, interstitial and endothelial cells was first
enunciated by Stokes (33).

Hormonal Stimulation of PGE_2 Formation by the Collecting Tubule

AVP and kinin derivatives such as lysyl-bradykinin stimulate PGE_2
formation by freshly microdissected rabbit cortical collecting tubules
(23,34; Fig. 1). The magnitudes of the responses are typically small--less
than twofold--and the responses vary with the age of the rabbits (34). In

contrast, kinins typically cause a five- to tenfold stimulation of PGE_2
synthesis by cultured collecting tubules derived from dog (35,36), rat (37)
and rabbit (31,38) renal cortex and papilla. Bradykinin, lysyl-bradykinin
and met-lysyl-bradykinin all appear to act via the same receptor to elicit
their responses (since their effects are nonadditive) in cultured rabbit
papillary collecting tubule (RPCT) cells (31); these epithelia are particularly responsive to kinins: 10^{-11} M bradykinin causes half-maximal stimulation of PGE_2 formation.

There is a sidedness to the response of collecting tubule epithelia to
kinins which appears to vary among species. For example, lysyl-bradykinin
appears only to function at the basolateral surfaces of the cell to cause
prostaglandin formation in microdissected rabbit cortical collecting tubules
(34), but bradykinin acts only from the apical side of cultured canine cortical collecting tubule (CCCT) cells to cause PGE_2 release (36).

The only cultured collecting tubule epithelia reported to exhibit AVP-
induced prostaglandin formation are those derived from the dog (30,35-37).
Interestingly, AVP stimulates PGE_2 release by a factor of two to three
when added to either the apical or the basolateral side of CCCT cells (35,
36), but AVP functions only from the basolateral surface of these cells to
elicit cAMP formation (36). These observations suggest that the action of
AVP to stimulate PGE_2 formation by CCCT cells is mediated through a receptor different from the V_2 receptor involved in AVP-induced cAMP formation.

Sidedness of PGE_2 Release from Collecting Tubules

Stokes (39) reported that PGE_2 added to the bath side, but not the
lumen side, of perfused rabbit cortical collecting tubules inhibits AVP-
induced water reabsorption. Since collecting tubules do not catabolize
PGE_2 (29), one can deduce from Stokes's observation that the collecting
tubule is essentially impermeable to PGE_2. Thus, one would reason that
for prostaglandins synthesized by the collecting tubule to inhibit the hydroosmotic effect of AVP *in vivo*, they must be released, at least in part,
on the basolateral side on the cell. In fact, PGE_2 can be released on
both the apical and basolateral sides of CCCT cells in response to bradykinin (36). Moreover, lysyl-bradykinin added to the bath side of the perfused rabbit collecting tubule causes a prostaglandin-mediated attenuation
of the hydroosmotic effect of AVP (34); these latter results of Schuster et
al. (34) are consistent with PGE_2 being released on the basolateral side
of the collecting tubule into the renal interstitium. Whether PGE_2 is

released exclusively on the basolateral surface of the rabbit collecting tubule is unknown. There is at least one example of sidedness to prostaglandin release: PGE_2 is released exclusively on the serosal side of colonic epithelia (40).

Finally, with regard to the sidedness of PGE_2 release from the collecting tubule, we mention again that biological membranes are not freely permeable to PGE_2 (36,41,42) and that there must exist carrier mechanisms to move prostaglandins from the interior of the cell--where most, if not all, biosynthesis occurs (3,42)--to the outside of the cell. There have been no direct studies of such mechanisms, but we conjecture that organic acid transport processes related to those that may carry prostaglandins across proximal tubular epithelia (43) and endothelial cells of the lung (44) are involved.

PGE_2 Inhibits the Hydroosmotic Effect of AVP in the Collecting Tubule

Grantham and Orloff were the first to demonstrate, nearly twenty years ago, that PGE_1 inhibits AVP-induced water reabsorption in the perfused rabbit cortical collecting tubule (45); and subsequent studies have indicated that this is a physiologically important effect of PGE_2 in vivo (46,47). In vitro, the inhibitory effect was observed with 10^{-9} M PGE_1; at higher concentrations (i.e., 10^{-7} M), PGE_1 in the presence of theophylline actually promoted water reabsorption (45). A curious feature of the inhibitory effect of PGE_1 was that it was observed with AVP, but not exogenous cAMP (45). Because it was recognized that cAMP mediated the hydroosmotic effect of AVP (46), these latter observations suggested that PGE_1 exerted its inhibitory effect by blocking AVP-induced cAMP formation. Interestingly, work occurring at about the same time on the inhibition by PGE_1 of hormone-induced lipolysis, another cAMP mediated event, suggested that PGE_2 can also inhibit hormone (epinephrine)-stimulated cAMP formation in adipocytes (48,49).

PGE_2 Prevents AVP-induced cAMP Accumulation in Collecting Tubule Cells

Considerable efforts have been directed during the past twenty years toward determining the mechanism by which PGE inhibits AVP-induced water reabsorption in the collecting tubule (46,47). The available data are consistent with the idea that PGE_2 does act to prevent, or at least attenuate, the magnitude of the increase in intracellular cAMP levels that occur in response to AVP; however, several mechanisms may exist involving effects

137

of PGE_2 (a) to stimulate cAMP phosphodiesterase via a $PGE-G_x$ receptor, (b) to cause heterologous desensitization of adenylate cyclase, and/or (c) to inhibit adenylate cyclase via a $PGE-G_i$ receptor (Fig. 1).

Edwards et al. (50) and Torikai and Kurokawa (51) have reported that PGE_2 inhibits accumulation of cAMP induced by AVP in rat papillary and cortical collecting tubules, respectively, but importantly, that the inhibitory effect of PGE_2 occurred only in the absence of cAMP phosphodiesterase inhibitors. These studies suggested that PGE_2 may exert its inhibitory effect on AVP-induced cAMP formation by stimulating a cAMP phosphodiesterase activity (Fig. 1). In fact, there is one report involving the P-815 mastocytoma line that PGE_2 can stimulate cAMP phosphodiesterase activity (52). In addition, PGE_2 can promote Ca^{++} mobilization in mesangial cells (53), and PGE_2 presumably stimulates Ca^{++} mobilization in acting via the "EP_1" smooth muscle receptor (54-55). Thus, one could envision PGE_2 acting in the collecting tubule to increase levels of intracellular Ca^{++} which could lead to activation of a Ca^{++}-calmodulin-dependent cAMP phosphodiesterase (56-58). There is, as yet, no direct evidence to support this possibility, but the question of whether PGE_2 can operate via a receptor coupled to G_x (designated as a $PGE-G_x$ receptor (59)) to stimulate a cAMP phosphodiesterase activity in the collecting tubule needs to be addressed. It should be noted that the putative $PGE-G_x$ receptor of collecting tubule cells is functionally analogous to the α_1-adrenergic receptor such as that reported to occur in cardiac myocytes (58).

A different type of inhibition than that seen with microdissected rat collecting tubules (50,51) is observed in cultured canine cortical collecting tubule (CCCT) cells. Garcia-Perez and Smith demonstrated that PGE_2 can inhibit AVP-induced cAMP formation in CCCT cells in the presence of the broad-specificity cAMP phosphodiesterase inhibitor isobutylmethylxanthine (36). Indeed, no appreciable hormone-stimulated cAMP accumulation is even detectable in CCCT cells in the absence of phosphodiesterase inhibitors. Somewhat surprisingly, half-maximal inhibition of AVP-induced cAMP formation in CCCT cells occurred with exceedingly low concentrations of PGE_2 (10^{-11} M), and inhibition was also measured with similarly low concentrations of $PGF_{2\alpha}$ (60). It is important to note that these inhibitory effects required pretreatment of the cells for a minimum of 10-20 min. with PGE_2 before the addition of AVP (36,60). An incubation period of this length is typically required to observe heterologous desensitization of the adenylate cyclase system by PGE_2 (61-63), and the results of Garcia-Perez and Smith (36) suggest that the adenylate cyclase system of the collecting

tubule may be under an analogous regulatory control. The net effect of this
type of control is to desensitize the collecting tubule to any hormone that
acts by stimulating cAMP synthesis when the cells have been chronically
exposed even to low concentrations of PGE_2. The biochemical mechanism for
this type of regulation is unknown (59,63).

A third type of inhibitory regulation of hormone-induced cAMP formation
by PGE_2 is a direct inhibition of adenylate cyclase involving the inhibitory guanine nucleotide regulatory protein G_i (64, Fig. 1). We have recently obtained evidence that there is a PGE receptor coupled to G_i in
rabbit cortical collecting tubule cells and that this receptor mediates a
direct inhibitory effect of PGE_2 on AVP-induced cAMP production. We initially characterized this receptor (designated as the PGE-G_i receptor) in
membranes prepared from canine outer medulla (64). The PGE-G_i receptor
shares several unique characteristics--most notably that GTP stimulates
PGE_2 binding--with a PGE binding activity from hamster adipocytes (65).
As noted earlier, PGE causes inhibition of epinephrine-induced cAMP synthesis in adipocytes (48,49). Pertussis toxin treatment of adipocytes blocks
the inhibitory effect of PGE_2 on epinephrine-induced cAMP formation (66),
and pertussis toxin treatment of the PGE-G_i receptor specifically elimi-
nates the stimulatory effect of GTP on binding (64). Anderson et al. (67)
have presented a preliminary report that pertussis toxin blocks the inhibition by PGE_2 of AVP-induced cAMP formation in rabbit cortical collecting
tubule epithelia cultured from microdissected tubules, and we have obtained
similar results with freshly isolated rabbit cortical collecting tubules
(unpublished observation). Thus, there appears to be an inhibitory PGE-G_i
receptor in collecting tubule epithelia that is functionally analogous to
the α_2-adrenergic receptor. This latter receptor mediates its action via
G_i and serves to inhibit adenylate cyclase (68-70). The molecular mechanism for the inhibitory effect of dissociated G_i is incompletely resolved
(70,71).

In summary, there may be two PGE receptors--PGE-G_x and PGE-G_i--
through which PGE_2 serves to limit the accumulation of cAMP in collecting
tubule epithelia (25-28; Fig. 1). In addition to these two inhibitory mechanisms, PGE_2 may also cause heterologous desensitization of the adenylate
cyclase system of the collecting tubule (36,60). It is not known whether
all these inhibitory pathways exist in the same cell or whether there is a
species specificity with, for example, the PGE-G_x receptor predominating
in rat (50,51), the PGE-G_i receptor predominating in rabbit (64,67), and
heterologous desensitization being most important in dog (36,60). There is,

however, a precedent for multiple inhibitory adenosine receptors being present in 3T3-L1 cells serving to limit epinephrine-induced cAMP formation (72).

AVP-PGE$_2$ Interactions in the Thick Ascending Limb

AVP can stimulate NaCl reabsorption in the thick ascending limb of Henle's loop (73-75), and cAMP appears to be the second messenger that mediates this effect (Fig. 1). In the mouse medullary thick ascending limb, AVP-induced NaCl reabsorption is attenuated by PGE$_2$ (73-75); in the rabbit thick ascending limb, the effects of AVP and PGE are variable (76). The inhibition of AVP-induced NaCl reabsorption by PGE$_2$ is blocked by pretreatment of mouse thick limb with pertussis toxin (75), suggesting that PGE$_2$ exerts its inhibitory action via a PGE-G$_i$ receptor (Fig. 1). Further support for this concept comes from the findings (a) that PGE$_2$ inhibits AVP-induced cAMP formation in rat medullary thick ascending limb, even in the presence of phosphodiesterase inhibitors (51), and (b) that the highest concentration of the PGE-G$_i$ receptor is present in the outer medulla (64) where approximately 25% of the cells are thick limb epithelia (27). Additional studies need to be performed to determine whether PGE$_2$ acts via a pertussis toxin sensitive mechanism to inhibit AVP-induced cAMP formation in medullary thick ascending limb cells and whether the PGE-G$_i$ receptor is actually present in these cells.

PGE-induced cAMP Formation by Collecting Tubule and Thick Ascending Limb Epithelia

At low concentrations (10^{-9} M), PGE$_1$ inhibits AVP-induced water reabsorption (45) and cAMP formation in rabbit cortical collecting tubule. At higher concentrations (ca. 10^{-7}), PGE$_2$ itself causes an antidiuresis. This is apparently due to the well-documented ability of higher concentrations of PGE$_2$ to stimulate cAMP formation (59), which then, in turn, overrides the acute inhibitory effect seen with low concentrations of PGE$_2$.

The cAMP stimulatory effect of PGE$_2$ is probably mediated through a stimulatory PGE receptor coupled to the stimulatory guanine nucleotide regulatory protein, G$_s$. This receptor is designated as a PGE-G$_s$ receptor, and it appears to be present throughout the collecting tubule and in the thin limb of Henle's loop (77). There is also evidence that the PGE-G$_s$ receptor is present in the rabbit and mouse (27), but not the rat (77), cortical and medullary thick ascending limb. Precedents for the stimulatory effects of PGE_2 on cAMP formation being mediated by a PGE receptor coupled directly to G_s comes from work on a frog PGE erythrocyte receptor (78), where PGE_2 binding is inhibited by GTP, and from work on the human platelet PGI_2/PGE_1 receptor, where PGE_1 stimulates GTPase activity (79). We have suggested that at high concentrations, PGE_2 may act via the $PGE-G_s$ receptor in a feedback loop to limit hormone-induced prostaglandin synthesis (60; Fig. 1), and there is some recent evidence to support this concept in papillary collecting tubule cells (80). Such a mechanism is predicated on compartmentalization of PGE_2-induced and AVP-induced cAMP formation. In fact, there is indirect evidence for a functional compartmentation of the $PGE-G_s$ and ß-adrenergic receptors in mouse 3T3 fibroblasts (81).

SUMMARY

PGE_2 formed by renal collecting tubules is an important factor in regulating NaCl and water reabsorption in the collecting tubule and medullary thick ascending limb. PGE_2 appears to act, depending on its ambient concentration, via several different receptors present in these renal epithelia to modulate cAMP turnover in both positive and negative directions. These putative PGE receptors form a family of receptors, all coupled to G proteins, and this family of PGE receptors is homologous to the adrenergic receptor family.

ACKNOWLEDGEMENTS

This work was supported in part by National Institutes of Health Grants Nos. DK22042, DK36485 and HL35731, a Grant-In-Aid from the American Heart Association of Michigan, and by an NIH Predoctoral Training Grant, HL-07404.

REFERENCES

1. M.J. Dunn, J.F. Liard, F. Dray, Basal and stimulated rates of renal secretion and excretion of prostaglandins E_2, F_α, and 13, 14- dihydro-15-keto F_α in the dog, Kidney Int. 13:136-143 (1978).
2. S.H. Ferreira and J.R. Vane, Prostaglandins: their disappearance from and release into the circulation, Nature 216:868-873 (1967).
3. W.L. Smith, Cellular and subcellular compartmentalization of prostaglandin and thromboxane synthesis, in: "Biochemistry of Arachidonic Acid Metabolism," W.E.M. Lands, ed., Martinus Nijhoff, Boston, pp. 79-93 (1985).
4. J.B. Lee, K. Crowshaw, B.H. Takman, K.A. Attrep, The identification of

prostaglandins E_2, $F_{2\alpha}$ and A_2 from rabbit kidney medulla, Biochem. J. 105:1251-1260 (1967).

5. E.G. Daniels, J.W. Hinman, B.E. Leach and E.E. Muirhead, Identification of prostaglandin E_2 as the principal vasodepressor lipid of rabbit renal medulla, Nature 215:1298-1299 (1967).

6. M. Hamberg, Biosynthesis of prostaglandins in the renal medulla of rabbit, FEBS Lett. 5:127-130 (1969).

7. K. Crowshaw, Prostaglandin biosynthesis from endogenous precursors in rabbit kidney, Nature New Biol. 231:240-242 (1971).

8. J.B. Lee, Renal homeostasis and the hypertensive state: a unifying hypothesis, in: "The Prostaglandins," Vol. I, P.W. Ramwell, ed., Plenum Press, New York, pp. 133-187 (1973).

9. M.J. Dunn and V.L. Hood, Prostaglandins and the kidney, Am. J. Physiol. 233:F169-F184 (1977).

10. E.E. Muirhead, G.B. Brown, G.S. Germain, B.E. Leach, The renal medulla as an antihypertensive organ, J. Lab. Clin. Invest. 76:641-649 (1970).

11. A.R. Whorton, M. Smigel, J.A. Oates, J.C. Frohlich, Regional differences in prostaglandin formation by the kidney: prostacyclin is a major prostaglandin of the renal cortex, Biochem. Biophys. Acta 529:176-180 (1978).

12. W.L. Smith, D.L. DeWitt, M.L. Allen, Bimodal distribution of the prostaglandin I_2 synthase antigen in smooth muscle cells, J. Biol. Chem. 258:5922-5926 (1983).

13. J.C. McGiff, Interactions of renal prostaglandins with the renin-angiotensin and kallekrein-kinin systems, in: "Prostaglandins in Cardiovascular and Renal Function," A. Scriabine, A.M. Lefer, and F.A. Kuehl, Jr., eds., Spectrum Publications, New York, pp. 387-398 (1980).

14. F.H.A. Janszen and A.H. Nugteren, A histochemical study of the prostaglandin biosynthesis in the urinary system of rabbit, guinea pig, goldhamster and rat, in: "Advances in the Biosciences," S. Bergstrom, ed., Pergamon, New York, pp. 287-292 (1973).

15. W.L. Smith and G.P. Wilkin, Immunochemistry of prostaglandin endoperoxide-forming cyclooxygenase: the detection of the cyclooxygenase in rat, rabbit and guinea pig kidneys by immunofluorescence, Prostaglandins 13:873-892 (1977).

16. W.L. Smith and T.G. Bell, Immunohistochemical localization of the prostaglandin-forming cyclooxygenase in renal cortex, Am. J. Physiol. 235:F451-F457 (1978).

17. W.L. Smith, F.C. Grenier, T.G. Bell, G.P. Wilkin, Cellular distribution

of enzymes involved in prostaglandin metabolism in the mammalian kidney, in: "Prostaglandins in Cardiovascular and Renal Function," A. Scriabine, A.M. Lefer and F.A. Kuehl, Jr., eds., Spectrum Publications, New York, pp. 71-91 (1980).

18. W.L. Smith, F.C. Grenier, D.L. DeWitt, A. Garcia-Perez, T.G. Bell, Cellular compartmentalization of the biosynthesis and function of PGE_2 and PGI_2 in the renal medulla, in: "Prostaglandins and the Kidney," M.J. Dunn and C. Patrono, eds., Plenum Publishing Co., New York, pp. 27-39 (1983).

19. M. Hemler, W.E.M. Lands, W.L. Smith, Purification of the cyclooxygenase that forms prostaglandins. Demonstration of two forms of iron in the holoenzyme, J. Biol. Chem. 251:5575-5579 (1976).

20. M.G. Currie and P. Needleman, Renal arachidonic acid metabolism, Ann. Rev. Physiol. 46:327-341 (1984).

21. N. Farman, P. Pradelles, J.P. Bonvalet, Determination of prostaglandin E_2 synthesis along rabbit nephron by enzyme immunoassay, Am. J. Physiol. 251:F238-F244 (1986).

22. M.A. Kirschenbaum, A.G. Lowe, W. Trizna, L.G. Fine, Regulation of vasopressin action by prostaglandins, J. Clin. Invest. 70:1193-1204 (1982).

23. D. Schlondorff, J.A. Satriano, V.W. Folkert, J. Eveloff, Prostaglandin synthesis by isolated collecting tubules from adult and neonatal rabbits, Am. J. Physiol. 248:F134-F144 (1985).

24. F.C. Grenier and W.L. Smith, Formation of 6-keto-$PGF_{1\alpha}$ by collecting tubule cells isolated from rabbit renal papillae, Prostaglandins 16: 759-772 (1978).

25. D. Schlondorff, R. Zanger, J.A. Satriano, V.W. Folkert, J. Eveloff, Prostaglandin synthesis by isolated cells from the outer medulla and from the thick ascending loop of Henle of rabbit kidney, J. Pharmacol. Exp. Therap. 223:120-124 (1982).

26. J. Sraer, J. Foidart, D. Chansel, P. Mahieu, B. Kourznetzova, R. Ardaillou, Prostaglandin synthesis by mesangial and epithelial glomerular cultured cells, FEBS Lett. 104:420-424 (1979).

27. M.L. Allen, W.K. Sonnenburg, M. Burnatowska-Hleden, W.S. Spielman, W.L. Smith, Immunodissection of cortical and medullary thick ascending limb cells from rabbit kidney, Am. J. Physiol. submitted.

28. A. Hassid, M. Konieczkowski, M. Dunn, Prostaglandin synthesis in isolated glomeruli, Proc. Nat. Acad. Sci. U.S.A. 76:1155-1159 (1979).

29. F.C. Grenier and W.L. Smith, Formation of 6-keto-$PGF_{1\alpha}$ by collecting tubule cells isolated from rabbit renal papillae, Prostaglandins 16:759-772 (1978).

30. F.C. Grenier, T.E. Rollins, W.L. Smith, Kinin-induced prostaglandin synthesis by renal papillary collecting tubule cells in culture, Am. J. Physiol. 241:F94-F104 (1981).

31. Y. Tanaka, S.L. Ward, W.L. Smith, Immunochemical and kinetic evidence for two different prostaglandin H-prostaglandin E isomerases in sheep vesicular gland microsomes, J. Biol. Chem. 262:1374-1381 (1987).

32. A.J. Marcus, B.B. Weksler, E.A. Jaffe, M.J. Broekman, Synthesis of prostacyclin from platelet-derived endoperoxide by cultured human endothelial cells, J. Clin. Invest. 66:979-983 (1980).

33. J.B. Stokes, Integrated actions of renal medullary prostaglandins in the control of water excretion, Am. J. Physiol. 240:F471-F480 (1981).

34. V.L. Schuster, J.P. Kokko, H.R. Jacobson, Interactions of lysyl-bradykinin and anti-diuretic hormone in the rabbit cortical collecting tubule, J. Clin. Invest. 73:1659-1669 (1984).

35. A. Garcia-Perez and W.L. Smith, Use of monoclonal antibodies to isolate cortical collecting tubule cells: AVP induces PGE release, Am. J. Physiol. 244:C211-C220 (1983).

36. A. Garcia-Perez and W.L. Smith, Apical-basolateral membrane asymmetry in canine cortical collecting tubule cells: bradykinin, arginine vasopressin, prostaglandin E_2 interrelationships, J. Clin. Invest. 74:63-74 (1984).

37. M. Sato and M.J. Dunn, Interactions of vasopressin, prostaglandins, and cAMP in rat renal papillary collecting tubule cells in culture, Am. J. Physiol. 247:F423-F433 (1984).

38. W.S. Spielman, W.K. Sonnenburg, M.L. Allen, L.J. Arend, K. Gerozissis, W.L. Smith, Immunodissection and culture of rabbit cortical collecting tubule cells, Am. J. Physiol. 251:F348-F357 (1986).

39. J.L. Stokes, Modulation of vasopressin-induced water permeability of the cortical collecting tubule by endogenous and exogenous prostaglandins, Mineral Electrolyte Metab. 11:240-248 (1985).

40. A.W. Cuthbert, P.V. Halushka, H.S. Margolius, J.A. Spayne, Mediators of the secretory response to kinins, Br. J. Pharmacol. 82:597-607 (1984).

41. W.L. Smith, Prostaglandin biosynthesis and its compartmentalization in vascular smooth muscle and endothelial cells, Ann. Rev. Physiol. 48:251-262 (1986).

42. L.A. Bito and R.A. Baroody, Impermeability of rabbit erythrocytes to prostaglandins, Am. J. Physiol. 229:1580-1584 (1975).

43. J.J. Irish, Secretion of prostaglandin E_2 by rabbit proximal tubules, Am. J. Physiol. 237:F268-F273 (1979).

44. S. Moncada, R. Korbut, S. Bunting, J.R. Vane, Prostacyclin is not a

circulating hormone, <u>Nature</u> 273:767-769 (1978).

45. J.J. Grantham and J. Orloff, Effect of prostaglandin E_1 on the permeability response of the isolated collecting tubule to vasopressin, adenosine 3',5'-monophosphate, and theophylline, <u>J. Clin. Invest</u>. 47:1154-1161 (1968).

46. J.S. Handler and J. Orloff, Antidiuretic hormone, <u>Ann. Rev. Physiol</u>. 43:611-624 (1981).

47. T.R. Beck and M.J. Dunn, The relationship of antidiuretic hormone and renal prostaglandins, <u>Mineral Electrolyte Metab</u>. 6:46-59 (1981).

48. D. Steinberg, M. Vaughan, P.J. Nestel, O. Strand, S. Bergstrom, Effects of prostaglandins on hormone-induced mobilization of free fatty acids, <u>J. Clin. Invest</u>. 43:1533-1540 (1963).

49. R.W. Butcher and C.E. Baird, Effects of prostaglandins on adenosine 3',5'-monophosphate levels in fat and other tissues, <u>J. Biol. Chem</u>. 243:1713-1717 (1968).

50. R.M. Edwards, B.A. Jackson, T.P. Dousa, ADH-sensitive cAMP system in papillary collecting duct: effect of osmolality and PGE_2, <u>Am. J. Physiol</u>. 240:F311-F318 (1981).

51. S. Torikai and K. Kurokawa, Effect of PGE_2 on vasopressin-dependent cell cAMP in isolated single nephron segments, <u>Am. J. Physiol</u>. 245:F58-F66 (1983).

52. K. Yatsunami, A. Ichikawa, K. Tomita, Accumulation of adenosine 3',5'-monophosphate induced by prostaglandin E_1 binding to mastocytoma P-815 cells, <u>Biochem. Pharmacol</u>. 11:1325-1332 (1981).

53. P. Menè, G.R. Dubyak, A. Scarpa, M.J. Dunn, Stimulation of cytosolic free calcium and inositol phosphates by prostaglandins in cultured rat mesangial cells, <u>Biochem. Biophys. Res. Comm</u>. 142(2):579-586 (1987).

54. R.A. Coleman and I. Kennedy, Characterization of the prostanoid receptors mediating contraction of guinea pig isolated trachea, <u>Prostaglandins</u> 29:363-375 (1985).

55. R.A. Coleman, P.P.A. Humphrey, I. Kennedy, Prostanoid receptors in smooth muscle: further evidence for a proposed classification, in: "Trends in Autonomic Pharmacology," Vol. 3, Kalsner, S., ed., Taylor and Francis Publishing, London, pp. 35-49 (1985).

56. D.A. Ausiello and J.V. Bonventre, Calcium and calmodulin as mediators of hormone action and transport events, <u>Sem. in Nephrol</u>. 4:134-143 (1984).

57. V.C. Manganiello, T. Yamamoto, M. Lin, M.L. Elks, M. Vaughan, Regulation of specific forms of cyclic nucleotide phosphodiesterases in cultured cells, <u>Adv. Cyclic Nucleotide Res</u>. 16:291-301 (1984).

58. I.L.O. Buxton and L.L. Brunton, Action of the cardiac α_1-adrenergic receptor. Activation of cAMP degradation, J. Biol. Chem. 260:6733-6737 (1985).

59. W.L. Smith, W.K. Sonnenburg, T. Watanabe, K. Umegaki, Mechanism of action of prostaglandin E_2 and prostaglandin $F_{2\alpha}$: PGE and PGE receptors, in: "Eicosanoids in the Cardiovascular and Renal Systems," P.V. Halushka, ed., MTP Press, Lancaster, in press.

60. W.L. Smith and A. Garcia-Perez, A two receptor model for the mechanism of action of prostaglandins in the renal collecting tubule, in: "Prostaglandins, Leukotrienes and Lipoxins," J.M. Bailey, ed., Plenum Publishing, New York, pp. 35-45 (1985).

61. B.B. Clark and R.W. Butcher, Desensitization of adenylate cyclase in cultured fibroblasts with prostaglandin E_1 and epinephrine, J. Biol. Chem. 254:9373-9378 (1979).

62. S. Kassis and P.H. Fishman, Different mechanism of desensitization of adenylate cyclase by isoproterenol and prostaglandin E_1 in human fibroblasts: role of regulatory components in desensitization, J. Biol. Chem. 257:5312-5318 (1982).

63. D.R. Sibley and R.J. Lefkowitz, Molecular mechanism of receptor desensitization using the ß-adrenergic receptor-coupled adenylate cyclase system as a model, Nature 317:124-129 (1985).

64. T. Watanabe, K. Umegaki, W.L. Smith, Association of a solubilized prostaglandin E_2 receptor from renal medulla with a pertussis toxin-reactive guanine nucleotide regulatory protein, J. Biol. Chem. 261:13430-13439 (1986).

65. R. Grandt, K. Aktories, K.H. Jakobs, Guanine nucleotides and monovalent cations increase agonist affinity of prostaglandin E_2 receptors in hamster adipocytes, Mol. Pharmacol. 22:320-326 (1982).

66. J.A. Garcia-Sainz, Decreased sensitivity to α_2 adrenergic amines, adenosine and prostaglandins in white fat cells from hamsters treated with pertussis toxin, FEBS Lett. 126:306-308 (1981).

67. R.J. Anderson, P.D. Wilson, M.A. Dillingham, R. Breckon, U. Schwertschlaf, J.A. Garcia-Sainz, Pertussis toxin reverses prostaglandin E_2 inhibition of arginine vasopressin (AVP) and forskolin in rabbit collecting tubular epithelium, (Abstract), Am. Soc. Nephrol. Mtg. 154, (1984).

68. J.W. Regan, H. Nakata, R.M. DeMarinis, M.G. Caron, R.J. Lefkowitz, Purification and characterization of the human platelet α_2-adrenergic receptor, J. Biol. Chem. 261:3894-3900 (1986).

69. R.A. Cerione, J.W. Regan, H. Nakata, J. Codina, J.L. Benovic, P. Gierschik, R.L. Somers, A.M. Spiegel, L. Birnbaumer, R.J. Lefkowitz,

M.G. Caron, Functional reconstitution of the α_2-adrenergic receptor with guanine nucleotide regulatory proteins in phospholipid vesicles, J. Biol. Chem. 261:3901-3909 (1986).

70. T. Katada, G.M. Bokoch, J.K. Northup, M. Ui, A.G. Gilman, The inhibitory guanine nucleotide-binding regulatory component of adenylate cyclase: properties and function of the purified protein, J. Biol. Chem. 259:3568-3577 (1984).

71. T. Katada, G.M. Bokoch, M.D. Smigel, M. Ui, A.G. Gilman, The inhibitory guanine nucleotide-binding regulatory component of adenylate cyclase: Subunit dissociation and the inhibition of adenylate cyclase in S49 cyc⁻ and wild type membranes, J. Biol. Chem. 260:3477-3483 (1985).

72. M.L. Elks, M. Jackson, V.C. Manganiello, M. Vaughan, Effect of N^6-(L-2-phenylisopropyl)adenosine and insulin on cAMP metabolism in 3T3-L1 adipocytes, Am. J. Physiol. 252:C342-C348 (1987).

73. R.M. Culpepper and T.E. Andreoli, Interactions among prostaglandin E_2, antidiuretic hormone, and cyclic adenosine monophosphate in modulating Cl⁻ absorption in single mouse medullary thick ascending limbs of Henle, J. Clin. Invest. 71:1588-1601 (1983).

74. S. Hebert and T.E. Andreoli, Control of NaCl transport in the thick ascending limb, Am. J. Physiol. 246:F745-F756 (1984).

75. R.M. Culpepper, Pertussis toxin blunts PGE_2 inhibition of ADH-stimulated Ve in mouse mTALH, Kidney Int. 27:255 (1985).

76. J.L. Stokes, Personal communication.

77. S. Torikai and K. Kurokawa, Distribution of prostaglandin E_2-sensitive adenylate cyclase along the rat nephron, Prostaglandins 21:427-438 (1981).

78. R.J. Lefkowitz, D. Mullikin, C.L. Wood, T.B. Gore, C. Mukherjee, Regulation of prostaglandin receptors by prostaglandins and guanine nucleotides in frog erythrocytes, J. Biol. Chem. 252:5295-5303 (1977).

79. H.A. Lester, M.L. Steer, A. Levitzki, Prostaglandin-stimulated GTP hydrolysis associated with activation of adenylate cyclase in human platelet membranes, Proc. Nat. Acad. Sci. U.S.A. 79:719-723 (1982).

80. I. Teitelbaum, J.N. Mansour, T. Berl, Effect of cAMP on prostaglandin E_2 production in cultured rat inner medullary collecting tubule cells, Am. J. Physiol. 251:F671-F677 (1986).

81. W.G. Tarpley, N.K. Hopkins, R.R. Gorman, Reduced hormone-stimulated adenylate cyclase activity in NIH-3T3 cells expressing the EJ human bladder *ras* oncogene, Proc. Nat. Acad. Sci. U.S.A. 83:3703-3707 (1986).

PEPTIDE HORMONES, CYTOSOLIC CALCIUM AND RENAL EPITHELIAL RESPONSE

Aubrey R. Morrison, Didier Portilla, Daniel Coyne

Departments of Medicine and Pharmacology
Washington University School of Medicine
St. Louis, Missouri 63110

INTRODUCTION

The response of renal cells to extracellular signals has recently attracted increasing experimental evaluation. The cellular response to a variety of peptide hormones, neurotransmitters and growth factors are fundamental to understanding how the signals mediated by circulatory substances, which interact with cell surface receptors, produce their effects intracellularly. The cellular responses to a wide variety of signal molecules are somewhat limited. Occupancy of receptors initiates the production of intracellular messengers including cAMP, cGMP and the second messenger molecules derived from phosphoinositides (1-3). The phosphoinositides constitute 5-8% of lipids in the cell membranes of eukaryotic cells and are essential for cell viability (4). These phosphoinositides are storage forms for the messenger molecules that transmit signals across the cell membrane and evoke responses to extracellular signals.

BIOCHEMISTRY

Current views on the metabolism of phosphoinositide-derived second messengers are illustrated in Figure 1. These metabolites are produced in response to a wide variety of agonists; and in the kidney, both epithelial and nonepithelial (in particular, mesangial smooth muscle) cells respond to agonists such as angiotensin II (5,6), bradykinin (7,8), vasopressin (9), and parathyroid hormone (10,11). The key enzyme that controls the formation of the phosphorylated inositols is phospholipase C which can be either a soluble enzyme (12-14) or membrane bound (15,16). It is presumably the

149

membrane associated form of the enzyme that is activated to initiate the hydrolysis of the phosphoinositides with receptor-mediated stimuli. The soluble enzymes have been better characterized and are inactive unless they are bound to the lipid bilayer containing the appropriate substrate. The PLC enzymes are specific for phosphatidylinositol and the polyphosphoinositides and appear to require Ca^{2+} for full activation. The phosphoinositides are substrates for phospholipase C and, when all three substrates are present, hydrolysis of the polyphosphoinositides are favored. Increasing the concentration of Ca^{2+}, however, shifts the substrate specificity in favor of phosphatidylinositol (17). Hydrolysis of phosphatidylinositol by the purified enzyme results in a mixture of inositol 1 phosphate and inositol cyclic 1,2 phosphate (Fig. 1) (18-20). Thus, the finding of two products suggests that either the 2-position hydroxyl on the inositol ring or free OH^- in solution may attack the phosphorus. It now appears that analogous cyclic compounds may be formed by PLC hydrolysis of the polyphosphoinositides (21-23). Thus, the activation of PLC through receptor-mediated stimuli will cause the initial hydrolysis of phosphatidylinositol bisphosphate with the formation of inositol 1,4,5 trisphosphate and cyclic 1,2 inositol 4,5 trisphosphate. Both inositol 1,4,5 trisphosphate and cyclic 1,2 inositol 4,5 trisphosphate (23) are active biologically and will mobilize Ca^{2+} from intracellular, nonmitochondrial stores to increase cytosolic calcium concentrations.

The metabolism of these inositol phosphates is very complex, and some of this complexity is demonstrated in Figure 1. It is now believed the inositol 1,4,5 trisphosphate can be further phosphorylated to inositol 1,3,4,5 tetrakisphosphate (25-27) by a 3-phosphokinase. The enzyme has been demonstrated in soluble fraction homogenates and has a K_m of approximately 0.6

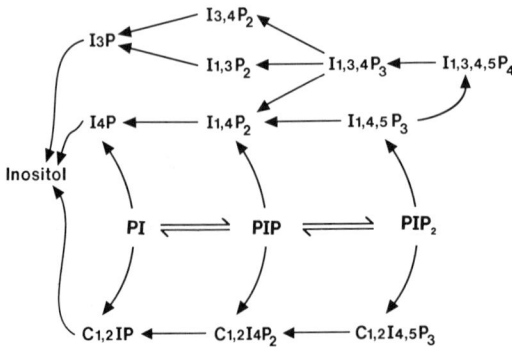

Fig. 1. Scheme of membrane phosphoinositols and breakdown products. PI, phosphatidylinositol; PIP, phosphatidylinositol monophosphate; PIP_2, phosphatidylinositol bisphosphate; C_nIP, cyclic inositol polyphosphates; I_nP, inositol polyphosphates.

uM for inositol (1,4,5) trisphosphate. The inositol tetrakisphosphate is a substrate for the 5-phosphomonoesterase which converts it to inositol 1,3,4 trisphosphate. Further metabolism can then yield a number of different isomers; for example, inositol (3,4) P_2 (28), inositol (1,3) P_2 (28,29) and inositol (1,4) P_2 (18,29,30). The biology of the tetrakisphosphate and the 1,3,4 trisphosphate isomers are currently being evaluated but, in nonrenal tissues, it has been suggested that the 1,3,4 isomer can alter membrane potential in neuroblastoma cells (31) and that the tetrakisphosphate (IP_4) may regulate Ca^{2+} gating across the plasma membrane of sea urchin eggs (32). In addition to the generation of the phosphorylated inositols, the phosphodiester cleavage of PIP_2 also generates diacylglycerol which is an activator of protein kinase C (34). Thus, a PLC activation generates two intracellular messengers, IP_3 (for cyclic $1,21^{4,5}P_3$) and diacylglycerol.

Indications are that the phospholipase C is not directly connected to the various receptors but interacts with them through a third membrane component, one of the family GTP binding proteins (35-39). Thus, the organization of the PI-linked receptor complex parallels that of the well-studied adrenergic receptor complex and is depicted in Figure 2. Here we show bradykinin as the agonist interacting with its cell surface receptor and coupled to the membrane-associated PLC through a GTP binding protein which is pertussis toxin inhibitable.

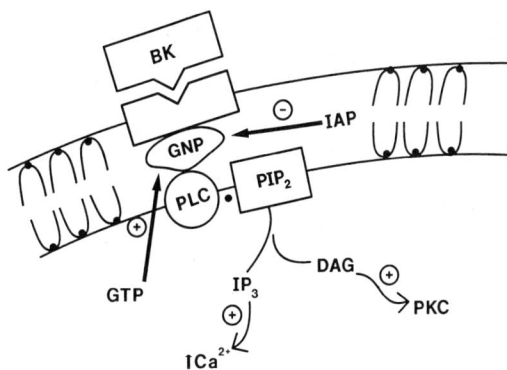

Fig. 2. Scheme of bradykinin (BK) activated, membrane-associated PIP_2 phosphodiesterase (phospholipase C). GNP, guanine nucleotide (GTP) binding protein; IAP, islet activating protein or pertussis toxin; PLC, phospholipase C; PIP_2, phosphatidylinositol bisphosphate; IP_3, inositol trisphosphate; DAG, diacylglycerol; PKC, protein kinase C.

Bradykinin

Rabbit papillary collecting tubule cells isolated by modification of Grenier et al. (40) placed in culture can be grown in inositol-free media and their endogenous phosphoinositide pool labeled within twenty-four hours by supplementing media with $[^3H]$inositol (7). Cells, when stimulated by bradykinin at $10^{-7}M$, produce a rapid increase in inositol trisphosphate labeling within 10 seconds which returns to control values after two minutes of stimulation with bradykinin. Dose response curves for IP_3 labeling in response to bradykinin shows an ED_{50} of about $5 \times 10^{-9}M$ bradykinin which is close to the ED_{50} observed for PGE_2 formation in these cells (40).

To measure intracellular CA^{2+} changes in response to the kinin, the RPCT cells were loaded with the fluorescent probe Quin 2 (41). The mean intracellular calcium concentration equaled 120 ± 13 nM. Bradykinin, when added directly to the stirred cell suspension, resulted in a rapid (less than 3 second) increase in Quin 2 fluorescence. This fluorescence peaked by 10 seconds and typically lasted 60 seconds or less. The rapid increase in Quin 2 fluorescence was similar to the time course of bradykinin-induced changes in inositol trisphosphate production previously reported in these cells. Further addition of bradykinin to the cell suspension was accompanied by the absence of a further calcium transient. A similar tachyphylactic response to bradykinin in prostaglandin E_2 production has been demonstrated in isolated perfused kidneys (42). Calculated changes in kinin-induced calcium transients demonstrated a dose dependence, and rapid increases in calcium transients which returned to baseline were observed at all effective doses, as was the desensitization to bradykinin. The cumulative data from several experiments assessing kinin-induced changes in Quin 2 fluorescence are displayed in Figure 3. The maximal change in intracellular calcium concentration expressed as a function of bradykinin concentration displays an ED_{50} of 5×10^{-8} M. This closely approximates that reported previously for inositol trisphosphate labeling. In order to assess the possible dependency of bradykinin on the presence of extracellular calcium, cells were preincubated with EGTA (1 mM) and subsequently stimulated with bradykinin $(10^{-7}M)$. The Quin 2 measured basal intracellular calcium concentration was unaffected by EGTA. The addition of bradykinin, however, did result in an increased, albeit attenuated, intracellular calcium concentration. Similar results have been obtained in MDCK cells which when stimulat-

Bradykinin-Induced Changes in $[Ca^{2+}]_i$

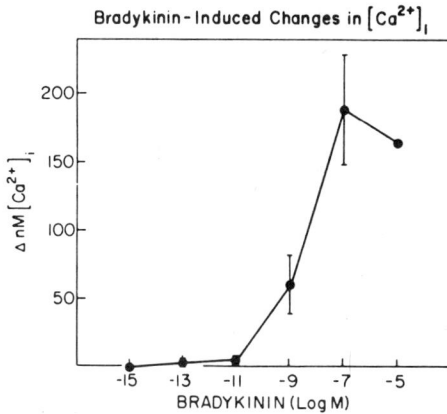

Fig. 3. Dose-dependent increase in cytosolic Ca^{2+} in response to bradykinin in rabbit papillary collecting duct cells.

ed with bradykinin results in a dose-dependent increase in IP_3 and cytosolic Ca^{2+} concentration.

Role of GTP Binding Protein in Bradykinin Response

Initial experiments were designed to evaluate the effects of IAP (pertussis toxin) on the formation of intracellular IP_3. After 4 hours of preincubation, MDCK monolayers washed x3 with Krebs buffer and stimulated with BK 10^{-7}M produced a marked increase in IP_3 formation, within 10 seconds, from a basal level of 46.2 picomoles/mg protein to 686.6 pmol/mg protein. Pretreatment with IAP (200 μg/ml for 4 hours) reduced this stimulation to 205.8 pmol/mg protein. These experiments demonstrated that the increase in PIP_2 hydrolysis with bradykinin stimulation was dependent on an intact GTP binding protein which was inhibitable by pertussis toxin (IAP).

ADP ribosylation of MDCK membranes was carried out with $[\alpha\ -^{32}P]$ NAD and the $[^{32}P]$ content of the membrane protein fractions was then analyzed by SDS polyacrylamide gel electrophoresis. As shown in Figure 4, a protein of Mr 41 kDa was labeled in vitro only when IAP was present in the reaction mixture. When IAP was replaced by cholera toxin in the incubation mixture, another protein with Mr of 45 kDa was also ADP ribosylated though less intensely than the 41 kDa protein with IAP. When MDCK membranes were prepared from IAP pretreated cells in culture, the IAP-induced ^{32}P labeling of the Mr 41 kDa protein was significantly decreased.

In addition to the hydrolysis of PIP_2 in both rabbit papillary collecting tubule cells and MDCK cells, we observed the formation of inositol

153

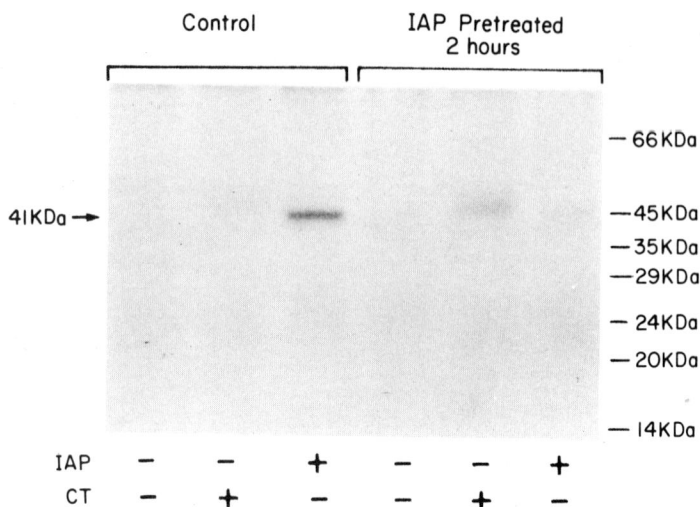

Fig. 4. Polyacrylamide gel electrophoresis of solubilized MDCK membranes subjected to ADP
ribosylation *in vitro*. IAP, islet
activating protein; CT, cholera toxin.

1,2 cyclic phosphate in rabbit papillary cells in response to bradykinin
(43). When renal papillary collecting tubule cells were labeled with myo-
[2-^{3}H]inositol and subsequently extracted with neutral chloroform and
methanol, a water-soluble metabolite that eluted with 0.1 M ammonium formate
on Dowex ion exchange chromatography was observed. This peak was not observed previously in cultures extracted with acidified methanol. To identify this metabolite, we applied a mixture of inositol monophosphates and synthetic myo-inositol 1,2-(cyclic)phosphate to an identical column. A component was observed to elute with 0.1 M ammonium formate as determined by
phosphate analysis. No compounds containing phosphate eluted with 0.1 M
ammonium formate when this mixture was exposed to 0.1 M HCl, conditions
known to hydrolyse myo-inositol 1,2-(cyclic) phosphate.

To confirm the structure of this metabolite, we subjected pentakis(trimethylsilyl) inositol 1,2-(cyclic)phosphate to GC-mass spectroscopy. The
mass spectrum of the pentakis(trimethylsilyl) myo-inositol 1,2-(cyclic) phosphate showed characteristic ions at m/z 587 $(M - Ch_3)^+$, m/z 497 (M -
CH_3-Me_3SiOH)+, m/z 422 $(M - 2x Me_3SiOH)^+$ m/z 398 (M - 2x Me$_3$
SiOCH)+, m/z 342 (M - Me$_3$SiOH-Me$_3$SiOPO$_3$H$_2$)+, m/z 315 [(Me$_3$SiO)$_3$
POH]+, m/z 299 [(Me$_3$SiO)$_2$(Me$_2$SiO)PO]$^+$, m/z 211[(Me$_2$SiO)$_2$
PO$_2$]+. Pooled cells (approx. 10^7) were extracted and subjected to
anion-exchange chromatography; the material eluting with 0.1 M ammonium

formate was similarly derivatized. When gas chromatographed, a distinct peak with an identical time retention to that of authentic derivatized myo-inositol 1,2-(cyclic) phosphate was observed. Selected ions at m/z 211, 227, 299 and 342 demonstrated the same ratios both for the cellular metabolite and for synthetic myoinositol 1,2-(cyclic)phosphate.

To ascertain whether hormonal stimulation alters intracellular levels of myo-inositol 1,2-(cyclic)phosphate, we labeled cultured cells with 8 uCi/well of myo-{2-^{3}H]inositol and subjected them to bradykinin (10^{-7}M) or to Krebs buffer alone. This concentration of bradykinin has been demonstrated previously to stimulate maximally both prostaglandin E_2 production and inositol trisphosphate labeling. When compared with control cultures, bradykinin-exposed cultures demonstrated a quantitative rise in myo-inositol 1,2-(cyclic)phosphate labeling. This increase peaked at 60 s and approached baseline by 5 min. (Fig. 5). This increase in cIP_1 over a time course different from inositol 1,4,5, P_3 formation could either indicate cIP_1 being formed as a metabolic degradation product of cyclic 1,2 inositol 4,5 P_3 (see Fig. 1) or represent a delayed hydrolysis of PI in the membrane. Our experiments do not distinguish between these two possibilities.

Vasopressin

Since the original observation that vasopressin stimulates prostaglandin E_2 synthesis and release in anuran membranes (44), much controversy has been generated by experiments carried out in anuran and mammalian

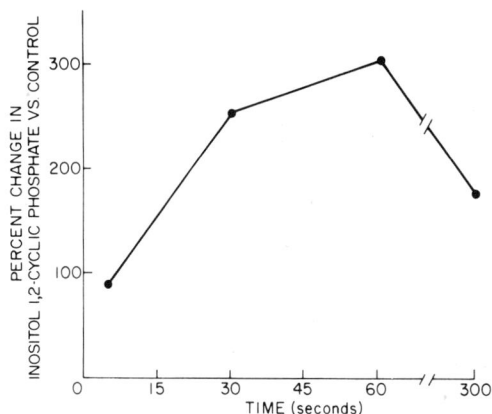

Fig. 5. Time course of [^{3}H]inositol 1,2 cyclic phosphate formation in rabbit papillary collecting tubule cells exposed to 10^{-7} M bradykinin.

systems concerning the potential for prostaglandin E_2 synthesis stimulated by vasopressin in the kidney. Bisordi et al. (45) and Forrest et al. (46) could not confirm the original observations of Zusman et al. (44) in anuran membranes, while Burch and Halushka (43) were able to demonstrate that vasopressin stimulated prostaglandin E_2 and thromboxane B_2 formation in toad urinary bladder. In mammalian systems, the same variability of results is observed. Grenier et al. (40) have been unable to demonstrate vasopressin-stimulated prostaglandins in papillary collecting duct cells of the rabbit in culture, while Sato and Dunn (48) have shown that papillary collecting duct cells in the rat supplemented with exogenous arachidonate when stimulated with vasopressin produces a modest increase in prostaglandin E_2 formation. In contrast, cortical collecting duct cells have been shown to respond to vasopressin with an increase in prostaglandin E_2 (49,50).

Vasopressin has been shown to have two distinct cellular effects mediated by different receptor types in mammalian cells (51,52). One receptor (V_2) is coupled to the activation of adenylate cyclase and mediates water transport (53), and the other receptor (V_1) mediates vasoconstriction (54,55). In some systems, for example the liver, vasopressin binding is coupled to the activation of polyphosphoinositide hydrolysis and release of inositol 1,4,5-trisphosphate (56). This response involves the activation of phospholipase C with phosphodiesteric cleavage of the polyphosphoinositides to release water-soluble inositols and diacylglycerol (57). Diacylglycerol can then be further metabolized by diacylglycerol lipase to release arachidonic acid (58). We have previously demonstrated that bradykinin increases formation of IP_3 (7), increases intracellular Ca^{2+} (41), and is associated with the synthesis and release of prostaglandin E_2 from rabbit papillary collecting tubule (RPCT) cells. With this background, we asked the question whether vasopressin stimulates the hydrolysis of $PI(4,5)P_2$, mobilizes intracellular Ca^{2+} and increases prostaglandin E_2 biosynthesis in RPCT cells. We, therefore, labeled the endogenous inositol phospholipid pool of RPCT cells in culture with [^{3}H]-myo-inositol. Vasopressin at 100 nM did not produce any significant changes in labeled inositol phosphates when compared with control over a 5 min. incubation time.

To assess whether the cell cultures under investigation could demonstrate active phosphatidylinositol metabolism, we preincubated cells for 15 min. with or without 10 mM lithium chloride. Lithium has been demonstrated previously to inhibit inositol-1-phosphatase (59), thereby resulting in the

Table 1. Accumulated Myo-[2-^{3}H]Inositol 1-Phosphate as a
Percentage of Total Myo-[2-^{3}H]Inositol

NaCl (10 mM)	8.2 ± 1.5
Bradykinin (10^{-7} M) + NaCl (10 mM)	10.1 ± 2.3
Lithium chloride (10 mM)	8.6 ± 0.8
Bradykinin (10^{-7} M)	
+ lithium chloride (10 mM)	20.5 ± 3.3*
Arginine vasopressin (10^{-7} M)	
+ NaCl (10 mM)	8.5 ± 1.2
Arginine vasopressin (10^{-7} M)	
+ lithium chloride (10 mM)	9.3 ± 2.5

Changes in distribution of inositol 1-phosphate as a percent of total incorporated myo-[2-3H]inositol. Cells were preincubated in Krebs buffer for 10 min to which was added either 10 mM NaCl as control or 10 mM LiCl. The cells were washed and then incubated with or without effectors under conditions as indicated above. * denotes P< 0.001,

accumulation of inositol 1-phosphate. As depicted in Table 1, increased distribution of [^{3}H] inositol into the inositol 1-phosphate fraction occurred only in cells exposed to bradykinin and then only in the presence of lithium chloride. No difference in cultures exposed to lithium chloride alone or in the presence of effectors alone were observed after 15 min. of incubation. Vasopressin did not increase the accumulation of inositol 1-phosphate in the presence of lithium chloride. Similarly, when the RPCT cells were loaded with the fluorescent probe Quin 2, we also could detect no increase in intracellular Ca^{2+} in response to 10^{-7} M and 10^{-6} M vaso-pressin. In unpublished observations, we have exposed MDCK cells, loaded with fura-2, to vasopressin and were unable to detect any increase in cyto-solic Ca^{2+} in response to arginine vasopressin. The MDCK cells are of canine origin and presumed to be derived from the cortical collecting duct. In our hands, we cannot demonstrate PGE$_2$ formation, IP$_3$ labeling (and presumably formation) and mobilization of intracellular calcium in both MDCK and RPCT cells in culture when exposed to vasopressin. Both of these cells, however, produce cAMP in response to vasopressin. Thus, they lack V$_1$ receptor coupling mechanisms and show V$_2$ receptor activation.

Parathyroid Hormone

PTH has long been known to activate adenylate cyclase and produce a rise in renal adenosine 3',5'-cyclic monophosphate (cAMP) (60). Receptors for PTH appear to be located on the basolateral membrane of several cortical segments, including the proximal convoluted tubule (PCT), cortical ascending limb, and the distal convoluted tubule (61). In the PCT, PTH produces a decrease in fluid, Na+, Ca^{2+}, and phosphate (62) reabsorption, while increasing gluconeogenesis (63) and hydroxylation of 25-hydroxycholecalciferol. Extensive studies have shown PTH inhibition of phosphate reabsorption to be mediated through cAMP (64) via cAMP-induced protein phosphorylation in the brush border membrane (65,66). However, even in this system, extracellular calcium (and by implication, increases in intracellular calcium) has been suggested as a modulator of PTH effect (67,68). Yanagawa and Jo have demonstrated that an increase in Ca^{2+} from 1 mM to 2 mM permitted PTH and 8-Bromo-cAMP to inhibit phosphate reabsorption in isolated perfused rabbit PCT (68). They have also shown that PTH is without effect on phosphate transport in brush border membranes prepared from proximal tubules, if the tubules were preincubated with PTH in the presence of the calcium antagonists, W7 or trifluoroperazine (67). The mechanisms of action of PTH on Na^{+} (64) and HCO_3^{-} (69) reabsorption have been less well worked out, though both are mimicked by cAMP or its analogues. Other effects of PTH seem clearly not to be mediated by cAMP (63,70). PTH stimulates gluconeogenesis in the proximal tubule, and this effect is mimicked by the calcium ionophore, A23187 (71), but not exactly by cyclic nucleotides (63). Hence, the inability to explain all of the effects of PTH via cAMP, the dependence of cAMP-mediated effects on extracellular Ca^{2+} and the blocking of these effects by calcium antagonists has led researchers to look for another mechanism of cell activation by PTH.

In a study by Hruska et al. (10), canine renal proximal tubule cells were grown in primary culture, and intracellular Ca^{2+} was measured in cell suspensions and monolayers grown on Nucleopore filters using Quin-2. Dose-dependent increases in intracellular Ca^{2+} were produced by 10^{-10} to 10^{-6} M bovine PTH (1-84). A rise in intracellular Ca^{2+} was seen within 1 minute, with peak levels occurring 20-30 minutes after addition of PTH. The dose response curve of cAMP production in response to PTH paralleled Ca^{2+} stimulation. Dibuturyl cAMP, IBMX (a phosphodiesterase inhibitor), and forskolin all failed to produce changes in intracellular Ca^{2+} over a one-to-five-minute period. These results are consistent with a preliminary report by Yanagawa (72) showing intracellular Ca^{2+} rises from 97.6 ± 14.8

158

nM to 221.9 $\pm$ 61.9 nM in isolated perfused rabbit PCT loaded with Fura-2,
and slightly larger increases, to 290 $\pm$ 88.3 nM, in perfused proximal
straight tubules. However, measurements using Quin-2, by Dolson et al.
(73), in a suspension of rabbit proximal tubules, showed an actual decline
in intracellular Ca^{2+} in response to PTH. This decline in Ca^{2+} by PTH
was mimicked by dibuturyl cAMP. Whether the buffering capacity of Quin-2
obscured a change in Ca^{2+}, or the different means of isolation and study
of these cells accounts for the discrepancy, is unknown. PTH stimulation of
rapid rises in cytosolic Ca^{2+} has been demonstrated in other cell types
(74).

Goligorsky et al. (11) performed single cell calcium measurements on
canine PCT cells grown on coverslips and loaded with Fura-2. Intracellular
Ca^{2+} rose from a basal level of 70.8 $\pm$ 7.4 nM to almost 400 nM in response
to 10^{-8} M PTH. Within three minutes of PTH washout, the intracellular
Ca^{2+} had returned to baseline. Zero extracellular calcium has been shown
to block the calcium transient of PTH, but not of angiotensin II which is
known to mobilize intracellular Ca^{2+} stores (11). Thus, PTH stimulates
Ca^{2+} transients by promoting calcium influx as well as through release of
intracellular stores. Repeated treatment with PTH at 5-minute, but not
20-minute, intervals led to a decrement in the calcium transients observed,
while cAMP production declined even with PTH stimulation at 20-minute intervals. Thus, there was a dissociation of calcium transients from cAMP
response, consistent with reports of the dissociation of anticalciuria from
cAMP tachyphylaxis to PTH in isolated perfused rabbit kidney (75).

Fewer data are available on how PTH stimulates increases in cytosolic
Ca^{2+}. The inositol (1,4,5)trisphosphate arm of phosphatidylinositol bisphosphate has been shown to produce cytosolic Ca^{2+} transients in a variety
of cells (2), and there is some evidence to support its involvement in mediating PTH stimulated cytosolic Ca^{2+} transients. Earlier investigations
have shown an effect of PTH on phosphoinositide metabolism (76-78) in renal
cortical tissue. A recent report by Hruska et al. (79) examined the opossum
kidney (OK) cell line in culture. A dose-dependent rise in IP^3 upon stimulation with PTH was noted at 10 and 30 seconds, and 5 minutes. Maximal
stimulation occurred at 10^{-7}M PTH. At the same dose of PTH, cytosolic
Ca^{2+} immediately rose by 40-50% and returned to basal levels within one
minute, as determined by the fluorescent probe Indo-1. Diacylglycerol was
measured in OK cells labeled with [^{14C}]arachidonic acid and at 30 and 60
seconds after PTH stimulation had increased to over twice control levels,
again suggesting breakdown of PIP_2. In summary, evidence is accumulating

that PTH stimulates a rapid rise in cytosolic Ca^{2+} in proximal tubular
cells, that this is a dose-dependent response that is not mimicked by cAMP,
and that this mechanism of PTH cell activation may be through the phosphoinositide second messenger pathway.

SUMMARY

We have reviewed the evidence that a number of hormones interact with
renal tubular epithelial cells. The evidence suggests that in the mammalian
renal tubule bradykinin and parathyroid hormone interact with cell surface
receptors to initiate the hydrolysis of PIP$_2$ leading to the formation of I
1,4,5P$_3$ and diacylglycerol in the distal and proximal tubule, respectively. The activation of this second messenger system leads to the mobilization of Ca^{2+} from intracellular stores. Vasopressin does not activate
this second messenger system in mammalian renal epithelial cells, and we
cannot demonstrate I 1,4,5P$_3$ formation and Ca^{2+} mobilization either in
the rabbit papillary collecting tubules or in MDCK cells. There is evidence
emerging, but not discussed here, that angiotensin II may also mediate some
of its effects on the mammalian proximal tubule via the inositol polyphos-
phate second messenger system.

ACKNOWLEDGEMENTS

This work was supported by PHS award DK30542 and AM38111. Dr. Morrison
is a recipient of the Burroughs Wellcome Clinical Pharmacology Award.

REFERENCES

1. Y. Nishizuka, Studies and perspectives of protein kinase C, Science
   233:305-312 (1986).
2. M.J. Berridge and R.F. Irvine, Inositol trisphosphate, a novel second
   messenger in signal transduction, Nature (London) 312:315-321 (1984).
3. P.N. Majerus, T.M. Connolly, H. Deckmyer, T.S. Ross, T.E. Bross, H.
   Ishii, V.S. Bansal, D.B. Wilson, The metabolism of phosphoinositide-
   derived messenger molecule, Science 234:1519-1526 (1986).
4. J. Esko and C.R.H. Raetz, Mutants of Chinese hamster ovary cells with
   altered membrane phospholipid composition, J. Biol. Chem. 255:4474-
   4480 (1980).
5. S. Shin, Y. Fujiwara, A. Wada, T. Takama, Y. Orita, T. Kamada, K.
   Tagawa, Angiotensin II-induced increase in inositol 1,4,5-trisphos-
   phate in cultured rat mesangial cells: evidence by refined High

Performance Liquid Chromatography, BBRC 142:70-77 (1987).

6. J.E. Benabe, L.A. Spry, A.R. Morrison, Effects of angiotensin II on phosphatidylinositol and polyphosphatidylinositol turnover in rat kidney, J. Biol. Chem. 257:7430-7434 (1982).

7. J.A. Shayman and A.R. Morrison, Bradykinin-induced changes in phosphatidylinositol turnover in cultured rabbit papillary collecting tubule cells, J. Clin. Invest. 76:978-984 (1985).

8. D. Portilla and A.R. Morrison, Bradykinin-induced changes in inositol trisphosphate mass in MDCK cells, BBRC 140:644-649 (1986).

9. D.A. Troyer, J.I. Kreisberg, D.W. Schwertz, M. Venkatachalam, Effects of vasopressin on phosphoinositide and prostaglandin production in cultured mesangial cells.

10. K.A. Hruska, M. Goligorsky, J. Schoble, M. Tsutsumi, S. Westbrook, D. Moskowitz, Effects of parathyroid hormone on cytosolic calcium in renal proximal tubule primary cultures, Am. J. Physiol. 251:F188-F198 (1986).

11. M.S. Goligorsky, D.J. Loftus, K.A. Hruska, Cytoplasmic calcium in individual proximal tubular cells in culture, Am. J. Physiol. 251:F938-F944 (1986).

12. S.L. Hofmann and P.W. Majerus, Purification and properties of phosphatidylinositol specific phospholipase C from sheep seminal vesicular glands, J. Biol. Chem. 257:6461-6467 (1982).

13. M.G. Low, R.C. Carroll, W.B. Weglicki, Multiple forms of phosphoinositide-specific phospholipase C of different relative molecular masses in animal tissue, Biochem. J. 221:813-820 (1984).

14. M.G. Low, R.C. Carroll, A.C. Cox, Characterization of multiple forms of phosphoinositide-specific phospholipase C purified from human platelets, Biochem. J. 237:139-145 (1986).

15. S. Cockcroft, The dependence on Ca^{2+} of the guanine-nucleotide-activated polyphosphoinositide phosphodiesterase in neutrophil plasma membrane, Biochem. J. 240:503-507 (1986).

16. S. Cockcroft, J.A. Taylor, Fluoroaluminates mimic guanosine 5'[γ-thio]-triphosphate in activating the polyphosphoinositide phosphodiesterase of hepatocyte membranes, Biochem. J. 241:409-414 (1987).

17. D.B. Wilson, T.E. Bross, S.L. Hoffmann, P.W. Majerus, Hydrolysis of polyphosphoinositides by purified sheep seminal vesicle phospholipase C enzymes, J. Biol. Chem. 259:11718-11724 (1984).

18. R.M. Dawson, N. Freinkel, F.B. Jungalwala, N. Clarke, The enzymatic formation of myoinositol 1,2 cyclic phosphate from phosphatidylinositol, Biochem. J. 122:605-607 (1971).

19. P.W. Majerus, T.M. Connolly, H. Deckmyn, T.S. Ross, T. E. Bross, H.

Ishii, V.S. Bansal, D.B. Wilson, The metabolism of phosphoinositide devoid messenger molecules, Science 234:1519-1526 (1986).

20. J.A. Shayman, R.J. Auchus, A.R. Morrison, Bradykinin-induced changes in myo-inositol 1,2 (cyclic) phosphate in rabbit papillary collecting tubule cells, Biochem. Biophys. Acta. 888:171-175 (1986).

21. D.B. Wilson, T.E. Bross, W.R. Sherman, R.A. Berger, P.W. Majerus, Inositol cyclic phosphates are produced by cleavage of phosphatidylphosphoinositols (polyphosphoinositide) with purified sheep seminal vesicle phospholipase C enzymes, Proc. Natl. Acad. Sci. 82:4013-4017 (1985).

22. D.B. Wilson, T. Connolly, T.E. Bross, P.W. Majerus, W.R. Sherman, A. Tyler, L.J. Rubin, J.E. Brown, Isolation and characterization of the inositol cyclic phosphate products of polyphosphoinositide cleavage by phospholipase C, J. Biol. Chem. 260:13496-13581 (1985).

23. R.F. Irvine, A.J. Letcher, D.J. Lander, M.S. Berridge, Specificity of inositol phosphate-stimulated Ca^{2+} mobilization from Swiss mouse 3T3 cells, Biochem. J. 240:301-304 (1986).

24. F.A. O'Rourke, S.P. Halenda, G.B. Zavoico, M.B. Feinstein, Inositol 1,4,5 trisphosphate releases Ca^{2+} from a Ca^{2+}-transporting membrane vesicle fraction derived from human platelets, J. Biol. Chem. 260:956-962 (1985).

25. I.R. Batty, S.R. Nahorski, R.F. Irvine, Rapid formation of inositol 1,3,4,5 tetrakis phosphate following muscarinic receptor stimulation of rat cerebral cortical slices, Biochem. J. 232:211-215 (1985).

26. R.F. Irvine, A.J. Letcher, J.P. Heslop, M.J. Berridge, The inositol tris/tetrakisphosphate pathway - demonstration of $Ins(1,4,5)P_3$3-kinase activity in animal tissues, Nature (London) 320:631-634 (1986).

27. C.A. Hansener, S. Mah, J.R. Williamson, Formation and metabolism of inositol 1,3,4,5 tetrakisphosphate in liver, J. Biol. Chem. 261: 8100-8103 (1986).

28. R.F. Irvine, A.J. Letcher, D.J. Lander, J.P. Heslop, M.J. Berridge, Inositol(3,4) bisphosphate and inositol(1,3) bisphosphate in GH_4 cells - evidence for complex breakdown of inositol(1,3,4) bisphosphate, BBRC 143:353-359 (1987).

29. R.C. Inhorn, V.S. Bansal, P.W. Majerus, Pathway for inositol 1,3,4 trisphosphate and 1,4 bisphosphate metabolism, Proc. Natl. Acad. Sci. 84: 2170-2174 (1987).

30. C.D. Downes, M.C. Mussat, R.H. Michell, The inositol trisphosphate phosphomonoesterase of the human erythrocyte membrane, Biochem. J. 203: 169-177 (1982).

31. G.J. Tertoolen, B.C. Tilly, R.F. Irvine, W.H. Moolenaar, Electrophysio-
    logical responses to bradykinin and microinjected polyphosphates in
    neuroblastoma cells. Possible role of inositol 1,3,4 trisphosphate
    in altering membrane potential, FEBS Lett. 214:365-369 (1987).

32. R.F. Irvine and R.M. Moor, Microinjection of inositol 1,3,4,5 tetrakis-
    phosphate activates sea urchin eggs by a mechanism dependent in exter-
    nal Ca^{2+}, Biochem. J. 240:917-920 (1986).

33. M.D. Honsay, Egg activation unscrambles a potential role for IP_4,
    TIBS 12:133-134 (1987).

34. Y. Takai, U. Kikkawa, Y. Kaibuchi, Y. Nishizuka, Membrane phospholipid
    metabolism and signal transduction for protein phosphorylation, Adv.
    Cyclic Nucl. Protein Phos. Res. 18:119-158 (1984).

35. A.M. Speigel, Signal transduction by guanine nucleotide binding pro-
    teins, Molecular and Cellular Endocrinology 49:1-16 (1987).

36. M. Oinuma, T. Katuda, M. Ui, A new GTP-binding protein in differentiated
    Human Leukenic (HL-60) cells serving as the specific substrate of
    islet activating protein pertussis toxin, J. Biol. Chem. 262:8347-
    8353 (1987).

37. I. Magnaldo, H. Talwar, W.D. Anderson, J. Pouyssegur, Evidence for a
    GTP-binding protein coupling thrombin receptor to PIP_2-phospho-
    lipase C in membranes of hamster fibroblasts, FEBS Lett. 210:6-10
    (1987).

38. S. Cockcroft, The dependence on Ca^{2+} of the guanine nucleotide-
    activated polyphosphoinositide phosphodiesterase in neutrophil plasma
    membranes, Biochem. J. 240:503-507 (1986).

39. G.M. Bokoch and A.G. Gilman, Inhibition of receptor-mediated release of
    arachidonic acid by pertussis toxin, Cell 39:301-308 (1984).

40. P.C. Grenier, T.E. Rollins, W.L. Smith, Kinin induced prostaglandin
    synthesis by renal papillary collecting tubule cells in culture, Am.
    J. Physiol. F94-F104 (1981).

41. J.A. Shayman, K. Hruska, A.R. Morrison, Bradykinin stimulates increased
    intracellular calcium in papillary collecting tubules of the rabbit,
    Biochem. Biophys. Res. Comm. 134:299-306 (1986).

42. P.C. Isakson, A. Raz, S.E. Denny, A. Wyche, P. Needleman, Hormonal stimu-
    lation of arachidonate release from isolated perfused organs: rela-
    tionship to prostaglandin biosynthesis, Prostaglandins 14:853-871
    (1977).

43. J.A. Shayman, R.J. Auchas, A.R. Morrison, Bradykinin-induced changes in
    myoinositol 1,2(cyclic) phosphate in rabbit papillary collecting
    tubule cells, Biochem. Biophys. Acta. 888:171-175 (1986).

44. R.M. Zusman, J.R. Keiser, J.E. Handler, Vasopressin-stimulated prosta-

glandin E biosynthesis in the toad urinary bladder, J. Clin. Invest. 60:1339-1347 (1977).

45. J.E. Bisordi, D. Schlondorff, R.M. Hayes, Interaction of vasopressin and prostaglandins in the toad urinary bladder, J. Clin. Invest. 66: 1200-1210 (1980).

46. J.M. Forrest, C.J. Schneider, D.B. Goodman, Role of prostaglandin E_2 in mediating the effects of pH on the hydrosomotic response to vasopressin in the toad urinary bladder, J. Clin. Invest. 69:499-506 (1982).

47. R.M. Burch and P.V. Halushka, Vasopressin stimulates prostaglandin and thromboxane synthesis in toad bladder epithelial cells, Am. J. Physiol. 243:F593-F597 (1982).

48. M. Sato and M. Dunn, Interactions of vasopressin, prostaglandins, and cAMP in rat renal papillary collecting tubule cells in culture, Am. J. Physiol. 247:F423-F433 (1984).

49. A. Garcia-Perez and W.L. Smith, Use of monoclonal antibodies to isolate cortical collecting tubule cells: AVP induces PGE release, Am. J. Physiol. 244:C211-C220 (1983).

50. M. Kirschenbaum, A.G. Lower, W. Trizma, L.G. Fine, Regulation of vasopressin action by prostaglandins, J. Clin. Invest. 70:1193-1204 (1982).

51. B.M. Altura, Selective microvascular constrictor actions of some neurohypophyseal peptides, Eur. J. Pharmacol. 24:43-60 (1973).

52. B.M. Altura and B.T. Altura, Actions of vasopressin, oxytocin, and synthetic analogs on vascular smooth muscle, Fed. Proc. 43:80-86 (1984).

53. J. Grantham and J. Orloff, Effect of prostaglandin E_1 on the permeability response of the isolated collecting tubule to vasopressin, adenosine 3'-5'-monophosphate, and theophylline, J. Clin. Invest. 47:1154-1161 (1968).

54. S.Z. Katasic, J.T. Shepherd, P.M. Van Loutte, Vasopressin causes endothelium-dependent relaxations of the canine basilar artery, Circ. Res. 55:575-579 (1984).

55. T. Nabika, P.A. Velletri, W. Lovenberg, M. Beaven, Increase in cytosolic calcium and phosphoinositide metabolism induced by angiotensin II and [Arg]vasopressin in vascular smooth muscle cells, J. Biol. Chem. 260:4661-4670 (1985).

56. D. Rhodes, V. Prpic, J.H. Exton, P.F. Blackmore, Stimulation of phosphatidylinositol 4,5-bisphosphate hydrolysis in hepatocytes by vasopressin, J. Biol. Chem. 258:2770-2773 (1983).

57. J.R. Williamson, R.H. Cooper, K.J. Suresh, A.L. Thomas, Inositol trisphosphate and diacylglycerol as intracellular second messengers in

liver, Am. J. Physiol. 248:C203-C216 (1985).

58. S.N. Prescott and P.W. Majerus, Characterization of 1,2-diacylglycerol hydrolysis in human platelets, J. Biol. Chem. 258:764-769 (1983).

59. L.M. Hallacher and W.R. Sherman, The effects of lithium ion and other agents on the activity of myoinositol-1-phosphatase from bovine brain, J. Biol. Chem. 255:10896-10901 (1980).

60. L.R. Chase and G.D. Aurbach, Parathyroid function and renal excretion of 3'5' adenylic acid, Proc. Natl. Acad. Sci. USA 58:518-525 (1967).

61. D. Charbardes, M. Imbert, A. Clique, M. Montegut, F. Morel, PTH-sensitive adenylate cyclase activity of the rabbit nephron, Pflugers Arch. 354:229 (1975).

62. Z.S. Agus, L.B. Gardner, L.H. Beck, M. Goldberg, Effects of parathyroid hormone on renal tubular reabsorption of calcium, sodium and phosphate, Am. J. Physiol. 224:1143-1148 (1973).

63. K. Kurokawa, T. Ohno, H. Rasmussen, Ionic control of renal gluconeogenesis II. Effects of Ca^{2+} and H^+ upon response to parathyroid hormone and cyclic AMP, Biochim. Biophys. Acta. 313:32-41 (1973).

64. Z.S. Agus, J.B. Puschett, D. Senesky, M. Goldberg, Mode of action of parathyroid hormone and cyclic adenosine 3'5'-monophosphate on renal tubular phosphate reabsorption in the dog, J. Clin. Invest. 50:617-626 (1971).

65. M.R. Hammerman and K.A. Hruska, Cyclic AMP-dependent protein phosphorylation in canine renal brush-border membrane vesicles is associated with decreased phosphate transport, J. Biol. Chem. 257:992-999 (1982).

66. M.R. Hammerman, V.A. Hansen, J.J. Morrissey, Cyclic AMP-dependent protein phosphorylation and dephosphorylation alter phosphate transport in canine renal brush border vesicles, Biochim. Biophys. Acta. 755:10-16 (1983).

67. N. Yanagawa and O.D. Jo, Possible role of calcium mediators in parathyroid hormone action on phosphate transport in rabbit renal brush border membrane, BBRC 128:278-284 (1985).

68. N. Yanagawa and O.D. Jo, Possible role of calcium in parathyroid hormone actions in rabbit renal proximal tubules, Am. J. Physiol. 250:F942-F948 (1986).

69. T.D. McKinney and P. Myers, PTH inhibition of bicarbonate transport by proximal convoluted tubules, Am. J. Physiol. 239:F127-F134 (1980).

70. S. Sabatini, Parathyroid hormone inhibits water flow in isolated toad bladder, Am. J. Physiol. 250:F532-F538 (1986).

71. P.A. Mennes, J. Yates, S. Klahr, Effects of ionophore A23187 and external calcium concentrations on renal gluconeogenesis, Proc. Soc. Exp.

Med. 157:168-174 (1978).

72. N. Yanagawa, Cytosolic free calcium in isolated perfused rabbit proximal tubules: effect of parathyroid hormone (Abstract), Kidney Int. 31:361(A) (1987).

73. G.M. Dolson, M.K. Hise, E.J. Weinman, Relationship among parathyroid hormone, cAMP, and calcium on proximal tubule sodium transport, Am. J. Physiol. 249:F409-F416 (1985).

74. C. Kleeman, D. Yamaguchi, S. Muallem, Regulation of parathyroid hormone-activated calcium channel by phorbol ester, Kidney Int. 31:351(A) (1987).

75. A. Besarab and J.W. Swanson, Tachyphylaxis to PTH in the isolated perfused rat kidney: resistance of anticalciuria, Am. J. Physiol. 247: F240-F245 (1984).

76. P. Bidot-Lopez, R.V. Farese, M.A. Sabiro, Parathyroid hormone and adenosine 3',5' monophosphate acutely increases phospholipids of the phosphatidate-polyphosphoinositide pathway in rabbit kidney cortex tubules in vitro by a cycloheximide-sensitive process, Endocrinology 108:2078-2081 (1981).

77. V. Metzler, S. Weinreb, E. Bellorin-Font, K.A. Hruska, Parathyroid hormone stimulation of renal phosphoinositide metabolism is a cyclic nucleotide-independent effect, Biochim. Biophys. Acta. 712:258-267 (1982).

78. H. Lo, D.C. Lehotay, D. Katz, G.S. Levey, Parathyroid hormone-mediated incorporation of ^{32}P-orthophosphate into phosphatidic acid and phosphatidylinositol in renal cortical slices, Endocrin. Res. Commun. 3(Suppl. 6):377-385 (1976).

79. K.A. Hruska, D. Moskowitz, P. Esbrit, R. Civitelli, S. Westbrook, M. Huskey, Stimulation of inositol triphosphate and diacylglycerol production in renal tubular cells by parathyroid hormone, J. Clin. Invest. 79:230-239 (1987).

PROSTAGLANDINS, THROMBOXANE AND LEUKOTRIENES

IN THE CONTROL OF MESANGIAL FUNCTION

Paolo Mené , Michael S. Simonson and Michael J. Dunn

Departments of Medicine and Physiology,
Case Western Reserve University School of Medicine
Division of Nephrology, University Hospitals of Cleveland
Cleveland, Ohio 44106

1. INTRODUCTION

Mesangial cells are intercapillary cells of the kidney glomerulus, serving both smooth muscle and immune effector functions (1-3). Since the first report of contraction of cultured rat mesangial cells by the vaso-active peptides, angiotensin II (ANG II) and arginine vasopressin (AVP) (2,4), considerable interest focused on the hypothesis that mesangial cells may regulate glomerular blood flow and filtration in synergism with afferent and efferent resistances, by altering capillary surface area and hence the ultrafiltration coefficient, K_f (5,6). Changes of capillary surface area might result from mechanical stretching and constriction of vessel walls by surrounding mesangial cells or intraglomerular shunting of blood flow as a result of segmental occlusion of certain loops. Although no definitive evidence of such function has been provided thus far, a number of observations link the mesangial cell to a regulatory role on glomerular hemodynamics, including reports of selective changes of K_f upon induction of glomerular immune injury (7) or infusion of vasoactive agents (8-10). Further support to the concept of mechanical properties of the mesangium comes from studies in freshly isolated rat and human glomer-uli, which are rapidly contracted by ANG II, as evaluated by different techniques (11,12). Inasmuch as mesangial cells are the most prominent contractile cell type of the glomerulus, it is reasonable to infer that mesangial contraction mediates volumetric changes of glomeruli in vitro

(6,13). Similar effects were also observed when analogues of the vasoconstrictor metabolite of arachidonic acid (AA), thromboxane A_2 (TxA_2)
(6,11,14), were added to glomerular suspensions, thereby prompting us to
hypothesize that impairment of renal hemodynamics following acute intra-
arterial administration of these compounds in experimental animals (15,16)
might be partially mediated by effects on the glomerular mesangium.

TxA_2 is a major product of platelet and leukocyte cyclooxygenation
of AA (17-19), released at the site of interaction with damaged capillary
walls or during inflammatory reactions such as glomerular immune injury
(20), ureteral obstruction (21), renal allograft rejection (22), and lupus
nephritis in man (23). A negative hemodynamic effect of TxA_2 has been
demonstrated in these disease models by improvement of renal function
after administration of various TxA_2-synthetase inhibitors or receptor
antagonists. A possible additional source of increased TxA_2 is local
synthesis by glomerular epithelial and mesangial cells, which also metabolize AA into TxA_2, albeit in relatively minor amounts as compared with
other prostaglandins (PG) such as PGE_2 and $PGF_{2\alpha}$ (24,25).

Leukotrienes (LT)C_4 and D_4 and various other lipoxygenase metabolites of AA are also potent constrictors of vascular smooth muscle and
chemotactic compounds (26,27), abundantly produced by leukocytes and possibly resident cells in various forms of infiltrative glomerulonephritis
(28,29). Glomerular cells are a potential target for the contractile
effects of LT, as they express specific binding sites for LTC_4 (30), and
mesangial contraction may contribute to the marked renal hemodynamic effects of intravenous infusion in the rat (31,32).

We therefore investigated the actions of TxA_2, $PGF_{2\alpha}$ and the sulfidopeptide leukotrienes on contraction of cultured rat mesangial cells,
simultaneously assessing effects of other major metabolites of AA such as
PGE_2 and PGI_2 or prostacyclin, a product of endothelial cells (33),
that display a functional antagonism of TxA_2 in other systems (34,35).
Specificity of responses to the various prostanoids as well as their putative mechanism(s) of action on mesangial cells were also analyzed by
monitoring the generation of intracellular second messengers that have
been implicated in the control of smooth muscle contractility, such as
inositol(poly)phosphates, cytosolic calcium and cyclic nucleotides (36,
37,38). Finally, pharmacologic manipulation with cyclooxygenase inhibitors was used in order to assess the functional relevance of endogenous
mesangial AA metabolism modulated by contractile stimuli.

2. MESANGIAL CELL CONTRACTION/RELAXATION

2.1. Effects of Endoperoxide Analogues, TxA$_2$ and PGF$_{2\alpha}$ on Mesangial Cell Contractility.

Multiple independent rat mesangial cell lines established from glomerular explants were used for morphometric experiments in passages 2 through
7. Mesangial origin and purity of the cultures were assessed by morphological and biochemical criteria as described in detail elsewhere (39,40).
Contractility was studied morphometrically by computer enhancement of
serial phase-contrast photomicrographs of the cells at 200 x. Cross-
sectional area (CSA) profiles for each individual cell were averaged to
obtain the time course of contraction in responding cells, defined as
cells undergoing a reduction CSA >8% of the average of 3 basal pictures
following addition of the agonist to be studied. This figure was obtained
by preliminary analysis of technical error and fluctuations of cell size
in resting cells during control incubations (39,40).

Increasing concentrations of the stable PGG$_2$/PGH$_2$/TxA$_2$ analogues
U-44069 (9,11-dideoxy-9α, 11α-epoxymethano-PGF$_{2\alpha}$) and U-46619 (9,11-
dideoxy-11α, 9α-epoxymethano-PGF$_{2\alpha}$, Upjohn, Kalamazoo, MI) stimulated a
progressively larger number of cultured rat mesangial cells to contract,
with a maximum of approximately 35% of the cells at 1 uM (Fig.1). In control experiments less than 8% of the cells displayed a detectable

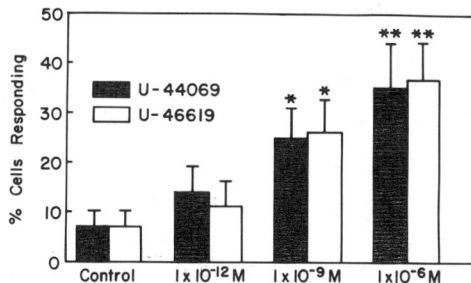

Fig. 1. Effects of increasing concentration of the thromboxane A$_2$/
endoperoxide analogues U-44069
and U-46619 on contraction of
cultured rat mesangial cells.
Percent cells contracting $\pm$ SE,
n=4-6 experiments each group.
* p <0.05, ** p <0.01 vs. controls, chi square tests. (From
Ref. 40, with permission.)

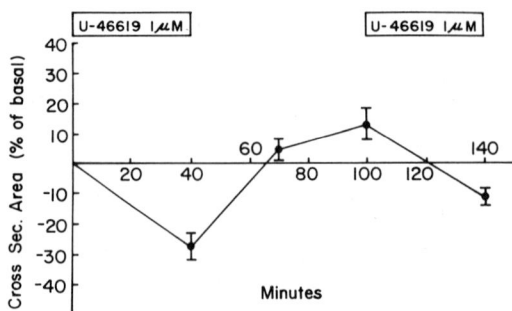

Fig. 2. Time course of cross-sectional
area (CSA) changes of cultured rat
mesangial cells, repeated 40-min
incubations with 1 uM U-46619.
Cells were rinsed and incubated
for 60 min at 37°C in RPMI 1640
medium without agonist following
the first incubation. Mean CSA
changes ± SE from basal and end of
second incubation, respectively,
n=7 experiments. (From Ref. 40,
with permission.)

reduction of CSA during 40 min incubations with the vehicle of the TxA_2
analogues (ethanol, final concentration < 0.01%). Contraction was time-
dependent, with detectable reductions of CSA as early as 5 min. after the
addition of each agonist, and a linear progression for 30 to 40 min., when
CSA stabilized or spontaneously increased in certain cells. The process
was fully reversible upon removal of the agonist(s), as summarized in the
experiments (Fig. 2), in which replacement of the medium at the end of 40
min. incubations with U-46619 was followed by a rapid return to prestimulation size of all responding cells. When U-46619 was added again 60 min.
later, the same cells that contracted initially underwent an almost identical reduction of CSA, indicating that morphologic changes are a specific
response to the agonists and the cells are not injured by exposure to vasoconstrictor prostanoids. Moreover, viability studies in trypsinized cells
showed that 87% of agonist-treated cells excluded Trypan Blue at the end
of 30 min. incubations, as opposed to 85.2% in control experiments, while
no staining was detectable on coverslips at the end of the incubations
with U-46619. All studies were performed at room temperature, but in selected experiments at 37°C the contractile responses were enhanced (not
shown). The effects of U-44069 and U-46619 were also significantly
blunted by the structurally different TxA_2-receptor antagonists EP-092
(a gift from Dr. R.L. Jones, University of Edinburgh, Scotland) (41,42)

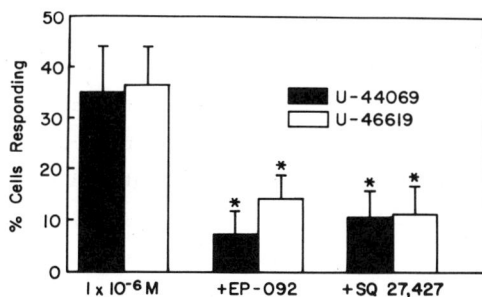

Fig. 3. Effects of the thromboxane
A$_2$-receptor antagonists EP
092 and SQ 27,427 (10 uM) on
U-44069/U-46619 (1 uM)-induced
contraction of cultured rat mesangial cells. Percent cells
contracting ± SE, n=3-5 experiments each group. * p <0.05
vs. agonists alone, chi-square
tests. (From Ref. 40, with
permission.)

and SQ 27,427 (a gift of Squibb, Princeton, NJ) (43), added 10 min. prior
to stimulation of the cells at a 10-fold molar excess of the agonists
(Fig. 3). The TxA$_2$-receptor antagonists had no intrinsic agonist activity in control incubations without endoperoxide analogues.

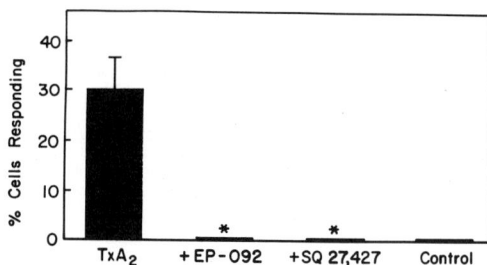

Fig. 4. Effects of thromboxane A$_2$
(TxA$_2$) synthesized by sheep
platelet microsomes incubated
30 sec. with arachidonic acid
on contraction of cultured rat
mesangial cells, alone or in
the presence of the TxA$_2$-re-
ceptor antagonists EP 092 and
SQ 27,427. Percent cells contracting ± SE, n=5 experiments
each group. * p <0.01 vs.
TxA$_2$, chi-square tests.
(From Ref. 40, with permission.)

171

Experiments with authentic TxA$_2$ generated by sheep platelet microsomes preincubated with AA (44,45) yielded results similar to the endoperoxide analogues, with approximately 30% of the cells contracted by concentrations of the agonist in the nanomolar range, as evaluated by measurement of total immunoreactive TxB$_2$ generation at the end of each experiment (Fig. 4). It should be noted that the initial concentration of
TxA$_2$ to which the cells were exposed is unknown, but it is conceivable
that 10-fold dilution of the microsomes/AA mixture at the time of addition
to the cells reduced TxA$_2$ synthesis, as 10 min. preincubation of the
same batch of microsomes yielded 2.42 nM TxB$_2$, compared with an average
recovery of 0.77±0.08 nM in the contraction experiment. TxB$_2$ had no
effect on contraction in 4 experiments at 1 nM and 1 uM. Similar to the
experiments with endoperoxide analogues, TxA$_2$-receptor antagonists completely blocked responses to the platelet microsomes, once again without
displaying any agonist activity even when used at concentrations 1,000-
fold in excess of TxA$_2$. A possible explanation for the different
effects of TxA$_2$-receptor antagonists on platelet microsome or endoperoxide analogue-induced mesangial contraction, that is, complete inhibition
of responses to TxA$_2$ (Fig. 4) and 7.3 to 14.2% residual responses to
U-44069 or U-46619 (Fig. 3), is binding of the latter compounds to receptors for contractile eicosanoids other than TxA$_2$. One such site could
be a receptor for PGF$_{2\alpha}$, which in separate experiments also stimulated
mesangial cell contraction (Fig. 5). The effects of PGF$_{2\alpha}$ were

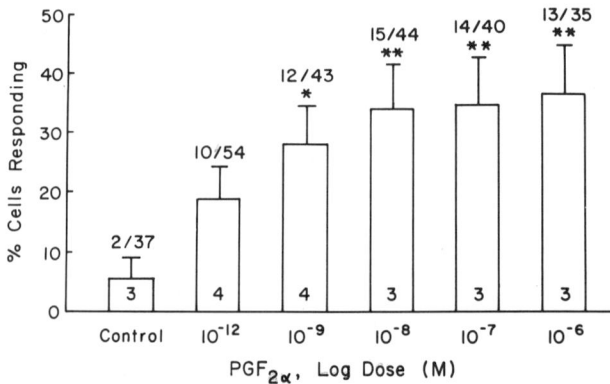

Fig. 5. Effects of PGF$_{2\alpha}$ on contraction of cultured rat mesangial cells. Percent
cells contracting ± SE. On top and bot-
tom of bars, number of cells contracting
out of total analyzed and number of experiments, respectively. * p <0.05, **
p <0.01 vs. control, chi-square tests.

detectable over a wide dose range, with a progressively larger fraction of
the cells contracting in response to increasing concentrations. Maximal
responses were obtained at 10 nM, with a time course and CSA reductions
similar to those seen with the endoperoxide analogues.

Immunoreactive PGE_2 synthesis in confluent monolayers was stimulated
by the endoperoxide analogues, with a plateau at 1 nM and significant increases still detectable at 1 uM (U-44069 100.9% and 47.9%, U-46619 141.6%
and 50.7% above basal synthesis during 15 min. incubations 1 nM and 1 uM,
respectively).

These experiments provide evidence that vasoconstrictor prostanoids,
similar to ANG II, AVP and other structurally unrelated compounds (2,4,
46,47,48,49,50), stimulate rapid, reversible contraction of the cells,
thereby suggesting that their effects on glomerular volume (11,14) are
probably mediated by the mechanical properties of the mesangium. The
apparent lack of responses in a large number of the cells studied is a
common finding in our and others' studies, and has been extensively discussed elsewhere (1,39,40,50,51). It should be remembered that, while
vascular smooth muscle cells do not generally retain contractility in
culture (52,53), mesangial cells display stable and reproducible responses
to agonists in early passages, that are rate-limited by adhesion to non-
flexible culture substrates. It is generally recognized that mechanical
resistances to contraction of adherent cultured cells may result in iso-
metric responses, that is, interaction of contractile proteins without
shortening of the cells (1,51). Work with mesangial cells grown of flexible surfaces, such as silicone rubber (50) or poly-HEMA (1,51) clearly
indicates agonist-induced contraction in a vast majority of the cells, as
expressed by either measurable surface changes or "wrinkling" of the sub-
strata, similar to earlier reports in fibroblasts by Harris et al. (54).
The dose-response relationship between concentration of the drug and
number of cells undergoing significant dimensional changes in our studies
may therefore be the result of increasing mechanical forces progressively
overcoming adhesion to plastic or glass (39,40).

Alternatively, uneven contraction may indicate variable expression of
receptors for the specific agent tested in a multiclonal population, subsets of cells with different functional characteristics (e.g., a modified
smooth muscle type as opposed to resident phagocytes), or defective pharma-
comechanical coupling with loss of contractile properties in vitro. It is
quite possible that subgroups of mesangial cells can be identified with

respect to the expression of receptors, an issue that will be addressed when techniques for the detection of biochemical responses or intracellular signals in single cells become available. Our experiments with non-cloned cells undoubtedly lend themselves to a certain variability, even though immunocytochemistry, electron microscopy and antigen markers all indicate substantial homogeneity of the populations studied. On the other hand, a major advantage of this approach, as opposed to work on cloned cells, is the maintenance of in vivo characteristics of the cells and a lower incidence of de-differentiation or uncontrolled transformation in culture. The possibility that certain functional responses or the expression of receptors may be related to phases of the cell cycle must also be kept in mind. Future work should also address the issue as to whether contractility is expressed selectively by cells in a certain phase of the cycle, therefore resulting in an irregular distribution of contractile cells in asynchronous populations. Cloned cells may display a more homogeneous behavior because of rapid and synchronized growth kinetics. Additional modulatory influences related to endogenous synthesis of relaxant AA metabolites will be discussed extensively in the following sections.

The experiments with selective receptor antagonists, vital staining of the cells and the lack of effect of the inactive metabolite TxB_2 rule out cell toxicity or a nonspecific effect of polyunsaturated fatty acids as an explanation for the dimensional changes observed in response to U-44069 and U-46619. No differences in potency between the two compounds and a similar degree of inhibition by TxA_2-receptor antagonists were observed, consistent with marked structural homology and evidence for binding of endoperoxide analogues to a TxA_2 receptor in platelets and smooth muscle (55). Continuous occupancy of this putative binding site seems to be required for sustained contraction, since removal of U-46619 after initial stimulation was accompanied by prompt return of the cells to basal shape and size. These considerations may also apply to native TxA_2, which, despite its unstable nature, may be "protected" from hydrolysis when complexed with a receptor molecule.

2.2. Effects of Sulfidopeptide Leukotrienes on Mesangial Cell Contractility

The sulfidopeptide LT C_4 and D_4, but not B_4, (gifts of Dr. J. Gleason of Smith, Kline and French Laboratories, Philadelphia, PA) had potent contractile effects on cultured rat mesangial cells, with maximal activity at nanomolar concentrations and an apparent EC_{50} in the picomolar range (Fig. 6, panel A). Interestingly, the maximum % of responding

174

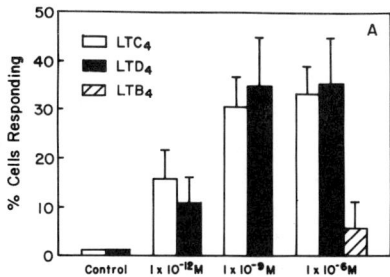

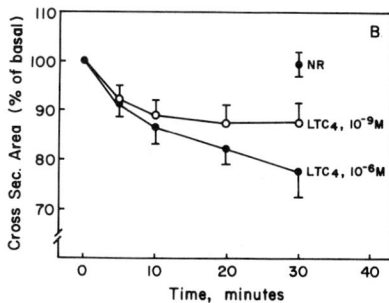

Fig. 6. Effects of leukotrienes C_4, D_4 and B_4 on contraction of cultured rat mesangial cells. Panel A, percent cells contracting $\pm$ SE. Panel B, time course of contraction, 1 nM/1 uM LTC_4, mean reductions of cross-sectional area $\pm$ SE vs. basal, n=5-6 experiments each group. NR, non-responding cells. (From Transactions of the American Clinical and Climatological Association, 98:71-79, 1987, with permission.)

cells was virtually identical to TxA_2, albeit such response was achieved at lower concentrations. Higher doses of LTC_4/LTD_4 failed to stimulate contraction in a larger fraction of the cells, but increased significantly the average CSA reductions in responding cells, as shown in Fig. 6, panel B. Similar to the dose-response curve to TxA_2 and its stable analogues, reductions of CSA in excess of 8% of the average of 3 basal determinations were already detectable within 5 min. of addition of both LTC_4 and LTD_4. Contraction induced by 1 nM LTD_4 was blocked by a large molar excess of the receptor antagonist, 4R,5S,6Z-2-nor-LTD_1 (Smith, Kline and French) (65.6% inhibition of response at 10 uM nor-LTD_1), without significant effects on the CSA changes of responding cells. Unlike ANG II, AVP or U-44069/U-46619, LTC_4 did not stimulate PGE_2 synthesis at concentrations between 1 pM and 1 uM. This observation is in agreement with the data of Barnett et al. (56), who also demonstrated the contractile activity of LTC_4 and D_4 on isolated rat glomeruli and mesangial cells. Similar to the experiments with TxA_2 analogues, blockade of LTD_4 receptors was not associated with reduced contractile forces in the fewer responding cells, as evaluated by CSA reductions. This suggests that once a threshold for contraction is attained, the process is rather independent of the number of receptors occupied, or that mechanical resistances offset any modulation of responses, shifting the cut-off point for isotonic changes (8% of initial CSA in our studies) to a smaller fraction of the cells.

175

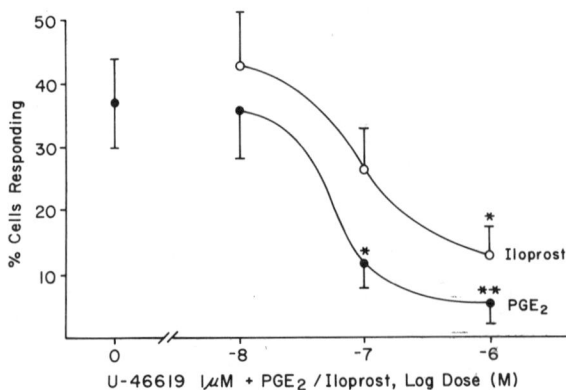

Fig. 7. Inhibition of U-46619-induced contraction of cultured rat mesangial cells
by PGE$_2$ and Iloprost. Percent
cells contracting ± SE. * p <0.05,
** p <0.001 vs. U-46619 alone, chi-
square tests. (From Circulation
Research 62:916-925, 1988, with permission.)

2.3. Modulatory Effects of Vasodilator PG and AA

Structurally unrelated contractile agents stimulate AA release and
PGE$_2$ synthesis in cultured rat mesangial cells, with the only notable
exception of LT, as previously discussed (25,46,56). We therefore evaluated whether this PGE$_2$ response serves any modulatory function on contraction induced by U-46619 and ANG II, by preincubating the cells with
increasing concentrations of PGE$_2$ for 10 min. prior to addition of
U-46619. The incubations were performed, as previously described, in the
continuous presence of PGE$_2$. PGE$_2$ dose-dependently inhibited U-46619-
induced contraction, an effect also shared by the PGI$_2$ analogue,
Iloprost, in the same range of concentrations (Fig. 7). PGE$_2$ did not
stimulate significant contraction by itself (7.2% of the cells during 40
min. incubations at 1 uM, vs. 5.4% in control experiments with the vehicle). Occasionally, PGE$_2$ and Iloprost triggered a "shape change" of the
cells, with the appearance of thin, tapered peripheral processes, accentu-
ating the stellate morphology of the cells. This observation is consistent with earlier reports by Kreisberg et al., who studied the effects of
cyclic adenosine monophosphate (cAMP) stimulation in these cells (1,57).
The surface changes did not markedly affect CSA, and occurred in a majority of the cells studied in a given experiment. The inhibitory effect of
PGE$_2$ on contraction was also evident when ANG II was used as an agonist,

176

as in the experiments described in Fig. 8. ANG II alone stimulated contraction of approximately 40% of the cells, whereas preincubation for 10 min. with 1 uM PGE_2 prior to addition of ANG II resulted in marked inhibition of contraction. If exogeneous PGE_2 has a modulatory effect on agonist-induced contraction, we reasoned that addition of AA as an excess substrate for cyclooxygenase should mimic the effects of added PGE_2 through stimulation of mesangial synthesis of the eicosanoid (25). 10 uM AA did, in fact, inhibit U-46619-induced contraction to 15.9% vs. 37.2% of the cells in control experiments (not shown). The number of responding cells, albeit significantly reduced in comparison with U-46619 alone, was larger than in experiments with 1 uM PGE_2, in which only 5% of the cells were contracted. A possible explanation is the simultaneous conversion of AA to other prostanoids with agonist activity on contraction, e.g., $PGF_{2\alpha}$ and TxA_2, the former particularly being produced in relatively larger amounts by cultured rat mesangial cells (25).

The observation that both PGE_2 and PGI_2 have inhibitory actions on contraction is in agreement with earlier data by Foidart and Mahieu in

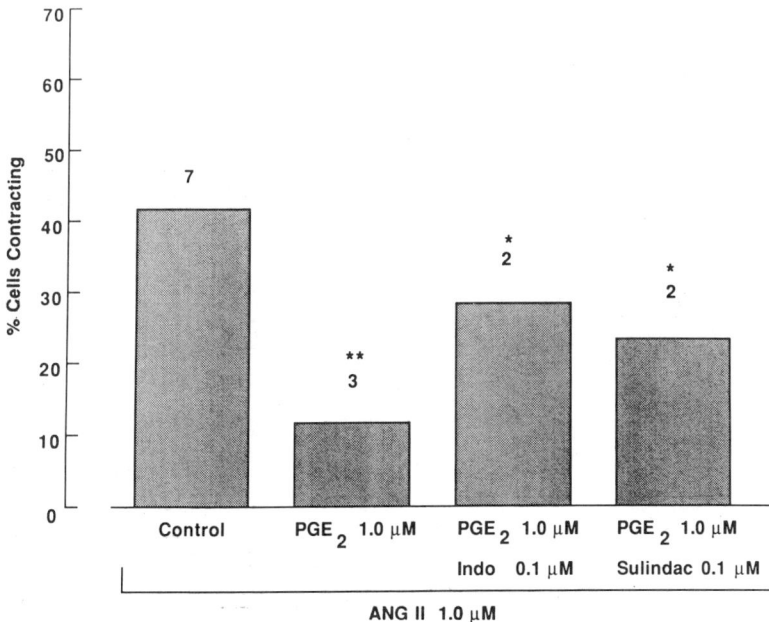

Fig. 8. Effects of PGE_2 and nonsteroidal antiinflammatory drugs on angiotensin II (ANG)-induced contraction of cultured rat mesangial cells. Percent cells contracting, n of experiments on top of bars. Indo, indomethacin. * p <0.05, ** p <0.001 vs. ANG II alone, chi-square tests. (From Ref. 59, with permission.)

confluent cultures of rat mesangial cells (58), and suggests that contractility in vivo may be under the tonic influence of endogenous PG synthesis. It is of interest that these cells both release and respond to eicosanoids, with the potential for autacrine control of cell function. Additionally, the mesangium may be a target for prostanoids with opposite biologic actions, locally released by infiltrating blood-borne cells or by other resident cells of the glomerulus.

2.4. Pharmacologic Manipulation of Mesangial Cell Contractility: Effects of NSAID

As exogenously added PGE_2 and Iloprost potently modulate contractile responses in vitro, we asked whether basal or stimulated endogenous production of PGE_2 has any influence on contractility, and particularly whether partial responses--that is, only approximately one-third of the cells contracted by a given agonist--could be explained, at least in part, by an internal negative feed-back mechanism. We therefore pretreated mesangial cells with one of five NSAID that inhibit cyclooxygenase,

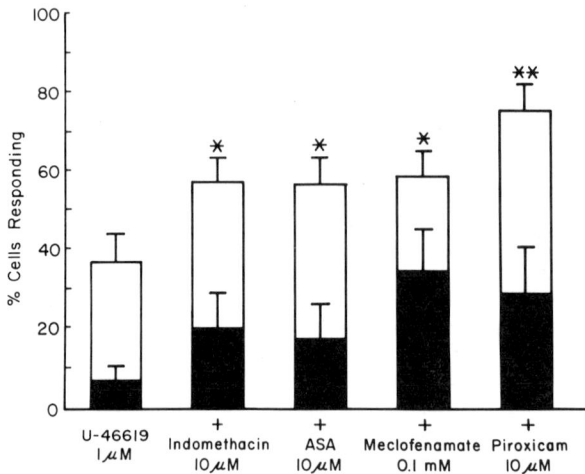

Fig. 9. Enhancement of U-46619-induced contraction of cultured rat mesangial cells by nonsteroidal antiinflammatory drugs (NSAID). Inset solid bars, control experiments with NSAID alone and/or vehicle of U-46619. Percent cells contracting $\pm$ SE. * $p < 0.05$, ** $p < 0.01$ vs. U-46619 alone, chi-square tests. ASA, acetylsalicylic acid. (From Circulation Research 62:916-925, 1988, with permission.)

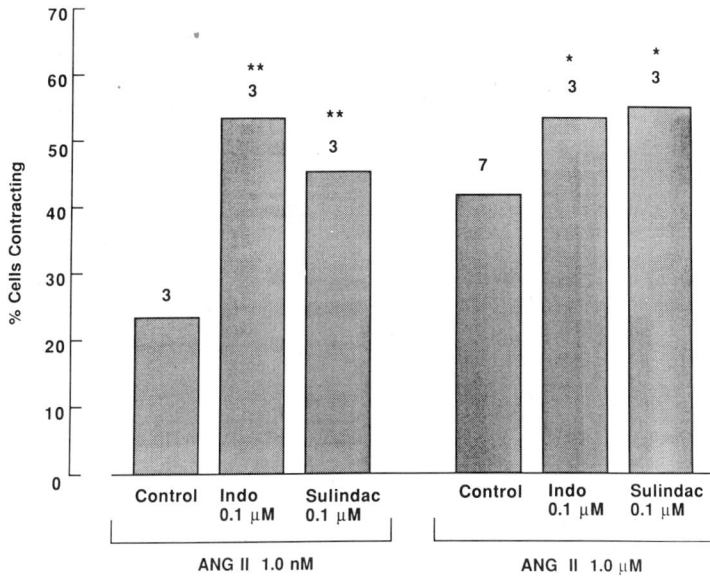

Fig. 10. Effects of nonsteroidal antiinflammatory drugs
on angiotensin II (ANG II)-induced contraction
of cultured rat mesangial cells. Percent cells
contracting, n of experiments on top of bars.
Indo, indomethacin. * p <0.05, ** p <0.01 vs.
ANG II alone, chi-square tests. (From Ref. 59,
with permission.)

thereby blocking PG synthesis. As indicated in Fig. 9 for U-44619 and in
Fig. 8 and 10 for ANG II, indomethacin, meclofenamate, ASA, sulindac and
piroxicam markedly enhanced contractile responses, as evaluated by either
number of cells responding or degree of CSA reduction (not shown). Notably, NSAID also amplified spontaneous contraction, suggesting that the
resting morphology of mesangial cells in culture is dependent on local
generation of PGE_2, consistent with its suggested role of "autacoid", a
compound whose receptors are in close proximity to the site of
production. The use of multiple, structurally different drugs in these
experiments ruled out a pharmacologic effect on biochemical pathways
independent of cyclooxygenase. The effects of each NSAID were tightly
linked to inhibition of immunoreactive PGE_2 production, as evaluated in
parallel dose-response biochemical studies. No signs of cytotoxicity were
evidenced by Trypan Blue viability assays at the concentrations of NSAID
used, chosen on the basis of effective inhibition of basal and AA-stimu-
lated PGE_2 synthesis.

Taken together with the effects of exogenous vasodilator PG, these
experiments indicate that contraction in vitro is regulated by endogenous

AA metabolism, in agreement with earlier reports by Foidart et al. (58) and Dunn et al. (13,59), underscoring a possible physiological role of PG in the modulation of mesangial and smooth muscle reactivity in the kidney. Our data with structurally unrelated stimuli such as TxA_2 and ANG II, and a study by Schlondorff et al. with platelet activating factor (46), point to a general anticontractile effect of endogenous and exogenous PG of the E and I series, which is independent of the nature of the agonist used.

PG inhibition in patients with chronic glomerular disease reduces renal perfusion and glomerular filtration (60-63), as well as proteinuria (64). It has been suggested that in these settings glomerular synthesis of PGE_2 and PGI_2 may serve a compensatory function, counteracting local vasoconstriction by ANG II or catecholamines and maintaining a relative increase of blood flow in residual nephrons to support ultrafiltration (62,63). Unopposed action of vasoconstrictors on vascular smooth muscle and, possibly, enhanced mesangial contraction may provide a basis for the understanding of this renal effect of cyclooxygenase inhibitors. It remains to be established whether this property of NSAID can be used therapeutically to prevent the postulated intraglomerular damage that follows compensatory vasodilation and hypertension (64,65).

3. INTRACELLULAR SIGNALS FOR MESANGIAL CONTRACTION/RELAXATION

3.1. <u>Effects of PG on $[Ca^{2+}]_i$ and Inositol Phosphates</u>

In an effort to elucidate the mechanisms of action of eicosanoids in mesangial cells, we directed our attention to putative intracellular "second messengers." Because TxA_2 and $PGF_{2\alpha}$ increase free cytosolic Ca^{2+} ($[Ca^{2+}]_i$) in platelets and vascular smooth muscle (66,67), respectively, and this event is followed by contractile and secretory responses in these cells, we studied the effects of various prostanoids on $[Ca^{2+}]_i$ of cultured rat mesangial cells, as measured by fluorometry (68) in confluent monolayers loaded with the fluorescent, Ca^{2+}-sensitive probe, fura-2 (69,70). $PGF_{2\alpha}$ and U-46619 triggered a rapid, dose-dependent rise of $[Ca^{2+}]_i$, followed by a reduction towards prestimulation levels, with a residual, sustained evaluation for >10 min. (Fig. 11). Of all the other eicosanoids tested, only PGE_2 had some weak but reproducible effects on $[Ca^{2+}]_i$ at concentrations above 10 nM. PGE_1 and the PGI_2 analogue, Iloprost, had no effects upon $[Ca^{2+}]_i$, even at

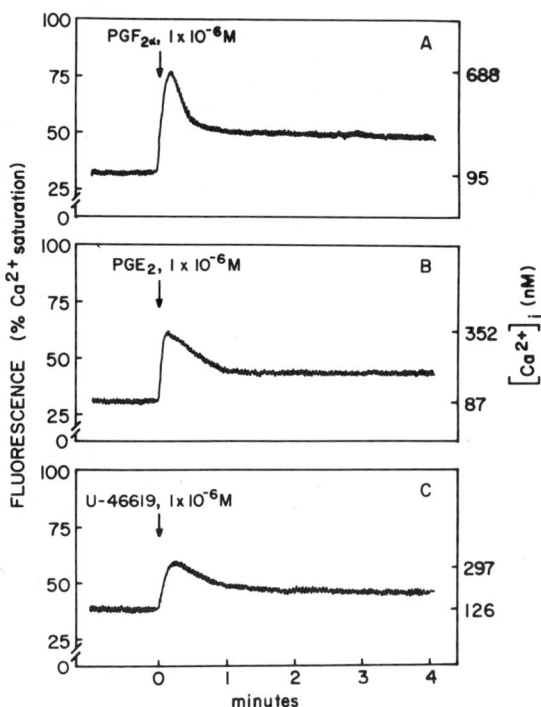

Fig. 11. Stimulation of free cytosolic
calcium ($[Ca^{2+}]_i$) by $PGF_{2\alpha}$,
PGE_2 and U-46619 in fura 2-
loaded monolayers of cultured rat
mesangial cells, representative
experiments. (From Ref. 70, with
permission.)

concentrations as high as 10 uM. $PGF_{2\alpha}$ displayed greater potency on a
molar basis and lower threshold than PGE_2 and U-46619, with responses
peaking at 10 uM. The rapid phase of the $[Ca^{2+}]_i$ transient was not
modified by chelation of extracellular Ca^{2+} with EGTA 3 mM, while the
long-lasting, tonic phase of responses was suppressed, with immediate
return of $[Ca^{2+}]_i$ to basal levels. Sequential additions of the same
agonist at increasing concentrations resulted in graded responses. On the
other hand, pretreatment with a maximal concentration of any PG desensi-
tized responses to a subsequent addition of the same or other PG and to
chemically unrelated stimuli, such as ANG II or AVP, for >10 min. (70).
The specificity of $[Ca^{2+}]_i$ responses to each prostanoid was therefore
evaluated using TxA_2-receptor antagonists (as shown in Fig. 12) and
Iloprost, whose effects on mesangial cells are not mediated by
$[Ca^{2+}]_i$. Pretreatment of the monolayers for 3 min. with SQ 27,427 and
(not shown) EP 092 at a 10-fold molar excess of the agonist inhibited

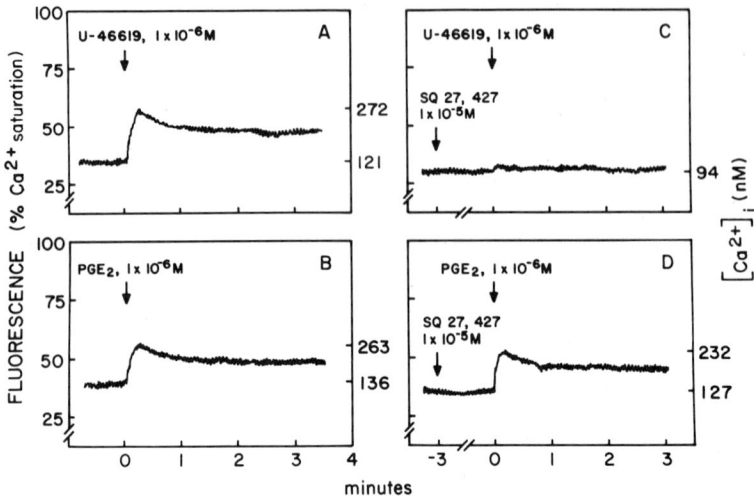

Fig. 12. Selective inhibition of U-46619-induced stimulation of free cytosolic calcium ($[Ca^{2+}]_i$) in fura 2-loaded monolayers of cultured rat mesangial cells by the thromboxane A_2-receptor antagonist SQ 27,427 (panels A and C). SQ 27,427 had no effects on responses to PGE_2 (panels B and D). (From Ref. 70, with permission.)

selectively the responses to U-46619 without affecting $PGF_{2\alpha}$ or PGE_2-induced $[Ca^{2+}]_i$ changes, consistent with the presence of a receptor for TxA_2 distinct from $PGF_{2\alpha}$ and PGE_2 binding sites. Iloprost, on the other hand, did not affect responses to $PGF_{2\alpha}$, PGE_2 or U-46619, even when used at a 10-fold molar excess, thereby ruling out the existence of a common binding site or a nonspecific effect of unsaturated fatty acids on mesangial cells.

The effects of $PGF_{2\alpha}$, PGE_2 and U-46619 on $[Ca^{2+}]_i$ were accompanied by dose-dependent accumulation of total inositol phosphates, measured after anion exchange chromatography in perchlorate-extracted confluent monolayers prelabeled for 36 hours with myo-[2-^{3}H(N)]inositol in serum-free, inositol-free RPMI 1640 medium (Table) (70). Together with the $[Ca^{2+}]_i$ studies, these results further confirm the specificity of the effects of TxA_2 and $PGF_{2\alpha}$, and are consistent with extensive observations that an elevation of $[Ca^{2+}]_i$ is an initial event in smooth muscle contraction (71,72). $PGF_{2\alpha}$ is the most potent PG stimulating $[Ca^{2+}]_i$ on a molar basis, although the magnitude of peak elevations at saturating concentrations is somewhat lower than with ANG II and AVP in our studies and earlier reports in mesangial cells (73,74). Responses to

Table. Effects of PG on Total Inositol Phosphates in
         Cultured Rat Mesangial Cells

	Conditions	Total inositol phosphates, cpm/well
Basal		$5,417 \pm 296$
$PGF_{2\alpha}$	10^{-8} M	$24,520 \pm 1,571$
	10^{-7} M	$27,704 \pm 2,178$
	10^{-6} M	$36,477 \pm 475$
PGE_2	10^{-7} M	$11,221 \pm 154$
	10^{-6} M	$36,151 \pm 3,664$
	10^{-5} M	$50,308 \pm 4,459$
U-46619	10^{-7} M	$5,637 \pm 357$
	10^{-6} M	$8,537 \pm 663$
	10^{-5} M	$21,099 \pm 1,854$

Values are mean $\pm$ SE of triplicate determinations, representative experiments. 15-min. incubations in the presence
of 10 mM LiCl, anion exchange chromatography of perchlorate-
extracted cells prelabeled for 36 hours with [^{3}H]-myoinosi-
tol (4 uCi/ml). (From Ref. 70, with permission.)

TxA$_2$ analogues and PGE$_2$ are weaker, although apparently specific, as
indicated by studies with selective TxA$_2$-receptor antagonists. The elevation of [Ca^{2+}]$_i$ is not a generic effect of unsaturated fatty acids,
since PGE$_1$ and Iloprost, which share structural similarities with
PGF$_{2\alpha}$, TxA$_2$ and PGE$_2$, are devoid of any agonist activity on
[Ca^{2+}]$_i$ while potently stimulating cAMP accumulation. Similarly, the
effect on [Ca^{2+}]$_i$ cannot be explained by ionophoretic properties of
the eicosanoids (75), since it also occurred in the virtual absence of
free extracellular Ca^{2+}, following chelation with EGTA. Simultaneous,
dose-dependent accumulation of total water-soluble inositol phosphates in
response to PGF$_{2\alpha}$, TxA$_2$ and PGE$_2$ suggests activation of phospholipase C (PLC), with breakdown of (poly)phosphoinositides. Phospha-
tidylinositol-4, 5-bisphosphate (PIP$_2$) is a membrane-associated phospholipid of the inositol phosphate cycle that is specifically acted upon by
PLC, thereby yielding inositol trisphosphate (IP$_3$) and diacylglycerol
(76,77). IP$_3$ is a mixture of two isomers, one of which, inositol
(1,4,5)-trisphosphate, is capable of releasing Ca^{2+} from intracellular
stores, when introduced in various cell types upon permeabilization (72,
78), and is therefore likely to mediate the initial, transient elevation
of [Ca^{2+}]$_i$ (76). Rapid return to near-basal [Ca^{2+}]$_i$ is believed
to result from homeostatic mechanisms, including Ca^{2+} extrusion by a
Ca^{2+}/H$^+$ ATPase or sequestration into cytoplasmic organelles (36,79).
Resetting of prestimulation levels to a slightly elevated steady state is

common in many cell types following stimulation of PLC (36), and may be critical to sustained functional responses, such as tonic contraction of mesangial cells in our studies. Removal of extracellular Ca^{2+} abolishes this long-term effect, suggesting its dependency on enhanced Ca^{2+} entry, possibly through divalent ion channels regulated by another metabolite of the inositol phosphate cycle, inositol (1,3,4,5)-tetrakisphosphate (80, 81). Increased $[Ca^{2+}]_i$ is believed to control cell function(s) through regulation of multiple Ca^{2+}-dependent enzymes, mediated by the Ca^{2+} binding protein, calmodulin (36). PG stimulation by agonists that increase $[Ca^{2+}]_i$ is the result of secondary, Ca^{2+}-dependent activation of phospholipase A_2 or diacylglycerol lipases that release AA from diacylglycerol. Free AA is subsequently metabolized to PG or TxA_2 by cyclooxygenase and specific synthetases. The Ca^{2+}-calmodulin complex has been shown to activate myosin light chain kinase and myosin ATPase, cytoskeletal enzymes catalyzing myosin/actin interactions and, therefore, cell contractility (82,83). The stability of the resulting actomyosin complexes may be controlled by $[Ca^{2+}]_i$ levels or other intracellular signal molecules, one of which is probably protein kinase C, the substrate for diacylglycerol (84). Activated, membrane-associated protein kinase C has been shown to phosphorylate multiple key proteins in the contraction-relaxation process (85), and may by itself trigger slow, sustained contraction of smooth muscle (36) and mesangial cells (86). In addition to agonist-induced phosphorylation of myosin light chain kinase, as shown by Kreisberg et al. in mesangial cells (87), the possibility should also be considered that contraction in vitro may in part result from cytoskeletal changes or reorganization of the intracellular actin network.

3.2. Effects of PG on cAMP

A well-characterized signal transduction system for PG of the E and I series in various cell types is cAMP (88-90). We therefore evaluated the action of these compounds on mesangial cell adenylate cyclase, in the belief that dose-dependent cAMP accumulation would explain the antagonism of contraction observed in previous studies. Fig. 13 shows the effects of PGE_2 and Iloprost on intracellular cAMP accumulation measured by radioimmunoassay at the end of 3-min. incubations in the presence of IBMX. The duration of exposure to the PG was determined on the basis of preliminary studies of the time course of cAMP responses, which peaked at 3 min. Both PG stimulated significant elevation of cAMP, with threshold at 10 nM/0.1 uM, and 4- to 5-fold higher potency of PGI_2 on a molar basis. $PGF_{2\alpha}$ and U-46619 did not stimulate cAMP at any of the concentrations tested.

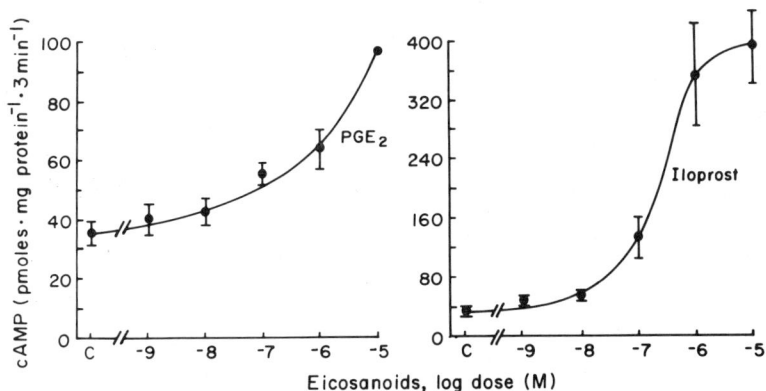

Fig. 13. Effects of PGE$_2$ and Iloprost on intracellular cAMP
of cultured rat mesangial cells. 3-min. incubations
in the presence of 0.1 mM isobutylmethylxanthine.
All values mean $\pm$ SE of duplicate determinations
from 6 experiments. (From the Proceedings of the
Xth International Congress of Nephrology, London,
1987, with permission.)

These data further differentiate the biologic actions of eicosanoids on
the glomerular mesangium, showing that opposite functional responses are
associated with distinct intracellular mediators. The molecular mechanisms of inhibition of contraction by cAMP-dependent protein kinase A have
been partially elucidated in smooth muscle (91), and the following scheme
may therefore apply to mesangial cells as well. The target of activated
protein kinase A is also myosin light chain kinase, whose phosphorylation
at a site different from protein kinase C inhibits its action on myosin,
thereby blocking contraction at the level of myosin/actin interaction
(82,91). Studies by Kreisberg et al. indicate that conformational changes
of cultured mesangial cells are induced by cAMP, resulting in dissolution
of "stress fibers," bundles of activated microfilaments, with ensuing
"shape change", a process accompanied by simultaneous dephosphorylation of
myosin light chain (1,57,87). Our morphologic observations are indeed consistent with this cytoskeletal effect, which should be kept distinct from
contraction, as indicated by inhibition of responses to ANG II (92) and
U-46619 in cells pretreated with agents that increase cAMP.

3.3. Manipulation of Intracellular Signals

To gain an insight into the mechanisms of control of intracellular
signals for PG and TxA$_2$, we evaluated the effects of protein kinase C
and A activation on $[Ca^{2+}]_i$ and contraction of cultured rat mesangial

cells. In the representative experiments shown in Fig. 14, panel A, preincubation of confluent monolayers with 17% FBS for 24 hours prior to an experiment markedly inhibited responses to $PGF_{2\alpha}$, U-46619 (not shown) and
other stimuli of $[Ca^{2+}]_i$. Since serum contains various activators of
protein kinase C, we stimulated serum-deprived monolayers with the specific protein kinase C activator, phorbol myristate acetate (PMA). As shown
in Fig. 14, panel B, PMA, but not the inactive phorbol, 4α-12,13-phorbol
didecanoate, mimicked the effects of serum, indicating that prior protein
kinase C activation, independent of the agonist, potently inhibits intracellular signals that arise from phosphoinositide breakdown. The potential of protein kinase C to regulate proximal and distal events leading to
mesangial cell contraction requires further evaluation.

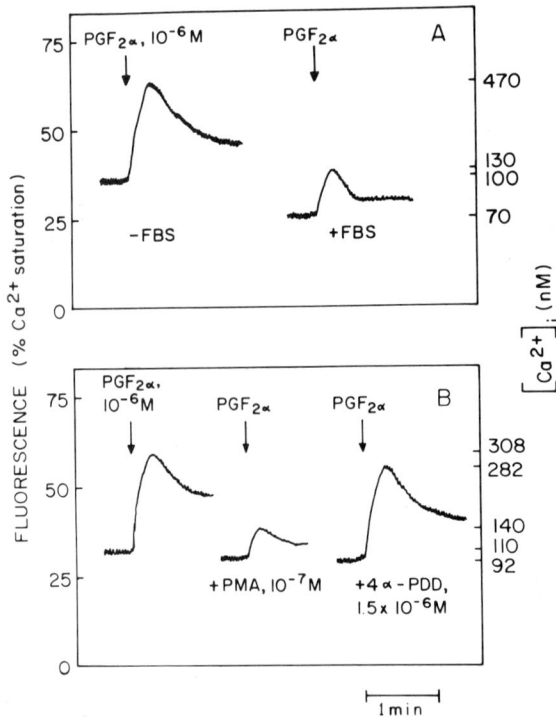

Fig. 14. Inhibition of $PGF_{2\alpha}$-induced
elevation of free cytosolic calcium
($[Ca^{2+}]_i$) in fura 2-loaded monolayers of cultured rat mesangial
cells by 17% fetal bovine serum
(FBS, panel A) or by 3-min. preincubation with phorbol myristate acetate (PMA, panel B), representative
experiments. 4α-PDD, 4α-phorbol
12,13-didecanoate.

Inasmuch as exposure of mesangial cells to PGE_2 and PGI_2 results in inhibition of contractility parallel to stimulation of intracellular cAMP, we asked whether inhibition of agonist-induced cAMP accumulation reverses the actions of these prostanoids. We therefore inhibited cAMP accumulation in response to PGE_2 and PGI_2 by preincubating the cells with the false substrate for adenylate cyclase, 2',5'-dideoxyadenosine (2',5'-DDA), which has been shown to block agonist-induced cAMP rises in several cell types (93,94). Ten-min. preincubations of the cells with 0.1 mM 2',5'-DDA, but not with the isomer 3',5'-DDA, significantly inhibited cAMP accumulation in response to 10 uM PGE_2 (PGE_2 79.6±16.6, 2'5'-DDA + PGE_2 40.1±4.3, 3',5'-DDA + PGE_2 90.8±26.3 pmol/mg prot/3 min. ± SE) and Iloprost (Iloprost 308.9±58.5, 2',5'-DDA + Iloprost 161.7±26.9, 3',5'-DDA + Iloprost 244.5±43.9 pmol/mg prot/3 min. ± SE). In cells pre-treated with 2',5'-DDA, PGE_2 failed to inhibit U-46619-induced contraction, displaying marked contractile activity by itself, different from control incubations in which it had no detectable effect (Fig. 15). More-over, 2',5'-DDA enhanced the action of U-46619, unmasking responses in approximately 40% of the cells that previously did not contract in response to the agonist alone.

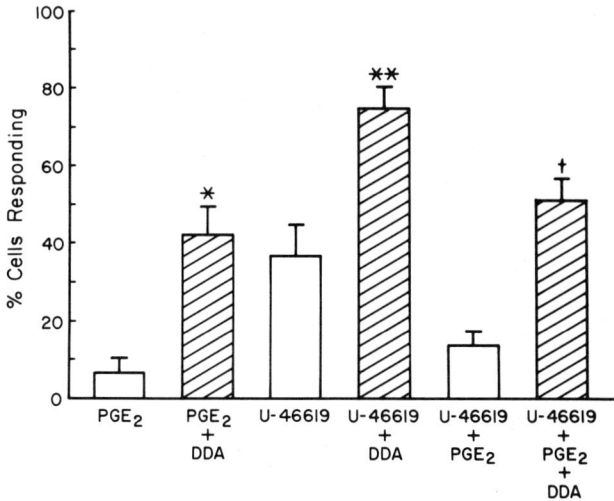

Fig. 15. Effects of 0.1 mM 2',5'-dideoxyadenosine (DDA) on contraction of cultured rat mesangial cells. PGE_2, 1 uM, and U-46619, 1 uM, added alone or in combination to cells preincubated for 10 min. with DDA. Percent cells contracting ± SE. * p <0.001 vs. PGE_2 alone; ** p <0.01 vs. U-46619 alone; + p <0.001 vs. U-46619 + PGE_2, chi-square tests. (From Circulation Research 62:916-925, 1988, with permission.)

These results support our belief that cAMP accumulation in response to PGE$_2$ functionally inhibits mesangial cells, preventing and/or antagonizing contraction induced by various agonists. Enhancement of contractile responses by 2',5'-DDA, that is, a larger number of cells displaying dimensional changes in response to U-46619, indicates an inhibitory action of cAMP stimulated by endogenously produced PGE$_2$. Suppression of PGE$_2$-stimulated cAMP also unmasked a contractile action of PGE$_2$ itself, as initially reported by Venkatachalam et al. (51) in cloned mesangial cells, an observation at variance with earlier studies by the same group (4) and .with our work in noncloned, subcultured cells. A likely explanation for this discrepancy is our finding of PLC activation by PGE$_2$, with elevation of [Ca^{2+}]$_i$ which, although of minor amplitude and somewhat inconstant, may account for contractility, unless modulated by a concomitant rise of cAMP. Cloning of a subtype of mesangial cell expressing a PGE$_2$ receptor selectively coupled to PLC may well explain the contractile effects observed by these investigators. Functionally distinct PGE$_2$ binding sites have been identified in smooth muscle and classified as EP$_1$ (constrictor) and EP$_2$ (dilator) receptors by Coleman et al. (95). Our observations are consistent with the presence of subclasses of PGE$_2$ receptors in cultured rat mesangial cells. Whether the cAMP and Ca^{2+} responses are a feature of independent subsets of mesangial cells or coexist in the same cells, indicating dual receptors or a common receptor coupled to distinct signal transduction systems, cannot be established with the current available techniques. PGE$_2$ could also have some agonist activity on a PGF$_{2\alpha}$ receptor, but the analysis of this possibility is complicated by the lack of specific PGF$_{2\alpha}$ receptor antagonists.

As discussed in 3.2, Iloprost did not inhibit [Ca^{2+}]$_i$ signals evoked by a PG or by ANG II or AVP, thus suggesting that the inhibitory effects of selective cAMP stimulation are distal to PLC activation and related initial biochemical responses. On the other hand, the inhibition of [Ca^{2+}]$_i$ responses to U-46619 or PGF$_{2\alpha}$ by PGE$_2$ is most likely related to its initial agonist activity on [Ca^{2+}]$_i$ (heterologous desensitization) rather than to cAMP accumulation.

4. CONCLUSIONS

Our evidence of functional receptors for various eicosanoids in the glomerular mesangium points to a possible hemodynamic role of AA metabolites at sites other than pre- and postcapillary resistances. These findings are in agreement with a general model of mesangial cells as

188

modified, intraglomerular smooth muscle, and provide an insight to the
intracellular signals that may control the interactions of cytoskeletal
and contractile components in these cells. We have also shown that the
contraction-relaxation responses to eicosanoids can be pharmacologically
manipulated at various levels, employing PG synthesis inhibitors, receptor
blockers or drugs that act on intracellular signaling pathways. Future
work should aim at correlating the results of studies in culture with
mesangial function in vivo and providing evidence of its regulatory role
on glomerular hemodynamics. In this context, the data reviewed here may
contribute to an appreciation of the interactions of PG and TxA_2 in the
control of the renal microcirculation under normal conditions and in
settings of enhanced production, such as inflammation (20,96,97). Other
biologic actions of these eicosanoids, including the regulation of cell
proliferation, secretion and processing of macromolecules, also demand
extensive investigation, with the potential for improved understanding of
glomerular disease and novel approaches to its pharmacologic treatment.

ACKNOWLEDGEMENTS

This work was supported by National Heart, Lung and Blood Institute
Grant HL-22563. P. Mene' was the recipient of Research Fellowships from
the American Heart Association, Northeast Ohio Affiliate and from
Consiglio Nazionale delle Ricerche of Italy. We are indebted to Drs.
George Dubyak and Antonio Scarpa for scientific advice and fruitful
discussion. The excellent technical support of Cheryl Subjoc, Edith
Hanzmann, Charles Rettberg and Guy McDermott is also gratefully
acknowledged.

REFERENCES

1. J.I. Kreisberg, M. Venkatachalam, D. Troyer, Contractile properties of
    cultured glomerular mesangial cells, Am. J. Physiol. 249 (Renal
    Fluid Electrolyte Physiol 18):F457-F463 (1985).
2. P.R. Mahieu, J.B. Foidart, C.H. Dubois, C.A. Dechenne, J. Deheneffe,
    Tissue culture of normal rat glomeruli: contractile activity of the
    cultured mesangial cells, Invest. Cell Pathol. 3:121-128 (1980).
3. G.E. Striker and L.J. Striker, Biology of disease: glomerular cell
    culture, Lab. Invest. 53:122-131 (1985).
4. D.A. Ausiello, J.I. Kreisberg, C. Roy, M.J. Karnovski, Contraction of
    cultured rat glomerular cells of apparent mesangial origin after

stimulation with angiotensin II and arginine vasopressin, <u>J. Clin. Invest</u>. 65:754-760 (1980).

5. B.M. Brenner, L.D. Dworken, I. Ichikawa, Glomerular ultrafiltration, in: "The Kidney", B.M. Brenner, F.C. Rector, eds., Philadelphia, W.B. Saunders, pp. 124-144 (1986).

6. L.A. Scharschmidt, E. Lianos, M.J. Dunn, Arachidonate metabolites and the control of renal function, <u>Federation Proc</u>.42:3058-3063 (1983).

7. D.A. Maddox, C.M. Bennett, W.M. Deen, R.J. Glassock, D. Knutson, T.M. Daughterty, B.M. Brenner, Determinants of glomerular filtration in experimental glomerulonephritis in the rat, <u>J. Clin. Invest</u>. 55: 305-318 (1975).

8. R.C. Blantz, K.S. Konnen, B.J. Tucker, Angiotensin II effects upon the glomerular microcirculation and ultrafiltration coefficient of the rat, <u>J. Clin. Invest</u>. 57:419-434 (1976).

9. I. Ichikawa and B.M. Brenner, Mechanism of action of histamine and histamine antagonists on the glomerular microcirculation in the rat, <u>Circ. Res</u>. 45:737-745 (1979).

10. N. Schor, I. Ichikawa, B.M. Brenner, Mechanisms of action of various hormones and vasoactive substances on glomerular ultrafiltration in the rat, <u>Kidney Int</u>. 20:442-451 (1981).

11. L.A. Scharschmidt, J.G. Douglas, M.J. Dunn, Angiotensin II and eicosa-noids in the control of glomerular size in rat and human, <u>Am. J. Physiol</u>. 250 (Renal Fluid Electrolyte Physiol.19):F348-F356 (1986).

12. V. Savin, In vitro effects of angiotensin II on glomerular function, <u>Am. J. Physiol</u>. 251 (Renal Fluid Electrolyte Physiol.20):F627-F634 (1986).

13. M.J. Dunn and L.A. Scharschmidt, Prostaglandins modulate the glomeru-lar actions of angiotensin II, <u>Kidney Int</u>. 31 (Suppl.20)S-95-S-101 (1987).

14. R. Loutzenhiser, M. Epstein, C. Horton, P. Sonke, Reversal of renal and smooth muscle actions of the thromboxane mimetic U-44069 by diltiazem, <u>Am. J. Physiol</u>. 250 (Renal Fluid Electrolyte Physiol.19):F619-F626 (1986).

15. L.P. Feigen, B.M. Chapek, J.E. Flemming, J.M. Flemming, P.J. Kadowitz, Renal vascular effects of endoperoxide analogs, prosta-glandins and arachidonic acid, <u>Am. J. Physiol</u>. 233 (Heart Circ. Physiol. 1):H573-H579 (1977).

16. J.G. Gerber, E. Ellis, J. Hollifield, A.S. Nies, Effect of prostaglan-din endoperoxide analogues on canine renal function, hemodynamics and renin release, <u>Eur. J. Pharmacol</u>. 53:239-246 (1979).

17. M. Hamberg, J. Svensson, B. Samuelsson, Thromboxanes: a new group of

biologically active compounds derived from prostaglandin endoperoxides, <u>Proc. Nat. Acad. Sci. USA</u> 72:2994-2998 (1975).

18. W.A. Scott, J.M. Zrike, A.L. Hamill, J. Kempe, Z.A. Cohn, Regulation
of arachidonic acid metabolism in macrophages, <u>J. Exp. Med.</u> 152:
324-355 (1980).

19. W.A. Scott, N.A. Pawlowski, M. Andreach, Z.A. Cohn, Resting macrophages produce distinct metabolites from exogenous arachidonic
acid, <u>J. Exp. Med.</u> 155:535-547 (1982).

20. E.A. Lianos, G.A. Andres, M.J. Dunn, Glomerular prostaglandin and
thromboxane synthesis in rat nephrotoxic serum nephritis. Effects
on renal hemodynamics, <u>J. Clin. Invest.</u> 72:1439-1448 (1983).

21. T. Okegawa, P.E. Jonas, K. DeSchryver, A. Kawasaki, P. Needleman,
Metabolic and cellular alterations underlying the exaggerated renal
prostaglandin and thromboxane synthesis in ureter obstruction in
rabbits, <u>J. Clin. Invest.</u> 71:81-90 (1983).

22. T.M. Coffman, W.E. Yarger, P.E. Klotman, Functional role of thromboxane production by acutely rejecting renal allografts in rats, <u>J.
Clin. Invest</u>. 75:1242-1248 (1985).

23. C. Patrono, G. Ciabattoni, G. Remuzzi, E. Gotti, S. Bombardieri, O.
DiMunno, G. Tartarelli, G.A. Cinotti, B.M. Simonetti, A. Pierucci,
Functional significance of renal prostacyclin and thromboxane A_2
production in patients with systemic lupus erythematosus, <u>J. Clin.
Invest</u>. 76:1011-1018 (1985).

24. J.I. Kreisberg, M.J. Karnovsky, L. Levine, Prostaglandin production by
homogeneous cultures of rat glomerular epithelial and mesangial
cells, <u>Kidney Int</u>. 22:355-359 (1982).

25. L.A. Scharschmidt and M.J. Dunn, Prostaglandin synthesis by rat glomerular mesangial cells in culture, <u>J. Clin. Invest</u>. 71:1756-1764
(1983).

26. R.A. Lewis and K.F. Austen, The biologically active leukotrienes. Biosynthesis, functions, and pharmacology, <u>J. Clin. Invest</u>. 73:889-
897 (1984).

27. S. Hammarstroem, Leukotrienes, <u>Ann. Rev. Biochem</u>. 52:355-377 (1983).

28. E.A. Lianos, M.A. Rahman, M.J. Dunn, Glomerular arachidonate lipoxygenation in rat nephrotoxic serum nephritis, <u>J. Clin. Invest</u>. 76:
1355-1359 (1985).

29. J.E. Stork, M.A. Rahman, M.J. Dunn, Eicosanoids in experimental and
human renal disease, <u>Am. J. Med</u>. 80(suppl 1A):34-45 (1986).

30. B.J. Ballermann, R.A. Lewis, E.J. Corey, K.F. Austen, B.M. Brenner,
Identification and characterization of leukotriene C_4 receptors
in isolated rat renal glomeruli, <u>Circ. Res</u>. 56:324-330 (1985).

31. A. Rosenthal, C.R. Pace-Asciak, Potent vasoconstriction of the iso-
    lated perfused rat kidney by leukotrienes C_4 and D_4, Can. J.
    Phys. Pharmacol. 61:325-328 (1983).

32. K. Badr, C. Baylis, J. Pfeffer, M. Pfeffer, R.J. Soberman, R.A. Lewis,
    K.F. Austen, E.J. Corey, B.M. Brenner, Renal and systemic hemody-
    namic responses to intravenous infusion of leukotriene C_4 in the
    rat, Circ. Res. 54:492-499 (1984).

33. S. Bunting, R. Gryglewski, S. Moncada, J.R. Vane, Arterial walls
    generate from prostaglandin endoperoxides a substance (prosta-
    glandin X) which relaxes strips of mesenteric and coeliac arteries
    and inhibits platelet aggregation, Prostaglandins 12:897-913
    (1976).

34. B.J.R. Whittle and S. Moncada, Pharmacologic interactions between
    prostacyclin and thromboxanes, Br. Med. Bull. 39:232-238 (1983).

35. S. Bunting, S. Moncada, J.R. Vane, The prostacyclin-thromboxane A_2
    balance: pathophysiological and therapeutic implications, Br. Med.
    Bull. 39:271-276 (1983).

36. H. Rasmussen and P.Q. Barrett, Calcium messenger system: an integrated
    view, Physiol. Rev. 64:938-984 (1984).

37. M.J. Berridge, Inositol trisphosphate and diacylglycerol as second
    messengers, Biochem. J. 220:345-360 (1984).

38. R.S. Adelstein and D.R. Hathaway, Role of calcium and cyclic adenosine
    3':5' monophosphate in regulating smooth muscle contraction, Am. J.
    Cardiol. 44:783-787 (1979).

39. M.S. Simonson and M.J. Dunn, Leukotriene C_4 and D_4 contract rat
    glomerular mesangial cells, Kidney Int. 30:524-531 (1986).

40. P. Mene' and M.J. Dunn, Contractile effects of TxA_2 and endoperoxide
    analogues on cultured rat glomerular mesangial cells, Am. J.
    Physiol. 251 (Renal Fluid Electrolyte Physiol.20):F1029-F1035
    (1986).

41. R.A. Armstrong, R.L. Jones, V. Peesapati, S.G. Will, N.H. Wilson,
    Competitive antagonism at thromboxane receptors in human platelets,
    Br. J. Pharmacol. 84:595-607 (1985).

42. R.A. Armstrong, R.L. Jones, N.H. Wilson, Ligand binding to thromboxane
    receptors in human platelets: correlation with biological activity,
    Br. J. Pharmacol. 79:953-964 (1983).

43. R. Greenberg, T. Steinbacher, M.F. Haslanger, Thromboxane receptor
    antagonist properties of SQ 27,427 in the anesthetized guinea pig,
    Eur. J. Pharmacol. 103:19-24 (1984).

44. G. Graff, Preparation of PGG_2 and PGH_2, in: "Methods in

Enzymology", W. Lands, W. Smith, eds., New York, Academic Press, pp. 376-385 (1982).

45. T. Yoshimoto, S. Yamamoto, M. Okuno, M. Hayaishi, Solubilization and resolution of thromboxane synthesizing systems from microsomes of bovine blood platelets, J. Biol. Chem. 252:5871-5874 (1977).

46. D. Schlondorff, J.A. Satriano, J. Hagege, J. Perez, L. Baud, Effect of platelet activating factor on prostaglandin E_2 synthesis, arachidonic acid release and contraction of cultured rat mesangial cells, J. Clin. Invest. 73:1227-1231 (1984).

47. J. Foidart, J. Sraer, F. Delarue, P. Mahieu, R. Ardaillou, Evidence for mesangial glomerular receptors for angiotensin II linked to mesangial contractility, FEBS Lett. 121:333-339 (1980).

48. T. Tanaka, Y. Fujiwara, Y. Orita, E. Sasaki, H. Kitamura, H. Abe, The functional characteristics of cultured rat mesangial cell, Japn. Circ. J. 48:1017-1029 (1984).

49. J.R. Sedor and H.E. Abboud, Histamine modulates contraction and cyclic nucleotides in cultured rat mesangial cells, J. Clin. Invest. 75: 1679-1689 (1985).

50. P.C. Singhal, L.A. Scharschmidt, N. Gibbons, R.M. Hays, Contraction and relaxation of cultured mesangial cells on a silicone rubber surface, Kidney Int. 30:862-873 (1986).

51. M.A. Venkatachalam and J.I. Kreisberg, Agonist-induced isotonic contraction of cultured mesangial cells after multiple passage, Am. J. Physiol. 249 (Cell Physiol.18):C48-C55 (1985).

52. J. Chamley-Campbell, G.R. Campbell, R. Ross, The smooth muscle cell in culture, Physiol. Rev. 59:1-61 (1979).

53. H.E. Ives, G.S. Schultz, R.E. Galardy, J.D. Jamieson, Preparation of functional smooth muscle cells from the rabbit aorta, J. Exp. Med. 148:1400-1413 (1978).

54. A.K. Harris, P. Wild, D. Stork, Silicone rubber substrates: a new wrinkle in the study of locomotion, Science 208:177-179 (1980).

55. D.E. Mais, D.L. Saussy, Jr., A. Chaikouni, P.J. Kochel, D.R. Knapp, N. Hamanaka, P.V. Halushka, Pharmacologic characterization of human and canine thromboxane A_2/prostaglandin H_2 receptors in platelets and blood vessels: evidence for different receptors, J. Pharmacol. Exp. Therap. 233:418-424 (1985).

56. R. Barnett, P. Goldwasser, L.A. Scharschmidt, D. Schlondorff, Effects of leukotrienes on isolated rat glomeruli and cultured mesangial cells, Am. J. Physiol. 250 (Renal Fluid Electrolyte Physiol. 19): F838-F844 (1986).

57. J.I. Kreisberg, M.A. Venkatachalam, P.Y. Patel, Cyclic AMP-associated

shape change in mesangial cells and its reversal by prostaglandin
E_2, <u>Kidney Int</u>. 25:874-879 (1984).

58. J.B. Foidart and P. Mahieu, Glomerular mesangial cell contractility in
vitro is controlled by an angiotensin-prostaglandin balance, <u>Mol.
Cell Endocrinol</u>. 47:163-173 (1986).

59. L.A. Scharschmidt, M. Simonson, M.J. Dunn, Glomerular prostaglandins,
angiotensin II, and nonsteroidal anti-inflammatory drugs, <u>Am. J.
Med</u>. 81 (suppl.2B):30-42 (1986).

60. R.P. Kimberly, R.E. Boden, H.R. Keiser, P.H. Plotz, Reduction of renal
function by newer nonsteroidal anti-inflammatory drugs, <u>Am. J. Med</u>.
64:804-807 (1978).

61. M.J. Dunn, Nonsteroidal antiinflammatory drugs and renal function.
<u>Ann. Rev. Med</u>. 35:411-428 (1984).

62. G. Ciabattoni, G.A. Cinotti, A. Pierucci, B.M. Simonetti, M. Manzi,
F. Pugliese, P. Barsotti, G. Pecci, F. Taggi, C. Patrono, Effects
of sulindac and ibuprofen in patients with chronic glomerular
disease. Evidence for the dependence of renal function on prostacyclin, <u>N. Eng. J. Med</u>. 310:279-283 (1984).

63. C. Patrono and M.J. Dunn, The clinical significance of inhibition of
renal prostaglandin synthesis, <u>Kidney Int</u>. 32:1-12 (1987).

64. R. Vriesendorp, A.J.M. Donker, D. de Zeeuw, P.E. de Jong, G.K. van der
Hem, J.R.H. Brentjens, Effects of nonsteroidal anti-inflammatory
drugs on proteinuria, <u>Am. J. Med</u>. 81 (suppl.2B):84-94 (1986).

65. B.M. Brenner, Nephron adaptation to renal injury or ablation, <u>Am. J.
Physiol</u>. 249 (Renal Fluid Electrolyte Physiol.19):F324-F337 (1985).

66. L.D. Brace, D.L. Venton, G.C. Le Breton, Thromboxane A_2/prostaglan-
din H_2 mobilizes calcium in human blood platelets, <u>Am. J.
Physiol</u>. 249 (Heart Circ. Physiol.18):H1-H7 (1985).

67. K. Fukuo, S. Morimoto, E. Koh, S. Yukawa, H. Tsuchiya, S. Imanaka, H.
Yamamoto, T. Onoshi, Y. Kumahara, Effects of prostaglandins on the
cytosolic free calcium concentration in vascular smooth muscle
cells, <u>Biochem. Biophys. Res. Comm</u>. 136:247-252 (1986).

68. G.R. Dubyak and M.B. De Young, Intracellular Ca^{++} mobilization
activated by extracellular ATP in Ehrlich ascites tumor cells, <u>J.
Biol. Chem</u>. 260:10653-10661 (1985).

69. G. Grynkiewicz, M. Poenie, R.Y. Tsien, A new generation of Ca^{2+}
indicators with greatly improved fluorescence properties, <u>J. Biol.
Chem</u>. 260:3440-3450 (1985).

70. P. Mene', G.R. Dubyak, A. Scarpa, M.J. Dunn, Stimulation of cytosolic
free calcium and inositol phosphates by prostaglandins in cultured

194

rat mesangial cells, <u>Biochem. Biophys. Res. Comm</u>. 142:579-586
(1987).

71. J.B. Smith, Angiotensin-receptor signaling in cultured vascular smooth
muscle cells, <u>Am. J. Physiol</u>. 250 (Renal Fluid Electrolyte
Physiol.19):F759-F769 (1986).

72. A.V. Somlyo, M. Bond, A.P. Somlyo, A. Scarpa, Inositol trisphosphate-
induced calcium release and contraction in vascular smooth muscle,
<u>Proc. Natl. Acad. Sci. USA</u> 82:5231-5235 (1985).

73. J.V. Bonventre, K.L. Skorecki, J.I. Kreisberg, J.Y. Cheung, Vasopressin increases cytosolic free calcium concentration in glomerular
mesangial cells, <u>Am. J. Physiol</u>. 251 (Renal Fluid Electrolyte
Physiol.20):F94-F102 (1986).

74. A. Hassid, N. Pidikiti, D. Gamero, Effects of vasoactive peptides on
cytosolic calcium in cultured mesangial cells, <u>Am. J. Physiol</u>. 251
(Renal Fluid Electrolyte Physiol. 20):F1018-F1028 (1986).

75. M. Deleers, P. Grognet, R. Brasseur, Structural considerations for
calcium ionophoresis by prostaglandins, <u>Biochem. Pharmacol</u>. 34:
3831-3836 (1985).

76. M.J. Berridge and R.F. Irvine, Inositol trisphosphate, a novel second
messenger in cellular signal transduction, <u>Nature</u> (London) 312:
315-321 (1984).

77. J.R. Williamson, Role of inositol lipid breakdown in the generation of
intracellular signals, <u>Hypertension</u> 8 [Suppl.II]:II-140-II-156
(1986).

78. S. Muallem, M. Schoeffield, S. Pandol, G. Sachs, Inositol trisphosphate modification of ion transport in rough endoplasmic reticulum,
<u>Proc. Natl. Acad. Sci. USA</u> 82:4433-4437 (1985).

79. B.C. Berk, T.A. Brock, M.A. Gimbrone, R.W. Alexander, Early agonist-
mediated ionic events in cultured vascular smooth muscle cells, <u>J.
Biol. Chem</u>. 262:5065-5072 (1987).

80. P.T. Hawkins, L. Stephens, C.P. Downes, Rapid formation of inositol
1,3,4,5-tetrakisphosphate and inositol 1,3,4-trisphosphate in rat
parotid glands may both result indirectly from receptor-stimulated
release of inositol 1,4,5-trisphosphate from phosphatidylinositol
4,5-bisphosphate, <u>Biochem. J</u>. 238:507-516 (1986).

81. R.F. Irvine and R.M. Moor, Micro-injection of inositol 1,3,4,5-tetra-
kisphosphate activates sea urchin eggs by a mechanism dependent on
external Ca^{2+}, <u>Biochem. J</u>. 240:917-920 (1986).

82. R.S. Adelstein and E. Eisenberg, Regulation and kinetics of the actin-
myosin-ATP interaction, <u>Ann. Rev. Biochem</u>. 49:921-956 (1980).

83. E.D. Korn, Actin polymerization and its regulation by proteins from

nonmuscle cells, <u>Physiol. Rev.</u> 62:672-737 (1982).

84. Y. Nishizuka, Studies and perspectives of protein kinase C, <u>Science</u>
233:305-312 (1986).

85. T. Hunter and J.A. Cooper, Protein-tyrosine kinases, <u>Ann. Rev.
Biochem.</u> 54:897-930 (1985).

86. D.A. Troyer, O.F. Gonzalez, J.G. Douglas, J.I. Kreisberg, Phorbol
ester inhibits arginine vasopressin activation of phospholipase C
and promotes contraction of and prostaglandin production by
cultured mesangial cells, <u>Biochem. J.</u> (in press) (1988).

87. J.I. Kreisberg, M.A. Venkatachalam, R.A. Radnik, P.Y. Patel, Role of
myosin light-chain phosphorylation and microtubules in stress fiber
morphology in cultured mesangial cells, <u>Am. J. Physiol.</u> 249 (Renal
Fluid Electrolyte Physiol. 18):F227-F235 (1985).

88. R.R. Gorman, S. Bunting, O.V. Miller, Modulation of human platelet
adenylate cyclase by prostacyclin (PGX), <u>Prostaglandins</u> 13:377-388
(1977).

89. N.K. Hopkins and R.R. Gorman, Regulation of endothelial cell cyclic
nucleotide metabolism by prostacyclin, <u>J. Clin. Invest.</u> 67:540-546
(1981).

90. D. Oliva, A. Noe', S. Nicosia, F. Bernini, R. Fumagalli, B.J.R.
Whittle, S. Moncada, J.R. Vane, Prostacyclin-sensitive adenylate
cyclase in cultured myocytes: differences between rabbit aorta and
mesenteric artery, <u>Eur. J. Pharmacol.</u> 105:207-213 (1984).

91. E.G. Krebs and J.A. Beavo, Phosphorylation-dephosphorylation of
enzymes, <u>Ann. Rev. Biochem.</u> 48:923-959 (1979).

92. R. Barnett, P.C. Singhal, L.A. Scharschmidt, D. Schlondorff, Dopamine
attenuates the contractile response to angiotensin II in isolated
rat glomeruli and cultured mesangial cells, <u>Cir. Res.</u> 59:529-533
(1986).

93. J.N. Fain, R.H. Pointer, W.F. Ward, Effects of adenosine nucleotides
on adenylate cyclase, phosphodiesterase, cyclic adenosine monophosphate accumulation and lipolysis in fat cells, <u>J. Biol. Chem.</u> 247:
6866-6872 (1972).

94. R.J. Haslam, M.M.L. Davidson, J.V. Desjardin, Inhibition of adenylate
cyclase by adenosine analogues in preparations of broken and intact
platelets, <u>Biochem. J.</u> 176:83-95 (1978).

95. R.A. Coleman and I. Kennedy, Characterization of the prostanoid receptors mediating contraction of guinea-pig isolated trachea,
<u>Prostaglandins</u> 29:363-375 (1985).

96. J.E. Stork and M.J. Dunn, Hemodynamic roles of thromboxane A_2 and

prostaglandin E_2 in glomerulonephritis, <u>J. Pharm. Exp. Ther</u>. 233:672-678 (1985).

97. M.A. Rahman, S.N. Emancipator, M.J. Dunn, Immune complex effects on glomerular eicosanoid production and renal hemodynamics, <u>Kidney Int</u>. 31:1317-1326 (1987).

INTERACTIONS OF PLATELET ACTIVATING FACTOR AND PROSTAGLANDINS

IN THE GLOMERULUS AND IN MESANGIAL CELLS

Detlef Schlondorff

Albert Einstein College of Medicine
New York

Platelet activating factor (PAF) has been recognized as a potent mediator of inflammation and identified as 1-alkyl-2-acetyl-sn-glycero-3-phospho-
choline (Fig. 1). For biological activity, PAF requires an alkyl chain
linked by an ether bond at the sn-1 position (most commonly a C_{16} length
saturated chain) and an acetyl group at the sn-2 position. Loss of the
acetyl group yielding lyso-PAF is associated with loss of biological activity (1-3). Many inflammatory cells have the capability to produce PAF upon
stimulation, including peripheral leucocytes, with the exception of lymphocytes (1). Furthermore, cultured endothelial cells and renal cells (see
below) can generate PAF. Stimuli that result in PAF formation cause an
increase in intracellular calcium and include calcium ionophore, receptor-
mediated phagocytosis or endocytosis, complement component C_{5a}, formyl -
methionyl - leucyl - phenyl - alanine and endotoxin (1-3). In general,
resting cells do not produce PAF but require prior stimulation. Upon stimulation, PAF formation occurs via deacylation by phospholipase A_2 of the
precursor molecule 1-alkyl-2-acyl-sn-glycero-3-phosphocholine to yield lyso
PAF (Fig. 2). Lyso PAF is subsequently acetylated in position 2 by a specific acetyl-coenzyme A transferase to yield the active PAF molecule (1). It
is of interest that, in many cells, the precursor molecule 1-alkyl-2-acyl-
sn-glycero-3-phosphocholine is highly enriched in position 2 with arachidonic acid (AA), so that stimulation of PAF formation will also result in
release of AA and subsequent conversion to eicosanoids (1,4). Another pathway for PAF formation may exist in some cells (1) that generate PAF from
1-alkyl-2-acetyl-diglyceride via addition of the polar choline group catalyzed by phosphocholine transferase (Fig. 2).

GLYCEROL ETHER   ALKYL

ACETYL   H₂C — O — (CH₂)₁₅₋₁₈ — CH₃

$$\text{ACETYL} \quad H_2C-O-(CH_2)_{15-18}-CH_3$$

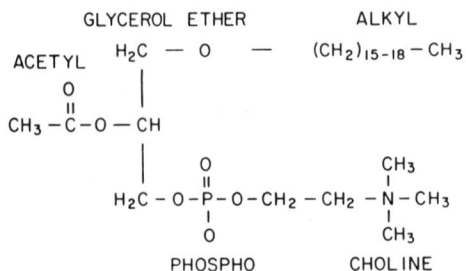

Fig. 1. Chemical formula of platelet-activating factors (1-alkyl-2-acetyl-sn-glycero-3-phosphocholine).

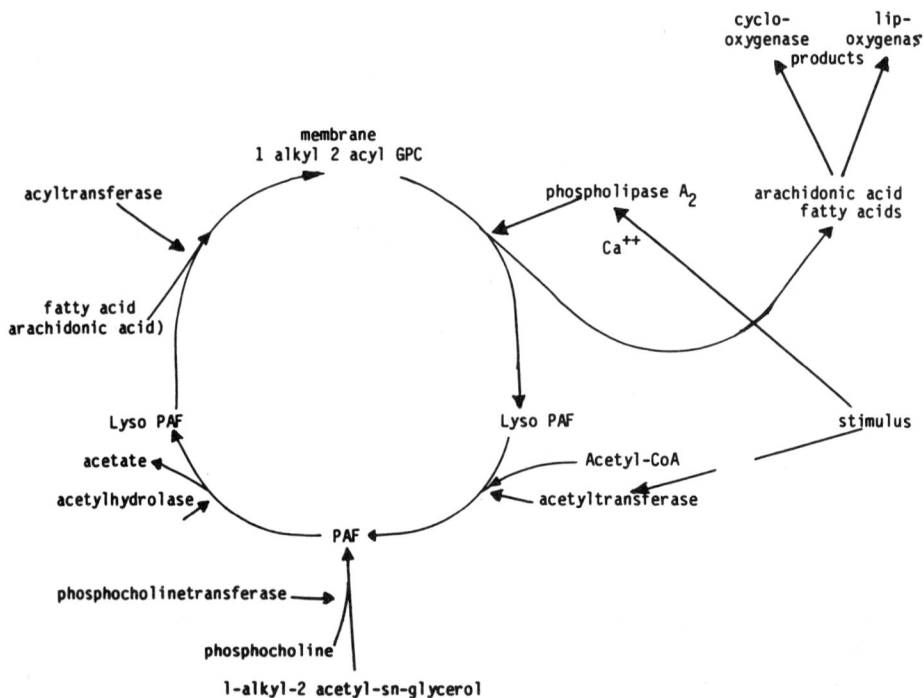

$$CH_3-\overset{O}{\overset{\|}{C}}-O-CH$$

$$H_2C-O-\overset{O}{\overset{|}{\underset{O}{\overset{\|}{P}}}}-O-CH_2-CH_2-\overset{CH_3}{\overset{|}{\underset{CH_3}{\overset{|}{N}}}}-CH_3$$

PHOSPHO   CHOLINE

Fig. 2. Schematic representation of pathways for generation and metabolism of platelet-activating factor. Reproduced from reference 5 with permission.

Labels in Fig. 2:
cyclo-oxygenase lip-oxygenase products
membrane 1 alkyl 2 acyl GPC
acyltransferase
phospholipase A₂
Ca^{++}
arachidonic acid fatty acids
fatty acid arachidonic acid)
Lyso PAF
Lyso PAF
stimulus
acetate
Acetyl-CoA
acetylhydrolase
acetyltransferase
PAF
phosphocholinetransferase
phosphocholine
1-alkyl-2 acetyl-sn-glycerol

EFFECTS OF PAF ON KIDNEY FUNCTION

Infusion of PAF into experimental animals results in profound changes of blood pressure, cardiac output, peripheral resistance, intravascular volume, etc. (1-3). Obviously, all of these have marked influences on renal function (5). Furthermore, many of the observed effects may be secondary to PAF-induced synthesis of leukotrienes, prostaglandins, and thromboxane (1-3). Several studies, however, have provided evidence that PAF can also directly reduce renal function and, specifically, glomerular filtration and blood flow (see Table 1). In dogs receiving PAF via direct intrarenal artery infusion, in doses that do not exert effects on the systemic blood

Table 1. Effects of PAF in The Kidney

Model	Effect	Reference
In _vivo_ intrarenal artery infusion in dogs	Dose-dependent decrease in renal plasma flow and glomerular filtration rate; further enhanced by PG-inhibition	Hebert et al 1987 Scherf et al 1986
In _vivo_ intrarenal artery infusion in rabbit	Loss of glomerular anionic charge. Intraglomerular accumulation of PMN and platelets. Transient and mild proteinuria	Camussi et al 1984
In _vitro_ isolated perfused rat kidney	Decreased renal vascular resistance Decrease in GFR Increased PG synthesis Increased vascular permeability Mild proteinuria	Schwertschlag et al 1986 and 1987 Weisman et al 1985 Pirotzky et al 1985

pressure or cardiac function, a dose-dependent and reversible decrease in renal blood flow has been observed (6,7). This is accompanied by a marked decline in glomerular filtration, urinary flow rate and sodium excretion. Pretreatment with indomethacin accentuated the changes in renal function. This suggests the production of renal vasodilatory prostaglandins that blunted the effect of the infused PAF.

Infusion of PAF into the isolated perfused rat kidney produces a some-what different picture (8,9). In this model PAF may have some vasodilatory action, yet causes a decline in glomerular filtration possibly by altering the ultrafiltration characteristics (9). PAF stimulates PG formation by the isolated kidney (8-10) and increases glomerular capillary permeability for macromolecules and albumin (9,11). A direct change of glomerular capillary characteristics has been reported with PAF perfusion of the isolated kidney (9). *In vivo* administration of PAF can result in mild proteinuria (12), though this may, in part, be secondary to activation of inflammatory cells in the glomerulus.

INVOLVEMENT OF PAF IN EXPERIMENTAL MODELS OF RENAL DISEASE

At present, involvement of PAF in the pathophysiology of experimental diseases must be inferred largely through the use of PAF receptor blocking agents. The importance of PAF has been recognized in a variety of nonrenal and renal diseases (1-3,5,13). Table 2 summarizes the experimental studies of PAF and renal disease. Bertani et al. (14) studied the influence of a PAF-receptor blocker on experimental nephrotoxic serum nephritis in rabbits. The PAF antagonists slightly improved proteinuria and partially prevented the decline in GFR. PAF blockage also ameliorated the more severe crescentic stage in the autologous phase of the disease. Similar observations have been reported in a rat model of nephrotoxic nephritis (15). Camussi et al. (16) examined the effect of PAF-receptor blockade on the course of acute serum sickness nephritis in rabbits, with similar beneficial results. More recently, Camussi et al. (17) have examined the influence of PAF on renal injury induced by glomerular *in situ* deposition of immune complexes. In these studies the inflammatory sequelae to this insult (pro-teinuria and glomerular cellular proliferation) were markedly decreased by administration of PAF blocker to the experimental rats. Circulating levels of PAF in plasma extracts were also decreased by the PAF blocker. The authors deduced that PAF is an important mediator of the *in situ* immune complex-mediated form of acute glomerulonephritis in the rat.

Table 2. Evidence for Involvement of
          PAF in Renal Injury

Model	Reference
Hyperacute allograft rejection in rabbits	Ito et al. 1984
Acute serum sickness in rabbits	Camussi et al. 1982
Immune complex nephritis in rats	Camussi et al. 1987
Nephrotoxic serum nephritis (anti GBM)	Hruby et al. 1986 Bertani et al. 1987
Endotoxin-induced acute renal insufficiency	Wang and Dunn 1987

Wang and Dunn (18) have studied the effect of PAF antagonist on an endotoxin-induced form of acute renal insufficiency. They reported that the
fall in glomerular filtration rate, renal plasma flow and filtration fraction seen in animals given a bolus of endotoxin was abolished by pretreatment of the animals with the PAF blocker. Infusion of PAF into the rats
resulted in findings similar to those of endotoxin administration and were
also antagonized by the PAF blockers. The authors, therefore, concluded
that PAF significantly contributes to the acute renal insufficiency of endotoxin. Part of the PAF generated in endotoxin shock may even originate from
the kidney, as Wang et al. subsequently demonstrated that endotoxin can stimulate cultured mesangial cells to produce PAF (19).

PAF PRODUCTION BY GLOMERULAR CELLS

    In view of the potential pathophysiological role of PAF in renal disease, the ability of kidney tissue to generate PAF is of interest. Camussi et al. first reported that rabbit kidney slices were capable of producing bioactive PAF (16). Subsequently, Pirotzky et al. demonstrated release of PAF bioactivity from the isolated perfused rat kidney stimulated by the calcium ionophore A23187 (20). These authors also demonstrated generation of bioactive PAF material from isolated glomeruli and suspensions of rat tubules and medullary cells upon stimulation (21). We confirmed PAF production by isolated rat (22) and human (Ardaillou and Schlondorff, unpublished) glomeruli and extended these studies to glomerular mesangial cells in culture (22) as a model for studying the control of platelet-activating factor production by glomerular cells. In these experiments, mesangial cells cultured from explanted rat glomeruli were used in their second to fifth subculture, exposed to the calcium ionophore A23187 for 30 minutes, and cells and buffer were extracted together, partially purified, and then tested for bioactivity, using the bioassay system of rabbit platelet aggregation (3). Mesangial cells were found to produce about 20 pmol PAF per mg protein over a 30 min. incubation with A23187. In order to demonstrate the structural identity of the bioactive PAF material, samples were further purified by reverse phase HPLC and subjected to fast atom bombardment spectrometry, then derivatized and examined by mass spectrometry after gas chromatography. The PAF produced was unequivocally identified as the 1-0 hexadecyl form (C-16 PAF). In the absence of calcium ionophore, mesangial cells did not produce PAF.

Subsequently, we examined the effect of angiotensin II on PAF formation, as angiotensin is a potent stimulus for phospholipase A2 activation and prostaglandin production in mesangial cells (23,24). As PAF formation requires phospholipase A2 activation and an increase in intracellular calcium, we thought that angiotensin II may stimulate PAF production by mesangial cells. In fact, angiotensin II stimulated PAF production in a dose-dependent manner, with a threshold at 10^{-9}M angiotensin. The amount of PAF produced was maximal at five minutes decreasing over the next 25 minutes (Fig. 3). The PAF produced remains mostly cell associated, with little bioactivity demonstrable in the buffer. The structural identity of the material as hexadecyl PAF was confirmed by negative ion chemical ionization mass spectrometry. The time course of PAF formation by mesangial cells in response to angiotensin II is of interest as it is reminiscent of that for PGE_2 formation (23). This probably indicates that agents such as angiotensin II, which cause a rapid increase in intracellular calcium and activation of phospholipase A_2, can simultaneously stimulate PGE_2 formation and PAF generation. The significance of the angiotensin II-induced PAF formation remains

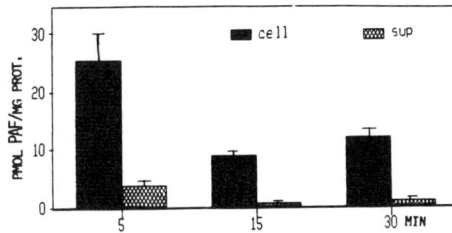

Fig. 3. Effect of angiotensin II
(10^{-8}M) on synthesis of platelet-
activating factor (PAF) by rat mesangial cells. Mesangial cells in
their 3rd to 5th subculture were incubated with or without angiotensin
II for 5, 15 or 30 min. in buffer.
At the end of each period, buffer
(sup) and cells (cell) were extracted separately, prepurified, and
samples assayed for PAF by rabbit
platelet aggregation bioassay. No
PAF was detectable in the control
incubations (not shown). Results
are expressed as pmol PAF generated
per mg cellular protein (Neuwirth
and Schlondorff, unpublished).

to be determined, as does the role of intracellular PAF. It is unclear
whether cell-associated PAF has an effect in mesangial cells. It is of
interest, however, that in cultured endothelial cells, intracellular PAF influences adhesion of leucocytes (25). If cell membrane-associated PAF in
mesangial cells is biologically active, it could be chemotactic and cause
local activation of leucocytes and platelets. Similarly, PAF generated and
released from mesangial cells after stimulation with high concentrations of
angiotensin II could contribute to angiotensin-related pathophysiological
states with local platelet activation, fibrin formation, and proteinuria.
Obviously, this is highly speculative at this point.

UPTAKE OF MACROMOLECULES AND PAF FORMATION BY MESANGIAL CELLS

The mesangium represents a site for deposition of macromolecules, including immune complexes, under a variety of experimental conditions and glomerular diseases (26,27). The potential for macromolecular uptake by cultured
mesangial cells had been previously examined by Baud et al. (28,29). These
authors were able to demonstrate that uptake of opsonized zymosan by mesangial cells was associated with generation of lipoxygenase products of AA and
release of reactive oxygen metabolites (28,29). Furthermore, opsonized

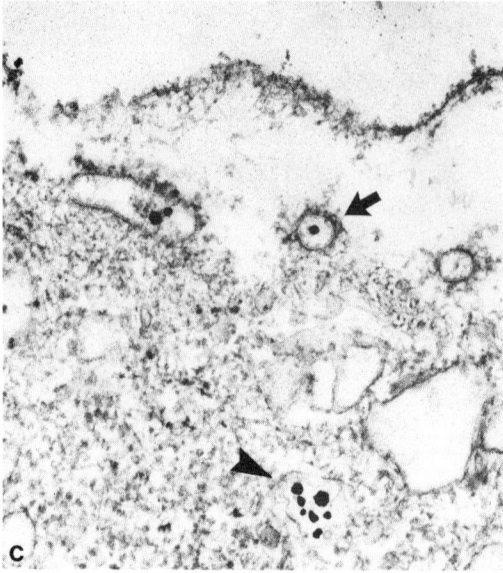

Fig. 4. Transmission electron micrograph of mesangial cells incubated with serum-coated colloidal gold particles for 30 min. Gold particles can be seen in coated pits and endosomes. Reproduced in part from reference 31 with permission.

zymosan stimulated PGE_2 production by cultured mesangial cells, and this was associated with stimulation of both phospholipases C and A_2 (30). We have recently extended these studies to examine in more detail uptake of serum-coated gold particles by mesangial cells in culture (31). Mesangial cells took up serum-coated gold particles by an active and specific process involving a coated pit-coated vesicle-endosome-lysosome pathway (Fig. 4). This was accompanied by stimulation of PGE_2 synthesis, which involves phospholipase A_2 activation. We therefore examined whether endocytosis of macromolecules would also result in PAF formation (32). In these studies we also sought to determine whether IgG immune complexes were potential ligands. Mesangial cells were therefore incubated with [^{195}Au] labeled, colloidal, gold particles that had been coated by either:

1. Bovine serum albumin (BSA) alone;
2. BSA followed by anti-BSA IgG, thus forming immune complexes;
3. BSA followed by the F(ab')2 fragment of the anti-BSA IgG, resulting in immune complexes lacking the Fc portion.

After 15-minute incubations with the respective gold particles, PGE_2 production was measured in the incubation buffer, PAF in lipid extracts of buffer and cells and uptake of immune complex-coated [^{195}Au] gold particles by scintillation spectrometry. Mesangial uptake of [^{195}Au], with BSA/anti-BSA IgG immune complexes, was significantly higher than that of [^{195}Au] with the F(ab')2 fragments or BSA only, and also resulted in stimulation of PGE_2 and PAF synthesis (Fig. 5). Most of the PAF (>90%) remained cell-associated, with less than 10% of PAF released into the

206

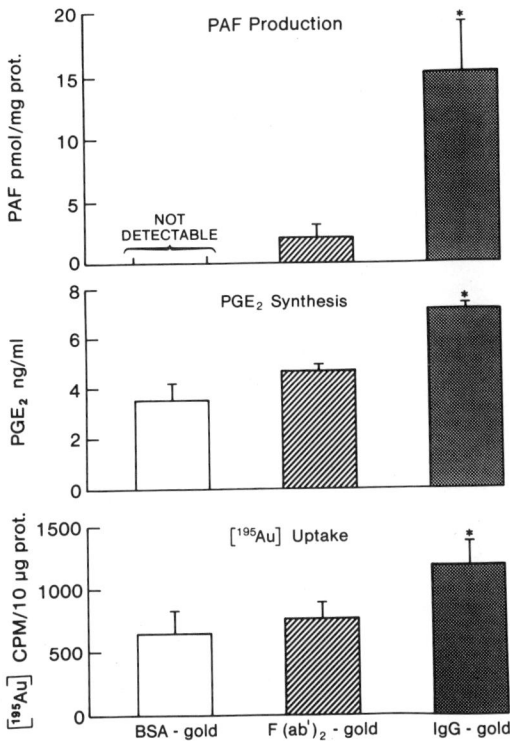

Fig. 5. Effects of different coatings of [^{195}Au] colloidal gold particles on uptake by mesangial cells (lowest panel), PGE$_2$ synthesis (middle panel), and production of platelet-activating factor (top panel). Mesangial cells were incubated for 15 min. with [^{195}Au] gold particles that had been coated either with only BSA (BSA), BSA followed by anti-BSA IgG antibody (BSA-IgG), or BSA followed by the F(ab')$_2$ fragment (BSA-F[ab']$_2$). * indicates $p<0.01$ compared with BSA or BSA-F(ab')$_2$ (Neuwirth et al., unpublished).

buffer. Mass spectrometry again confirmed these results and identified the PAF as exclusively of the hexadecyl species i.e. containing a saturated 16 carbon chain in position 1. These experiments also indicate that cultured mesangial cells contain Fc receptors for IgG. The binding and uptake of IgG immune complexes by Fc receptors of mesangial cells has also been shown by Sedor et al. (33) and is associated with the generation of reactive oxygen metabolites and an increase in intracellular calcium (33,34). Thus, it appears that mesangial cells can bind immune complexes by specific receptors, which results in PAF formation, eicosanoid production, and generation of reactive oxygen metabolites. This illustrates the close interrelationship of these various mediators of inflammation and also the potential roles that they may have in glomerular immune injury.

METABOLISM OF PAF BY MESANGIAL CELLS

Given its potent biological activities, it would appear imperative for cells to inactivate PAF rapidly (1). Metabolism of PAF by mesangial cells is catalyzed by an acetylhydrolase enzyme with consequent production of lyso-PAF (Fig. 2), and acylation of lyso-PAF to alkyl-acyl-glycerolphospho-

MEDIUM   CELLS

— PC

— C_{16}-PAF
— C_{18}-PAF

C_{16}-Lyso-PAF —
C_{18}-Lyso-PAF —

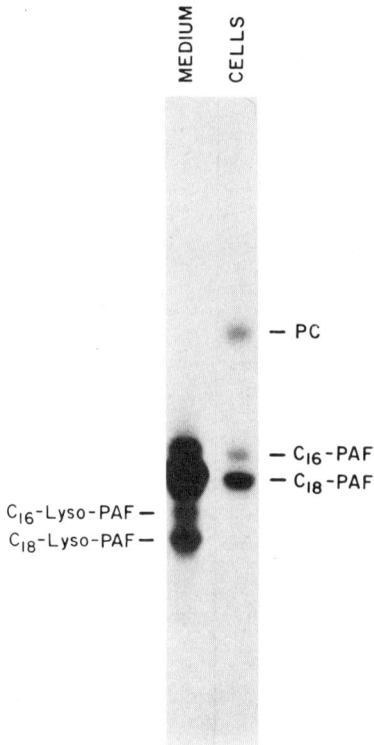

Fig. 6. Autoradiography of metabolites formed after incubation of mesangial cells
with [^{3}H] PAF for 15
min. The [^{3}H] PAF was
mostly of C_{18} and less
C_{16} alkyl species. Incubation medium and cells
were extracted separately
and analyzed by thin layer
radiochromatography. Position of authentic standards
is indicated. Note that in
the incubation buffer,
[^{3}H] PAF is mostly metabolized to [^{3}H] lyso-PAF
whereas, intracellularly,
[^{3}H] lyso-PAF is barely
apparent. Intracellularly,
some label migrates with
phosphatidylcholine (PC)
and probably represents
conversion to 1-alkyl-2-
acyl-sn-3 phosphocholine.

choline then follows. We have found that in mesangial cells the acetylhydrolase activity is located predominantly at the outer surface of the cell
membrane (Fig. 6) (35). This activity can be released into the buffer after
treatment of cells with trypsin, suggesting that the acetylhydrolase enzyme
may represent an ectoenzyme anchored in the plasma membrane but directed
towards the outside. This location of the PAF acetylhydrolase may represent
a protective mechanism whereby cells rapidly inactivate PAF released from or
presented to the cells, thus limiting the pathophysiologic consequences.

EFFECTS OF PAF ON MESANGIAL CELLS: PROSTAGLANDIN PRODUCTION

In cultured mesangial cells, addition of PAF stimulates PG synthesis
(30,36,37), generates reactive oxygen species (29,38), and causes cell contraction (30,36). In general, many effects of PAF are closely interrelated
with eicosanoids. For example, several consequences of *in vivo* PAF administration may be secondary to PAF-induced generation of eicosanoids
(1-3). The effect of PAF on PG production has been examined in cultured
mesangial cells from rat (30,37) and human glomeruli (36). In both cell

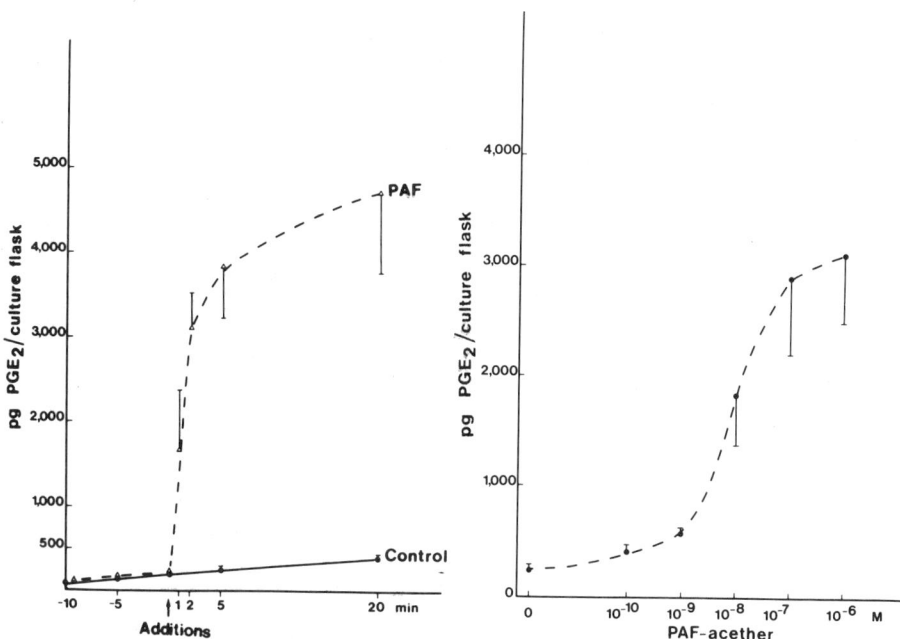

Fig. 7. Effect of PAF on PGE$_2$ production by mesangial cells. Panel on left shows the time course and panel on right the dose response curve. Reproduced from reference 30 with permission.

types, PAF markedly increased PG synthesis with a very rapid time course and in a dose-dependent manner with threshold stimulation occurring at 10^{-10}M PAF (Fig. 7). The PAF-induced stimulation of PGE$_2$ synthesis can be prevented by a number of structurally unrelated PAF antagonists (Fig. 8) (39,40). These PAF antagonists are specific for PAF as they did not inhibit the stimulation of PGE$_2$ synthesis by either angiotensin II or A23187 (39).

The release of AA from cellular phospholipids is generally considered to be the major control factor for PG synthesis. We therefore determined the effect of PAF alone or in combination with PAF antagonists on [^{14}C] AA release from prelabeled mesangial cells (39). PAF stimulated the release of [^{14}C] AA, an effect inhibitable by PAF antagonists (Fig. 9). These results also indicate that the PAF antagonists interfere with early steps of PAF's action. The knowledge of the specificity of PAF antagonists in mesangial cells should allow a more educated use of these blockers in future studies involving the kidney.

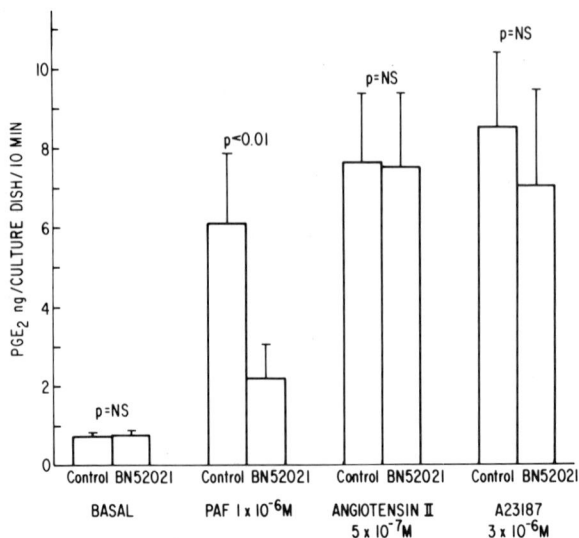

Fig. 8. Effect of PAF antagonist BN52021 (10^{-5}M) on PGE$_2$ synthesis by mesangial cells under basal conditions and, in response to PAF, angiotensin II or A23187. Reproduced from reference 39 with permission.

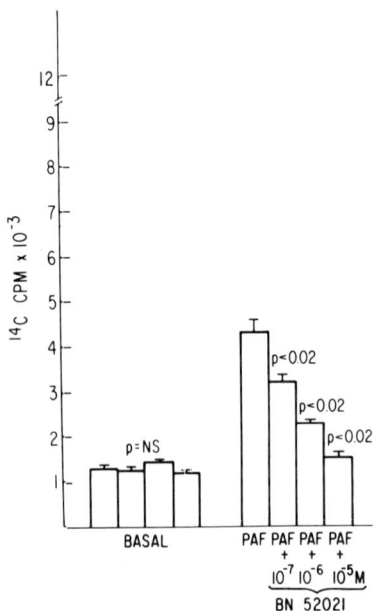

Fig. 9. Effect of PAF antagonist BN52021 (10^{-5}M) on [^{14}C] release from mesangial cells prelabeled with [^{14}C] arachidonic acid and then incubated either under basal conditions or with PAF (10^{-6}M). Reproduced in part from reference 39 with permission.

210

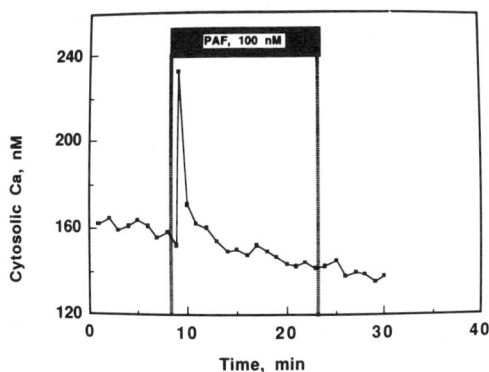

Fig. 10. Determination of intracellular
calcium by Fura 2 in mesangial
cells during control incubations and with addition of
PAF. Unpublished results kindly provided by A. Hassid.

ACTIVATION OF PHOSPHOLIPASE C AND A_2 BY PAF

Examination of the cellular lipids from [^{14}C] AA prelabeled mesangial
cells indicated that PAF activated both phospholipase C and phospholipase
A_2 (30). Phospholipase C activation was suggested by the increase in
[^{14}C]-labeled diacylglycerol and phospholipase A_2 activation by the
decrease in [^{14}C]-labeled phospholipids. The phospholipase C that is stimulated by PAF appears to be specific for phosphatidylinositol 4,5 bisphosphate, rapidly releasing inositol-trisphosphate (IP_3) and diacylglycerol
(37,41). The IP_3 so generated has been shown to release calcium from
intracellular stores in a number of cell systems (42). In experiments
carried out in collaboration with Dr. A. Hassid, the response of mesangial
cells to PAF was evaluated by determining intracellular calcium concentration by the use of the fluorescent indicator, Fura II (Fig. 10). PAF induced a very rapid rise in intracellular calcium which was followed by moderate elevation before calcium returned to the control level. Similar results
have been observed by others and have also been correlated with the appearance of IP_3 (37,41,43).

Taken together, these results demonstrate that PAF rapidly activates
phospholipase C for phosphatidylinositol 4,5 bisphosphate, thereby generating IP_3. The IP_3, in turn, causes an increase in calcium by release
from intracellular stores. Recent evidence suggests that GTP binding pro-
teins--G proteins--may also be involved in receptor-effector interactions

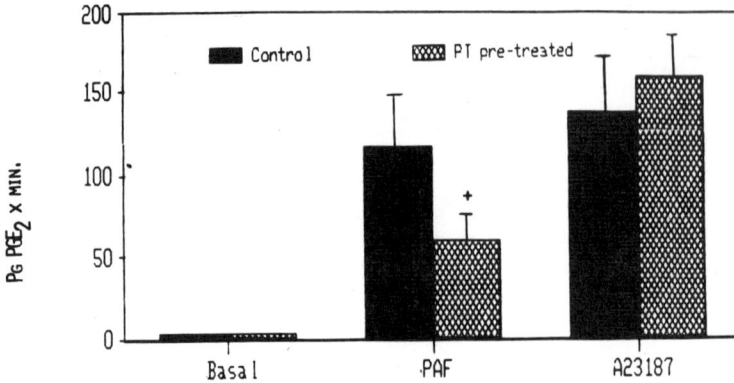

Fig. 11. Effect of 2 hours preincubation of mesangial cells with pertussis toxin (PT) on subsequent PGE_2 production under basal conditions and after stimulation with either angiotensin II or A23187. Cross indicates $P<0.05$ compared to PAF only (Schlondorff et al., unpublished).

that activate phospholipase C (44). Such G proteins may be similar to the inhibitory subunit (G_i) of the adenylate cyclase complex, in that they can be inactivated by ADP-ribosylation catalyzed by pertussis toxin. Pretreatment with pertussis toxin has been reported to inhibit the release of AA in PAF-stimulated 3T3 fibroblasts (45) and the activation of phospholipase C in mesangial cells (46,47). It appears, however, that pertussis toxin pretreatment may not inhibit phospholipase C activation in all cell systems (44). Furthermore, there may be multiple, and even cell-specific, G proteins that

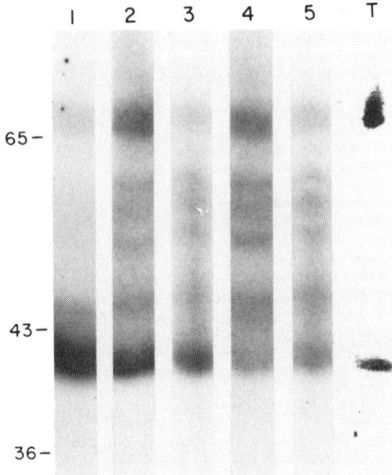

Fig. 12. Autoradiography of SDS-PAGE for pertussis toxin-catalyzed radiolabeling of mesangial membrane preparations. Mesangial cells were pretreated for 2 hours in the absence (lane 1) or presence of increasing concentrations of pertussis toxin (2nd lane: 25 ng/ml; 3rd lane: 50 ng/ml; 4th lane: 100 ng/ml; 5th lane: 200 ng/ml). Membranes were then prepared and [^{32}P] NAD ribosylated with activated pertussis toxin. Note doublet labeled band around 39-41 K Da. Lane T shows immunoblot with antibody directed against N terminal of the alpha subunit of G^1 protein. Schlondorff et al., unpublished.

can be ADP ribosylated by pertussis toxin and that may be different from G_i of adenylate cyclase (44). We therefore attempted to identify the role of such G proteins in mesangial cells. Pretreatment with pertussis toxin decreased the subsequent response to PAF in terms of [^{14}C] AA release and PGE_2 formation but not in response to A23187 (47). The results for PGE_2 are illustrated in Fig. 11. Pretreatment of mesangial cells with increasing doses of pertussis toxin decreased the subsequent in vitro ADP ribosylation, evaluated as [^{32}P] radiolabeling of protein bands on SDS gel (47). We were able to resolve two adjacent bands, which are ADP ribosylated in the presence of pertussis toxin (Fig. 12). When an immunoblot was performed with an antibody reacting with G_i, only one of the two ribosylated bands showed immunoreactivity (Fig. 12). Thus, it remains to be determined which of the two G proteins is involved with activation of phospholipase C by PAF. At this point it also remains unclear if and how activation of phospholipase C and A_2 interrelate. Two major possibilities to be considered are:

1. Phospholipase C activation with formation of IP_3 and diacylglycerol increase intracellular calcium which, in turn, stimulates phospholipase A_2 resulting in release of AA and PG formation.

2. Phospholipase A_2 is activated independently from phospholipase C. This could involve interaction of specific receptors with phospholipase A_2, perhaps also controlled by a GTP binding protein. Elucidation of these questions will require future work which will be complicated by the very rapid time course of the cellular events and the multitude of G proteins emerging.

EFFECT OF PAF ON MESANGIAL CELL CONTRACTION

PAF has been demonstrated to cause smooth muscle contraction in a number of systems (2). PAF also contracts isolated rat glomeruli (48), a phenomenon that has been related to the action of the smooth muscle-like mesangial cells. In fact, PAF has been shown to affect cultured mesangial cells directly in a manner consistent with cell contraction (30,36). This PAF-induced cell contraction is inhibited by PAF blockers (39), consistent with a PAF receptor-mediated event. It is likely that the increase in tension of the mesangial cells is related to phospholipase C activation and the increase in intracellular calcium. Contraction of mesangial cells in response to PAF is enhanced by inhibition of the endogenous PG formation and is mitigated by addition of exogenous PGE (30). Thus, PAF stimulates PG synthesis by mesangial cells which, in turn, modulates the contraction of mesangial cells. This feedback mechanism is similar to that observed for angiotensin

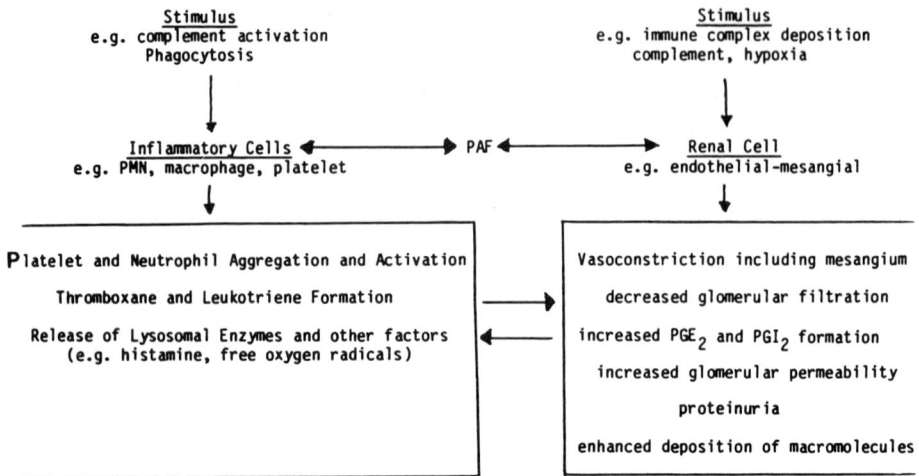

```
         Stimulus                                              Stimulus
 e.g. complement activation                          e.g. immune complex deposition
       Phagocytosis                                      complement, hypoxia

            │                                                   │
            ▼                                                   ▼
   Inflammatory Cells  ◄───────────►  PAF ◄───────────►   Renal Cell
 e.g. PMN, macrophage, platelet                        e.g. endothelial-mesangial
            │                                                   │
            ▼                                                   ▼
```

Platelet and Neutrophil Aggregation and Activation	Vasoconstriction including mesangium
Thromboxane and Leukotriene Formation	decreased glomerular filtration
Release of Lysosomal Enzymes and other factors (e.g. histamine, free oxygen radicals)	increased PGE_2 and PGI_2 formation
	increased glomerular permeability
	proteinuria
	enhanced deposition of macromolecules

Fig. 13. Potential role for PAF during interactions between inflammatory and renal cells. Reproduced from reference 5 with permission.

II (49). It is also operational in the intact kidney, as inhibition of PG formation aggravates the decrease in GFR in response to PAF administration (6,7). On the other hand, it has to be considered that PAF-induced mesangial cell contraction and PGE_2 formation could act synergistically to enhance glomerular vascular permeability resulting in proteinuria.

POTENTIAL ROLE FOR PAF IN GLOMERULAR INJURY

Based on our present knowledge of PAF, it appears reasonable to add PAF to the list of potential mediators of glomerular injury, as already supported by a number of experimental studies (Table 2; for review see ref. 13). A dual role for PAF, involving both resident glomerular cells and infiltrating inflammatory cells (Fig. 13), can be envisioned. (a) PAF could be generated by glomerular endothelial and mesangial cells, and potential stimuli for such intraglomerular PAF formation could be local deposition or formation of antigen-antibody complexes, with or without complement activation, mesangial uptake of immune complexes, endotoxin or local stimulation by angiotensin II. Such PAF synthesis by glomerular cells could cause adhesion and activation of leucocytes and platelets passing through the glomerulus with further generation of mediators of inflammation, such as thromboxane, leukotrienes, histamine, reactive oxygen species, etc.. (b) PAF could be released from inflammatory cells within the glomerulus, resulting in activation of platelets and leukocytes. In addition, PAF would cause glomerular mesangial contraction, decreasing glomerular filtration and possibly enhancing glomerular

214

capillary permeability, thus producing proteinuria. Finally, all of these factors may also contribute to the eventual outcome of the glomerular injury by influencing local cell proliferation and fibrosis. While much of this remains hypothetical at present, it appears likely that PAF will have to be added to the list of local mediators of glomerular injury.

ACKNOWLEDGEMENTS

This work was supported by grant AM-22036 from the National Institutes of Health and a grant-in-aid from the New York Heart Association.

REFERENCES

1. F. Snyder, Chemical and biochemical aspects of platelet activating factor: a novel class of acetylated ether-linked choline-phospholipids, Med. Res. Rev. 5:107-140 (1985).

2. R.N. Pinckard, L.M. McManus, D.J. Hanahan, Chemistry and biology of acetyl glyceryl ether phosphorylcholine (platelet-activating factor), Adv. Inflamm. Res. 4:147-180 (1982).

3. B.B. Vargaftig, M. Chignard, J. Benveniste, J. Lefort, F. Wal, Background and present status of research on platelet-activating factor (PAF-acether), Ann. NY Acad. Sci. 370:119-137 (1981).

4. F.H. Chilton, J.M. Ellis, S.C. Olson, R.L. Wykle, 0-Alkyl-2:sn-glycero-3-phosphocholine. A common source of platelet-activating factor and arachidonate in human polymorphonuclear leukocytes, J. Biol. Chem. 257:5402-5407 (1982).

5. D. Schlondorff and R. Neuwirth, Platelet-activating factor and the kidney, Am. J. Physiol. 251:F1-F11 (1986).

6. R.L. Hebert, P. Sirois, P. Braquet, G.E. Plante, Hemodynamic effects of PAF-acether on the dog kidney, Prostaglandins, Leukotrienes and Med. 26:189-202 (1987).

7. H. Scherf, A.S. Nies, U. Schwertschlag, M. Hughes, J.G. Gerber, Hemodynamic effects of platelet-activating factor in the dog kidney in vivo, Hypertension 8:737-741 (1986).

8. U. Schwertschlag, H. Scherf, J.G. Gerber, M. Mathias, A.S. Nies, L-platelet-activating factor (L-PAF) induces changes on renal vascular resistance, vascular reactivity and renin release in the isolated perfused rat kidney, Circulation Res. 60:534-539 (1987).

9. U.S. Schwertschlag and V.W. Dennis, Renal hemodynamic and functional effects of platelet-activating factor, Clin. Res. 34:608A (1986).

10. S.M. Weisman, D. Felsen, E.D. Vaughan, Platelet-activating factor is a

potent stimulus for renal prostaglandin synthesis: possible significance in unilateral ureteral ligation, J. Pharmacol. Exp. Ther. 235:
10-15 (1985).

11. E. Pirotzky, C. Page, J. Morley, J. Bidault, J. Benveniste, Vascular permeability induced by PAF-acether (platelet-activating factor) in the
isolated perfused rat kidney, Agents Actions 16:1-2 (1985).

12. G. Camussi, C. Tetta, R. Coda, G.P. Segoloni, A. Vercellone, Platelet-
activating factor-induced loss of glomerular anionic charges, Kidney
Int. 25:73-81 (1984).

13. G. Camussi, Potential role of platelet-activating factor in renal pathophysiology, Kidney Int. 29:469-477 (1986).

14. T. Bertani, M. Livio, D. Macconi, M. Mongi, G. Bisogno, C. Patrono, G.
Remuzzi, Platelet-activating factor (PAF) as a mediator of injury in
nephrotoxic nephritis, Kidney International 31:1248-1256 (1987).

15. Z. Hruby, R.P. Lowry, R.D.C. Forbes, D. Blais, Immune reactivity and
immunosuppressive intervention in experimental nephritis. IV. Effects
of PAF (AGEPC) inhibitor CV3988 on albuminuria and histopathology in
the accelerated autologous form of nephrotoxic serum nephritis, in:
"Proceedings of the First Sandoz Research Symposium. New Horizons in
Platelet-Activating Factor Research." C.M. Winslow, ed., Wiley,
London (1986).

16. G. Camussi, C. Tetta, C. Deregitus, F. Bussolino, G. Segolomi, A.
Vercellone, Platelet-activating factor (PAF) in experimentally-
induced rabbit acute serum sickness: role of basophil-derived PAF in
immune complex deposition, J. Immunol. 128:86-94 (1982).

17. G. Camussi, I. Pawlowski, R. Saunders, J. Brentjens, G. Andres, A receptor antagonist of platelet-activating factor inhibits inflammatory
injury induced by in situ formation of immune complexes in renal
glomeruli and in skin, Lab. Invest., in press.

18. J. Wang and M.J. Dunn, Platelet-activating factor mediates endotoxin-
induced acute renal insufficiency in rats, Am. J. Physiol. 253 (Renal
Fluid Electrolyte Physiol. 22):F1283-F1289, 1987.

19. J. Wang, M. Kester, M.J. Dunn, Endotoxin stimulates platelet-activating
factor synthesis in cultured rat glomerular mesangial cells, Clin.
Res. 35:638A (1987).

20. E. Pirotzky, J. Bidault, C. Burtin, M.C. Gubler, J. Benveniste, Release
of platelet-activating factor, slow reacting substance and vasoactive
species from isolated rat kidneys, Kidney Int. 25:404-410 (1984).

21. E. Pirotzky, E. Ninio, J. Bidault, P. Pfister, J. Benveniste, Biosynthesis of platelet-activating factor. VI. Precursor of platelet-activating factor and acetyl transferase activity in isolated rat kidney

cells, Lab. Invest. 51:567-572 (1984).

22. D. Schlondorff, P. Goldwasser, R. Neuwirth, J.A. Satriano, K. Clay, Production of platelet-activating factor in glomeruli and cultured glomerular mesangial cells, Am. J. Physiol. 250:F1123-1127 (1986).

23. D. Schlondorff, J. Perez, J.A. Satriano, Differential stimulation of PGE$_2$ synthesis in mesangial cells by angiotensin and A23187, Am. J. Physiol. 248:C119-C126 (1985).

24. D. Schlondorff, S. DeCandido, J.A. Satriano, Angiotensin stimulates both phospholipase A$_2$ and C in mesangial cells, Am. J. Physiol., in press.

25. G.A. Zimmerman, T.M. McIntyre, S.M. Prescott, Thrombin stimulates the adherence of neutrophils to human endothelial cells in vitro, J. Clin. Invest. 76:2235-2246 (1985).

26. M.G. Farquhar and G.E. Palade, Functional evidence for the existence of a third cell type in the renal glomerulus, J. Cell Biol. 13:55-87 (1962).

27. A.F. Michael, W.F. Keane, L. Raij, R.L. Vernier, S.M. Mauer, The glomerular mesangium, Kidney Int. 17:141-154 (1980).

28. L. Baud, J. Hagege, E. Straer, E. Rondeau, J. Perez, R. Ardaillou, Reactive oxygen production by cultured rat glomerular mesangial cells during phagocytosis is associated with stimulation of lipoxygenase activity, J. Exp. Med. 158:1836-1852 (1983).

29. L. Baud, J. Perez, R. Ardaillou, Dexamethasone and hydrogen peroxide production by mesangial cells during phagocytosis, Am. J. Physiol. 250:F596-F604 (1986).

30. D. Schlondorff, J.A. Satriano, J. Hagege, J. Perez, L. Baud, Effect of platelet-activating factor and serum-treated zymosan on prostaglandin E$_2$ synthesis, arachidonic acid release and contraction of cultured rat mesangial cells, J. Clin. Invest. 73:1227-1231 (1984).

31. P.C. Singhal, G.H. Ding, S. DeCandido, N. Franki, R.M. Hays, D. Schlondorff, Endocytosis by cultured mesangial cells and associated changes in prostaglandin E$_2$ synthesis, Am. J. Physiol. 252:F627-F634 (1987).

32. R. Neuwirth, P. Singhal, A. Sinha, R.M. Hays, D. Schlondorff, Macromolecular uptake by mesangial cells is enhanced by Fc portion of IgG and is associated with production of PGE$_2$ and platelet-activating factor, in: "Proceedings of the Xth. Int. Congress of Nephrology," London, (1987).

33. J.R. Sedor, S.W. Carey, S.N. Emancipator, Immune complexes bind to cultured rat glomerular mesangial cells to stimulate superoxide release: evidence for an Fc receptor, J. Immunol. 138:3751-3757, 1987.

34. P. Menè, S.A. Ricanati, G.R. Dubyak, S.N. Emancipator, M.J. Dunn, Stimulation of cytosolic free calcium and contraction by immune complexes in cultured rat mesangial cells, Clin. Res. 35:662A (1987).

35. R. Neuwirth, P. Braquet, D. Schlondorff, Metabolism of [^{3}H] platelet-activating factor by cultured mesangial cells, in: "Proc. of the 2nd Int. on Platelet-activating Factor," Gatlinburg (1986).

36. N. Ardaillou, J. Hagege, M.-P. Nivez, R. Ardaillou, D. Schlondorff, Vasoconstrictor-evoked prostaglandin synthesis in cultured human mesangial cells, Am. J. Physiol. 248:F240-246 (1985).

37. J. Pfeilschifter, A. Kurtz, C. Bauer, Role of phospholipase C and protein kinase C in vasoconstrictor-induced prostaglandin synthesis in cultured rat renal mesangial cells, Biochem. J. 234:125-130 (1986).

38. J.R. Sedor and H.E. Abboud, Platelet-activating factor stimulates oxygen radical release by cultured rat mesangial cells (abstract), Kidney Int. 27:222A (1985).

39. R. Neuwirth, P. Singhal, J.A. Satriano, P. Braquet, D. Schlondorff, Effect of platelet-activating factor antagonists on cultured rat mesangial cells, J. Pharm. Exp. Therapeut., in press.

40. J.E. Stork, T.Y. Shen, M.J. Dunn, Stimulation of prostaglandin E$_2$ and thromboxane B$_2$ production in cultured rat mesangial cells by platelet-activating factor: inhibition by a specific receptor antagonist (abstract), Kidney Int. 27:267A (1985).

41. J.V. Bonventre and P.C. Weber, Effect of platelet-activating factor and platelet-derived growth factor on cytosolic free calcium concentration and phospholipase A$_2$ and C activation in glomerular mesangial cells, Kidney Int. 31:161A (1987).

42. J.J. Berridge and R.F. Irvine, Inositol trisphosphate, a novel second messenger in cellular signal transduction, Nature 312:315-321 (1984).

43. M. Kester, P. Menè, G.R. Dubyak, M.J. Dunn, Platelet-activating factor elevates cytosolic free calcium concentration in cultured rat mesangial cells, Clin. Res. 35:550A (1987).

44. A.M. Spiegel, Signal transduction by guanine nucleotide binding proteins, Mol. Cell Endocrinol. 49:1-16 (1987).

45. T. Murayama and M. Ui, Receptor-mediated inhibition of adenylate cyclase and stimulation of arachidonic acid release in 3T3 fibroblasts, J. Biol. Chem. 260:7226-7233 (1985).

46. J. Pfeilschifter and C. Bauer, Pertussis toxin abolished angiotensin II-induced phosphoinositide hydrolysis and prostaglandin synthesis in rat renal mesangial cells, Biochem. J. 236:289-294 (1986).

47. D. Schlondorff, J.A. Satriano, S. DeCandido, Different concentrations of pertussis toxin have opposite effects on agonist-induced PGE$_2$

formation in mesangial cells, <u>Biochem. Biophys. Res. Comm.</u> 141:39-45 (1986).

48. R. Barnett, P. Goldwasser, L.A. Scharschmidt, D. Schlondorff, Effects of leukotrienes on isolated rat glomeruli and cultured mesangial cells, <u>Am. J. Physiol</u>. 250:F838-844 (1986).

49. L.A. Scharschmidt, M.S. Simonson, M.J. Dunn, Glomerular prostaglandins, angiotensin II, and nonsteroidal antiinflammatory drugs, <u>Am. J. Med.</u> 21:30-42 (1986).

EICOSANOIDS AND PLATELET ACTIVATING FACTOR AS POSSIBLE MEDIATORS OF INJURY

IN EXPERIMENTAL NEPHROPATHIES

Giuseppe Remuzzi, M.D.

Mario Negri Institute for Pharmacological Research,
Via Gavazzeni 11, Bergamo and Division of Nephrology and
Dialysis, Ospedali Riuniti de Bergamo, Italy

INTRODUCTION

That secondary mediators of inflammation such as arachidonate (AA) metabolites and platelet activating factor (PAF) determine morphological and clinical expression of toxic or immune-mediated nephropathies has been the subject of extensive investigations in the last few years. More recently the demonstration that, beside circulating inflammatory cells, resident renal cells can generate prostaglandins and thromboxane A_2 (TxA_2) (1,2) as well as PAF (3) has opened new perspectives to clarify the mechanisms of tissue damage in the above-mentioned conditions. Most interestingly, both AA metabolites and PAF may have a common precursor (4,5), and both can be formed as a consequence of phospholipase A_2 activation (6,7). Thus, a new field of investigation in renal diseases will be to clarify the relation between the two pathways as determinant of tissue injury.

This chapter considers (a) adriamycin (ADR) and puromycin (PA) nephrosis in the rat as an example of a toxic nephropathy mainly characterized by an increased glomerular permeability to proteins with many similarities with human minimal change nephrosis; (b) antiglomerular basement membrane glomerulonephritis in the rabbit as an example of immune-mediated glomerulopathy characterized by intra- and extracapillary proliferation with massive infiltration of inflammatory cells and fibrin deposits; the disease is the animal equivalent of human rapidly progressive glomerulonephritis; (c) CyA nephrotoxicity in the rat as an example of toxic tubulo-interstitial disease with macrophage accumulation and progressively deteriorating renal function.

In these conditions the evidence that eicosanoids and/or PAF mediate
tissue damage will be reviewed, and the possible interrelationship between
the two pathways will be discussed.

ADRIAMYCIN AND PUROMYCIN NEPHROSIS IN THE RAT

During the last years, many studies have been devoted to investigate the
factors that, in the glomerular capillary wall, oppose the passage of circulating macromolecules into the urinary space. Here I will briefly review
the subject first, then describe the most employed animal models of chronic
proteinuria, and lastly focus on the possible role of eicosanoids and PAF as
mediators of altered glomerular capillary permeability to proteins.

Studies with particulate tracers, cationic probes, and immunohistochemical markers have now estabished that the glomerular capillary wall restricts
the passage of circulating macromolecules by two major determinants: the
size and the charge selectivity properties of the glomerular basement membrane (GBM) (8-10). The size selective properties of GBM have been clearly
demonstrated with the use in experimental animals of neutral dextrans of different molecular weight (11-14), whereas the charge selectivity properties
have been documented employing ferritin with different isoelectric point
(10,15). In 1979 Kanwar and Farquhar (16,17) isolated proteoglycans from
the GBM and found that the anionic sites of the GBM consisted of heparan sulfate proteoglycans located in the laminae rarae interna and externa. The
role of proteoglycans seems to be important both in maintaining the GBM
charge selective barrier and in preserving the size selective properties
(18).

Conditions of chronic proteinuria in experimental animals and humans
often precede the development of glomerular focal and segmental sclerosis, a
lesion that is believed to be the main cause of progressive deteriorating
renal function, possibly due to hemodynamic changes in residual glomeruli.
Recent advances in understanding the mechanisms of proteinuria and the subsequent development of focal and segmental sclerotic lesions derive from the
studies of two rat models: ADR and PA-induced nephrosis. In both models the
early phase of the disease is characterized by abnormal glomerular permeability to proteins, which precedes the development of glomerular sclerosis and
renal function deterioration.

Adriamycin nephrosis is induced in the rat by a single intravenous (IV)
injection of the drug, which causes proteinuria (within a few days), hypoal-

buminemia and increased serum lipid levels. Proteinuria reaches its maximum values after 14-16 days from the ADR injection. Ultrastructural examination of the kidney shows extensive damage to the glomerular epithelial cell (which appears as early as 36-48 hours from ADR injection) with diffuse swelling and foot process fusion. In the cytoplasm of proximal tubular cells, protein reabsorption droplets are detected with casts in the distal tubules. Adriamycin nephrosis is a chronic disease of persisting protein-uria and progressive glomerulosclerosis (19).

Puromycin nephrosis also can be induced by a single injection of amino-nucleoside of PA (80 mg/kg). As in the ADR model, proteinuria develops in a few days and reaches maximal values in 14-16 days. As in ADR nephrosis, ani-mals develop hypoalbuminemia and increased serum lipids (20). Kidney ultra-structural alterations are also very similar in both conditions (21,22). However, at variance with the ADR model, animals treated with a single IV injection of PA recover after 4 weeks of nephrotic syndrome (21). A chronic and persistent proteinuria can be obtained in PA-treated animals only after repeated injections (23) (Table).

In studies of the renal metabolism of AA in ADR nephrosis (24), it has been found that 14 days after ADR injection, in the presence of marked pro-teinuria, both the generation of TxA_2, measured as immunoreactive thrombox-ane B_2 by isolated glomeruli, and urinary excretion of TxB_2 were signifi-cantly increased. Trials with thromboxane inhibitors in ADR nephrosis showed a reduction in glomerular production and urinary excretion of TxB_2 with a concomitant decrease in urinary protein excretion (24,25). Results obtained in a unilateral model of ADR nephrosis (26) excluded the possibil-ity that this abnormality was a consequence of nephrotic syndrome.

The source of the increased urinary excretion of TxB_2 remains an open issue. Since thromboxane is known to be released by white cells and plate-lets (27-29), the high levels measured in isolated glomeruli might derive from polymorphonuclear cells, macrophages or platelets, trapped in the glo-merular preparation. That platelet activation might play a role in the de-velopment of proteinuria derives from the studies of Tetta and coworkers (30) who showed that the localization of platelet-derived cationic proteins in the glomerular capillary wall was concomitant with the loss of fixed ani-onic charges. Whether the increased thromboxane excretion in experimental nephrosis is a marker of platelet activation is a possibility for further investigation. If this is the case, thromboxane inhibitors may reduce pro-tein excretion, reducing the degree of platelet activation and the release

Table. Characteristics of the Animal Model

Adriamycin (ADR) nephrosis	Puromycin (PA) nephrosis
* Rat injected i.v. with ADR (7.5 mg/Kg)	* Rat injected i.v. with PA (80 mg/Kg)

* Development of proteinuria in few days
* Massive proteinuria at 14-16 days
* Light microscopy ---> no significant changes
* Electron microscopy ---> glomerular epithelial cell
  - foot process fusion
  - intracytoplasmic inclusions
  - focal detachment of cytoplasm from the basement membrane.

Adriamycin (ADR) nephrosis	Puromycin (PA) nephrosis
* Persisting proteinuria and progressive glomerulosclerosis	* Complete recovery after 4 weeks
	* Chronic and persisting proteinuria only after PA repeated injections

of platelet-derived cationic proteins. Another possibility is that resident glomerular cells, rather than circulating cells, generate an excessive amount of TxA_2 in ADR nephrosis. In support of the latter hypothesis is the consideration that ultrastructural examination of glomeruli from animals with ADR nephrosis failed to reveal a significant number of platelets or other inflammatory cells entrapped in the capillary lumens (19,24). However, the possibility that blood-borne cells, particularly platelets, contribute to the increased thromboxane production in the ADR model still remains an attractive possibility.

Some of the studies made in the ADR model have recently been repeated in PA nephrosis. Thus, Goto and coworkers (31) assessed the daily urinary excretion rate of proteins and the *in vitro* synthesis of TxA_2, measured as TxB_2, from isolated glomeruli of rats with PA nephrosis. A significant correlation was found between the *in vitro* TxB_2 synthetic rate and the urinary protein excretion rate. Moreover, when PA nephrosis animals were given a selective TxA_2 inhibitor, a significant decrease in urinary protein excretion was found with respect to PA animals that did not receive the compound. Thus, for both the currently available models of experimental nephrosis, there is now quite convincing evidence that the increased glomerular permeability to proteins is associated with an increased renal synthesis of TxA_2 and that pharmacological blocking of the excessive TxA_2 generation results in a reduction of proteinuria.

Appropriate studies are necessary to ascertain whether an inhibition of TxA$_2$ synthesis accounts for the favorable effect on proteinuria repeatedly reported in humans with nephrotic syndrome given indomethacin. Adriamycin and PA nephrosis in rats have remarkable similarities with human minimal change glomerulopathy with respect to the morphological expression and evolution of the disease. Of most interest is the recent observation that a TxA$_2$ synthase inhibitor effectively decreased urinary protein excretion in patients with minimal change nephrotic syndrome and that one of these patients, resistant to previous therapy, underwent remission with the TxA$_2$ synthase inhibitor alone (32). The therapeutic effect of the TxA$_2$ synthase inhibitor was associated with a reduction in the exaggerated urinary excretion of both TxB$_2$ and 2,3-dinor-TxB$_2$, thus suggesting that both resident glomerular cells and platelets participate in the increased formation of thromboxane in this condition.

Two questions arise from the above-mentioned studies. (a) Does the increased renal thromboxane contribute to the evolution of the disease to glomerulosclerosis and, if so, can pharmacological manipulations of TxA$_2$ synthesis or activity protect against the development of chronic renal failure in experimental nephrosis? (b) Is the increased TxA$_2$ synthesis a primary phenomenon in experimental nephrosis, or is it mediated by other system(s) within the kidney which promote membrane phospholipase activation, increasing the release and the consequent metabolism of AA?

Regarding the first question, the rationale for looking at the possible protective effect of TxA$_2$ inhibitors on the evolution of experimental nephrosis to glomerulosclerosis and renal failure rests on the observation of Purkerson and coworkers (33), who showed that chronic administraton of an inhibitor of TxA$_2$ synthesis in rats with a remnant kidney decreased protein and TxB$_2$ urinary excretion, improved renal histology and ameliorated renal function parameters. The mechanism responsible for the favorable effect of inhibiting TxA$_2$ synthesis on the subsequent evolution of renal disease in animals with subtotal renal ablation is not clear. On the basis of para-aminohippuric acid (PAH) and inulin clearances, the authors calculated theoretical values for single nephron plasma flow and single nephron glomerular filtration rate (GFR) and found both parameters to be increased in rats with remnant kidney as compared with normal rats, in agreement with previous data. However, the TxA$_2$ synthase inhibitor significantly prevented the progression of the disease despite increasing hyperfiltration. The authors concluded that factors other than hyperperfusion and hyperfiltration are responsible for the progression in this model. ADR and PA nephrosis progress to glomerulosclerosis without apparent changes in perfusion and capillary pressure. It would be of great interest in the near future to evaluate in these models whether pharmacological manipulation of TxA_2 synthesis can protect against glomerular sclerotic lesions and renal function deterioration.

The second question, whether the increased renal TxA_2 is a primary phenomenon or depends upon the activation of other mediator(s) within the kidney, is also worth investigating. In this context a recent paper by Egido and coworkers (34) opened interesting perspectives for future studies. These authors demonstrated that PAF receptor antagonists protect ADR-treated rats from the development of proteinuria. That PAF can play a role in glomerular damage in experimental and human glomerulopathies has been suggested by several lines of evidence in recent years. Thus, in acute serum sickness the glomerular deposition of circulating immune complexes is associated with platelet activation and degranulation of IgE-sensitized basophils with PAF release (35,36). Moreover, PAF is released during hyperacute allograft rejection (37). Finally, basophil degranulation and PAF release are found in patients in the acute phase of systemic lupus (38).

In all the above-mentioned conditions of glomerular damage, PAF is thought to be released from circulating cells infiltrating the glomerular tuft. In contrast, in the ADR model there is no visible infiltration of inflammatory cells into glomeruli (19,24). It is therefore tempting to speculate that the PAF receptor blocking in ADR nephrosis may antagonize the deleterious effect of an excessive amount of PAF, probably generated by resident glomerular cells. A possible link between PAF release at renal level and the excessive generation of thromboxane, either in experimental animals or in humans, is provided by studies showing that PAF causes a rapid and reversible increase of cytosolic free calcium (39-41), possibly through a phospholipase-dependent signal transduction system. Since the enhanced intracellular calcium induced by PAF is most likely responsible for activation of phospholipase A_2, resulting in the release of AA and subsequent metabolism to cyclooxygenase or lipoxgenase products (43-46), some of the effects obtained by the use of PAF receptor blocking may be attributable to reduced AA conversion to vasoactive and pro-inflammatory eicosanoids. In the case of ADR nephrosis, the possibility that PAF receptor blocking prevents PAF-mediated generation of excessive amounts of TxA_2 would be an important issue for further investigation. However, on the basis of the available results, it is worth mentioning that the TxA_2 synthase inhibitors studied so far in experimental nephrosis only partially protect animals from the development of

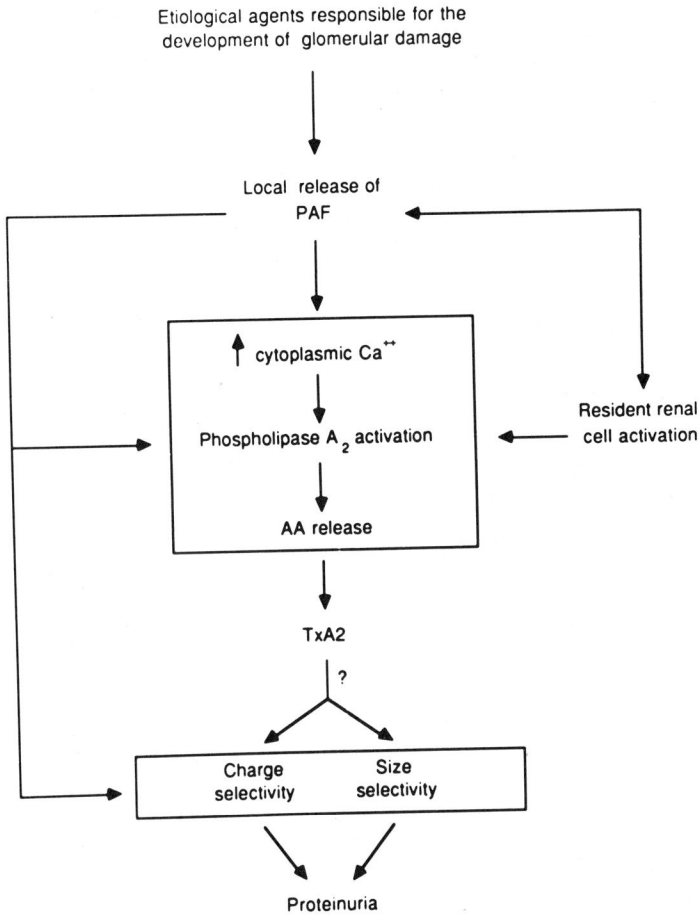

Fig. 1. Hypothetical role of PAF as possible mediator of changes in glomerular permselectivity to proteins in experimental nephrosis.

proteinuria (24,25), whereas the PAF receptor antagonist seems to preserve completely the permeability properties of glomerular capillaries in ADR-treated animals (34).

If confirmed, these results would indicate that the activity of PAF receptor antagonist is only partially due to its effects on AA metabolism, thus suggesting that the formation of PAF at renal level may play an important role in the altered glomerular permeability to proteins. That PAF can increase glomerular permeability to proteins, inducing the release of circulating cell cationic proteins in the glomerular capillary wall and loss of fixed anionic charges, has been demonstrated by Camussi and coworkers (47)

227

after infusing rabbits with PAF. Since, in experimental nephrosis, platelet or leukocyte deposition cannot be demonstrated at the glomerular level, it is difficult to establish whether the above-mentioned mechanism can contribute to the increased glomerular permeability to proteins in ADR-treated animals.

An alternative mechanism by which PAF might directly contribute to the increased glomerular permeability to proteins has been recently proposed. Actually, Perico and coworkers (48) have found that PAF, unlike 2-lyso-PAF, induced a dose-dependent progressive increase in urinary protein excretion in a model of isolated kidney perfused with a cell-free medium. Of interest is the observation that a specific PAF receptor antagonist inhibited the effect of PAF on glomerular permeability to proteins. In addition, the effect of PAF on glomerular permeability to proteins in isolated perfused kidney was not mediated by cyclooxygenase or lipoxygenase products, nor was it the result of oxygen-free radical generation. A direct effect of PAF in increasing glomerular permeability to protein can better explain the result obtained in ADR animals given PAF receptor antagonists (Fig. 1).

ANTIGLOMERULAR BASEMENT MEMBRANE GLOMERULONEPHRITIS IN THE RABBIT

Rabbits injected with sheep anti-rabbit nephrotoxic serum (NTS) have a proliferative glomerulonephritis with glomerular fibrin degradation and progressively deteriorating renal function (49). The disease develops in two subsequent phases. The first phase (heterologous phase) develops within a few hours of the injection of antiserum. It is due to the antibody binding to kidney GBM and is complement and polymorphonuclear cell (50,51) mediated. During the heterologous phase rabbits have heavy proteinuria and renal failure. Then proteinuria decreases, and renal function recovers until the development of the delayed phase (autologous phase) which starts 7-10 days after the injection of antiserum and is due to the host's immune response to heterologous antibody bound to the GBM. In this phase of the disease, proteinuria and renal insufficiency are associated with macrophage infiltration, fibrin formation and extracapillary proliferation (52-55).

The mechanism responsible for the formation of fibrin is not clear. It has been postulated that an increase in glomerular capillary permeability due to interruption of the GBM favors the passage of fibrinogen in extravascular space (56); then circulating cells infiltrating the glomerular tufts are likely to play an important role in promoting fibrin formation (53,54, 57-59). In this context recent studies have indicated that the deposition

228

of fibrin in Bowman's capsule can result from macrophage activation which generates procoagulant activities (60,61). The signal(s) for macrophage accumulation, which is complement independent, is not known.

Because PAF has been implicated as a mediator of damage in inflammatory diseases of the kidney (36,37) and given the fact that PAF activates macrophages (62) and increases glomerular permeability to proteins (47,63), several recent studies have addressed the possibility that PAF participates in the development of glomerular damage in nephrotoxic nephritis (NTN). So far, determination of PAF activity is only possible by bioassays based either on platelet aggregation (36) or serotonin release (64). These methods are inadequate to study conditions such as NTN in that the massive infiltration of inflammatory cells within the kidney leads to a release of more than one factor with potential platelet aggregating properties. However, the availability of specific and potent PAF receptor antagonists in recent years has provided indirect information on the potential role of PAF as mediator of damage in NTN (Fig. 2).

Actually, it has been found that PAF receptor antagonists reduce proteinuria, ameliorate renal function parameters and decrease histological lesions in both rat (65) and rabbit (66) NTN. These results suggest that PAF is likely to be released at the renal level in extracapillary glomerulonephritis and participate in tissue injury. At the moment there are no data to clarify whether PAF derives from circulating cells infiltrating the glomerular tuft or is generated by resident renal cells.

The effect of PAF receptor blocking on proteinuria supports the concept derived from many recent studies that PAF can increase glomerular permeability to proteins either because it causes release of cationic proteins from circulating cells (47) or directly, as has been shown in isolated perfused kidney models (48,63). Moreover, the favorable effect of PAF receptor antagonists on renal functional parameters implies that PAF can contribute to renal function deterioration in NTN. This possibility is consistent with experimental data showing that PAF can contract mesangial cells *in vitro* (46,67). If this mechanism can operate *in vivo*, this would result in a decrease in glomerular surface area with a consequent reduction in GFR. Since PAF causes vasoconstriction in isolated hydronephrotic kidneys (68), this can be regarded as a possible additional mechanism by which PAF mediates a decline in GFR.

The protective effect of the PAF receptor blocking on histopathological

Immune complexes

Complement activation

Deposition at endothelial and mesangial cell level

↑ cytoplasmic Ca^{++}

Macrophage recruitment → PAF generation

Phospholipase A_2 activation

Macrophage activation

Mesangial cell contraction

Arachidonic acid release

PMN and platelet-derived cationic proteins

Procoagulant activity

↓ Glomerular surface area

TxA_2 generation

Proteinuria

Fibrin formation

↓ GFR

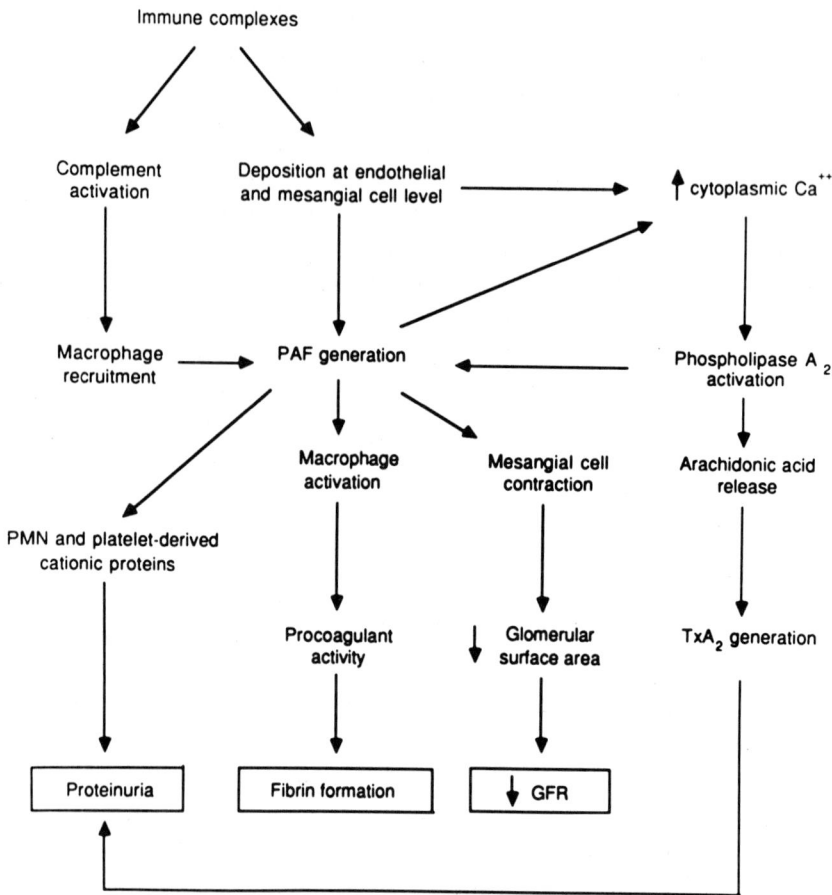

Fig. 2. Potential role of PAF as mediator of glomerular injury in nephrotoxic nephritis.

abnormalities and particularly on fibrin deposition suggests that PAF release in NTN interferes with processes of fibrin formation. In this context the possibility must be considered that PAF, by promoting macrophage accumulation and activation, determines the expression of macrophage procoagulant activity, which in turn favors fibrin deposition (60,61). Thus, PAF receptor antagonist may limit fibrin formation in NTN by reducing macrophage infiltration. In this context the recent demonstration that PAF directly stimulates inflammatory cells to express procoagulant activity (69) strongly supports the possibility that PAF-receptor antagonists inhibit fibrin formation by limiting PAF-induced macrophage activation.

That PAF can indeed be regarded as a new mediator of damage in NTN is also consistent with some recent preliminary data showing that PAF bioactivity can be documented in short-term culture of glomeruli from rats with NTN but not from control rats (64). It has been documented that the formation of PAF is initiated by activation of phospholipase A_2 (7). In the kidney following immune complex formation, phospholipase A_2 can be activated by an increased intracellular calcium at endothelial (41) or mesangial cell level (42). It is of interest that AA can be released by membrane phospholipids upon phospholipase A_2 activation (6); thus, the formation of cyclooxygenase and lipoxygenase metabolites can parallel PAF release. It follows that some of the effects attributed to PAF can actually result from the simultaneous formation of eicosanoids with inflammatory or vasoactive properties. Thus, in evaluating the results obtained by pharmacological blocking of PAF-receptor, it would be of interest to discriminate between effects directly due to the inhibition of PAF bioactivity and those depending on a reduction of PAF-dependent activation of AA metabolism.

That AA metabolism can be altered in NTN has been consistently demonstrated by several studies. Lianos and coworkers (70) have shown that the generation of TxB_2 is increased in glomeruli isolated from rats in the heterologous phase of NTN. Moreover, pretreatment of animals with thromboxane synthase inhibitor partially prevented the acute decrease in GFR and renal plasma flow (RPF) observed in this model. Of interest in the rat model, despite a marked increase in glomerular TxB_2 in the autologous phase, concomitant with a fall in GFR and RPF, the thromboxane synthase inhibitor did not influence renal function parameters in this phase of the disease (71). In the rabbit model glomerular TxB_2 production was found markedly increased over the normal values in both the heterologous and the autologous phase, whereas PGE_2 and 6-keto-$PGF_{1\alpha}$ were either not modified (heterologous phase) or reduced (6-keto-$PGF_{1\alpha}$ in the autologous phase) (66). The use of a selective PAF-receptor antagonist normalized the excessive glomerular thromboxane synthesis in the heterologous phase and markedly decreased it in the autologous phase of the disease (72).

These *in vivo* findings are consistent with *in vitro* results showing that, in primary cultures of rat (46) or human (67) glomerular mesangial cells, PAF causes a rapid and dose-dependent stimulation of AA metabolism that can be inhibited by specific PAF receptor antagonists (73,74). Thus, it is conceivable that in rabbit NTN the beneficial effects of PAF receptor blocking is at least partly mediated by the suppression of PAF-mediated increase in glomerular thromboxane synthesis. The observation that the use of a selective thromboxane synthase inhibitor in NTN reduced proteinuria, prevented renal function deterioration and ameliorated the morphological

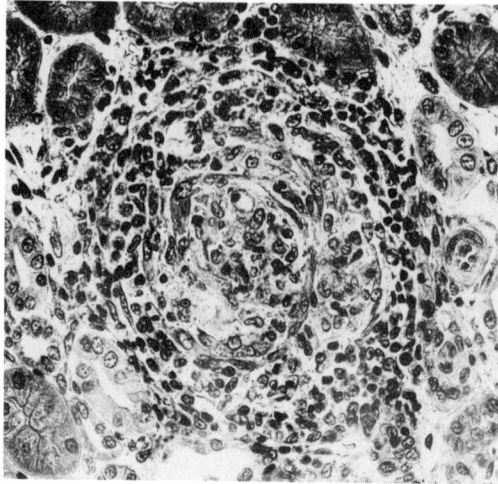

Fig. 3. Photomicrograph of a glomerulus from a rabbit on day 10 after induction of anti-GBM glomerulonephritis, demonstrating a severe extracapillary proliferation with crescent formation (Masson's Trichrome x 250).

expression of the disease supports this possibility (Figs. 3 and 4). However, the effect of PAF receptor blocking had a better protective effect on proteinuria in the autologous phase of the disease with respect to thromboxane synthase inhibition (72), thus indicating that PAF has probably some additional direct damaging effect on the glomerulus in NTN.

NEPHROTOXICITY OF CyA IN THE RAT

The use of cyclosporin (CyA) to prevent graft rejection is associated with major problems of nephrotoxicity (75-77). Extensive investigation has been performed in this area, but the cause(s) of CyA nephrotoxicity remains poorly understood. Most transplant patients on CyA develop episodes of acute renal failure, rapidly reversible upon reduction of CyA dose (76,78). A chronic type of CyA nephrosis has also been reported in patients given CyA to prevent graft rejection and is characterized by progressively deteriorating renal function (79,80). At pathological examination CyA induces proximal tubular cell damage, both in experimental animals (81,82) and in humans (83), which nevertheless does not appear to correlate with renal function deterioration (82).

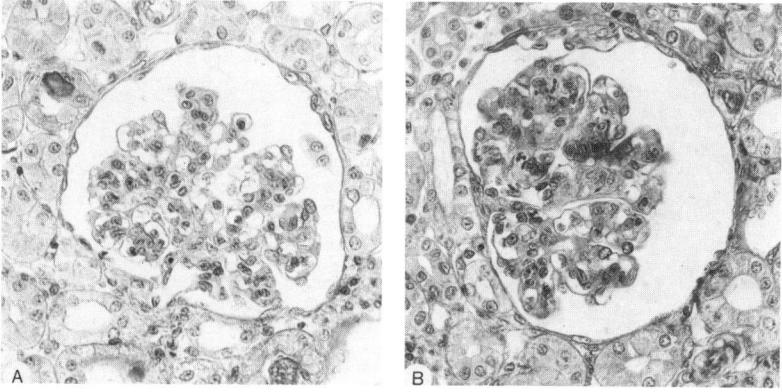

Fig. 4. Photomicrograph of a glomerulus from a rabbit on day 10
after the induction of anti-GBM glomerulonephritis and
treated with a thromboxane synthase inhibitor (panel A)
or with a PAF-receptor antagonist (panel B) (Masson's
Trichrome x 250). Note the absence of extracapillary
proliferation after both treatments.

Thus, several explanations have been attempted for CyA-induced decrease
in GFR, including stimulation of the renin-angiotensin system and activation
of tubuloglomerular feedback. A stimulation of renin-angiotensin axis has
been documented in some studies (84-86), but not in others (87,88). Recently it has been documented that the renal vasoconstriction and the fall in
GFR and RPF induced by CyA in the rat can be prevented by captopril, suggesting that angiotensin II may play an important role in the pathogenesis of
renal vasoconstriction produced by CyA (86). The possibility that reduction
in GFR induced by CyA results from the activation of the tubuloglomerular
feedback mechanism has also been considered (89). However, conclusive evidence in favor of this possibility is lacking.

That CyA can induce renal function deterioration, altering the formation
of AA metabolites at the renal level, derives from the observation that CyA
alters the metabolism of AA in several cellular systems, including blood monocytes (90) and smooth muscle cells in culture (91). Kawaguchi et al. (92)
first demonstrated that the administration of CyA to nonoperated Fischer
rats as well as to rats receiving a heterotopic cardiac isograft induced an
increase in urinary excretion of TxB_2. In the above-mentioned study, CyA
was administered at the dose of 15 mg/kg/day for 14 days. These findings
have been subsequently confirmed with other schedules of short-term CyA administration to normal rats (20 mg/kg/day for 20 days and 25 mg/kg/day for
45 days) (85,93).

The latter studies also showed that changes in urinary excretion of TxB_2 were rather selective, since prostaglandin E_2 (PGE_2) and prostacyclin (PGI_2) urinary excretion did not increase, and preceded the increase in serum creatinine and the decrease in GFR. In the same animals, short-term CyA administration was also associated with a stimulation of the renin-angiotensin system. In physiological conditions angiotensin, in addition to its direct effect of glomerular hemodynamics (94,95), stimulates PGE_2 and PGI_2 (96) which minimize the vasoconstrictory effect of angiotensin on the glomerular capillary network. It is possible that in CyA-treated animals the concomitant effect of excessive TxA_2 synthesis and stimulation of renin-angiotensin axis, not counterbalanced by a parallel increase in PGE_2 and PGI_2, is the cause of CyA-induced decrease in GFR.

The mechanisms by which an enhanced renal TxA_2 synthesis may decrease the GFR has been recently investigated. Since TxA_2 is a potent vasoconstrictor, an increased tone in efferent and afferent arteriole may explain a decrease in GFR due to a marked increase in glomerular vascular resistance, which in turn causes a decrease in renal blood flow (97). Alternatively TxA_2 may reduce the glomerular surface area stimulating mesangial cell constriction (98).

A more careful evaluation of the relationship between decreased GFR and excretion of TxB_2 is provided in another recent study performed in normal rats given CyA (40 mg/kg every other day) as a chronic administration (3 months) (99). This study showed that chronic treatment with CyA was associated with a progressive decrease in GFR, which rapidly normalizes after CyA withdrawal; TxB_2 urinary excretion increased in CyA-treated animals and normalized after CyA withdrawal. Of interest is the statistically significant correlation ($r=0.82$, $p<0.01$) found in this study between urinary excretion of TxB_2 and GFR measured as inulin clearance. No statistical correlation has been found between the exaggerated urinary excretion of TxB_2 and RPF. Taken together, the correlation of urinary TxB_2 with GFR but not with RPF suggests that TxA_2, by reducing the glomerular surface area, reduces the ultrafiltration coefficient. That increased renal TxA_2 synthesis has a functional significance in rats given CyA is further supported by the finding that the administration of a selective TxA_2 synthase inhibitor at a dose effectively inhibiting urinary excretion of TxB_2 was associated with a significant increase in GFR. However, GFR increased but did not normalize following the thromboxane-synthase inhibitor, thus suggesting that factor(s) other than increased TxA_2 synthesis must also play a role in CyA-induced renal insufficency (Fig. 5).

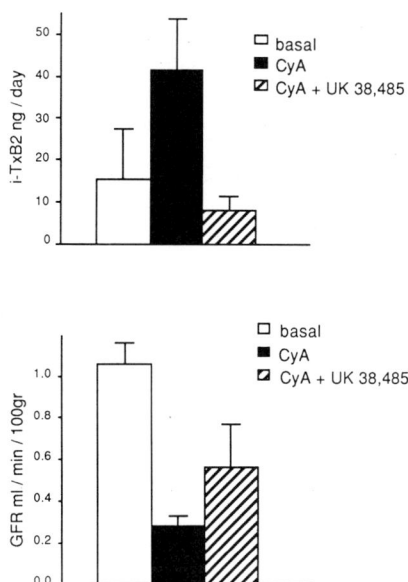

Fig. 5. Effect of UK-38,485, a thromboxane synthase inhibitor, on urinary excretion of TxB_2 and glomerular filtration rate in rats chronically given CyA as oral administration.

The cellular origin of the increased urinary TxB_2 in CyA-treated animals has not been identified so far. Since platelets represent the major source of thromboxane in various animal species, a platelet activation occurring in the systemic circulation has to be excluded before considering the possible renal origin of the increased urinary TxB_2. Alternatively, the increase in urinary excretion of TxB_2 could be due to locally activated blood-borne cells (platelets and macrophages) which may either infiltrate the glomerular tuft or be activated during their passage through the kidney. Finally, the possibility that resident glomerular or tubular cells participate in the increased urinary excretion of TxB_2 in CyA-treated animals has to be considered. Concerning the first possibility, Benigni et al. (100) have been unable to observe an increase in serum TxB_2 generation in rats given CyA compared with control animals. Moreover, rat platelet-rich plasma (PRP), exposed to CyA *in vitro* and then challenged with AA or ADP, generated a normal amount of TxB_2.

None of these approaches, however, can rule out conclusively the possibility that an extrarenal activation occurs in CyA-treated animals that can account for the excessive urinary excretion of TxB_2. To discriminate between the other two possibilities of interpreting the results of urinary TxB_2 excretion in CyA-treated animals, two different approaches have been attempted. First, the relative amounts of 2,3-dinor-TxB_2 versus TxB_2 in the urine of CyA-treated animals have been studied using gas chromatography-negative ion chemical ionization mass spectrometry (100). Although studies in rats are lacking, human studies have suggested that the major urinary metabolite of TxB_2, 2,3-dinor-TxB_2, is a sensitive marker of *in vivo* platelet activation, whereas urinary TxB_2 reflects the renal synthesis of the parent compound (101). Thus, urinary 2,3-dinor-TxB_2 is increased in severe atherosclerosis (102,103) and in other syndromes of platelet activation (104,105).

In the above-mentioned study (100), CyA-treated animals had significantly higher urinary excretion of both TxB_2 and 2,3-dinor-TxB_2 than animals given the vehicle alone. Thus, these data do not allow us to establish the cell origin of the increased urinary excretion of TxB_2 in CyA-treated animals. It is possible that in the rat TxB_2 generated by resident renal cells follows a different metabolic pathway than in humans. Alternatively, cells infiltrating the glomerular tuft generate an increased amount of TxA_2 which only partially returns to systemic circulation and is metabolized before being excreted as 2,3-dinor-TxB_2 (Fig. 6).

In this context it is of interest that CyA nephrotoxicity in humans is associated with a chronic type of arteriolar damage (83). To analyze further the possibility that platelets and macrophages are recruited at the renal circulation as a consequence of CyA-induced endothelial damage and possibly release TxA_2 locally, Benigni et al. (100) performed ultrastructural studies on kidney tissue of CyA-treated animals. This work indeed documented a mild endothelial damage with focal loss of endothelial fenestrae and doubling of the basement membrane. Moreover, at the site of injury, a marked infiltration of blood-borne cells, mainly of monocyte-macrophage type, has been found in the glomerular tuft. Further studies addressing the biosynthesis, metabolism and renal handling of TxA_2 in rats will contribute to defining the cellular origin of the increased urinary TxB_2 in rats given CyA.

The above-mentioned data on urinary excretion of TxB_2 in CyA-treated rats are in agreement with the recent findings of Coffman et al. (107), who found that in a post-ischemic denervated rat model, a 12-14 day treatment with CyA resulted in decreased renal function associated with increased urinary excretion of TxB_2 and 2,3-dinor-TxB_2. Very recently, Kuhn et al. found that 2,3-dinor-TxB_2 urinary excretion is increased also in humans given CyA (108). These authors have studied 40 patients with multiple sclerosis undergoing CyA as a 2-year treatment (5 mg/kg/day). Serum creatinine was significantly elevated during CyA treatment, and urinary excretion of

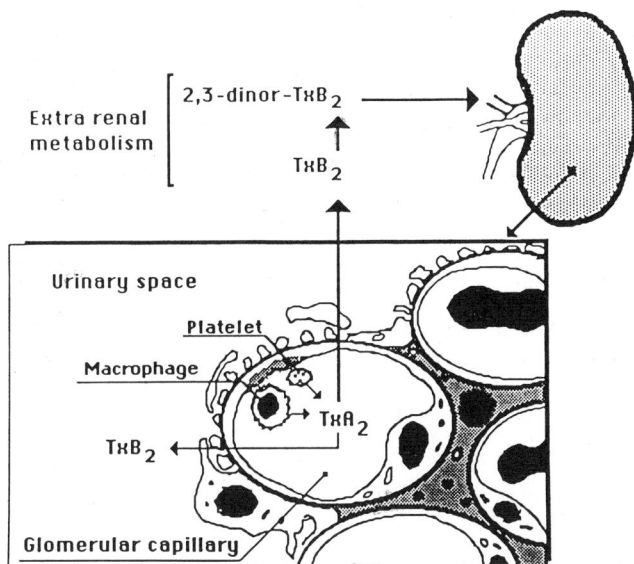

Fig. 6. Hypothetical metabolic consequences of
TxA$_2$ generated by inflammatory cells infiltrating the glomerular tuft in CyA-
treated rat. TxA$_2$ released by circulating cells at renal level may in part escape
β-oxidation and be excreted as TxB$_2$ and
in part return to the systemic circulation,
be metabolized and excreted as 2,3-dinor-
TxB$_2$.

2,3-dinor-TxB$_2$ was 60 + 25% greater under CyA than at 3 months after cessa-
tion of the treatment, when renal function normalized.

CONCLUSIONS

   In the last few years, considerable evidence has been accumulating to
suggest that eicosanoids and PAF are formed in excessive amounts within the
kidney in both toxic and immune-mediated nephropathies. Given the potent
pro-inflammatory and vasoactive properties of these compounds, the possibility that both AA metabolites and PAF are involved as mediators of tissue injury has generated enormous interest in the nephrological community and
served as a basis for hypothesis and new pathogenetic theories. Our knowledge is still limited, however, and the definitive proof that either AA metabolites or PAF actually causes renal damage is lacking. The biochemical
data already available and the results of inhibitor trials are interesting
enough to encourage future investigation in the field.

The next steps will include studies aimed at defining the cellular origin of the increased eicosanoids and PAF in renal diseases. In particular, it will be important to discriminate between circulating cells (possibly infiltrating the diseased kidney) and resident renal cells as sources of potential mediators of damage in renal diseases. The significance of changes in urinary excretion of eicosanoids and possibly of PAF in disease states will be studied. With respect to eicosanoids, efforts will be devoted to clarifying the origin of the material found in excessive amounts in the urine. Regarding PAF, it will be important to develop methods that can establish whether PAF is excreted in experimental animal and human urine by chemical identification and determine the relative amount excreted in normal and disease conditions.

The mechanism(s) involved in toxic stimuli or immune complex-induced eicosanoid or PAF formation in the kidney have also to be extensively investigated. The sophisticated methodology available for studying receptor distribution will bring about new information and help to design new molecules that can antagonize eicosanoid and PAF actions on the kidney. The possible interactions between eicosanoid and PAF pathways are also an open field for future studies. One of the main issues to be clarified is whether PAF induces renal damage directly or whether the pathophysiologic consequences of PAF formation at the renal level are expressed via the intrarenal formation of thromboxane. It is hoped that more selective and specific antagonists than those now available will contribute to defining the role of eicosanoids and PAF as mediators of renal damage and their precise pathophysiological role in renal diseases.

REFERENCES

1. W.L. Smith and G.P. Wilkin, Immunochemistry of prostaglandin endoperoxide-forming cyclo-oxygenase: The detection of the cyclooxygenases in rat, rabbit and guinea pig kidneys by immunofluorescence, Prostaglandins 13:873-892 (1977).
2. A. Hassid, M. Konieczkowski, M.J. Dunn, Prostaglandin synthesis in isolated rat kidney glomeruli, Proc. Natl. Acad. Sci. USA 76:1155 (1979).
3. D. Schlondorff, P. Goldwasser, R. Neuwirth, et al., Production of platelet-activating factor in glomeruli and cultured glomerular mesangial cells, Am. J. Physiol. 19:F1123-F1127 (1986).
4. D.H. Albert and F. Snyder, Release of arachidonic acid from 1-alkyl-2-acyl-sn-glycero-3-phosphocholine, a precursor of platelet-

activating factor, in rat alveolar macrophages, <u>Biochem. Biophys.</u> <u>Acta</u> 796:92- 101 (1984).

5. F.H. Chilton, J.M. Ellis, S.C. Olson, et al., 1-0-alkyl-2-arachidonoyl-sn-glycero-3-phosphocholine. A common source of platelet-activating factor and arachidonate in human polymorphonuclear leukocytes, <u>J. Biol. Chem</u>. 259:12014-12019 (1984).

6. E.G. Lapetina, Regulation of arachidonic acid production: Role of phospholipases C and A_2, <u>Trends in Pharmacological Sciences</u> 3:115-118 (1982).

7. D.H. Albert and F. Snyder, Biosynthesis of 1-Alkyl-2-acetyl-sn-glycero-3-phosphocholine (Platelet-activating Factor) from 1-Alkyl-2-sn-glycero-3-phosphocholine by rat alveolar macrophages. Phospholipase A_2 and acetyltransferase activities during phagocytosis and ionophore stimulation, <u>J. Biol. Chem</u>. 258:97-102 (1983).

8. B.M. Brenner, T.H. Hostetter, H.D. Humes, Molecular basis of proteinuria of glomerular origin, <u>N. Engl. J. Med</u>. 298:826-833 (1978).

9. M.G. Farquhar, The glomerular basement membrane--a selective macromolecular filter, <u>in</u>: "The Cell Biology of the Extracellular Matrix," E.D. Hay, ed., Plenum Press, New York, p. 335 (1981).

10. H.G. Rennke, R.S. Cotran, M.A. Venkatachalam, Role of molecular charge in glomerular permeability. Tracer studies with cationized ferritin, <u>J. Cell Biol</u>. 67:638-646 (1975).

11. J.P. Caulfield and M.G. Farquhar, The permeability of glomerular capillaries to graded dextrans, <u>J. Cell Biol</u>. 63:883-903 (1974).

12. G. Wallenius, Renal clearance of dextran as a measure of glomerular permeability, <u>Acta Soc. Med. Upsal</u>. 59:1-5 (1954).

13. R.L.S. Chang, I.F. Ueki, J.L. Troy, et al., Permselectivity of glomerular capillary wall to macromolecules. II. Experimental studies in rats using neutral dextran, <u>Biophys. J</u>. 15:887-906 (1975).

14. H.G. Rennke, M.A. Venkatachalam, Y. Patel, Glomerular permeability of macromolecules: Effect of molecular configuration on the fractional clearance of uncharged dextran and neutral horseradish peroxidase in the rat, <u>J. Clin. Invest</u>. 63:713-726 (1979).

15. H.G. Rennke and M.A. Venkatachalam, Glomerular permeability: In vivo tracer studies with polyanionic and polycationic ferritins, <u>Kidney Int</u>. 11:44-53 (1977).

16. Y.S. Kanwar and M.G. Farquhar, Presence of heparan sulfate in the glomerular basement membrane, <u>Proc. Natl. Acad. Sci. USA</u> 76:1303-1307 (1979).

17. Y.S. Kanwar and M.G. Farquhar, Isolation of glycosaminoglycans (heparan sulfate) from glomerular basement membranes, <u>Proc. Natl. Acad. Sci.</u>

USA 76:4493-4497 (1979).

18. Y.S. Kanwar, Biology of disease. Biophysiology of glomerular filtration and proteinuria, Lab. Invest. 51:7-21 (1984).

19. T. Bertani, G. Rocchi, G. Sacchi, et al., Adriamycin-induced glomerulosclerosis in the rat, Am. J. Kidney Dis. 7:12-19 (1986).

20. J. Grond, J.J. Weening, J.D. Elema, Glomerular sclerosis in nephrotic rats. Comparison of the long-term effects of adriamycin and aminonucleoside, Lab. Invest. 51:227-285 (1984).

21. G.B. Ryan and M.J. Karnovsky, An ultrastructural study of the mechanisms of proteinuria in aminonucleoside nephrosis, Kidney Int. 8:219-232 (1975).

22. T. Bertani, A. Poggi, R. Pozzoni, et al., Adriamycin-induced nephrotic syndrome in rats. Sequence of pathologic events, Lab. Invest. 46:16-23 (1982).

23. J.A. Velosa, R.J. Glasser, T.E. Nevins, et al., Experimental model of focal sclerosis. II. Correlation with immunopathologic changes, macromolecular kinetics, and polyanion loss, Lab. Invest. 36:527-534 (1977).

24. G. Remuzzi, L. Imberti, M. Rossini, et al., Increased glomerular thromboxane synthesis as a possible cause of proteinuria in experimental nephrosis, J. Clin. Invest. 75:94-101 (1985).

25. C. Ferti, L. Pierucci, G. Corsi, et al., A new TxA_2 synthetase inhibitor reduces adriamycin-induced nephrotic syndrome in rats, International Symposium on Renal Eicosanoids, Capri, June 9-11, p. 23 (1987).

26. T. Bertani, M. Abbate, G. Mecca, et al., Adriamycin-induced epithelial cell disease in the rat, in: "Drugs and Kidney," T. Bertani, G. Remuzzi, S. Garattini, eds., Raven Press, New York, pp. 1-14 (1986).

27. I.M. Goldstein, C.L. Malmsten, H. Kindahl, et al., Thromboxane generation by human peripheral blood polymorphonuclear leukocytes, J. Exp. Med. 901:787-792 (1978).

28. K. Brune, M. Glatt, H. Kalin, et al., Pharmacological control of prostaglandin and thromboxane release from macrophages, Nature 274:261-263 (1978).

29. P.O. Needleman, S. Moncada, S. Buntin, et al., Identification of an enzyme in platelet microsomes which generates thromboxane A_2 from prostaglandin endoperoxides, Nature 261:558-560 (1976).

30. C. Tetta, R. Coda, G. Camussi, Human platelet cationic probes bind to rat glomeruli, induce loss of anionic charges and increase glomerular permeability, Agents and Actions 16:24-26 (1985).

31. T. Goto, M. Mune, K. Matoba, et al., Effects of selective thromboxane

A$_2$ synthetase inhibitor of aminonucleoside induced nephrotic rats, Xth International Congress of Nephrology, London, July 26-31, p. 227 (1987).

32. T. Niwa, Y. Ozawa, T. Nomura, et al., Thromboxane A$_2$ metabolism and clinical effects of selective thromboxane A$_2$ synthetase inhibitor in chronic glomerulonephritis, Xth International Congress of Nephrology, London, July 26-31, p. 24 (1987).

33. M.L. Purkerson, J.H. Joist, J. Yates, et al., Inhibition of thromboxane synthesis ameliorates the progressive kidney disease of rats with sub-total renal ablation, Proc. Natl. Acad. Sci. 82:193-197 (1985).

34. J. Egido, A. Robles, A. Ortiz, et al., Role of platelet-activating fac-tor in adriamycin-induced nephropathy in rats, Eur. J. Pharmacol. 138:119-123 (1987).

35. J. Benveniste, J. Egido, V. Gutierrez-Millet, Evidence for the involve-ment of IgE basophils system in acute serum sickness of rabbits, Clin. Exp. Immunol. 26:449 (1976).

36. G. Camussi, C. Tetta, M.C. Deregibus, et al., Platelet-activating factor (PAF) in experimentally-induced rabbit acute serum sickness: Role of basophil-derived PAF in immune complex deposition, J. Immunol. 128: 86-94 (1982).

37. S. Ito, G. Camussi, C. Tetta, et al., Hyperacute renal allograft rejec-tion in the rabbit. The role of platelet-activating factor and of cat-ionic proteins derived from polymorphonuclear leukocytes and from platelets, Lab. Invest. 51:148-161 (1984).

38. G. Camussi, C. Tetta, R. Coda, et al., Release of platelet-activating factor in human pathology. I. Evidence for the occurrence of baso-phil degranulation and release of platelet-activating factor in sys-temic lupus erythematosus, Lab. Invest. 44:241-251 (1981).

39. T.J. Hallam, A. Sanchez, T.J. Rink, Stimulus-response coupling in human platelets. Changes evoked by platelet-activating factor in cytoplas-mic free calcium monitored with the fluorescent calcium indicator quin 2, Biochem. J. 218:819-827 (1984).

40. P.H. Naccache, M.M. Molski, E.L. Volpi, et al., Unique inhibitory pro-file of platelet-activating factor-induced calcium mobilization, poly-morphonuclear turnover and granule enzyme secretion in rabbit neutro-phils towards pertussis toxin and phorbol ester, Biochem. Biophys. Res. Commun. 130:677-684 (1985).

41. F. Bussolino, M. Aglietta, F. Sanavio, et al., Alkyl-ether phosphoglycer-ides influence calcium fluxes into human endothelial cells, J. Immunol. 135:2748-2753 (1985).

42. M. Kester, P. Mene', G.R. Dubyak, et al., Platelet activating factor

elevates cytosolic free calcium concentration in cultured rat mesangial cells (Abstract), <u>Clin. Res</u>. 35:550A (1987).

43. D. Schlondorff and R. Neuwirth, Platelet-activating factor and the kidney, <u>Am. J. Physiol</u>. 251:F1-F11 (1986).

44. J.O. Shaw, S.J. Klusick, D.J. Hanahan, Activation of rabbit platelet phospholipase and thromboxane synthesis by 1-0-hexadecyl-octadecyl-2-acetyl-sn-glyceryl-3-phosphorylcholine (platelet activating factor), <u>Biochem. Biophys. Acta</u>. 663:222-229 (1981).

45. F.H. Chilton, J.T. O'Flaherty, C.E. Walsh, et al., Platelet activating factor: Stimulation of the lipoxygenase pathways in polymorphonuclear leukocytes by 1-0-Alkyl-2-0-acetyl-sn-glycero-3-phosphocholine, <u>J. Biol. Chem</u>. 257:5402-5407 (1982).

46. D. Schlondorff, J.A. Satriano, J. Hagege, et al., Effects of platelet-activating factor and serum-treated zymosan on prostaglandin E_2 synthesis, arachidonic acid release, and contraction of cultured rat mesangial cells, <u>J. Clin. Invest</u>. 73:1227-1231 (1984).

47. G. Camussi, C. Tetta, R. Coda, et al., Platelet-activating factor-induced loss of glomerular anionic charges, <u>Kidney Int</u>. 25:73-81 (1984).

48. N. Perico, F. Delaini, M. Tagliaferri, et al., The effect of platelet-activating factor and its specific receptor antagonist on glomerular permeability to proteins in isolated perfused rat kidney, (submitted for publication).

49. E.R. Unanue and F.J. Dixon, Experimental glomerulonephritis: Immunological events and pathogenetic mechanisms, <u>Adv. Immunol</u>. 6:1-90 (1967).

50. E.R. Unanue and F.J. Dixon, Experimental glomerulonephritis. IV. Participation of complement in nephrotoxic nephritis, <u>J. Exp. Med</u>. 119:965-982 (1964).

51. C.G. Cochrane, E.R. Unanue, F.J. Dixon, A role of polymorphonuclear leukocytes and complement in nephrotoxic nephritis, <u>J. Exp. Med</u>. 122:99-116 (1965).

52. S.R. Holdsworth, T.J. Neale, C.B. Wilson, Abrogation of macrophage-dependent injury in experimental glomerulonephritis in the rabbit. Use of antimacrophage serum, <u>J. Clin. Invest</u>. 68:686-698 (1981).

53. G.F. Schreiner, R.S. Cotran, V. Pardo, et al., A mononuclear cell component in experimental immunological glomerulonephritis, <u>J. Exp. Med</u>. 147:369-384 (1978).

54. N.M. Thomson, S.R. Holsworth, E.F. Glasgow, et al., The macrophage in the development of experimental crescentic glomerulonephritis, <u>Am. J. Pathol</u>. 94:223-235 (1979).

55. S.R. Holdsworth and T.J. Neale, Macrophage-induced glomerular injury.

Cell transfer studies in passive autologous antiglomerular basement membrane antibody-initiated experimental glomerulonephritis, Lab. Invest. 51:172-180 (1984).

56. F.G. Silva, J.R. Hoyer, C.L. Pirani, Sequential studies of glomerular crescent formation in rats with antiglomerular basement membrane-induced glomerulonephritis and the role of coagulation factors, Lab. Invest. 51:404-415 (1984).

57. S.R. Holdsworth, N.M. Thomson, E.F. Glasgow, et al., Tissue culture of isolated glomeruli in experimental crescentic glomerulonephritis, J. Exp. Med. 147:98-109 (1978).

58. S.R. Holdsworth, N.M. Thomson, E.F. Glasgow, et al., The effect of defibrination on macrophage participation in rabbit nephrotoxic nephritis: Studies using glomerular culture and electron microscopy, Clin. Exp. Immunol. 37:38-44 (1979).

59. V. Cattell and S.W. Jamieson, The origin of glomerular crescents in experimental nephrotoxic serum nephritis in the rabbit, Lab. Invest. 39:584-590 (1978).

60. S.R. Holdsworth and P.G. Tipping, Macrophage-induced glomerular fibrin deposition in experimental glomerulonephritis in the rabbit, J. Clin. Invest. 76:1367-1374 (1985).

61. R.C. Wiggins, A. Glatfelter, J. Brukman, Procoagulant activity in glomeruli and urine of rabbits with nephrotoxic nephritis, Lab. Invest. 53: 156-165 (1985).

62. H.P. Hartung, M.J. Parnham, J. Winkelmann, et al., Platelet-activating factor (PAF) induces the oxidative burst in macrophages, Int. J. Immunopharmacol. 5:115-121 (1983).

63. E. Pirotzky, C. Page, J. Morley, et al., Vascular permeability induced by Paf-acether (platelet-activating factor) in the isolated perfused rat kidney, Agents and Actions 16:17-18 (1985).

64. C.A. Desmopoulos, R.N. Pinckard, D.J. Hanaham, Platelet-activating factor. Evidence for 1-0-alkyl-2-acetyl-sn-glyceryl-3-phosphorylcholine as the active component (a new class of lipid chemical mediators), J. Biol. Chem. 254:9355-9358 (1979).

65. Z. Hruby, R.P. Lowry, D. Blais, Effect of platelet activating factor antagonist CV3988 in an autologous nephrotoxic nephritis model, Clin. Res. 34:602A (1986).

66. T. Bertani, M. Livio, D. Macconi, et al., Platelet-activating factor (PAF) as a mediator of injury in nephrotoxic nephritis, Kidney Int. 31:1248-1256 (1987).

67. N. Ardaillou, J. Hagege, M.P. Nivez, et al., Vasoconstrictor-evoked prostaglandin synthesis in cultured human mesangial cells, Am. J.

Physiol. 248:F240-F246 (1985).

68. S.M. Weisman, D. Felsen, E.D. Vaughan, Jr., Platelet-activating factor is a potent stimulus for renal prostaglandin synthesis: Possible significance in unilateral ureteral obstruction, J. Pharmacol. Exp. Ther. 235:10-15 (1985).

69. R.N. Pinckard and P.M. Henson, Activation of procoagulant activity in rabbit platelets by basophil-derived platelet activating factor (PAFB) (Abstract), Fed. Proc. 36:1329 (1977).

70. E.A. Lianos, G.A. Andres, M.J. Dunn, Glomerular prostaglandin and thromboxane synthesis in rat nephrotoxic serum nephritis. Effects on renal hemodynamics, J. Clin. Invest. 72:1439-1448 (1983).

71. J.E. Stork and M.J. Dunn, Hemodynamic roles of thromboxane A_2 and prostaglandin E_2 in glomerulonephritis, J. Pharmacol. Exp. Ther. 233: 672-678 (1985).

72. D.M. Macconi, A. Benigni, M. Morigi, et al., Enhanced glomerular thromboxane A_2 mediates some pathophysiologic effect of platelet-activating factor in rabbit nephrotoxic nephritis: Evidence from biochemical measurements and inhibitor trials (submitted for publication).

73. R. Neuwirth, P. Singhal, J.A. Satriano, et al., Stimulation of PGE_2 synthesis by platelet-activating factor (PAF) in mesangial cells is inhibited by BN52021 and kadsurenone (Abstract), 6th International Conference on Prostaglandins and Related Compounds, Florence, June 3-6 (1986).

74. J.E. Stork, T.Y. Shen, M.J. Dunn, Stimulation of prostaglandin E_2 and thromboxane B_2 production in cultured rat mesangial cells by platelet activating factor: Inhibition by a specific receptor antagonist (Abstract), Kidney Int. 27:267A (1985).

75. G.B.G. Klintmalm, S. Iwatsuki, T.E. Starzl, Nephrotoxicity of cyclosporin A in liver and kidney transplant patients, Lancet 1:470-471 (1981).

76. S.M. Flechner, G. van Buren, R.H. Herman, et al,. The nephrotoxicity of cyclosporine in renal transplant recipients, Transplant Proc. 15: 2689-2694 (1983).

77. S.O. Bohman, G. Klintmalm, O. Rindgen, et al., Interstitial fibrosis in human kidney grafts after 12 to 46 months of cyclosporine therapy, Transplant Proc. 17:1168-1171 (1985).

78. E. von Willebrand and P. Hayry, Cyclosporin-A deposits in renal allografts, Lancet 2:189-192 (1983).

79. B.D. Myers, J. Ross, L. Newton, et al., Cyclosporine-associated chronic nephropathy, N. Engl. J. Med. 311:699-705 (1984).

80. B.D. Myers, Cyclosporine nephrotoxicity, Kidney Int. 30:964-974 (1986).

81. A.W. Thomson, P.H. Whiting, J.G. Simpson, Cyclosporine: Immunology, toxicity and pharmacology in experimental animals, <u>Agents Actions</u> 15: 306-327 (1984).

82. T. Bertani, N. Perico, M. Abbate, et al., Renal injury induced by long-term administration of cyclosporin A to rats, <u>Am. J. Pathol</u>. 127:569-579 (1987).

83. M.J. Mihatsch, G. Thiel, H.P. Spichtin, et al., Morphological findings in kidney transplants after treatment with cyclosporine, <u>Transplant Proc</u>. 15:2821-2835 (1983).

84. H. Siegl, B. Ryffel, R. Petric, et al., Cyclosporine, the renin-angiotensin-aldosterone system, and renal adverse reactions, <u>Transplant Proc</u>. 15: Suppl.1, 2719-2725 (1983).

85. N. Perico, C. Zoja, A. Benigni, et al., Effect of short-term cyclosporine administration in rats on renin-angiotensin and thromboxane A_2: Possible relevance to the reduction in glomerular filtration rate, <u>J. Pharmacol. Exp. Ther</u>. 239:229-235 (1986).

86. E.J.G. Barros, M.A. Boim, H. Ajzen, et al., Glomerular hemodynamics and hormonal participation on cyclosporine nephrotoxicity, <u>Kidney Int</u>. 32:19-25 (1987).

87. J.F. Gerkens, S.B. Bhagwandeen, P.J. Dosen, et al., The effect of salt intake on cyclosporine-induced impairment on renal function in rats, <u>Transplantation</u> 38:412-417 (1984).

88. J.P. Bantle, K.A. Nath, D.E.R. Sutherland, et al., Effects of cyclosporine on the renin-angiotensin-aldosterone system and potassium excretion in renal transplant recipients, <u>Arch. Intern. Med</u>. 1456: 505-508 (1985).

89. P.H.A. Whiting, A.W. Thomson, J.T. Blair, et al., Experimental cyclosporin A nephrotoxicity, <u>Br. J. Exp. Pathol</u>. 63:88-94 (1982).

90. R.L. Whisler, J.A. Lindsey, K.V.M. Proctor, et al., Characteristics of cyclosporine induction of increased prostaglandin levels from human peripheral blood monocytes, <u>Transplantation</u> 38:377-381 (1984).

91. J.A. Lindsey, N. Morisaki, J.M. Stitts, et al., Fatty acid metabolism and cell proliferation: IV. Effect of prostanoid biosynthesis from endogenous fatty acid release with cyclosporin A, <u>Lipids</u> 18:566-569 (1983).

92. A. Kawaguchi, M.H. Goldman, R. Shapiro, et al., Increase in urinary thromboxane B_2 in rats caused by cyclosporine, <u>Transplantation</u> 40: 214-216 (1985).

93. N. Perico, A. Benigni, E. Bosco, et al., Acute cyclosporine A nephrotoxicity in rats: Which role for renin-angiotensin system and glomerular prostaglandins, <u>Clin. Nephrol</u>. 25:Suppl. 1, S83-88 (1986).

94. B.M. Brenner, N. Schor, I. Ichikawa, Role of angiotensin II in the physiologic regulation of glomerular filtration, Am. J. Cardiol. 49: 1430-1433 (1982).

95. D.A. Ausiello, J.I. Kreisberg, C. Roy, et al., Contraction of cultured rat glomerular cells of apparent mesangial origin after stimulation with angiotensin II and arginine vasopressin, J. Clin. Invest. 65: 754-760 (1980).

96. D. Schlondorff, S. Roczniak, J.A. Satriano, et al., Prostaglandin synthesis by isolated rat glomeruli: Effect of angiotensin II, Am. J. Physiol. 238:F486-495 (1980).

97. M.J. Dunn, The role of arachidonic acid metabolites in glomerulonephritis, in: "Glomerular Injury: 300 years after Morgagni," T. Bertani and G. Remuzzi, eds., Wichtig Editore, Milan, pp. 75-88 (1983).

98. L.A. Scharschmidt, E. Lianos, M.J. Dunn, Arachidonate metabolites and the control of glomerular function, Fed. Proc. 42:3058-3063 (1983).

99. N. Perico, A. Benigni, C. Zoja, et al., Functional significance of the exaggerated renal thromboxane A_2 synthesis induced by cyclosporin A, Am. J. Physiol. 251:F581-F587 (1986).

100. A. Benigni, C. Chiabrando, A. Piccinelli, et al., Cyclosporin A nephrotoxicity: Significance of the increased urinary excretion of thromboxane B_2 and its metabolite 2,3-dinor-TxB_2 (submitted for publication).

101. G.A. FitzGerald, A.K. Pedersen, C. Patrono, Analysis of prostacyclin and thromboxane biosynthesis in cardiovascular disease, Circulation 67:1174-1177 (1983).

102. G.A. FitzGerald, B. Smith, A.K. Pedersen, et al., Prostacyclin biosynthesis is increased in patients with severe atherosclerosis and platelet activation, N. Engl. J. Med. 310:1065-1068 (1984).

103. F. Catella, J. Nowak, G.A. FitzGerald, Measurement of renal and nonrenal eicosanoid synthesis, Am. J. Med. 81:Suppl. 2B, 23-29 (1986).

104. I.A.G. Reilly, J.B. Doran, B. Smith, et al., Increased thromboxane biosynthesis in a human preparation of platelet activation: Biochemical and functional consequences of selective inhibition of thromboxane synthase, Circulation 73:1300-1309 (1986).

105. I.A.G. Reilly, L. Roy, G.A. FitzGerald, Biosynthesis of thromboxane in patients with systemic sclerosis and Raynaud's phenomenon, Br. Med. J. 292:1037-1039 (1986).

106. C. Zoja, L. Furci, F. Ghilardi, et al., Cyclosporin-induced endothelial cell injury, Lab. Invest. 55:455-462 (1986).

107. T.M. Coffman, D.R. Carr, W.E. Yarger, et al., Evidence that renal prostaglandin and thromboxane production is stimulated in chronic

cyclosporine nephrotoxicity, <u>Transplantation</u> 43:282-285 (1987).

108. K. Kuhn, U. Forstermann, J.C. Frolich, et al., Effect of cyclosporine A (CyA) on blood pressure and prostacyclin and thromboxane A_2 production, Xth International Congress of Nephrology, London, July 26-31, p. 231, (1987).

EICOSANOIDS: ROLE IN EXPERIMENTAL RENAL DISEASE

Saulo Klahr and Mabel L. Purkerson

Renal Division, Department of Medicine
Washington University School of Medicine
St. Louis, Missouri 63110
U.S.A.

Endogenous prostaglandin biosynthesis modulates such renal functions as regional blood flow (1), salt and water transport (2), renin secretion (3,4) and neurotransmitter release (5). More recently, the vasoconstrictor thromboxane A_2, which normally is quantitatively a minor product of arachidonic acid metabolism in the kidney under resting or basal conditions, has been shown to be increased in several models of renal injury and may be, in part, responsible for some of the pathophysiological derangements (6-9). Two models, ureteral obstruction and subtotal renal ablation, in which thromboxane seems to play a pathogenetic role, are described in more detail in this chapter. In addition, the role of prostaglandins, thromboxane and dietary fatty acids on kidney function and structure in several immunological models of renal disease and hypertension are discussed.

URETERAL OBSTRUCTION

In the rabbit, unilateral ureteral obstruction increases the basal synthesis of prostaglandin E_2 (PGE_2) in the experimental kidney perfused in vitro and the responsiveness of PGE_2 synthesis to stimulatory agents, such as bradykinin, angiotensin II, and norepinephrine (10). In these experiments, indomethacin treatment increased basal perfusion pressure, indicating that basal vascular resistance of the experimental kidney perfused in vitro was dependent, in part, on the synthesis of a vasodepressor prostaglandin. There is also enhanced prostaglandin production in response to stimulatory agents in the MRC/H strain of rats with congenital hydronephrosis. Thus, in both the surgical and genetic models of hydronephrosis, there is an increase in vasodepressor prostaglandin (probably PGE$_2$) production.

Although the production of vasodepressor prostaglandins may explain the fall in renal vascular resistance seen in the initial phases of acute ureteral obstruction, it does not explain the progressive fall in renal blood flow and glomerular filtration rate (GFR) observed with prolonged obstruction (11). This reduction in renal blood flow and GFR is probably due to preglomerular vasoconstriction (12). Furthermore, vasoconstriction in the postobstructed kidney is most severe in the outer cortex, an area rich in renin-containing glomeruli. Thus, the renin-angiotensin system may be involved in the increase in vascular resistance. However, some studies (13) suggest that angiotensin II does not play a significant role in this hemodynamic response. Nevertheless, a dynamic interplay of vasoconstrictor and vasodilatory forces is believed to modulate renal blood flow in mammals in a variety of experimental circumstances. Thromboxane A$_2$, a powerful vasoconstrictor, has attracted much attention following the demonstration by Morrison et al. (14) that the rabbit kidney subjected to ureteral obstruction had an increased capacity for thromboxane biosynthesis and that the major site of this synthesis in ureteral obstruction appears to be in the renal cortex (6).

The release of thromboxane A$_2$ by bradykinin in the obstructed kidney of the rabbit perfused in vitro is associated with an increase in resistance (15). The hemodynamic effects of bradykinin on the rabbit kidney can be reversed by the selective thromboxane synthetase inhibitor OKY-1581 (sodium-3-4,3-pyridylmethyl)phenyl-2-methyacrylate (16,17).

Early attempts to determine the sites along the nephron where these metabolites of arachidonic acid were produced and where functional changes were occurring utilized the isolation, by sieving techniques, of glomeruli and tubules (18). These experiments did not show the marked alterations of arachidonic acid metabolism by glomeruli and tubules that may have been anticipated from the results obtained in the isolated perfused kidney studies. However, they did demonstrate a population of cells that passed easily through the 80-micron nylon mesh which had significant synthetic capacity for PGE$_2$, thromboxane B$_2$ (TxB$_2$) and 6-keto-PGF$_{1\alpha}$. This finding was intriguing in view of previous reports (18,19) demonstrating that complete unilateral obstruction of 24 hr. in the rabbit produced a widening of the cortical interstitial space and an increase in fibroblasts and mononuclear cells. This latter observation was confirmed by Okegawa et al. (20). The potential interaction of mononuclear cells and fibroblasts through cell interactions or through factor(s) secreted by one of these cells has attracted much attention. Leibovitch and Ross (32) previously showed that reducing the number of circulating monocytes by antimacrophage serum and steroids delayed the appearance of fibroblasts at the site of injury. Thus, the macrophages accumulating at the site of injury may modulate fibroblast appearance and metabolism by secreting a soluble factor that stimulates PGE_2 biosynthesis (21). Macrophages have a high capacity to metabolize arachidonic acid and have been demonstrated to synthesize PGE_2, thromboxane A_2, prostacyclin (PGI_2) and a number of lipoxygenase products. Thus, the exaggerated prostaglandin synthesis observed following ureteral obstruction in the rabbit may be, in part, related to the invasion of the renal cortex by mononuclear cells and fibroblasts.

The importance of monocyte-macrophage infiltration in the rabbit kidney during unilateral ureteral obstruction has been investigated by Lefkowith et al. (22). Administration of endotoxin, a macrophage agonist, resulted in dramatic increases in the measured synthesis of eicosanoids by the perfused kidney. Nitrogen mustard administration, presumably by depletion of macrophages, blocked this effect. These studies suggest that renal injury produced by ureteral obstruction is followed by interstitial infiltration with monocyte-macrophages which can produce substantial amounts of thromboxane A_2 and also may stimulate renal synthesis of eicosanoids (22). These invading macrophages also may be responsible for the potent stimulatory effect of platelet activating factor (PAF) on PGE_2 and thromboxane B_2 release from the hydronephrotic kidney. Although normal kidneys respond with increased eicosanoid release to the administration of PAF, hydronephrotic kidneys have an accentuation of this response (23).

It has been shown previously that after 24 hours of bilateral ureteral obstruction, single nephron glomerular filtration rate declines to 40% of preobstruction levels in rats (24). Of interest, GFR remains markedly depressed several hours after release of obstruction due primarily to profound vasoconstriction of renal arterioles (12). The mechanism underlying this vasoconstrictive response after ureteral obstruction may be related to at least two vasoactive compounds, angiotensin II and thromboxane A_2 (12).

Pharmacologic blockade of thromboxane A_2 production has been shown by Yarger et al. (25) to significantly improve the depressed levels of GFR and renal plasma flow observed after release of unilateral ureteral obstruction. We have examined the effects of changes in dietary protein intake on the renal response to bilateral ureteral obstruction in rats (26). Rats were fed either a low- or a high-protein diet for 4 weeks prior to bilateral ureteral obstruction. After 24 hours of obstruction, one ureter was released and renal function was assessed by measurements of GFR, renal plasma flow, and also by examining the determinants of single nephron filtration rate (SNGFR) using micropuncture techniques. We found that values for inulin and PAH clearances were remarkably depressed in both groups after release of obstruction but to a greater extent in high-protein-fed rats, averaging less than 60% of values measured in low-protein-fed animals. Captopril, an inhibitor of the angiotensin I converting enzyme, increased inulin and PAH clearances remarkably but comparably, in high- or low-protein-fed rats. Micropuncture studies performed after unilateral release of bilateral ureteral obstruction in another group of rats fed a high- or a low-protein diet revealed lower values of glomerular plasma flow rate (Q_A) and SNGFR in rats fed a high-protein diet. Values for afferent and efferent arteriolar resistances in the kidney were nearly twofold greater in rats fed a high-protein diet when compared with low-protein-fed animals. Infusion of OKY 1581, an inhibitor of thromboxane A_2 synthetase, increased both Q_A and SNGFR, decreased arteriolar resistances, and increased glomerular capillary ultrafiltration coefficient (Kf) in high- but not in low-protein-fed rats (see Fig. 1). Urinary excretion of thromboxane B_2, per ml of GFR, was greater after release of BUO in rats fed a high-protein diet than in those fed a low-protein diet. These results suggest (a) that an increase in protein intake causes greater renal vasoconstriction following ureteral obstruction and (b) that the augmented renal vasoconstriction with a high-protein diet appears to be mediated by enhanced production and/or action of thromboxane A_2 during the acute injury.

Preliminary studies from our laboratory indicate that the degree of macrophage infiltration of the kidney that occurs during ureteral obstruction can be conditioned by the quantity of protein ingested in the diet prior to the onset of obstruction. Animals fed a low-protein diet demonstrated a lesser number of macrophages in the interstitium than animals fed a high-protein diet. In addition, in preliminary experiments we have shown that lethal bone marrow radiation 24 hours prior to the onset of obstruction results in almost complete absence of invading macrophages in the renal parenchyma. Concomitantly, studies of renal function, after unilateral release of bilateral ureteral obstruction, revealed that radiated rats, compared with control nonradiated animals, had higher values for renal plasma

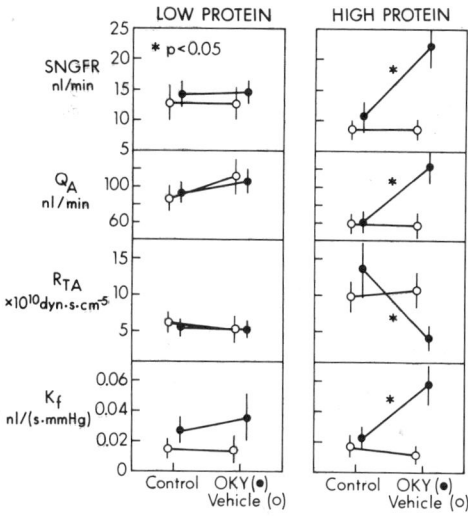

Fig. 1. Effects of OKY-1851 (●) or vehicle (o) infusion on glomerular dynamics in low (n=12 rats) or high (n=12 rats) protein-fed Munich-Wistar rats. These studies were performed after unilateral release of bilateral ureteral obstruction. After control observations were obtained, either OKY-1581 or vehicle was given, and measurements were repeated. Asterisk denotes a significant change (p<0.05). SNGFR, single nephron glomerular filtration rate; Q_A, glomerular capillary plasma flow; R_{TA}, total renal arteriolar resistance; K_f, glomerular capillary ultrafiltration coefficient. (Reproduced by permission from reference 26.)

flow and GFR. In addition, there was a significant decrease in the excretion of thromboxane B_2 in the urine after release of obstruction in radiated rats compared with nonradiated animals. Thromboxane excretion after release of obstruction was about 50% less in radiated rats compared with nonradiated rats. These studies strongly suggest that increased synthesis of thromboxane by invading macrophages is responsible in great part for the profound vasoconstriction that occurs in ureteral obstruction. In addition, the role of protein intake in conditioning the degree of vasoconstriction (26) seems to depend on the ability of the high-protein diet to increase the number of invading macrophages in the renal parenchyma.

Thromboxane may decrease the glomerular ultrafiltration coefficient (Kf) not only through a marked decrease in renal plasma flow but also through a potential effect of this compound on glomerular mesangial cells. Receptors for thromboxane A_2, a substance that presumably can induce mesangial contraction and lead to a decrement in the ultrafiltration coefficient by reduction in filtration surface area, have been demonstrated in mesangial cells (27).

Role of Thromboxane A_2 in the Progression of Renal Failure in Rats with
Subtotal Renal Ablation

Partial infarction of one kidney and contralateral nephrectomy (subtotal renal ablation) in the rat is associated with hypertension, proteinuria and progressive renal failure (28,29), which usually leads to death within 6 to 12 weeks. Light, electron, and immunofluorescence microscopy reveal extensive glomerular deposition of platelets and fibrin, mesangial expansion and sclerosis and fibrosis of glomerular capillaries (30). Studies by Purkerson et al. (31) have demonstrated that heparin retards development and progression of renal damage and hypertension, and prolongs the lifespan of rats with subtotal renal ablation. Because heparin had such a dramatic influence on the morbidity and mortality of these rats, it was postulated that platelets may play a role in the pathogenesis of the progression of renal failure; conversely, this suggested the possibility that administration of an inhibitor of thromboxane A_2 formation and platelet aggregation may be beneficial in this experimental model.

Normal rats and rats with 1-5/6 nephrectomy were studied. Rats with 1-5/6 nephrectomy were given either vehicle (saline) or OKY-1581, 20 mg/kg body wt daily for 26 to 35 days (9). Figure 2 shows the effects of OKY-1581 administration on mean arterial blood pressure and the ratio of heart weight

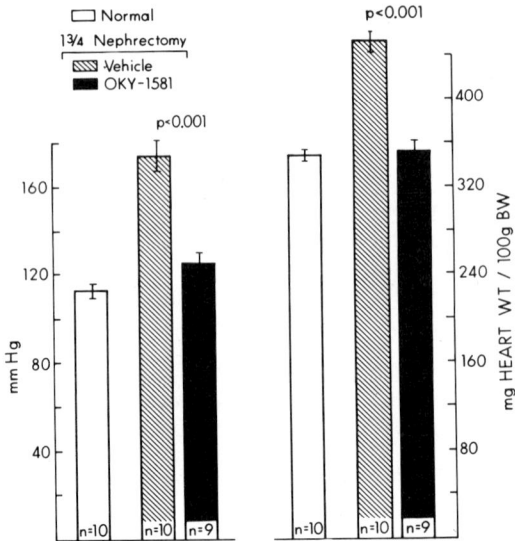

Fig. 2. Values for mean arterial blood pressure and the ratio of heart weight in normal and 1-5/6 nephrectomized rats receiving vehicle (normal saline) or OKY-1581, 20 mg/kg body weight twice daily for 5 weeks. Values for both blood pressure and the ratio of heart weight to body weight are significantly greater in 1-5/6 nephrectomized rats receiving vehicle. Values of 1-5/6 nephrectomized rats given OKY-1581 did not differ from values obtained in normal rats. (Reproduced by permission from reference 9.)

to body weight in the treated group. The inhibitor of thromboxane synthesis decreased the values for blood pressure and the ratio of heart weight to body weight to values comparable with those seen in normal rats. There was also a greater inulin clearance (5.34 ± 0.66 ml/min./kg body wt) in the OKY-1581-treated group when compared with the vehicle-treated group (2.34 ± 0.15 ml/min./kg body wt). Similarly, PAH clearances were lower (6.25 ± 0.15 ml/min./kg body wt) in the vehicle-treated rats than in the OKY-1581-treated rats (10.92 ± 0.91; $p < 0.001$).

The urinary excretion of thromboxane B_2, the stable metabolite of thromboxane A_2, was greater in 1-5/6 nephrectomized rats than in normal animals (Table I). The excretion rates of thromboxane B_2 were significantly greater, even when expressed as picograms per minute, despite the fact that rats with 1-5/6 nephrectomy had a reduced renal mass when compared with normal rats. Rats with 1-5/6 nephrectomy, given OKY-1581 by gavage daily, had decreased excretion rates of thromboxane B_2 in the urine when compared with rats with a remnant kidney given vehicle alone. The urinary thromboxane excretion rates of rats with reduced renal mass receiving OKY-1581 were not different from the excretion rates observed in normal rats. In contrast, no significant differences were observed in the excretion of 6-keto-PGF$_1$ between normals and 1-5/6 nephrectomized rats. In the latter group, administration of OKY-1581 did not significantly decrease the excretion of 6-keto-PGF$_{1\alpha}$ when compared with a group given vehicle alone (9).

Table I. Urinary Excretion of Thromboxane B_2 and 6-keto-PGF$_{1\alpha}$ in Normal Rats and 1-5/6 Nephrectomized Rats Given Vehicle or OKY-1581 for 5 Weeks

Thromboxane (pg/min/kg body wt)			6-keto-PGF$_{1\alpha}$ (pg/min/kg body wt)		
Normal Rats	1-5/6 Nephrectomized Rats		Normal Rats	1-5/6 Nephrectomized Rats	
	Vehicle	OKY-1581		Vehicle	OKY-1581
105	294	87.1	1060	1121.3	890.9
$\pm$ 40	$\pm$ 83.2	$\pm$ 16.1	$\pm$ 360	$\pm$ 247.6	$\pm$ 319.3
	< 0.05	< 0.05		N.S.	N.S.

The values given represent the mean results obtained in 5 rats in each group. In each rat the results of three clearance periods were averaged.

N.S., not significant.

The site of origin of the increased thromboxane excretion in the urine of 1-5/6 nephrectomized rats has not been established. It may be produced by renal cells per se or it could originate from platelets and/or infiltrative cells in the viable portion of the remnant kidney. Although we are not able to distinguish between these possibilities, it is of interest that the acute intravenous administration of the thromboxane inhibitor OKY-1581 resulted in an increase in PAH and inulin clearance values only in 1-5/6 nephrectomized rats and not in normal rats. These data suggest that the increased intrarenal thromboxane, synthesized either by the kidney per se or by cells infiltrating the kidney, had physiological effects on renal blood flow and glomerular filtration rate in animals with reduced renal mass. These results indicate that the increased synthesis of thromboxane has important physiological effects in rats with reduced renal mass. Recent preliminary studies from our laboratory indicate that there is infiltration by macrophages of the viable portion of the remnant kidney. In addition, preliminary evidence suggests that the amount of dietary protein may condition the degree of infiltration of the kidney by macrophages. It is tempting to speculate that the increased excretion of thromboxane in the urine of rats with a remnant kidney may be the result, at least in part, of synthesis of this eicosanoid by invading macrophages.

Effects of Changes in Dietary Fatty Acids on the Progression of Renal Disease in Rats with Subtotal Ablation

Barcelli, Weiss and Pollak (32) examined the effects of administering diets with a high or normal linoleic acid content to three-fourths nephrectomized and sham-operated rats. Linoleic acid is a precursor of both eicosa-8,11,15-trienoic (dihomo-gamma-linolenic) and eicosa-5,8,11,15-tetraenoic (arachidonic) acids which can be converted to prostaglandins, thromboxanes, prostacyclins, or leukotrienes. Subtotally nephrectomized rats fed the normal linoleic acid diet had progressive deterioration of renal function with serum creatinine levels rising to 1.55 mg/dl by week 20. By contrast, partially nephrectomized rats fed the high linoleic acid diet maintained stable renal function (mean serum creatinine 0.97 mg/dl at week 20). Urinary protein excretion was significantly lower and glomerular sclerosis was prevented in the rats fed the high linoleic acid diet. No changes were observed in the levels of blood pressure, serum cholesterol, or serum triglycerides as an effect of the diet. Increased PGE_2 production in the renal cortex of the rats fed the high linoleic acid diet may have had, according to the authors, a protective effect on renal function in this model of renal disease. We have confirmed the observations of Barcelli et al. (32) but found that a high linoleic acid diet decreased blood pressure in rats with subtotal nephrectomy when compared with rats fed a low linoleic acid diet (33). Similar findings have been reported by Izumi et al. (34). It has also been reported that rats with subtotal renal ablation fed a diet rich in eicosapentaenoic acid (EPA) had lower PGE_2 excretion, lower creatinine clearance and higher mortality than rats fed a standard diet (35).

Platelets and coagulation may play a role in the progression of the renal disease in subtotally nephrectomized rats (30,31). It is of interest that studies in rats fed different concentrations of dietary linoleic acid (from 0 to 6%) have shown that increased amounts of linoleic acid were associated with decreased susceptibility of platelets to thrombin-induced aggregation and prolongation of the clotting time of platelet-rich plasma (36). These studies in rats confirm and extend previous work showing that dietary saturated and polyunsaturated fatty acids had opposite effects on platelet function in humans.

IMMUNOLOGICAL MODELS OF RENAL DISEASE

Lupus Nephritis

In mice with lupus nephritis, Zurier et al. (37) reported that administration of PGE_1, daily or twice daily, increased survival. This effect occurred without alterations in the production of antinuclear antibodies. Hurd et al. (38) examined the effects of diets rich or deficient in essential fatty acids on the progression of lupus nephritis in NZB/NZW hybrid mice. These mice develop several immunologic abnormalities, as well as circulating immune complexes and immunoglobulin deposition in the renal glomerulus (39,40). The severity of the glomerulonephritis peaks at 9 months of age and causes death of most animals by 12 months of age (39-41). In Hurd's study, a group of mice was fed safflower oil which contains 78% linoleic acid. A second group of animals was fed a normal chow diet. A third group of animals was fed the chow diet and was injected with PGE_1. The fourth group received a diet deficient in essential fatty acids (< 1% linoleic acid). Animals fed the latter diet had remarkably prolonged life, reduced severity of glomerulonephritis, and lower levels of antinuclear antibodies and anti-DNA antibodies. Safflower oil had no apparent beneficial effect on the severity of glomerulonephritis or survival. Treatment with PGE_1 prolonged survival but was less effective than an essential-fatty-acid-deficient-diet in reducing or preventing autoantibody production or the development of glomerulonephritis. Although these studies did not directly address the biochemical basis for the beneficial effects of the essential-fatty-acid-deficient diets on this model of lupus, it is possible that one or more metabolites of arachidonic acid may play a role in the pathogenesis of this disease. Products of the lipoxygenase pathway (12-hydroxy-eicosate-traenoic acid, 12-hydroperoxy-eicosatetraenoic acid, leukotrienes, etc.) or cyclooxygenase pathway (prostacyclins, thromboxanes, prostaglandins, etc.) could be involved in mediating murine lupus by virtue of their biological activity in chemotaxis, vascular permeability and other factors involved in inflammation (42,43).

Selective stimulation of prostaglandin and thromboxane synthesis by feeding their precursors has been utilized to elucidate the potential role of the cyclooxygenase metabolites in chronic renal failure. Menhaden and other oils from fish or mammals are rich in eicosapentaenoic and docosahexaenoic, omega-3 fatty acids that are the precursors of the 3-series prostaglandins. Whereas prostaglandins of the 3-series appear to retain vasodilatory properties as compared with those of the 2-series, it appears that thromboxane A_3 does not have the same potency as a vasoconstrictor or platelet-aggregating agent as thromboxane A_2. Prickett, Robinson and Steinberg (44) demonstrated that enrichment of the diet with a polyunsaturated fatty acid, eicosapentaenoic acid, prevented proteinuria and prolonged survival in NZB/NZW F_1 mice. Dietary eicosapentaenoic acid (EPA, C20:5), a fatty acid analog of arachidonic acid (C20:4), has been shown to impair platelet aggregation in humans apparently through inhibition of the synthesis of prostaglandins and thromboxanes from arachidonic acid. Animals from 4 weeks of age were fed diets containing 25% lipid supplied either as beef tallow or menhaden oil with fatty acid analysis of less than 0.05 and 14.4% eicosapentaenoic acid, respectively. By 13 months of age, all mice on the beef tallow diet developed proteinuria, and the majority (six of nine) had died, with renal histologic examination revealing severe glomerulonephritis. In contrast, none of ten mice fed menhaden oil had developed proteinuria or died at this time. In a second experiment using 50 mice of each dietary group, 56% of the beef tallow group versus none of the menhaden oil group had developed proteinuria at 9 months of age. These results demonstrate that a diet high in eicosapentaenoic acid prevents the development of glomerulonephritis in NZB/NZW F_1 mice.

Kelley and collaborators, using the MRL-1pr murine model of lupus erythematosus have confirmed the beneficial effects on the immunologic presentation of the disease, both renal and extrarenal, of increased dietary fish oil. Fish oil treatment reduced renal dienoic prostaglandins, especially

PGE$_2$ and thromboxane B$_2$, and may have enhanced trienoic prostaglandin production (45). It is unknown whether the beneficial effects of fish oil in murine models of lupus are secondary to reduction of dienoic thromboxane and possibly leukotrienes or whether these changes are coincidental to other biochemical alterations. Steinhauer et al., using the NZB/NZW F$_1$ hybrid mouse, demonstrated that dietary supplementation with histidine or zinc substantially reduced the development of glomerulonephritis and was accompanied by reductions of renal PGE$_2$ and thromboxane B$_2$ (46).

To determine whether the amount of cyclooxygenase metabolites correlates with the development of lupus nephritis, Kelley et al. (47) measured intrarenal eicosanoid production in autoimmune mice. Disease progression was related to the renal biosynthesis of prostaglandin (PGE$_2$), prostacyclin (6 keto PGF$_{1\alpha}$), and thromboxane (TXB$_2$) using the MRL-1pr and NZB/NZW F1 hybrid mouse strains with predictably progressive forms of renal disease that mimic the human illness. Mice were evaluated for renal disease by measuring urinary protein excretion and renal immunopathological conditions, and these features were related to renal eicosanoid production. These studies show that: (a) intrarenal synthesis of TXB$_2$ rose incrementally in MRL-1pr and NZB x NZW F1 hybrid mice as renal function deteriorated and renal pathologic changes progressed (see Figure 3); (b) there were no consistent increases in the levels of two other cyclooxygenase metabolites, PGE$_2$ or 6-keto-PGF$_{1\alpha}$; (c) increased TXB$_2$ production occurred in the renal medulla, cortex, and within enriched preparations of cortical glomeruli; (d) when renal disease was prevented by pharmacologic doses of PGE$_2$, intrarenal TXB$_2$ did not increase; (e) administration of a dose of ibuprofen (9 mg/kg), a cyclooxygenase inhibitor capable of reducing 90% of platelet TXB$_2$ without affecting intrarenal levels, did not retard the progression of renal damage. Taken together, these data indicate that the intrarenal level of TXB$_2$ rises in relation to the severity of murine lupus nephritis. Furthermore, because of the potentially deleterious effects of TXA$_2$, enhanced production of this eicosanoid may be an important mediator of renal injury.

In experimental lupus nephritis it appears that decreased production of one or more metabolites of arachidonic acid has a beneficial effect on the disease. This effect may be related to the action of products of arachidonic acid on immunological and/or inflammatory processes. By contrast, in other experimental animal models of renal disease, increased synthesis of certain arachidonic acid metabolites seems to play a beneficial role. Whether or not changes in the synthesis of vasodilatory prostaglandins are responsible for these effects is not known. Of interest, however, is the

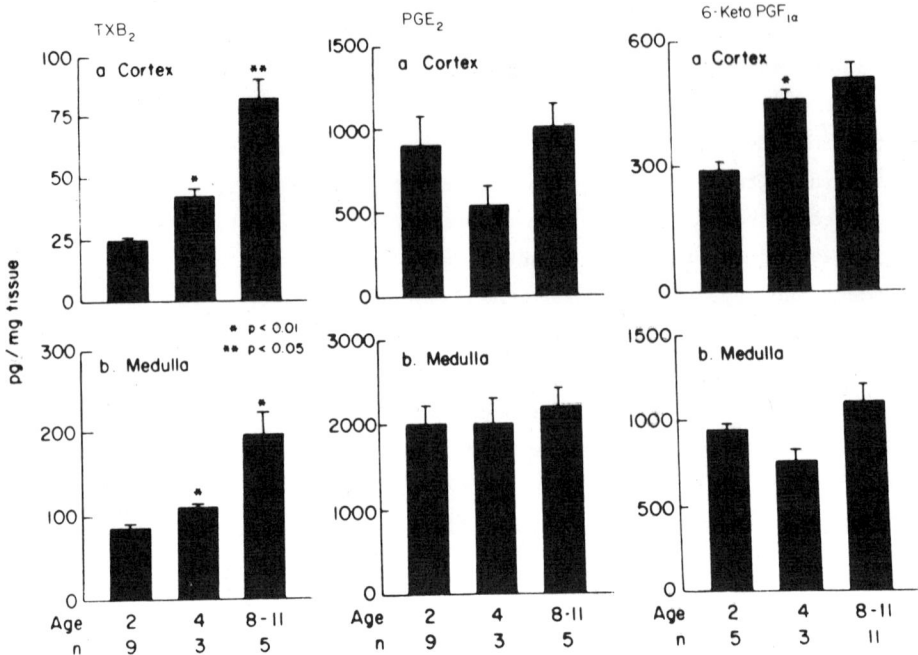

Fig. 3. TxB_2, PGE_2 and 6-keto-$PGF_{1\alpha}$ synthesis in renal tissues of NZBxW mice of 2,4 and 8-11 months of age. These eicosanoid measurements (pg/mg of tissue) were analyzed by radioimmunoassays. Mean renal cortex (a) and medulla (b) values are indicated by bars ± SEM. (Reproduced by permission from reference 47.)

fact that exogenous administration of prostaglandins may affect the course of renal disease in several animal models (48,49).

Heymann Nephritis

Stahl et al. (50) have determined the effects of induction of immune-mediated glomerular injury on the formation of cyclooxygenase products by glomerular cells. These authors measured PGE_2 and thromboxane B_2 formation in isolated glomeruli of rats with passive Heymann nephritis (PHN). PHN is a model of membranous nephropathy mediated by antibody and complement, independent of inflammatory cells. Five days after the induction of PHN by injection of heterologous antibody to rat proximal tubule brush border antigen (FX 1A), rats developed proteinuria, 36.5 ± 34 mg/day as compared with 3.8 ± 1 mg/day in controls. Treatment with cobra venom factor, which depleted complement C3 levels to less than 10% of baseline, prevented the development of proteinuria (6.9 ± 2 mg/day). The development of subepithelial glomerular immune complex deposits and proteinuria was associated with a significant stimulation of glomerular PGE_2 (87%) and thromboxane

B_2 (183%) formation. This increment in glomerular prostanoid biosynthesis was significantly inhibited (PGE_2 increased 22%, thromboxane B_2 increased 75%) in animals that were complement depleted with cobra venom factor (Fig. 4). Cobra venom factor had no effect on glomerular prostanoid formation in normal rats. In additional experiments the authors tested the hypothesis that thromboxane A_2 may mediate the proteinuria in PHN. They utilized a thromboxane synthetase inhibitor which reduced glomerular TXB_2 formation by 80% without influencing the glomerular deposition of ^{125}I labeled antibody and did not alter the levels of urine protein in rats with PHN. Using this inhibitor (UK 38485), proteinuria was 39 ± 24 mg/day as compared with 42 ± 21 mg in control animals. These data demonstrate that glomerular injury in rats with PHN is associated with increased thromboxane B_2 and PGE_2 production which is probably derived, in vivo, from glomer-

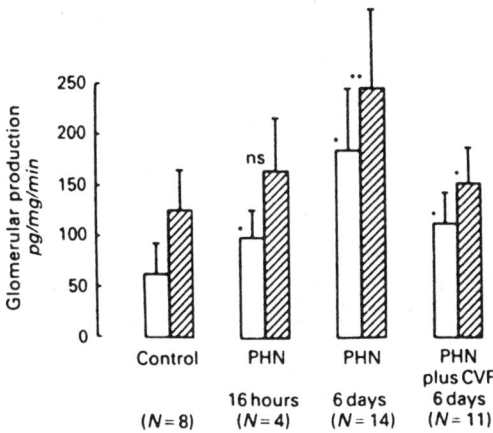

Fig. 4. Glomerular TxB_2 ($\square$) and PGE_2 ($\blacksquare$) production by control rats and animals with passive Heymann nephritis (PHN). PHN rats were studied 16 hours and 6 days after anti-Fx1A antibody injection. Glomerular prostanoid formation in PHN rats that received cobra venom factor (CVF) were also determined at day 6 after Fx1A application (PHN + CVF). Data are the means $\pm$ SE. Glomerular prostanoid formation was significantly higher in PHN rats when compared with control animals (* p < 0.05, ** p < 0.001). CVF reduced TxB_2 and PGE_2 formation significantly in PHN rats when studied at day 6 (PHN vs. PHN + CVF at day 6, p < 0.01). (Reproduced by permission from reference 50.)

261

ular cells. This stimulatory effect of glomerular immune deposit formation on prostanoid synthesis is inhibited by complement depletion. On the other hand, utilizing an inhibitor of thromboxane synthesis, the authors found no pathogenetic role for thromboxane B_2 in the mediation of proteinuria in this animal model. In this model relatively smaller increases in prostanoid synthesis were seen as compared with other models in which extraglomerular cells probably contribute significantly to the results observed (47,51-53).

Nephrotoxic Serum Nephritis

Stork and Dunn (51) examined the glomerular synthesis of PGE_2 and thromboxane A_2 and the effects of inhibitors of thromboxane synthesis or of cyclooxygenase on renal plasma flow and GFR in rats with nephrotoxic serum nephritis (51). Contrary to their previous studies, which showed a beneficial effect of inhibition of thromboxane A_2 within 3 hours of induction of nephrotoxic serum nephritis (52), these authors found that 24 hours and 14 days after initiation of nephrotoxic serum injury, inhibition of thromboxane A_2 with two different inhibitors did not increase GFR or renal plasma flow. Pharmacologic blockade of the thromboxane receptor did not unmask any pathophysiologic role of thromboxane A_2. Despite a significant increase in the glomerular synthesis of thromboxane A_2, two weeks after the administration of glomerular basement membrane antibodies, GFR was normal and renal plasma flow was increased in association with a tenfold rise in glomerular PGE_2 production. Inhibition of the cyclooxygenase with either indomethacin or meclofenamate resulted in a 50% fall in both renal plasma flow and GFR in these animals. The authors concluded that despite 10-15-fold increment in the synthesis of glomerular PGE_2 and thromboxane A_2 in rats with nephrotoxic serum nephritis, the major hemodynamic effects were mediated by PGE_2 and not by thromboxane A_2 (51).

Lianos et al. (53) have also examined the metabolism of eicosanoids by glomeruli in rats with nephrotoxic serum nephritis. They found a substantial increase of lipoxygenation of arachidonic acid to 12-hydroxy-eicosatetraenoic acid, a fatty acid that has chemotactic and pro-inflammatory properties which could be important in the mediation of glomerular immune injury (53). DuBois et al. (54), using the nephrotoxic serum nephritis model in rats, found that an essential-fatty-acid-deficient diet did not change the course of the disease, but the group of rats fed the essential-fatty-acid-deficient diet had more fibrin deposition in their glomeruli than did rats fed the control diet. The synthesis of prostaglandins or thromboxane was not measured in these experiments.

262

Models of Immune Complex Glomerulonephritis

In a murine model of acute immune complex glomerulonephritis induced by
the intraperitoneal injection of 1 ml of human plasma, Kelley and
Winkelstein (55) examined the effects of subcutaneous prostaglandin injection. They gave either PGE_2 or $PGF_{2\alpha}$ twice daily and found that rats
injected with PGE_2 had less proteinuria and developed fewer glomerular
pathological changes, despite more circulating antibodies to human plasma,
than did controls. Injection of $PGF_{2\alpha}$ failed to produce beneficial
effects. Similar results were reported by McLeish et al. (49) who injected
PGE_1 twice daily into mice with chronic serum sickness. In these studies
PGE_1 apparently reduced glomerular immune complex deposition; the immunological response apparently was not affected. The same mouse model was used
by McLeish (56) to demonstrate a protective effect on glomerular damage,
with the precursors of the dienoic prostaglandins originating from arachidonic acid. In addition, the levels of anti-apoferitin antibodies were decreased, and immune complexes were deposited in the mesangium and not in
capillary loops. Kher and coworkers (57) also have used an alternative
model of immune complex glomerulonephritis induced by the intraperitoneal
injection of apoferitin. Apoferitin-injected animals develop chronic glomerular changes due to immune complex deposition within the glomeruli and glomerular cell proliferation. When dietary linoleic acid was supplemented,
the high linoleic acid diet resulted in a decrease in proteinuria and a re-
tardation in the development of glomerulonephritis. Whether these changes
can be attributed to increased production of dienoic prostaglandin synthesis
as a consequence of the supplementation with linoleic acid could not be established. In a model of immune complex glomerulonephritis induced by the
injection of bovine serum albumin in rabbits, Saito et al. reported that
inhibition of thromboxane synthesis by the injection of benzyl imidazole,
given to rabbits throughout the course of immunization, resulted in reduced
proteinuria, decreased glomerular infiltration with polymorphonuclear leukocytes and monocytes and less glomerular fibrin deposition (58). In these
experiments it is difficult to separate the potential effects of thromboxane
A_2 inhibition in platelets and circulating whole blood cells from the
potential effect of the inhibitor of thromboxane synthesis on the glomerular
production of thromboxane A_2. These renal and extrarenal effects of thromboxane synthesis may not be mutually exclusive, and both may be important in
the production of renal disease.

The administration of adriamycin to rats results in the development of
experimental proteinuria and the nephrotic syndrome. This is associated

with an increased synthesis of thromboxane B_2 by glomeruli, 14 and 30 days after the induction of the disease (59). Treatment of these animals with a thromboxane A_2 synthetase inhibitor resulted in a decrease in proteinuria of about 50%. Urinary thromboxane B_2 excretion, which was increased in these nephrotic rats, was reduced to normal levels after the administration of the inhibitor of thromboxane synthesis. Since the lesions produced by adriamycin appear not to require infiltrative cells, the putative role of thromboxane may be ascribed to increased glomerular synthesis of thromboxane A_2 and a potential *in situ* effect of this vasoactive compound on protein permeability of the glomerular capillary wall.

Experimental Interstitial Nephritis

It has been shown that administration of PGE_1, from the time of immunization, inhibits the development of immune-mediated interstitial nephritis without demonstrable effects on levels of antibodies against α renal tubular basement membrane antigens (60). Immune lymphocytes from mice receiving PGE_1 do not exhibit a delayed hypersensitivity reaction response to tubular antigen, which is a response that correlates well with the presence of nephritogenic effector T cells. Inhibition of delayed hypersensitivity was a nonspecific effect of PGE_1 administration since the PPD response was also inhibited. *In vitro* studies indicated that PGE_1-mediated suppression of interstitial disease correlated with a dose-dependent and reversible inhibition of effector T cell differentiation. This inhibition, in turn, was mediated by PGE_1-induced soluble spleen products (nonspecific suppressor lymphokines). This suppression could be overcome by recombinant interleukin I, which suggests a mechanism related to either diminished interleukin I secretion or target cell sensitivity to interleukin I. It should be pointed out that recent studies suggest an inhibitory effect of endogenous prostaglandins on interleukin I production (61,62).

ROLE OF PROSTAGLANDINS AND THROMBOXANE IN THE PATHOPHYSIOLOGY OF HYPERTENSION

The role of prostaglandins and thromboxane in the pathophysiology of systemic hypertension has been a subject of extensive investigation. A spontaneously hypertensive strain of rats (Wistar Kyoto) has been studied in detail as an animal model of essential hypertension in man. The exact pathogenesis of hypertension in this strain of rats has not been clearly established, although several abnormalities have been identified, including an increase in sympathetic nerve activity, a rise in renal vascular resistance and enhanced renal vascular sensitivity to angiotensin II and norepinephrine (63). It appears also that the renal production of prostaglandins is increased in this strain of rats when compared with normotensive animals (64-66). Isolated glomeruli from young spontaneously hypertensive rats (SHR) also exhibit increased production of thromboxane as well as PGE_2, $PGF_{2\alpha}$ and 6-keto-$PGF_{1\alpha}$ (67). It has also been shown that prostacyclin (PGI_2) release by aortic strips obtained from spontaneously hypertensive rats is enhanced (68). The mechanisms underlying the renal and peripheral abnormalities described above remain unclear, but the observation that indomethacin administration in these rats causes a rise in systemic blood pressure suggests that one or more metabolites of arachidonic acid exert an effect in the control of blood pressure in this particular strain (69).

It is known that the kidney plays a primary role in the development of hypertension. In the developmental phase of hypertension, spontaneously hypertensive rats retain sodium, and this may contribute to the development of hypertension (70,71). This reabsorption of sodium in young SHR may be caused by an increase in renal sympathetic nerve activity (72,73). Shibouta et al. (74,75) demonstrated enhanced thromboxane A_2 synthesis in kidneys isolated from 6-week-old SHR in response to either angiotensin II or arachidonic acid administration. Shibouta et al. (76) also reported that in 6-week-old SHR, indomethacin and pinane thromboxane A_2, a thromboxane A_2 antagonist as well as a thromboxane A_2 synthetase inhibitor, resulted in natriuresis accompanied by an increase in the clearances of both inulin and para-aminohippurate. On the other hand, pinane thromboxane administration did not alter renal function in either 6-week-old normotensive Wistar Kyoto rats (WKY) or 18-week-old SHR. These studies suggest that increased synthesis of thromboxane, a powerful vasoconstrictor, may play a role in the hypertension that develops in SHR. In addition, Uderman et al. (77) have shown an attenuation of the development of hypertension in SHR after administration of the thromboxane synthetase inhibitor 4(imidazole 1-yl)-acetophenonone.

Grone et al. (78) treated SHR with UK-38485, an inhibitor of thromboxane synthesis and also administered a thromboxane A_2 receptor antagonist, EP-092 (78). Despite greater than 75% inhibition of thromboxane B_2 synthesis by glomeruli in both acute and chronic studies, these agents did not reduce blood pressure or increase renal plasma flow or GFR in SHR rats. Renal vascular resistance and sodium excretion were not altered by thromboxane A_2 synthetase inhibition. No endoperoxide shunting, as measured by changes of glomerular PGI_2 or PGE_2 synthesis, could be demonstrated.

As a result of these studies, Grone et al. (78) concluded that enhanced renal or extrarenal thromboxane synthesis does not contribute to the disordered blood pressure regulation in young SHR.

We have reported contrasting results to those of Grone et al. (78). We examined the effects of administration of OKY 046, an inhibitor of thromboxane synthesis, for 100 days on systemic blood pressure and renal function in SHR and in normotensive control rats (79). Untreated SHR had higher values for thromboxane excretion in the urine and higher values for blood pressure than did normotensive control (WKY) rats. Administration of OKY 046 decreased systolic and mean arterial blood pressure and urinary excretion of thromboxane and protein in SHR. Administration of OKY 046 decreased thromboxane excretion in the urine of WKY rats but had no effect on blood pressure or protein excretion. Renal function, as assessed by the clearances of inulin and para-aminohippuric acid, was greater in SHR rats treated with OKY 046 than in those receiving vehicle alone. In WKY rats, OKY 046 administration did not affect renal function. These results suggest that increased renal synthesis of thromboxane may play a role in the pathogenesis of the elevated blood pressure of SHR rats (79). The results of these studies support the conclusions from Shibouta et al. (76) and Uderman et al. (77) that thromboxane may be involved in the development of hypertension in SHR. However, it is possible that OKY 046 might possess antihypertensive actions completely independent of thromboxane synthesis inhibition such as direct vasodilatation or sympatholytic activity. Another possibility is that the observed antihypertensive effect was mediated either by the shunting of prostaglandin endoperoxides into pathways of prostacyclin biosynthesis or by increased vascular sensitivity to prostacyclin in the absence of thromboxanes. Although the increased excretion of thromboxane A_2 in the urine suggests that this thromboxane is of renal origin, other possibilities should be considered. If platelet activation is involved in renal malfunction at some stage in the development of hypertension in SHR rats, the increased capacity of the platelets of this strain to produce thromboxane A_2 (80) as compared with WKY may explain the differences in the excretion of this compound in the urine; hence, the urinary excretion of thromboxane may be related to intrarenal differences in the behavior of platelets in SHR as compared with WKY. However, even if this were the origin of the increased excretion of thromboxane in the urine, the functional data after thromboxane blockade indicate that the increase in blood flow and glomerular filtration rate must have been related to an increase in perfusion and vasodilatation of the afferent arteriole. Hence, thromboxane of renal origin or platelet origin or both may be acting on the afferent arteriole to produce vasoconstriction in SHR but not in WKY.

Martineau et al. (81) have also searched for alterations of renal and extrarenal prostaglandin production in SHR. Using urinary PGE_2 as a measure of renal PGE_2 synthesis and urinary 2,3-dinor-6-keto-$PGF_{1\alpha}$ as a measure of extrarenal PGI_2 synthesis, these investigators found reduced urinary excretion of PGE_2 in SHR and defective increments of the PGI_2 metabolite during salt loading in SHR compared with normotensive controls. These results may point to an important role of renal PGE_2 synthesis and extrarenal role of PGI_2 synthesis in the capacity to handle a sodium load and regulate blood pressure responses to increases of sodium intake (81).

SUMMARY

Because of their vasodilator and vasoconstrictor properties, vasoactive prostaglandins and thromboxane A_2 have been proposed as modulators of the hemodynamic changes that occur in experimental models of renal disease. Increased synthesis of vasodilatory prostaglandins (PGE_2) and perhaps prostaglandin I_2 (PGI_2) play a role in the maintenance of renal blood flow and GFR during states of impaired perfusion. In contrast, thromboxane A_2 has been implicated as the vasoconstrictor responsible for the reduction of renal blood flow and GFR in certain animal models of experimental renal disease. These products and other metabolites of arachidonic acid may also participate in the immunological events underlying the onset and/or progression of experimental renal disease. It is evident that the pathophysiologic role of eicosanoids in experimental renal disease is not fully understood. Additional studies and further understanding of the many other potential roles of eicosanoids on immunological events, hemodynamic states, mesangial cell physiology, etc. are needed to comprehend more fully the extent of the participation of eicosanoids in the pathogenesis and pathophysiology of renal disease.

ACKNOWLEDGMENTS

The original work from our laboratory described in this article was supported by U.S.P.H.S. NIDDK grants DK-09976 and DK-07126.
We would like to thank Ms. Pat Verplancke for her help in the preparation of this manuscript.

267

REFERENCES

1. K. Herbaczynska-Cedro and J.R. Vane, Contribution of intrarenal generation of prostaglandins to autoregulation of renal blood flow in the
   dog, Circ. Res. 23:428-436 (1973).
2. J.J. Grantham and J. Orloff, Effect of prostaglandin E_1 on the permeability response of the isolated collecting tubule to vasopressin,
   adenosine 3'-5' monophosphate and theophylline, J. Clin. Invest. 47:
   1154-1161 (1968).
3. C. Larsson, P. Weber, E. Anggard, Arachidonic acid increases and indomethacin decreases plasma renin activity in the rabbit, Eur. J.
   Pharmacol. 28:391-394 (1974).
4. A.J. Vander, Direct effects of prostaglandin on renal function and
   renin release in anesthetized dogs, Am. J. Physiol. 214:218-221
   (1968).
5. K.V. Malik and J.C. McGiff, Modulation by prostaglandins of adrenergic
   transmission in the isolated perfused rabbit and rat kidney, Circ.
   Res. 36:599-609 (1975).
6. A.R. Morrison, K. Nishikawa, P. Needleman, Thromboxane A_2 biosynthesis in the ureter obstructed isolated perfused kidney of the rabbit,
   J. Pharmacol. Exp. Ther. 205:1-8 (1978).
7. A.R. Morrison, F. Thornton, A. Blumberg, E. Darracott-Vaughn, Thromboxane A_2 is the major arachidonic acid metabolite of human cortical
   hydronephrotic tissue, Prostaglandins 21:471-481 (1981).
8. R. Zipser, S. Myer, P. Needleman, Exaggerated prostaglandin and thromboxane synthesis in the rabbit with renal vein constriction, Circ.
   Res. 47:231-237 (1980).
9. M.L. Purkerson, J.H. Joist, J. Yates, A. Valdes, A. Morrison, S. Klahr,
   Inhibition of thromboxane synthesis ameliorates the progressive kidney disease of rats with subtotal renal ablation, Proc. Natl. Acad.
   Sci. 82:193-197 (1985).
10. K. Nishikawa, A.R. Morrison, P. Needleman, Exaggerated prostaglandin bio-
    synthesis and its influence on renal resistance in the isolated hydro-
    nephrotic rabbit kidney, J. Clin. Invest. 59:1143-1150 (1977).
11. T.E. Moody, E.D. Vaughn Jr., J.Y. Gillenwater, Relationship between
    renal blood flow and ureteral pressure during eighteen hours of total
    unilateral occlusion. Implications for changing sites of renal resist-
    ance, Invest. Urol. 13:246-251 (1975).
12. S. Klahr, J. Buerkert, A. Morrison, Urinary tract obstruction, in: "The
    Kidney," 3rd edition, Brenner B.M. and Rector F.C. Jr., eds., W.B.
    Saunders, Philadelphia, pp. 1443-1490 (1986).

13. T.E. Moody, E.D. Vaughn Jr., A.T. Wyler, The role of intrarenal angiotensin II in the hemodynamic response to unilateral obstructive uropathy, Invest. Urol. 14:390-397 (1977).

14. A.R. Morrison, K. Nishikawa, P. Needleman, Unmasking of thromboxane A_2 synthesis by ureter obstruction in the rabbit kidney, Nature 267:259-260 (1977).

15. A.R. Morrison and P. Needleman, Biochemistry and pharmacology of renal prostaglandins: Hormonal function and the kidney, in: "Contemporary Issues in Nephrology," B.M. Brenner and J.H. Stein, eds., Churchill Livingstone, New York, pp. 68-88 (1979).

16. T. Hiyamoto, K. Taniguchi, T. Tanonchi, F. Hirata, Selective inhibitor of thromboxane synthetase. Pyridine and its derivatives, in: "Advances in Prostaglandin Research," Vol. 6, B. Samuelsson, P. Ramwell, R. Paoletti, eds., Raven Press, New York, pp. 443-445 (1980).

17. A. Kawasaki and P. Needleman, Contribution of thromboxane to renal resistance changes in the isolated perfused hydronephrotic rabbit kidney, Circ. Res. 50:486-490 (1982).

18. R.B. Nagle, R.E. Bulger, R.E. Culter, H.R. Jervis, E.P. Benditt, Unilateral obstructive nephropathy in the rabbit. I. Early morphologic, physiologic and histochemical changes, Lab. Invest. 28:456-467 (1973).

19. R.B. Nagle, M.E. Johnson, H.R. Jervis, Proliferation of renal interstitial cells following injury induced by ureteral obstruction, Lab. Invest. 35:18-22 (1976).

20. T. Okegawa, P.E. Jonas, K. DeSchryver, A. Kawasaki, P. Needleman, Metabolic and cellular alterations underlying the exaggerated renal prostaglandin and thromboxane synthesis in ureter obstruction in rabbits, J. Clin. Invest. 71:81-90 (1983).

21. S.J. Leibovitch and R. Ross, A macrophage-dependent factor that stimulates the proliferation of fibroblasts in vitro, Am. J. Pathol. 84: 501-513 (1976).

22. J.B. Lefkowith, T. Okegawa, K. DeSchryver-Kecskemetei, P. Needleman, Macrophage-dependent arachidonate metabolism in hydronephrosis, Kidney Int. 26:10-17 (1984).

23. S.M. Weisman, D. Felser, E.D. Vaughan Jr., Platelet-activating factor is a potent stimulus for renal prostaglandin synthesis: possible significance in unilateral ureteral obstruction, J. Pharmacol. Exp. Ther. 235:10-15 (1985).

24. J. Buerkert, M. Head, S. Klahr, Effects of acute bilateral ureteral obstruction on deep nephron and terminal collecting duct function in

the young rat, J. Clin. Invest. 59:1055-1065 (1977).

25. W.E. Yarger, D.D. Schocker, R.H. Harris, Obstructive nephropathy in the rat: possible roles for the renin-angiotensin system, prostaglandins and thromboxanes in post-obstructive renal function, J. Clin. Invest. 65:400-412 (1980).

26. I. Ichikawa, M.L. Purkerson, J. Yates, S. Klahr, Dietary protein intake conditions the degree of renal vasoconstriction in acute renal failure caused by ureteral obstruction, Am. J. Physiol. 249:F54-F61 (1985).

27. L.A. Scharschmidt, J.G. Douglas, M.J. Dunn, Angiotensin II and eicosanoids in the control of glomerular size in the rat and human, Am. J. Physiol. 250:F348-F356 (1986).

28. A. Chanutin and E. Ferris, Experimental renal insufficiency produced by partial nephrectomy; control diet, Arch. Intern. Med. 49:767-787 (1932).

29. T. Shimamura and A.B. Morrison, A progressive glomerulosclerosis occurring in partial five-sixths nephrectomized rats, Am. J. Pathol. 79:95-106 (1975).

30. M.L. Purkerson, P.E. Hoffsten, S. Klahr, Pathogenesis of the glomerulopathy associated with renal infarction in rats, Kidney Int. 9:407-417 (1976).

31. M.L. Purkerson, J.R. Joist, D. Greenberg, D. Kay, P.E. Hoffsten, S. Klahr, Inhibition by anticoagulant drugs of the progressive hypertension and uremia associated with renal infarction in rats, Thromb. Res. 26:227-240 (1982).

32. U.O. Barcelli, M. Weiss, V.E. Pollak, Effects of a dietary prostaglandin precursor on the progression of experimentally-induced chronic renal failure, J. Lab. Clin. Med. 100:786-797 (1982).

33. M. Heifets, J.J. Morrissey, M.L. Purkerson, A.R. Morrison, S. Klahr, Effect of dietary lipids on renal function in rats with subtotal nephrectomy, Kidney Int., in press.

34. Y. Izumi, T.W. Weiner, R. Franco-Saenz, P.J. Mulrow, Effects of dietary linoleic acid on blood pressure and renal function in subtotally nephrectomized rats (42404), Proc. Soc. Expl. Biol. Med. 183:193-198 (1986).

35. R. Hirschberg, D. Herrath, H. Klaus, W. Hofer, C. Schuster, H. Rottka, K. Schaefer, Effect of diets containing varying concentrations of essential fatty acids and triglycerides on renal function in uremic rats and NZB/NZW F_1 mice, Nephron 38:233-237 (1984).

36. L. McGregor, R. Morazain, S. Renaud, Effect of dietary linoleic acid on platelet function in the rat, Thromb. Res. 20:499-507 (1980).

270

37. R.B. Zurier, I. Damjanov, D.M. Sayadoff, N.B. Rothfield, Prostaglandin
    E_1 treatment of NZB/NZW F_1 hybrid mice. II. Prevention of glomer-
    ulonephritis, Arthritis Rheum. 20:1449-1456 (1977).

38. E.R. Hurd, J.M. Johnston, J.R. Okita, P.C. MacDonald, M. Ziff, J.N.
    Gilliam, Prevention of glomerulonephritis and prolonged survival in
    New Zealand Black/New Zealand White F_1 hybrid mice fed an essential
    fatty acid-deficient diet, J. Clin. Invest. 67:476-485 (1981).

39. J.B. Howie, B.J. Helyer, T.P. Casey, I. Aarons, Renal disease in auto-
    immune strains of mice, in: Proc. Third Int. Cong. Nephr., Basel, S.
    Karger 2:150-163 (1967).

40. E.L. Dubois, R.E. Horowitz, H.B. Demopoulos, R. Teplitz, NZB/NZW mice
    as a model of systemic lupus erythematosus, J. Am. Med. Assoc. 195:
    285-289 (1966).

41. E.R. Hurd and M. Ziff, Quantitative studies of immunoglobulin deposi-
    tion in the kidney, glomerular cell proliferation and glomerulo-
    sclerosis in NZB/NZW F_1 hybrid mice, Clin. Exp. Immunol. 26:261-
    268 (1976).

42. J.R. Vane, Prostaglandins as mediators of inflammation, in: "Advances in
    Prostaglandin and Thromboxane Research," Samuelsson, B. and Paoletti,
    R., Raven Press, New York 2:791-801 (1976).

43. E.J. Goetzl, Mediators of immediate hypersensitivity derived from arachi-
    donic acid, N. Engl. J. Med. 303:822-825 (1980).

44. J.D. Prickett, D.R. Robinson, A.D. Steinberg, Dietary enrichment with
    the polyunsaturated fatty acid eicosapentaenoic acid prevents pro-
    teinuria and prolongs survival in NZB x NZW F_1 mice, J. Clin.
    Invest. 68:556-559 (1981).

45. V.E. Kelley, A. Ferretti, S. Izui, T.B. Strom, A fish oil diet rich in
    eicosapentaenoic acid reduces cyclooxygenase metabolites and suppres-
    ses lupus in MRL-1pr mice, J. Immunol. 134:1914-1919 (1985).

46. H.B. Steinhauer, S. Batsford, P. Schollmeyer, R. Kluthe, Studies on
    thromboxane B_2 production in the course of murine autoimmune dis-
    ease: Inhibition by oral histidine and zinc supplementation, Clin.
    Nephrol. 24:63-68 (1985).

47. V.E. Kelley, S. Sneve, S. Musinski, Increased renal thromboxane produc-
    tion in murine lupus nephritis. J. Clin. Invest. 77:252-259 (1986).

48. V.E. Kelley, A. Winkelstein, S. Izui, Effect of prostaglandin E on
    immune complex nephritis in NZB/W mice, Lab. Invest. 41:531-537
    (1979).

49. K.R. McLeish, A.F. Gohara, W.T. Gunning, III, D. Senitzer, Prostaglandin
    E_1 therapy of murine chronic serum sickness, J. Lab. Clin. Med. 96:
    470-479 (1980).

50. R.A.K. Stahl, S. Adler, P.J. Baker, Y.P. Chen, P.M. Pritzl, W.G. Couser, Enhanced glomerular prostaglandin formation in experimental membranous nephropathy, Kidney Int. 31:1126-1131 (1987).

51. J.E. Stork and M.J. Dunn, Hemodynamic roles of thromboxane A_2 and prostaglandin E_2 in glomerulonephritis, J. Pharm. Exp. Ther. 233:672-678 (1985).

52. E.A. Lianos, G.A. Andres, M.J. Dunn, Glomerular prostaglandin and thromboxane synthesis in rat nephrotoxic serum nephritis. Effects on renal hemodynamics, J. Clin. Invest. 72:1439-1448 (1983).

53. E.A. Lianos, M.A. Rahman, M.J. Dunn, Glomerular arachidonate lipoxygenation in rat nephrotoxic serum nephritis, J. Clin. Invest. 76:1355-1359 (1985).

54. C.H. Dubois, J.B. Foidart, C.A. Dechene, P.R. Mahieu, Effects of a diet deficient in essential fatty acids on the glomerular hypercellularity occurring in the course of nephrotoxic serum nephritis in rats, Kidney Int. 21 (suppl.11):S-39-S-45 (1982).

55. V.E. Kelley and A. Winkelstein, Effect of prostaglandin E_1 treatment on murine acute immune-complex glomerulonephritis, Clin. Immunol. Immunolpathol. 16:316-323 (1980).

56. K.R. McLeish, A.F. Gohara, L.J. Johnson, D.L. Sustarsic, Alteration in immune-complex glomerulonephritis by arachidonic acid, Prostaglandins 23:383-389 (1982).

57. V. Kher, U. Barcelli, M. Weiss, V.E. Pollak, Effects of dietary linoleic acid enrichment on induction of immune complex nephritis in mice, Nephron 39:261-266 (1985).

58. H. Saito, T. Ideura, J. Takeuchi, Effects of a selective thromboxane A_2 synthetase on immune complex glomerulonephritis, Nephron 36: 38-45 (1984).

59. G. Remuzzi, L. Imberti, M. Rossini, C. Morelli, C. Carminati, G.M. Cattaneo, T. Bertani, Increased glomerular thromboxane synthesis as a possible cause of proteinuria in experimental nephrosis, J. Clin. Invest. 75:94-101 (1985).

60. C.J. Kelly, R.B. Zurier, K.A. Krakauer, N. Blanchard, E.G. Nielson, Prostaglandin E_1 inhibits effector T cell induction and tissue damage in experimental murine interstitial nephritis. J. Clin. Invest. 79:782-789 (1987).

61. O. Abehsura-Amar, C. Damars, M. Parait, L. Chedid, Strain dependence of muramyl dipeptide-induced LAF (IL-1) release by murine-adherent peritoneal cells, J. Immunol. 134:365-368 (1985).

62. S.L. Kunkel, S.W. Chensue, S.H. Phan, Prostaglandins as endogenous mediators of interleukin I production, J. Immunol. 136:186-192 (1986).

63. G.D. Fink and M.J. Brody, Renal vascular resistance and reactivity in the spontaneously hypertensive rat, Am. J. Physiol. 237:F128-F132 (1979).

64. C. Limas and C.J. Limas, Enhanced renomedullary prostaglandin synthesis in spontaneously hypertensive rats: Role of a phospholipase A_2, Am. J. Physiol. 236:H65-H72 (1979).

65. C.R. Pace-Asciak, Decreased renal prostaglandin catabolism precedes onset of hypertension in the developing spontaneously hypertensive rat, Nature 263:510-512 (1976).

66. C.J. Limas and C. Limas, Prostaglandin metabolism in the kidneys of spontaneously hypertensive rats, Am. J. Physiol. 233:H87-H92 (1977).

67. M. Konieczkowski, M.J. Dunn, J.E. Stork, A. Hassid, Glomerular synthesis of prostaglandins and thromboxane in spontaneously hypertensive rats, Hypertension 5:446-452 (1983).

68. C.R. Pace-Asciak, M.C. Carrara, G. Rangaraj, K.C. Nicolaou, Enhanced formation of PGI_2, a potent hypotensive substance, by aortic rings and homogenates of the spontaneously hypertensive rats, Prostaglandins 15:1005-1012 (1978).

69. J.V. Levy, Changes in systolic arterial blood pressure in normal and spontaneously hypertensive rats produced by acute administration of inhibitors of prostaglandin biosynthesis, Prostaglandins 13:153-160 (1977).

70. N. Farman and J.P. Bonvalet, Abnormal relationship between sodium excretion and hypertension in spontaneously hypertensive rats, Pflugers Arch. 354:39-53 (1975).

71. R. Dietz, A. Schomig, H. Haebara, J.F.E. Mann, W. Rascher, J.B. Lüth, N. Grünherz, F. Gross, Studies on the pathogenesis of spontaneous hypertension in rats, Circ. Res. 43 (suppl.I):I98-I111 (1978).

72. R.L. Kline, P.M. Kelton, P.F. Mercer, Effect of renal denervation on the development of hypertension in spontaneously hypertensive rats, Can. J. Physiol. Pharmacol. 56:818-822 (1978).

73. S.R. Winternitz, R.E. Katholi, S. Oparil, Role of the renal sympathetic nerves in the development and maintenance of hypertension in the spontaneously hypertensive rat, J. Clin. Invest. 66:971-978 (1980).

74. Y. Shibouta, Y. Inada, Z. Terashita, K. Nishikawa, S. Kikuchi, K. Shimamoto, Angiotensin-II-stimulated release of thromboxane A_2 and prostacyclin (PGI_2) in isolated, perfused kidneys of spontaneously hypertensive rats, Biochem. Pharmacol. 28:3601-3609 (1979).

75. Y. Shibouta, Z-I. Terashita, Y. Inada, K. Nishikawa, S. Kikuchi, Enhanced thromboxane A_2 biosynthesis in the kidney of spontaneously

hypertensive rats during development of hypertension, Eur. J. Pharmacol. 70:247-256 (1981).

76. Y. Shibouta, Z-I. Terashita, Y. Inada, K. Kato, K. Nishikawa, Renal effects of pinane-thromboxane A$_2$ and indomethacin in saline volume-expanded spontaneously hypertensive rats, Eur. J. Pharmacol. 85:51-59 (1982).

77. H.D. Uderman, R.J. Workman, E.K. Jackson, Attenuation of the development of hypertension in spontaneously hypertensive rats by the thromboxane synthetase inhibitor, 4'-(imidazol-1-YL)acetophenone, Prostaglandins 24:237-244 (1982).

78. H.J. Grone, R.S. Grippo, W.J. Arendshorst, M.J. Dunn, Role of thromboxane in control of arterial pressure and renal function in young spontaneously hypertensive rats, Am. J. Physiol. 250:F488-F496 (1986).

79. M.L. Purkerson, K.J. Martin, J. Yates, J.M. Kissane, S. Klahr, Thromboxane synthesis and blood pressure in spontaneously hypertensive rats, Hypertension 8:1113-1120 (1986).

80. F. DeClerck, L. VanGorp, B. Xhonneux, Y. Somers, L. Wouters, Enhanced platelet turnover and prostaglandin production in spontaneously hypertensive rats, Thromb. Res. 27:243-249 (1982).

81. A. Martineau, M. Robillard, P. Falardeau, Defective synthesis of vasodilator prostaglandins in the spontaneously hypertensive rat, Hypertension 6:1161-1165 (1984).

ABNORMALITIES OF GLOMERULAR EICOSANOID METABOLISM

IN STATES OF GLOMERULAR HYPERFILTRATION

Morris Schambelan, Burl R. Don, George A. Kaysen
and Susan Blake

Medical Service, San Francisco General Hospital Medical
Center, San Francisco, California 94110; Department of
Medicine, University of California, San Francisco
California 94143; Department of Medicine, Martinez
Veterans Administration Medical Center, Martinez
California 94553; Department of Internal Medicine
University of California, Davis, California 95817

INTRODUCTION

The glomerulus is a complex structure that functions primarily to
produce an ultrafiltrate of plasma. Micropuncture studies in the Munich-
Wistar rat have identified the major physical determinants of the glomerular filtration rate (GFR): glomerular plasma flow; afferent protein concentration; transcapillary hydraulic-pressure difference; and the glomerular capillary ultrafiltration coefficient (K_f) (1). Studies in this
model also indicate that a number of circulating hormones (e.g., parathyroid hormone, vasopressin), as well as other biological messengers that
are produced in the kidney and that act locally through an autacrine or
paracrine mechanism (prostaglandins, angiotensin II, histamine, and bradykinin), can affect the GFR (2). In addition to vasoactive effects on the
afferent and efferent renal arterioles with subsequent alteration of renal
vascular resistance, many of these substances affect the filtration
process by a reduction in K_f, an action that appears to be mediated by
contraction of the glomerular mesangial cell and reduction of the glomerular capillary surface area. This may be due to a direct contractile
effect of the agonist on the mesangial cell, as in the case of angiotensin
II and vasopressin, or by secondary activation of the renin-angiotensin
system (2-3). Prostaglandins, which are secreted by whole glomeruli (4-6)

and by cultured glomerular mesangial cells (7-9), can also influence GFR by modifying the vasoconstrictor effects of angiotensin II (10) as well as the glomerular contractile response to this agonist (11). Other metabolites of arachidonic acid, such as thromboxane A_2 (12), can directly stimulate glomerular contraction. Numerous studies in the past decade have defined a complex picture in which vasoactive autacoids and circulating hormones interact to control glomerular filtration.

As discussed elsewhere in this volume, it is evident that renal eicosanoid metabolism is altered in a variety of pathophysiologic settings in which the GFR is reduced. Enhanced glomerular synthesis and/or urinary excretion of vasoconstrictor eicosanoids such as TxA_2 has been reported in diverse lesions such as nephrotoxic serum nephritis (13), adriamycin nephrosis (14), following subtotal nephrectomy (15) and bilateral ureteral obstruction (16). A functional role for these abnormalities is suggested by the improvement in GFR that occurs when inhibitors of TxA_2 synthesis are administered (13-16). In these models, increased production of vasodilatory prostaglandins may serve to offset the vasoconstrictor effects of TxA_2 since recovery of GFR is attenuated in animals pretreated with inhibitors of cyclooxygenase (13). A compensatory role for vasodilatory prostaglandins is also suggested by the reduction in GFR that occurs frequently when inhibitors of prostaglandin synthesis are administered to patients with chronic renal diseases (17) and disorders characterized by ineffective circulating volume (18).

Recently, there has been considerable interest in the observation that early in the course of some renal diseases the GFR, rather than being reduced, is actually supernormal. This so-called "glomerular hyperfiltration" and the associated increase in the glomerular transcapillary hydraulic pressure difference ("glomerular hypertension") occurs in disorders such as diabetes mellitus (19-23) and in remnant nephrons following subtotal nephrectomy (24), and is modified by alterations in certain dietary constituents, particularly protein (24,25). In the following sections we will review the evidence that indicates that abnormalities of glomerular eicosanoid metabolism occur in various states characterized by glomerular hyperfiltration and will consider the potential role of these biochemical findings in the control of glomerular hemodynamics in these pathophysiologic settings.

DIABETES MELLITUS

Abnormalities of Glomerular Function in Diabetes Mellitus

Abnormalities of glomerular function are prominent features of diabetes mellitus. Approximately 40% of patients with type 1 (insulin-
dependent) diabetes and a somewhat smaller percentage of individuals with
type 2 (noninsulin-dependent) diabetes will develop diabetic nephropathy,
characterized by proteinuria, loss of renal function and rapid progression
to end-stage renal failure. Diabetic nephropathy is now the most common
cause of end-stage renal disease in the United States. The clinical features of this important cause of chronic renal insufficiency have been
reviewed previously (26-28).

Numerous studies have indicated that, prior to the onset of overt
clinical diabetic nephropathy and progressive renal insufficiency, many
diabetics, particularly type 1 diabetics studied at the time of initial
presentation of their disease, have a supernormal GFR (Fig. 1) and renal
plasma flow (RPF) (19-21,29,30). Retrospective analysis of the subsequent
course in such patients suggests that such hemodynamic abnormalities,
together with the presence of microalbuminuria and systemic hypertension,
may herald the subsequent development of diabetic nephropathy (31).

That some component of the metabolic derangement in diabetes is responsible for the alterations in glomerular hemodynamics is evidenced by the
observation that "tight control" with insulin will prevent or correct the
hyperfiltration both in patients with clinical diabetes mellitus (32-34)
and in animals in which diabetes is induced experimentally with streptozotocin (35). Structural hypertrophy, typically seen in kidneys of diabetic

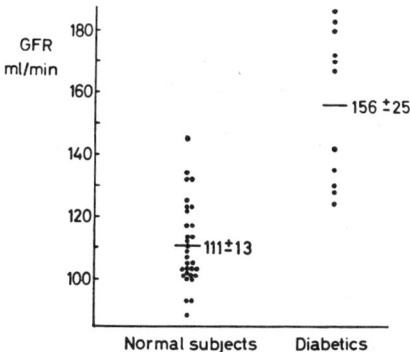

Fig. 1. Comparison of measurements of glomerular fil-
tration rate in 11
patients with untreated
type 1 diabetes melli-
tus and 31 normal sub-
jects. The difference
between groups was high-
ly significant
($p<0.001$). (Reproduced
from reference 20 with
permission).

patients and animals (36), may account in part for these findings. Indeed the increased GFR and RPF correlate with renal size in type 1 diabetics (29). In the diabetic rat a marked increase in renal weight and cellular hypertrophy can be detected within the first 24 hours after the onset of hyperglycemia (36).

Hyperglycemia per se may play a role as evidenced by the increase in GFR, documented to occur with infusion of glucose in both normal subjects and in insulin-dependent diabetics (37), and by the reduction in the elevated levels of GFR and RPF in type 1 diabetics that results when glucose levels are normalized by acute insulin therapy (38). In addition to hyperglycemia, the diabetic state is characterized by increased circulating levels of gluco-regulatory hormones such as glucagon and growth hormone, both of which can induce an increase in GFR (39,40).

The pathogenesis and pathophysiologic significance of glomerular hyperfiltration/hypertension in diabetic nephropathy has been elucidated further in studies performed in rats with experimentally-induced diabetes mellitus. Similar to the findings in humans with type 1 diabetes, whole kidney GFR is increased 30-40% above control in rats made diabetic with the beta cell toxin streptozotocin (22,41). Assessment of the determinants of GFR in this model have been made using the technique of free-flow micropuncture (22). These micropuncture studies indicate that, in diabetic rats, single nephron GFR (SNGFR) and glomerular capillary plasma flow are increased as a consequence of renal vasodilatation (Table 1). In addition, the transglomerular capillary pressure difference is increased. That such glomerular and/or systemic hypertension plays a critical role in the initiation and progression of diabetic nephropathy is suggested by the observation that, prior to the induction of diabetes mellitus in the rat, a unilateral reduction in renal perfusion pressure due to placement of a clip across one renal artery attenuated the subsequent development of diabetic nephropathy in the "clipped" kidney (42). Similarly, the occurrence of a unilateral atherosclerotic plaque in a renal artery of a diabetic patient protected the ipsilateral kidney from diabetic nephropathy, whereas the contralateral kidney with the patent renal artery demonstrated Kimmelstiel-Wilson lesions (43). Furthermore, despite persistent hyperglycemia, glomerular damage and proteinuria in the rat can be prevented by maintaining intraglomerular pressures at normal levels utilizing either dietary (25) or pharmacologic means (44) (vida infra). These data suggest that hemodynamic rather than metabolic abnormalities play a more important

Table 1. Glomerular Hemodynamics in Experimental Diabetes Mellitus [*]

Group	Glucose	GFR	SNGFR	Q_A	$\overline{\Delta P}$	R_A	R_E
	mg/dl	ml/min	nl/min		mmHg	dynes.sec.cm^{-5}	
Control	115	1.10	48.9	142	35.2	3.0	2.1
(n=8)	±10	±0.06	±3.8	±11	±1.3	±0.3	±0.2
Diabetic	375	1.47	69.0	240	44.4	1.9	1.6
(n=6)	±23	±0.13	±8.0	±34	±1.6	±0.3	±0.3
P values	<0.001	<0.02	<0.05	<0.01	<0.005	<0.01	NS

[*]Values are mean ± SEM; R_A and R_E x 10^{10}. Micropuncture studies were performed 2-15 weeks following injection of streptozotocin, 60 mg/kg body weight intravenously, or vehicle. Diabetic rats were treated with 2 U NPH insulin daily to maintain them in a state of moderate hyperglycemia. Abbreviations: GFR, glomerular filtration rate; SNGFR, single nephron GFR; Q_A, glomerular plasma flow; $\overline{\Delta P}$, mean glomerular filtration pressure; R_A, afferent arteriolar resistance; R_E, efferent arteriolar resistance. (Adapted from Table 1 in Reference 22 with permission.)

role in the pathogenesis of diabetic nephropathy and led us to consider whether glomerular eicosanoid metabolism is altered in diabetes.

Eicosanoid Metabolism in Diabetes Mellitus - General Considerations

There is ample evidence that eicosanoid metabolism is abnormal in a variety of tissues obtained both from patients with clinical diabetes mellitus and in animals with experimental diabetes. Increased release of PGE_2 and PGI_2 by adipocytes in rats with diabetic ketoacidosis has been inferred by the observation that elevated plasma levels of derivatives of these compounds are suppressed by both insulin and 5-methyl-pyrazole-3-carboxylic acid, two potent and structurally unrelated antilipolytic compounds (45). Abnormalities of eicosanoid metabolism by platelets are also well described in diabetes. Increased synthesis of TXA_2, as gauged by levels of its stable metabolite TXB_2, occurs in platelets obtained from humans with diabetes (46) and from rats with both streptozotocin-induced (47) and spontaneous (BB Wistar) diabetes (48). The finding of increased platelet aggregation and release of malonyldialdehyde in newborn infants and their diabetic mothers compared with control mother-neonate pairs provides further evidence of enhanced thromboxane synthesis activity (49). In contrast to increased synthesis of these proaggregatory eicosanoids by platelets, release of the potent antiaggregatory eicosanoid prostacyclin was decreased in arteries (50) and veins (51) of patients with diabetes mellitus, in umbilical arteries of neonates of mothers with either gestational or type 1 diabetes (52,53), and in aortic tissues obtained from rats (54,55) and swine (56) with streptozotocin-induced diabetes. In the pancreas a possible role for abnormalities of prostaglandin metabolism in the pathophysiology of impaired insulin secretion in diabetics has been inferred from the observation that PGE infusion inhibits glucose-induced acute insulin responses in normal human subjects and infusion of sodium salicylate partially restores the acute insulin response to glucose infusion in type 2 diabetics (57). Similarly, changes in prostaglandin metabolism in lung (58-60), heart (61), and seminal vesicles (62) may also contribute to altered function of these organs in diabetics.

Effect of Diabetes Mellitus on Renal Eicosanoid Metabolism

Studies of eicosanoid metabolism in diabetic animals have also been performed using homogenates obtained from renal tissue. In the rat microsomal preparations from whole kidney homogenates obtained in the first two weeks of both alloxan- and streptozotocin-induced diabetes mellitus released less PGE-like material when incubated with arachidonic acid than those from control animals (63). Production of PGI_2 was also decreased in homogenates of renal cortex obtained from rats studied 1-3 months after administration of streptozotocin (64). It should be noted, however, that results of studies performed on either crude homogenates or microsomal preparations of whole cortex may reflect abnormalities in prostaglandin metabolism in structures other than the glomerulus. Inasmuch as prostaglandins act locally within the cells or tissues in which they are synthesized, conclusions about the potential pathophysiologic role of eicosanoids in the glomerular hemodynamic abnormalities in diabetes mellitus require direct measurements of eicosanoid production by glomeruli isolated from diabetic animals.

Glomerular Eicosanoid Metabolism in Diabetes Mellitus

To evaluate directly the effect of diabetes mellitus on glomerular eicosanoid metabolism, we performed studies in the early stages of an experimental model of diabetes mellitus in the rat (65). Diabetes mellitus was induced in male Sprague-Dawley rats (150-250 g body wt) by intravenous administration of streptozotocin, 60 mg/kg body wt. Control rats, which were matched for age and weight at the time of streptozotocin administration, received an equal volume of the vehicle. The rats were maintained on standard rat chow and water ad libitum throughout the study.

The onset of diabetes was manifested by the development of polyuria and polydypsia within 2-3 days of administration of streptozotocin. Only those rats in which plasma glucose levels exceeded 250 mg/dl were included in the diabetic group.

Studies of prostaglandin biosynthesis were performed 9 to 23 days after administration of streptozotocin. Glomeruli were isolated by mechanical sieving and incubated in the presence of $CaCl_2$. Eicosanoid production was evaluated both by direct radioimmunoassay and by identification of labeled products by HPLC after incubation with [^{14}C]-arachidonic acid. When measured under basal conditions, glomerular prostaglandin production was increased in diabetic rats. In studies performed in individual rats, the mean production rates of both PGE_2 and $PGF_{2\alpha}$, the major prostaglandins produced by rat glomeruli, were approximately two-fold greater in the rats with diabetes whereas that of TxB_2 was not significantly greater than control (Fig. 2). Production of PGE_2 was also measured on the supernatant obtained from incubation of pools of glomeruli from diabetic and control rats. In these studies, mean basal PGE_2 production was also approximately two-fold greater in glomeruli from diabetic animals.

Similar increases in the synthesis of PGE_2 by glomeruli isolated from diabetic rats have also been reported by Kreisberg and Patel (66), Brown et al. (67), Chaudhari and Kirschenbaum (68), Craven et al. (69), and Moel et al. (70). Rogers and Larkin (71) also reported that the release of both 6-keto-$PGF_{1\alpha}$ and PGE_2 was increased in diabetic rats,

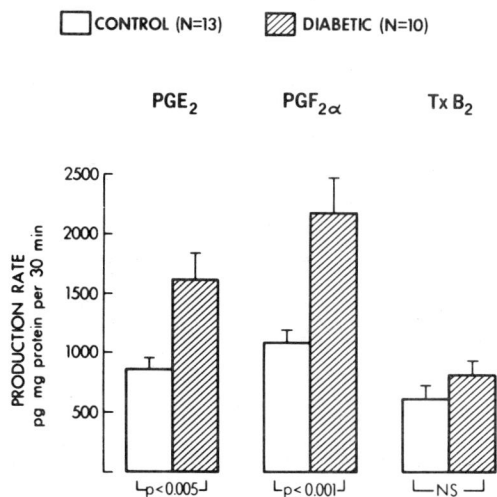

Fig. 2. Comparison of eicosanoid production by glomeruli isolated from rats with streptozotocin-induced diabetes mellitus and control animals. (Adapted from data in Table 1 in reference 65 with permission.)

281

but the differences did not reach the level of statistical significance. Taken together with the results of our study, it seems evident that, at least in the early stages of the disorder, glomerular prostaglandin synthesis is increased in diabetic rats.

In our study, the increased rate of prostaglandin production did not appear to be due to a nonspecific effect of streptozotocin inasmuch as glomerular prostaglandin production was not increased significantly in streptozotocin-treated rats made euglycemic with insulin therapy (Fig. 3). Furthermore, PGE_2 production by normal glomeruli was not affected by addition of streptozotocin to the incubation media, using a range of concentrations (1-10 mM) that have been demonstrated to inhibit glucose-stimulated proinsulin biosynthesis by (72) and to be cytotoxic to (73) pancreatic islet cells in vitro.

The increased rate of prostaglandin production also did not appear to be related directly to the severity of the diabetic state as reflected by the degree of hyperglycemia at the time of death. In fact, the rates of glomerular prostaglandin production in the individual diabetic animals correlated inversely with the plasma glucose concentration ($r= -0.67$, $p<0.05$ for PGE_2; $r= -0.88$, $p <0.001$ for $PGF_{2\alpha}$). Furthermore, in separate studies we determined that PGE_2 production by normal glomeruli was not influenced by the glucose concentration in the incubation media over a range of 1-40 mM. Thus, hyperglycemia per se did not appear to account for the abnormalities in eicosanoid metabolism.

Simply increasing the external concentration of glucose may bear little resemblance to the diabetic state, in which intracellular glucose and ATP concentrations may be decreased as a consequence of insulin deficiency. That such hypoglycemic conditions might stimulate prostaglandin production was suggested by studies in which release of both PGE_2 and $PGF_{2\alpha}$ by renal papillae was stimulated by reduction of the concentration of glucose in the media (74). However, in our studies utilizing pools of glomeruli obtained from normal rats, glomerular PGE_2 production was not increased during short-term incubations in media in which the concentration of glucose was reduced to levels as low as 1 mM. Release of PGE_2 and $PGF_{2\alpha}$ from rat papillae has also been reported to increase when this tissue is incubated in the presence of inhibitors of glycolysis (75), experimental conditions that would be expected to result in a reduction in availability of glucose intracellularly. Accordingly, we incubated pools of glomeruli obtained from normal rats with inhibitors of

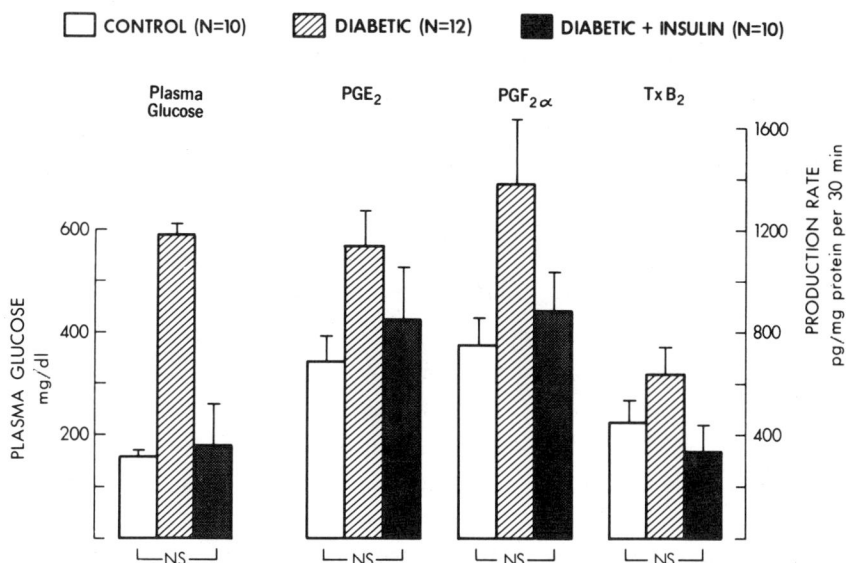

Fig. 3. Comparison of blood glucose concentrations and glomerular
eicosanoid production in untreated diabetic rats, diabetic
rats that received insulin, and control animals. Beginning
on the third day after injection of streptozotocin, a sub-
group of diabetic rats received NPH insulin in doses (2-5 U,
subcutaneously, each evening) sufficient to render them
euglycemic. Animals were sacrificed 10-14 days later, and
glomeruli were incubated as described in the legend of Fig.
1. (Adapted from data in reference 65 with permission.)

glycolysis at concentrations similar to those used with the papillary
preparations: D-L-glyceraldehyde, 100 mg/dl; sodium fluoride, 10 mM; and
iodoacetate, 50 mg/dl (75). Of these three inhibitors, only iodoacetate
appeared to result in increased release of PGE_2 and $PGF_{2\alpha}$. This
occurred in a dose dependent fashion and only in the absence of added
arachidonic acid, suggesting a stimulatory effect on phospholipase activity. However, since iodoacetate has many effects on cell function other
than inhibition of glycolysis, we repeated these studies using 2-deoxy-D-
glucose and D-L-glyceraldehyde in doses ranging from 0.1 to 100 mM. No
effect was seen with these agents, in either the absence or presence of
added arachidonic acid. Thus, we are reluctant to ascribe the stimulatory
effect noted with iodoacetate to its suggested effect to inhibit specifically glyceraldehyde-3-phosphate dehydrogenase (75).

Cyclooxygenase versus Phospholipase Activity

In our study the addition of arachidonic acid (10 μM) increased the production of both PGE_2 and $PGF_{2\alpha}$ to five to ten times the basal values in both the diabetic and control preparations. Although in the presence of arachidonic acid the mean values were 20-40% higher in the diabetic group, these values did not differ significantly from those in control animals. Such findings are often interpreted as evidence against the presence of increased glomerular cyclooxygenase activity (76), inasmuch as intraglomerular concentrations of arachidonic acid are presumed to be non-rate limiting under such conditions. Indeed, in their recent report Craven et al. (69) interpreted the absence of a difference in eicosanoid production between diabetic and control glomeruli in the presence of excess arachidonic acid as evidence against an increase in cyclooxygenase activity. These investigators therefore concluded that augmented eicosanoid production in diabetic glomeruli was a consequence of increased phospholipase activity.

To examine further the possibility that glomerular cyclooxygenase activity might be altered in diabetic rats, we performed additional studies in which glomeruli were incubated with an excess of labeled arachidonic acid. When conversion of $[^{14}C]$-arachidonic acid into labeled prostaglandins was examined using a reverse phase HPLC system, no qualitative differences in the profiles obtained from diabetic and control glomeruli were noted (Fig. 4): the patterns of elution in the diabetic animals were nearly identical to those obtained in the control animals and were similar to those reported previously using glomeruli obtained from normal rats (77). Whereas no differences in the general patterns of prostaglandin production were noted, incorporation of $[^{14}C]$-arachidonic acid into the major prostaglandins was increased in the diabetic glomeruli (Fig. 4, Table 2). When conversion of $[^{14}C]$-arachidonic acid into lipoxygenase products and other eicosanoids was studied using a straight phase system, there were again no differences in the patterns of $[^{14}C]$ products between control and diabetic preparations. At the high substrate concentration used in these experiments, glomeruli from normal rats form predominantly 12- and 15-HETE (78). No significant difference in the rate of formation of these lipoxygenase products was observed (Table 2). An additional major peak was identified as the cyclooxygenase product, hydroxyheptadecatrienoic acid (HHT). Similar to the prostaglandins, conversion to HHT was increased in diabetic glomeruli (Table 2). These results indicate that glomerular cyclooxygenase but not lipoxygenase activity was increased

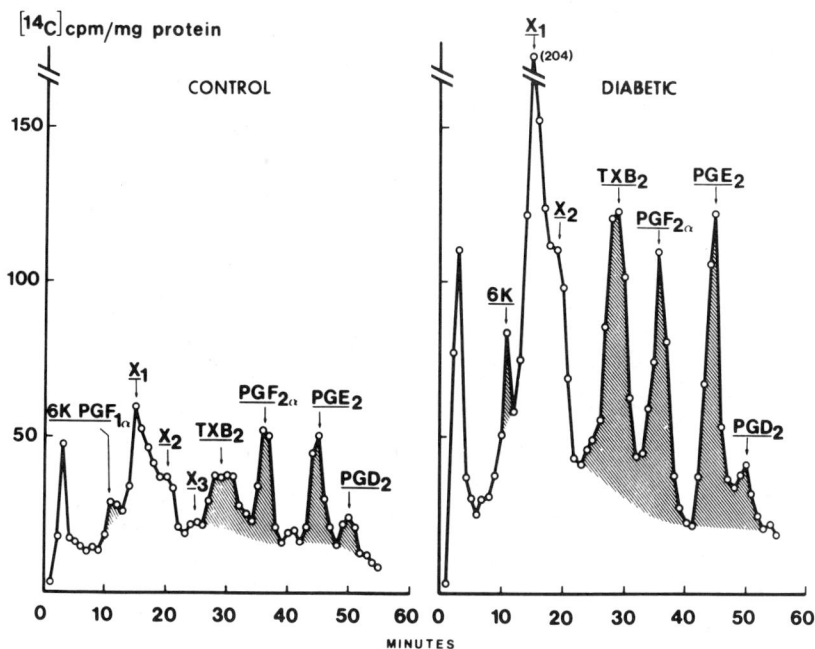

Fig. 4. Reverse phase HPLC of [^{14}C]-arachidonic acid products formed by incubation of glomeruli isolated from pools of control (left panel) and diabetic (right panel) rats. After an initial peak that represents an auto-oxidation product of [^{14}C]-arachidonic acid, five peaks (indicated by the shaded areas) with retention times of 11, 30, 36, 45, and 50 min. were identified by comigration with [^{3}H] standards as 6-keto-PGF$_{1\alpha}$, TXB$_2$, PGF$_{2\alpha}$, PGE$_2$, and PGD$_2$, respectively. Two large peaks (X1 and X2) eluting between 6-keto-PGF$_{1\alpha}$ and TXB$_2$, were also present as noted previously for rat glomeruli (77). On the right panel, the figure in parentheses next to peak X1 indicates the magnitude of this product.

in the diabetic animals. These results do not exclude a concomitant increase in glomerular phospholipase activity. More definitive studies utilizing isolated plasma membranes (phospholipase activity) and microsomes (cyclooxygenase activity) will be required before the precise biochemical events underlying the alterations in glomerular eicosanoid metabolism that occurs in diabetic animals can be established.

Pathogenesis of Altered Glomerular Eicosanoid Metabolism in Diabetes Mellitus

The results of our study did not identify the mechanisms responsible for the increase in prostaglandin production by glomeruli obtained from diabetic animals. Inasmuch as prostaglandins may function as modulators of the action of a variety of vasoactive substances, the increased prostaglandin synthesis we observed could represent a response to the systemic hemodynamic abnormalities and the changes in circulating vasoactive hormones (e.g., the renin-angiotensin system) that occur in severe diabetes. Although plasma volume per se is generally normal or increased in this setting (22), salt depletion, attending osmotic diuresis in severe diabetes, could conceivably be present independently of an alteration in intravascular volume. However, when salt depletion was induced in the rat by dietary restriction of sodium chloride (79), the pattern of prostaglandin synthesis by isolated glomeruli, namely an increase in $PGF_{2\alpha}$ and a decrease in PGE_2, was clearly different from that observed in our study. Furthermore, the highest prostaglandin production rates that we observed occurred in the rats with the least severe degrees of hyperglycemia, suggesting that the abnormalities were not related quantitatively to the disturbance of osmolality.

Alternatively, the abnormalities in arachidonic acid metabolism could result from the profound metabolic changes that occur in diabetics. As discussed previously, a direct stimulatory effect of hyperglycemia seems unlikely in view of the negative correlation between blood glucose levels

Table 2. Incorporation of [^{14}C] C20:4 into Labeled Eicosanoids by Glomeruli from Diabetic and Control Rats[*]

Study group	Glomerular Protein	Prostaglandins				12- and 15- HETE	HHT
		Total	TXB$_2$	PGF$_{2\alpha}$	PGE$_2$		
	mg/ml			cpm/mg protein			
Diabetic	3.44 ±0.52	1889 ±229	448 ±99	349 ±55	294 ±23	3843 ±815	2019 ±327
Control	3.56 ±0.40	908 ±119	168 ±36	145 ±22	132 ±10	3192 ±459	981 ±114
P value	NS	<0.01	<0.05	<0.01	<0.01	NS	<0.01

[*]Results are expressed as cpm/mg protein per 30-min. incubation. The area under the curve for each product was calculated as the sum of the cpm for each tube comprising that peak minus the baseline radioactivity level. Values obtained in five separate experiments were compared by the Mann-Whitney test. Abbreviations: HETE, hydroxyeicosatetraenoic acid; HHT, hydroxyheptadecatrienoic acid. (Adapted from Table IV in reference 65 with permission.)

and the rate of prostaglandin synthesis and the absence of a stimulatory effect when the concentration of glucose was increased in the incubation media in vitro (66). In tissues such as vascular endothelium, the interaction between glucose concentration and prostaglandin release is complex, with stimulation occurring at concentrations between 10 and 30 mM with a subsequent decrease at higher concentrations (80). Production of prostaglandins may be increased by activation of lipolysis, as evidenced by the reduction of elevated plasma prostaglandin levels that occurred when antilipolytic agents were administered to rats with diabetic ketoacidosis (45). That any one or more of the many metabolic abnormalities that constitute the diabetic milieu in vivo could conceivably influence prostaglandin production is suggested by the observation that prostacyclin production by human endothelial cells in culture was reduced when cultured in diabetic serum (81).

Pathophysiologic Role of Glomerular Eicosanoids in Glomerular Hyperfiltration

That the augmented production of glomerular prostaglandins observed in diabetic animals may play a role in the altered glomerular hemodynamics seen characteristically in this model has been suggested by studies employing cyclooxygenase inhibitors as pharmacologic probes. Treatment with either indomethacin or acetylsalicylate prevented glomerular hyperfiltration when administered in the early stages of experimental diabetic nephropathy in the rat (69,70) and reduced intraglomerular hypertension (82) and prevented the progressive decrease in GFR characteristic of the chronic stages of this model (70). Although studies evaluating the effect of chronic treatment with cyclooxygenase inhibitors in humans with glomerular hyperfiltration have not been reported, an acute infusion of lysine acetylsalicylate significantly reduced supernormal GFR and RPF in patients with type 1 diabetes mellitus (83).

Considerable caution must be exercised in interpreting the effects of cyclooxygenase inhibitors on glomerular hemodynamics, however, particularly when such studies are performed in anaesthetized animals or in states such as diabetes mellitus in which volume depletion can occur rapidly when the animals do not have full access to water. The failure to demonstrate reversal of glomerular hyperfiltration with indomethacin in a recent preliminary report (84) perhaps may be explained by differences in volume repletion during preparative surgery.

Abnormalities of Glomerular Function Following a Reduction in Renal Mass

As renal mass is reduced, the remaining nephrons increase in size and function. Studies in dogs (85) and rats (86) have demonstrated that glomerular hyperfiltration occurs following partial renal ablation, and the degree of this functional adaptation correlates with the amount of renal mass removed. Deen et al. (87) found that single nephron GFR (SNGFR) was increased in remnant nephrons from rats with partial renal ablation. Similar to the glomerular hemodynamic profile that occurs in rats with streptozotocin-induced diabetes mellitus, the determinants of this elevated SNGFR were noted to be an increase in the glomerular trans-capillary hydraulic pressure gradient and glomerular plasma flow rate, occurring as a consequence of glomerular arteriolar vasodilatation. Such altered glomerular hemodynamics result in progressive azotemia, protein-uria and glomerulosclerosis. In 1932, Chanutin and Ferris (88) reported that proteinuria, hypertension and progressive dysfunction occurred in rats subjected to three-quarters renal ablation. Studies by Shimamura and Morrison (89), and Olson et al. (90), have documented that progressive histologic changes occur in remnant nephrons following subtotal nephrec-tomy, which include fusion of epithelial foot processes, expansion of the mesangium, collapse of the capillary lumen and subsequent focal and seg-mental glomerular sclerosis. Ultrastructural changes have been demonstrat-ed as early as one week after partial renal ablation (90). Thus, it has been suggested that functioning remnant nephrons, which augment their fil-tration and plasma flow as an apparent compensatory measure, undergo an acceleration of their own destruction following renal injury (24). How-ever, a recent preliminary report in which serial micropuncture studies were performed in remnant nephrons following subtotal nephrectomy sug-gested that the degree of glomerular sclerosis might not be related to the degree of glomerular hemodynamic abnormalities measured at an earlier stage (91). In view of these recent findings, it must be concluded that the attractive hypothesis that links glomerular hyperfiltration and hyper-tension to the progression of renal insufficiency remains unproven.

Effect of Partial Renal Ablation on Glomerular Eicosanoid Metabolism

Although studies of glomerular eicosanoid metabolism in rats with par-tial renal ablation are limited in comparison with those in diabetes melli-tus, the available data appear to indicate that eicosanoid metabolism is augmented in this model as well. Stahl et al. (92) reported that glomerular production of PGE_2, $PGF_{2\alpha}$, and TXB_2 is greater in rats with partial renal ablation in comparison with sham-operated controls. Similar augmentation in glomerular synthesis of these eicosanoids in remnant nephrons obtained from rats that had undergone prior subtotal nephrectomy has been observed in our laboratory (Fig. 5). That the augmented prostaglandin production may contribute to the glomerular hemodynamic abnormalities in this model is suggested by the observation that inhibition of cyclooxygenase activity with indomethacin significantly reduced GFR in rats with partial renal ablation, but not in control rats (92). Similarly, recent micropuncture studies have shown that reducing renal prostaglandin production with indomethacin attenuated the augmented SNGFR in the subtotally nephrectomized rat, whereas this therapy had no effect on the glomerular microcirculation in sham controls (93). These studies at both the whole kidney and single nephron level suggest that renal prostaglandin production plays an important role in the augmented renal hemodynamic response to partial renal ablation. Thus, like diabetes mellitus, rats with partial renal ablation may share a common dependency on renal eicosanoid biosynthesis in mediating glomerular hyperfiltration.

DIETARY PROTEIN INTAKE

Role of Dietary Protein Intake in the Progression of Renal Disease

An increasing body of evidence supports an important role for dietary protein intake in the pathogenesis of glomerular hyperfiltration and the progression of renal insufficiency (94). Several investigators in the 1920-1940s noted the deleterious effect of a high-protein intake in accelerating the progression of renal disease and recommended a reduction in protein intake in patients with chronic renal disease for the purpose of reducing the "workload" of the functioning nephrons (95). Although restriction of dietary protein intake had long been recognized to reduce the symptoms of uremia and thus to play an important role in the conservative management of this disorder, with the advent of dialysis in the 1950s and 1960s, less attention was paid to the possible role of dietary therapy in retarding the progression of renal disease. Recently, Hostetter et al. (24) demonstrated an important link between dietary protein intake, altered glomerular hemodynamics and subsequent renal damage. Following partial renal ablation, rats fed a low-protein (6%) diet failed to develop glomerular hyperfiltration and hypertension, whereas animals fed standard rat chow (containing 21% protein) showed typical elevations in SNGFR and

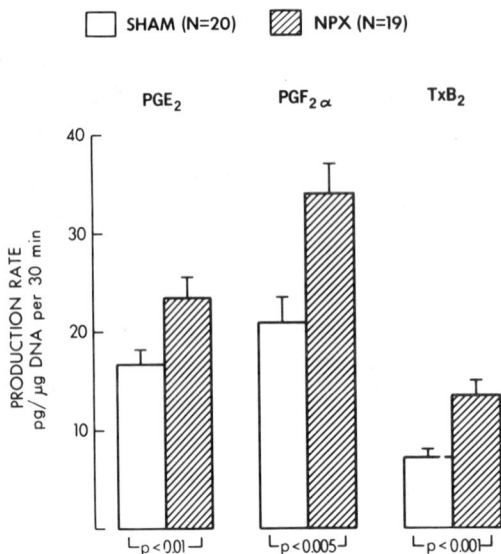

SHAM (N=20) NPX (N=19)

Fig. 5. Comparison of eicosanoid production by glomeruli isolated from rats with partial renal ablation and sham-operated control animals. Animals were studied 10-14 days following a total right nephrectomy and left heminephrectomy, performed as a one-stage procedure. Pools of glomeruli from two such animals and preparations obtained from control animals were incubated as described in the legend of Fig. 1.

glomerular hydraulic pressure. Additionally, prevention of glomerular hyperfiltration and hypertension by protein restriction was associated with preservation of normal glomerular histology. This important effect of dietary protein on glomerular hemodynamics has also been observed in normal rats: Ichikawa (96) noted a decrease in glomerular capillary plasma flow in rats fed a low-protein diet. Furthermore, acute and chronic protein loads increase GFR and renal plasma flow in both normal human subjects (97,98) and in experimental animals (99). The clinical relevance of these observations has been supported by the results of several studies that have evaluated the role of dietary protein restriction on the progression of renal disease in humans (100,101). Although still somewhat anecdotal, these studies suggest that restriction of dietary protein intake with or without dietary supplementation with essential amino acids or alpha-keto analogues may slow the rate of progression of renal insufficiency as predicted by a plot of the reciprocal of the serum creatinine concentration.

Effect of Variation in Dietary Protein on Glomerular Eicosanoids

We have undertaken studies to determine the effect of variation in dietary protein intake on glomerular eicosanoid production in rats with Heymann nephritis (102). This is an immunologically mediated renal disease with many similarities to human membranous glomerulopathy (103). Heymann nephritis was induced by the intraperitoneal injection of FX1A

antiserum. Two days following induction of the model, an equal number of
rats from each paradigm were randomized to receive either a high (40%
casein) or low (8.5% casein) protein diet. The two diets were rendered
isocaloric by adjustment of the sucrose and dextrin contents. Dietary
protein intake modulated glomerular prostaglandin production in these
animals. Production rates of PGE_2, $PGF_{2\alpha}$ and TXB_2 were all significantly greater in glomeruli isolated from the rats with Heymann nephritis
fed the high-protein diet in comparison with rats fed the low-protein diet
(Fig. 6). The degree of albuminuria was also greater in the rats ingesting the high-protein intake. Comparable increases in glomerular eicosanoid biosynthesis with a high-protein diet have also been reported to
occur in rats following partial renal ablation (104). In preliminary
studies performed recently in our laboratory, a similar modulation of
glomerular eicosanoid production by dietary protein intake was found in
rats with partial renal ablation, as well as in diabetic rats and in
normal animals (105).

Several lines of evidence implicate renal eicosanoids in the increases
in GFR and RPF that occur with protein intake in both humans and experimental animals. Renal production of eicosanoids is augmented following
protein loading, as reflected in the increased urinary excretion of PGE_2
with amino acid infusions in normal humans (106) and the elevated urinary
PGE_2 and 6-keto-$PGF_{1\alpha}$ excretion following the ingestion of a high-protein diet in patients with chronic glomerular diseases (107). The

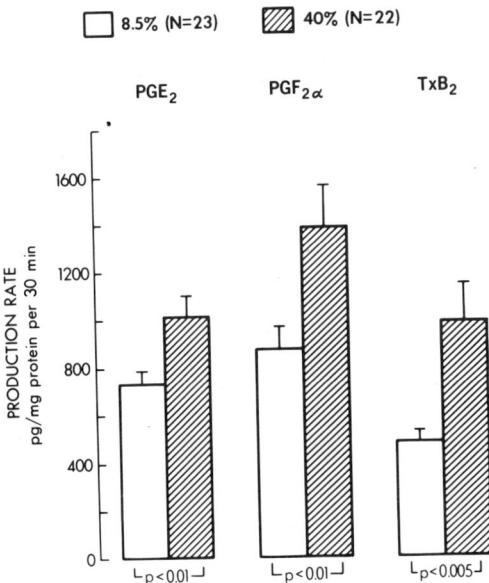

Fig. 6. Effect of varia-
tion of dietary
protein intake on
eicosanoid pro-
duction by glomer-
uli isolated from
rats with Heymann
nephritis. Two
days after injec-
tion of anti-FX1A
antiserum, the
rats were random-
ized to receive
either a high
(40%) or low
(8.5%) protein
diet. (Data are
adapted from
Table 1 in refer-
ence 102 with per-
mission).

increase in GFR and RPF that occurs in response to infusions of arginine
(108) or amino acid mixtures (106), or ingestion of a meat meal (109) in
normal human subjects can be blunted by pretreatment with cyclooxygenase
inhibitors. Similar observations have been noted in rats with experimental renal disease and in normal animals. For example, urinary excretion
of 6-keto-PGF$_{1\alpha}$ and PGE$_2$ was increased in normal rats that chronically
ingested a high-protein diet (110). Furthermore, pretreatment with indomethacin blocked both the increase in GFR and glomerular production of
PGE$_2$ and 6-keto-PGF$_{1\alpha}$ that occurred in response to an acute protein
load (111). Indomethacin treatment also prevented both the acute increase
in GFR following a protein bolus in uremic rats (111), as well as the
parallel increases in GFR and urinary excretion, and glomerular production
of 6-keto-PGF$_{1\alpha}$ associated with chronic ingestion of a high-protein diet
in rats with adriamycin nephrosis (112).

Interrelationship of Renal Autacoids in States of Glomerular Hypertension

In the foregoing discussion we have focused on the potential role of
abnormalities of glomerular eicosanoid metabolism in states of glomerular
hyperfiltration/hypertension. Other vasoactive regulatory systems may
also be altered in these pathophysiologic settings. In particular, the
renin-angiotensin system has been implicated in the alteration in glomerular function that occurs in animals with various experimentally induced
renal diseases characterized by glomerular hypertension. This implication
is based nearly entirely on indirect evidence derived from studies employing inhibitors of angiotensin converting enzyme (ACE) activity. For
example, treatment with the ACE inhibitor enalapril lowered the glomerular
transcapillary hydraulic pressure and limited the development of proteinuria and glomerular lesions in rats with partial renal ablation (113). A
similar beneficial effect of ACE inhibitor therapy in reducing glomerular
capillary pressure and preventing renal damage occurs in diabetic rats
(44). Treatment with ACE inhibitors also reduced established proteinuria
in rats with Heymann nephritis (114) and in patients with diabetes mellitus (115) and appeared to retard the rate of deterioration of GFR in
patients with diabetic nephropathy (116).

Whereas the salutary effect of treatment with ACE inhibitors may
simply reflect the benefit known to occur with a reduction in systemic
arterial pressure, it has been suggested that this class of drugs may pref-
erentially decrease efferent arteriolar resistance (117), presumably by
reversing the vasoconstrictor effect of angiotensin II on the efferent

arteriole (118). This would result in a reduction of glomerular capillary hydraulic pressure. Despite the "normalization" of glomerular pressures and protection against renal damage, therapy with enalapril did not lower the SNGFR or glomerular plasma flow rate in rats with partial renal ablation and diabetes mellitus. These results have been interpreted as evidence that glomerular hypertension rather than hyperfiltration is the factor responsible for glomerular injury.

The results of recent studies that indicate that dietary protein intake modifies the activity of the renin-angiotensin system, as reflected by levels of renin in the systemic circulation, may provide an additional link between this vasoregulatory system and paradigms of glomerular hypertension. In normal rats ingesting a high-protein diet, plasma renin activity was approximately three-fold greater than in low-protein-fed animals (110). Similar results have been observed in patients with chronic glomerular diseases. In such patients, plasma renin activity was two-fold greater while they ingested a high-protein diet than when they were maintained on a low-protein diet (107) and increased acutely following the ingesting of a large protein meal (119). That the renin-angiotensin system plays a role in the modulation of glomerular function by dietary protein is suggested by the observation that the salutary effect of a low-protein diet in reducing both glomerular capillary pressure and renal injury in experimental renal disease can be duplicated by ACE inhibitor therapy (44,113). This is in accord with our recent observation that treatment with enalapril completely prevented the increment in urinary albumin excretion that ordinarily occurs when rats with Heymann nephritis consume a high-protein diet (114). Because dietary protein restriction or ACE inhibitor therapy can induce a similar glomerular hemodynamic effect, Paller and Hostetter (110) have suggested that reduced dietary protein alters renal hemodynamics by lowering intrarenal renin and angiotensin II levels.

Inasmuch as renal hyperfiltration results from reduced arteriolar vascular resistance and increased renal plasma flow, it is difficult to ascribe the hemodynamic profile in states of glomerular hyperfiltration solely to an alteration in the glomerular action of angiotensin II. For example, infusion of angiotensin II in the rat results in an increase in renal vascular resistance and a decrease in the glomerular capillary ultrafiltration coefficient with a resultant decrease in glomerular capillary plasma flow rate and SNGFR (120). Since hyperfiltration is associated with an increase in glomerular capillary plasma flow rate and SNGFR, the presence of a renal vasodilator is strongly implied. Although direct infusion of "vasodilatory" prostaglandins in the rat have variable effects on glomerular hemodynamics (3), we suggest that the glomerular vasodilatation and hyperfiltration induced by diabetes mellitus, dietary protein intake and reduced renal mass could be mediated, at least in part, by an augmented glomerular production of vasodilatory prostaglandins. Angiotensin may contribute to this hemodynamic effect by increasing postglomerular arteriolar tone so that, while total renal vascular resistance is decreased, a relatively greater vasodilatation in the afferent renal arteriole could account for the increased intraglomerular pressure that is characteristic in this setting. Thus, there is probably a critical interplay between renal eicosanoids, the renin-angiotensin system and perhaps other humoral mediators in modulating glomerular hyperfiltration and hypertension. This interaction can be predicted to be complex, since, on the one hand, angiotensin II stimulates prostaglandin production by whole glomeruli (6,7) and cultured mesangial cells and, on the other, renin secretion and subsequent angiotensin II production is in part mediated by prostaglandins (121,122). Whether changes in one or both of these vasoregulatory systems in the primary event in states of glomerular hyperfiltration or whether these changes are occurring in response to alterations in other humoral factors (e.g., glucagon, growth hormone) remains to be determined.

ACKNOWLEDGMENTS

This work was supported by U.S. Public Health Service Research Grant HL-11046 from the National Heart, Lung, and Blood Institute; by a grant from the Research Evaluation and Allocation Committee, University of California, San Francisco; and from the Research Service of the United States Veterans Administration. Dr. Don was the recipient of a fellowship from the National Kidney Foundation.

REFERENCES

1. B.M. Brenner and H.D. Humes, Mechanics of glomerular ultrafiltration, N. Engl. J. Med. 297:148-154 (1977).

2. L.D. Dworkin, I. Ichikawa, B.M. Brenner, Hormonal modulation of glomerular function, Am. J. Physiol. 244:F95-F104 (1983).

3. N. Schor, I. Ichikawa, B.M. Brenner, Mechanisms of action of various hormones and vasoactive substances on glomerular ultrafiltration in the rat, Kidney Int. 20:442-451 (1981).

4. J. Sraer, J-D. Sraer, D. Chansel, F. Russo-Marie, B. Kouznetzova, R.

Ardaillou, Prostaglandin synthesis by isolated rat renal glomeruli, Mol. Cell Endocrinol. 16:29-37 (1979).

5. A. Hassid, M. Konieczkowski, M.J. Dunn, Prostaglandin synthesis in isolated rat kidney glomeruli, Proc. Natl. Acad. Sci. USA 76: 1155-1159 (1979).

6. D. Schlondorff, S. Roczniak, J.A. Satriano, V.W. Folkert, Prostaglandin synthesis by isolated rat glomeruli: effect of angiotensin II, Am. J. Physiol. 238:F486-F495 (1980).

7. J. Sraer, J. Foidart, D. Chansel, P. Mahieu, B. Kouznetzova, R. Ardaillou, Prostaglandin synthesis by mesangial and epithelial glomerular cultured cells, FEBS Lett. 104:420-424 (1979).

8. J.I. Kreisberg, M.J. Karnovsky, L. Levine, Prostaglandin production by homogeneous cultures of rat glomerular epithelial and mesangial cells, Kidney Int. 22:355-359 (1982).

9. L.A. Scharschmidt and M.J. Dunn, Prostaglandin synthesis by rat glomerular mesangial cells in culture--Effects of angiotensin II and arginine vasopressin, J. Clin. Invest. 71:1756-1764 (1983).

10. C. Baylis and B.M. Brenner, Modulation by prostaglandin synthesis inhibitors of the action of exogenous angiotensin II on glomerular ultrafiltration in the rat, Circ. Res. 43:889-898 (1978).

11. L.A. Scharschmidt, J.G. Douglas, M.J. Dunn, Angiotensin II and eicosanoids in the control of glomerular size in the rat and the human, Am. J. Physiol. 250:F348-356 (1986).

12. P. Mene' and M.J. Dunn, Contractile effects of TxA_2 and endoperoxide analogues on cultured rat glomerular mesangial cells, Am. J. Physiol. 251:F1029-F1035 (1986).

13. E.A. Lianos, G.A. Andres, M.J. Dunn, Glomerular prostaglandin and thromboxane synthesis in rat nephrotoxic serum nephritis, J. Clin. Invest. 72:1439-1448 (1983).

14. G. Remuzzi, L. Imberti, M. Rossini, C. Morelli, C. Carminati, G.M. Cattaneo, T. Bertani, Increased glomerular thromboxane synthesis as a possible cause of proteinuria in experimental nephrosis, J. Clin. Invest. 75:94-101 (1985).

15. M.L. Purkerson, J.H. Joist, J. Yates, A. Valdes, A. Morrison, S. Klahr, Inhibition of thromboxane synthesis ameliorates the progressive kidney disease of rats with subtotal renal ablation, Proc. Natl. Acad. Sci. USA 82:193-197 (1985).

16. I. Ichikawa, M.L. Purkerson, J. Yates, S. Klahr, Dietary protein intake conditions the degree of renal vasoconstriction in acute renal failure caused by ureteral obstruction, Am. J. Physiol. 249: F54-61 (1985).

17. G. Ciabattoni, G.A. Cinotti, A. Pierucci, B.M. Simonetti, M. Manzi, F. Pugliese, P. Barsotti, G. Pecci, F. Taggi, C. Patrono, Effects of sulindac and ibuprofen in patients with chronic glomerular disease, N. Engl. J. Med. 310:279-283 (1984).

18. M.J. Dunn and E.J. Zambraski, Renal effect of drugs that inhibit prostaglandin synthesis, Kidney Int. 18:609-622 (1980).

19. J. Ditzel and M. Schwartz, Abnormally increased glomerular filtration rate in short-term insulin-treated diabetic subjects, Diabetes 16:264-267 (1967).

20. C.E. Mogensen, Kidney function and glomerular permeability to macromolecules in early juvenile diabetes, Scand. J. Clin. Lab. Invest. 28:79-90 (1971).

21. J.S. Christiansen, J. Gammelgaard, M. Frandsen, H-H. Parving, Increased kidney size, glomerular filtration rate and renal plasma flow in short-term insulin-dependent diabetics, Diabetologia 20: 451-456 (1981).

22. T.H. Hostetter, J.L. Troy, B.M. Brenner, Glomerular hemodynamics in experimental diabetes mellitus, Kidney Int. 19:410-415 (1981).

23. P.K. Jensen, J.S. Christiansen, K. Steven, H-H. Parving, Renal function in streptozotocin-diabetic rats, Diabetologia 21:409-414 (1981).

24. T.H. Hostetter, J.L. Olson, H.G. Rennke, M.A. Venkatachalam, B.M. Brenner, Hyperfiltration in remnant nephrons: a potentially adverse response to renal ablation, Am. J. Physiol. 241:F85-F93 (1981).

25. R. Zatz, T.W. Meyer, H.G. Rennke, B.M. Brenner, Predominance of hemodynamic rather than metabolic factors in the pathogenesis of diabetic glomerulopathy, Proc. Natl. Acad. Sci. USA 82:5963-5967 (1985).

26. H.C. Knowles Jr., Magnitude of the renal failure problem in diabetic patients, Kidney Int. 4 (Suppl. 1):52-57 (1974).

27. A.S. Krowlewski, J.H. Warram, A.R. Christlieb, E.J. Busick, C.R. Kahn, The changing natural history of nephropathy in type I diabetes, Am. J. Med. 78:785-794 (1985).

28. R. Omachi, The pathogenesis and prevention of diabetic nephropathy, West. J. Med. 145:222-227 (1986).

29. C.E. Mogensen and M.J.F. Andersen, Increased kidney size and glomerular filtration rate in early juvenile diabetes, Diabetes 22:706-712 (1973).

30. C.E. Mogensen, Renal function changes in diabetes, Diabetes 25:872-879 (1976).

31. C.E. Mogensen and C.K. Christensen, Predicting diabetic nephropathy in

insulin-dependent patients, N. Engl. J. Med. 311:89-93 (1984).

32. J.S. Christiansen, J. Gammelgaard, B. Tronier, P.A. Swendsen, H-H. Parving, Kidney function and size before and during initial insulin treatment, Kidney Int. 21:683-688 (1982).

33. C.E. Mogensen and M.J.F. Andersen, Increased kidney size and glomerular function rate in untreated juvenile diabetes: normalization by insulin-treatment, Diabetologia 11:221-224 (1975).

34. M.J. Wiseman, A.J. Saunders, H. Keen, G.C. Viberti, Effect of blood glucose control on increased glomerular filtration rate and kidney size in insulin-dependent diabetes, N. Engl. J. Med. 312:617-621 (1985).

35. T.H. Hostetter, T.W. Meyer, H.G. Rennke, B.M. Brenner, Influence of strict control of diabetes on intrarenal hemodynamics, Kidney Int. 23:215 (1983).

36. K. Seyer-Hansen, Renal hypertrophy in experimental diabetes mellitus, Kidney Int. 23:643-646 (1983).

37. J.S. Christiansen, M. Frandsen, H-H. Parving, Effect of intravenous glucose infusion on renal function in normal man and in insulin-dependent diabetics, Diabetologia 21:368-373 (1981).

38. J.S. Christiansen, M. Frandsen, H-H. Parving, The effect of intravenous insulin infusion on kidney function in insulin-dependent diabetes mellitus, Diabetologia 20:199-204 (1981).

39. J. Corvilain and M. Abramow, Some effects of human growth hormone on renal hemodynamics and on tubular phosphate transport in man, J. Clin. Invest. 41:1230-1235 (1962).

40. H-H. Parving, J.S. Christiansen, I. Noer, B. Tronier, C.E. Mogensen, The effect of glucagon infusion on kidney function in short-term insulin-dependent juvenile diabetes, Diabetologia 19:350-354 (1980).

41. S.L. Carney, N.L.M. Wong, J.H. Dirks, Acute effects of streptozotocin diabetes on rat renal function, J. Lab. Clin. Med. 93:950-961 (1979).

42. S.M. Mauer, M.W. Steffes, S. Azar, S.K. Sandberg, D.M. Brown, The effects of Goldblatt hypertension on development of glomerular lesions of diabetes mellitus in the rat, Diabetes 27:738-744 (1978).

43. J. Berkman and H. Rifkin, Unilateral nodular diabetic glomerulo-sclerosis (Kimmelstiel-Wilson): report of a case, Metabolism 22: 715-722 (1973).

44. R. Zatz, B.R. Dunn, T.W. Meyer, S. Anderson, H.G. Rennke, B.M. Brenner, Prevention of diabetic glomerulopathy by pharmacologic

amelioration of glomerular capillary hypertension, J. Clin. Invest.
77:1925-1930 (1986).

45. L. Axelrod and L. Levine, Plasma prostaglandin levels in rats with
    diabetes mellitus and diabetic ketoacidosis, Diabetes 31:994-1001
    (1982).

46. P.V. Halushka, R.C. Rogers, C.B. Loadholt, H. Wohltman, R. Mayfield,
    S. McCoy, J.A. Colwell, Increased platelet prostaglandin and
    thromboxane synthesis in diabetes mellitus, Horm. Metab. Res. 11
    (Suppl. 1):7-11 (1981).

47. J.M. Gerrard, M.J. Stuart, G.H.R. Rao, M.W. Steffes, S.M. Mauer, D.M.
    Brown, J.G. White, Alteration in the balance of prostaglandin and
    thromboxane synthesis in diabetic rats, J. Lab. Clin. Med. 95:950-
    958 (1980).

48. M.T.R. Subbiah and D. Deitemeyer, Altered synthesis of prostaglandins
    in platelet and aorta from spontaneously diabetic Wistar rats,
    Biochem. Med. 23:231-235 (1980).

49. M.J. Stuart, H. Elrad, J.E. Graeber, D.O. Hakanson, S.G. Sunderji,
    M.K. Barvinchak, Increased synthesis of prostaglandin endoperoxides
    and platelet hyperfunction in infants of mothers with diabetes
    mellitus, J. Lab. Clin. Med. 94:12-17 (1979).

50. M. Johnson, H.E. Harrison, A.T. Raftery, J.B. Elder, Vascular prosta-
    cyclin may be reduced in diabetes in man, Lancet 1:325-326 (1979).

51. K. Silberbauer, G. Schernthaner, H. Sinzinger, H. Piza-Katzer, M.
    Winter, Decreased vascular prostacyclin in juvenile-onset diabetes,
    N. Engl. J. Med. 300:366-367 (1979).

52. M.J. Stuart, S.G. Sunderji, J.B. Allen, Decreased prostacyclin produc-
    tion in the infant of the diabetic mother, J. Lab. Clin. Med. 98:
    412-416 (1981).

53. C. Dadak, A. Kefalides, H. Sinzinger, G. Weaver, Reduced umbilical
    artery prostacyclin formation in complicated pregnancies, Am. J.
    Obstet. Gynecol. 144:792-795 (1982).

54. H.E. Harrison, A.H. Reece, M. Johnson, Decreased vascular prostacyclin
    in experimental diabetes, Life Sci. 23:351-356 (1978).

55. S.P. Rogers and R.G. Larkins, Production of 6-oxo-prostaglandin $F_{1\alpha}$
    by rat aorta: influence of diabetes, insulin treatment, and caloric
    deprivation, Diabetes 30:935-939 (1981).

56. K. Silberbauer, P. Clopath, H. Sinzinger, G. Schernthaner, Effect of
    experimentally induced diabetes on swine vascular prostacyclin
    (PGI_2) synthesis, Artery 8:30-36 (1980).

57. R.P. Robertson and M. Chen, A role for prostaglandin E in defective
    insulin secretion and carbohydrate intolerance in diabetes melli-

tus, J. Clin. Invest. 60:747-753 (1977).

58. W.C. Lubawy and M. Valentovic, Streptozotocin-induced diabetes decreases formation of prostacyclin from arachidonic acid in intact rat lungs, Biochem. Med. 28:290-297 (1982).

59. M.Y. Tsai, L.E. Schallinger, M.W. Josephson, D.M. Brown, Disturbance of pulmonary prostaglandin metabolism in fetuses of alloxan-diabetic rabbits, Biochim. Biophys. Acta. 712:395-399 (1982).

60. I.S. Watts, J.T. Zakrzewski, Y.S. Bakhle, Altered prostaglandin synthesis in isolated lungs of rats with streptozotocin-induced diabetes, Thromb. Res. 28:333-342 (1982).

61. P. Rosen and K. Schror, Increased prostacyclin release from perfused hearts of acutely diabetic rats, Diabetologia 18:391-394 (1980).

62. A. Kandil, S. Fouad, S. Samaan, Diabetes versus prostaglandin activity in the seminal vesicles, in: "Proc. of Vth International Conf. Prostaglandins," Florence, p. 113 (1982).

63. J.R.S. Hoult and P.K. Moore, Prostaglandin synthesis and inactivation in kidneys and lungs of rats with experimental diabetes, Clin. Sci. 59:63-66 (1980).

64. H.E. Harrison, A.H. Reece, M. Johnson, Effect of insulin treatment on prostacyclin in experimental diabetes, Diabetologia 18:65-68 (1980).

65. M. Schambelan, S. Blake, J. Sraer, M. Bens, M-P. Nivez, F. Wahbe, Increased prostaglandin production by glomeruli isolated from rats with streptozotocin-induced diabetes mellitus, J. Clin. Invest. 75:404-412 (1985).

66. J.E. Kreisberg and P.Y. Patel, The effects of insulin, glucose and diabetes on prostaglandin production by rat kidney glomeruli and cultured mesangial cells, Prostaglandins Leukotrienes Med. 11:431-442 (1983).

67. D.M. Brown, J.M. Gerrard, J. Peller, G.H.R. Rao, J.B. White, Glomerular prostaglandin metabolism in diabetic rats, Diabetes 29 (Suppl): 55 (1980).

68. A. Chaudhari and M.A. Kirschenbaum, Effect of experimental diabetes mellitus (DM) on eicosanoid biosynthesis in isolated rat glomeruli, Kidney Int. 25:326 (1984).

69. P.A. Craven, M.A. Gaines, F.R. DeRubertis, Sequential alterations in glomerular prostaglandin and thromboxane synthesis in diabetic rats: relationship to the hyperfiltration of early diabetes, Metabolism 36:95-103 (1987).

70. D. Moel, R.L. Safirstein, R.C. McEvoy, W. Hsueh, The effect of aspirin on experimental diabetic nephropathy, J. Lab. Clin. Med. 110:300-

307 (1987).

71. S.P. Rogers and R.G. Larkins, Production of 6-oxo-prostaglandin $F_{1\alpha}$ and prostaglandin by isolated glomeruli from normal and diabetic rats, Br. Med. J. 284:1215-1217 (1982).

72. S. Sandler and A. Andersson, The partial protective effect of the hydroxyl radical scavenger dimethyl urea on streptozotocin-induced diabetes in the mouse in vivo and in vitro, Diabetologia 23:374-378 (1982).

73. H. Kromann, M. Christy, A. Lernmark, M. Nedergaard, J. Nerup, The low dose streptozotocin murine model of type 1 (insulin-dependent) diabetes mellitus: studies in vivo and in vitro of the modulating effect of sex hormones, Diabetologia 22:194-198 (1982).

74. J. Tannenbaum, B.J. Sweetman, A.S. Nies, K. Aulsebrook, J.A. Oates, The effect of glucose on the synthesis of prostaglandins by the renal papilla of the rat in vitro, Prostaglandins 17:337-350 (1979).

75. J.S. Tannenbaum, Prostaglandins: a study of their effects on the kidney and factors controlling their release, Ph. D. Thesis, Washington University School of Medicine, St. Louis, MO. (1980).

76. J-D. Sraer, L. Moulonguet-Doleris, F. Delarue, J. Sraer, R. Ardaillou, Prostaglandin synthesis by glomeruli isolated from rats with glycerol-induced acute renal failure, Circ. Res. 49:775-783 (1981).

77. J. Sraer, W. Siess, L. Moulonguet-Doleris, J-P. Oudinet, F. Dray, R. Ardaillou, In vitro prostaglandin synthesis by various rat renal preparations, Biochim. Biophys. Acta. 710:45-52 (1982).

78. J. Sraer, M. Rigaud, M. Bens, H. Rabinovitch, and R. Ardaillou, Metabolism of arachidonic acid via the lipoxygenase pathway in human and murine glomeruli, J. Biol. Chem. 258:4325-4330 (1983).

79. P. Chaumet-Riffaud, J-P. Oudinet, J. Sraer, C. Lajotte, R. Ardaillou, Altered PGE_2 and $PGF_{2\alpha}$ production by glomeruli and papilla of sodium-depleted and sodium-loaded rats, Am. J. Physiol. 241:F517-524 (1981).

80. J.Y. Jeremy, D.P. Mikhailidis, P. Dandona, Simulating the diabetic environment modifies in vitro prostacyclin synthesis, Diabetes 32:217-221 (1983).

81. R.C. Paton, R. Guillot, P.H. Passa, J. Canivet, Prostacyclin production by human endothelial cells cultured in diabetic serum, Diabete. Metab. 8:323-328 (1982).

82. P.K. Jensen, K. Steven, H. Blaehr, J.S. Christiansen, H-H. Parving, Effects of indomethacin on glomerular hemodynamics in experimental diabetes, Kidney Int. 29:490-495 (1986).

83. E. Estamajes, M.R. Fernandez, I. Halperin, J. Camps, J. Gaya, V. Arroyo, F. Rivera, D. Figuerola, Renal hemodynamic abnormalities in patients with short term insulin-dependent diabetes mellitus: role of renal prostaglandins, J. Clin. Endocrinol. Metab. 60:1231-1236 (1986).

84. N. Bank, M.A.G. Lahorra, H.S. Aynedjian, D. Schlondorff, Role of vaso-active hormones in hyperfiltration of diabetes, Kidney Int. 31:259 (1987).

85. N.S. Bricker, S. Klahr, R.E. Rieselbach, The functional adaptation of the diseased kidney--I. Glomerular filtration rate, J. Clin. Invest. 43:1915-1921 (1964).

86. A.B. Morrison and R.M. Howard, The functional capacity of the hyper-trophied nephrons--Effect of partial nephrectomy on the clearance of inulin and PAH in the rat, J. Exp. Med. 123:829-844 (1966).

87. W.M. Deen, D.R. Maddox, C.R. Robertson, B.M. Brenner, Dynamics of glomerular hyperfiltration in the rat--VII. Response to reduced renal mass, Am. J. Physiol. 227:556-562 (1974).

88. A. Chanutin and E. Ferris, Experimental renal insufficiency produced by partial nephrectomy: I. Control diet, Arch. Intern. Med. 49:767-787 (1932).

89. T. Shimamura and A.B. Morrison, A progressive glomerulosclerosis occurring in partial five-sixths nephrectomized rats, Am. J. Pathol. 79:95-101 (1975).

90. J.L. Olson, T.H. Hostetter, H.G. Rennke, B.M. Brenner, M.A. Venkatachalam, Altered glomerular permeability and progressive sclerosis following ablation of renal mass, Kidney Int. 22:112-126 (1982).

91. Y. Yoshida, A. Fogo, H. Shinaga, A. Glick, I. Ichikawa, Serial micro-puncture analysis of single nephron function in subtotal renal ablation, Kidney Int. 33:855-867 (1988).

92. R.A.K. Stahl, S. Kudelka, M. Paravicini, P. Schoolmeyer, Prostaglan-din and thromboxane formation in glomeruli from rats with reduced renal mass, Nephron 42:252-257 (1986).

93. K.A. Nath, D.H. Chmilewski, T.H. Hostetter, Regulatory role of prosta-noids in glomerular microcirculation of remnant nephrons, Am. J. Physiol. 21:F829-F837 (1987).

94. S. Klahr, J. Buerkert, M.L. Purkerson, Role of dietary factors in the progression of chronic renal disease, Kidney Int. 24:579-587 (1983).

95. T. Addis, "Glomerular Nephritis: Diagnosis and Treatment," Macmillan, New York, (1948).

96. I. Ichikawa, M.L. Purkerson, S. Klahr, J.L. Troy, M. Martinez-Maldanado, B.M. Brenner, Mechanism of reduced glomerular filtration rate in chronic malnutrition, J. Clin. Invest. 65:982-988 (1980).

97. T.N. Pullman, A.S. Alving, R.J. Dern, M. Landowne, The influence of dietary protein on specific renal function in normal man, J. Lab. Clin. Med. 44:320-332 (1954).

98. J.P. Bosch, S. Lew, S. Glabman, A. Lauer, Renal hemodynamic changes in humans: response to protein loading in normal and diseased kidneys, Am. J. Med. 81:809-815 (1986).

99. R.F. Pitts, The effect of infusing glycine and of varying the dietary protein intake on renal hemodynamics in the dog, Am. J. Physiol. 142:355-365 (1944).

100. G. Maschio, L. Oldrizzi, N. Tessitore, A. D'Angelo, E. Valvo, A. Lupo, C. Loschiavo, A. Fabris, L. Gammaro, C. Rugio, G. Panzetta, Effects of dietary protein and phosphorus restriction on the progression of early renal failure, Kidney Int. 22:371-376 (1982).

101. W.E. Mitch, Conservative management of chronic renal failure, in: "Contemporary Issues In Nephrology--Chronic Renal Failure," Vol. 7, B.M. Brenner and J.H. Stein, Eds., Churchhill-Livingston, New York, pp. 116-152, (1981).

102. F.N. Hutchinson, B.R. Don, G.A. Kaysen, S. Blake, M. Schambelan, Dietary protein intake modulates glomerular eicosanoid production in nephrotic rats, Adv. in Prostaglandins Thromboxane and Leukotriene Res. 17:725-728 (1987).

103. W.G. Couser, D.R. Steinmuller, M.M. Stilmant, D.J. Salant, L.J. Lowenstein, Experimental glomerulonephritis in the isolated per-fused rat kidney, J. Clin. Invest. 62:1275-1287 (1978).

104. R.A.K. Stahl, S. Kudelka, U. Helmchen, High protein intake stimulates glomerular prostaglandin formation in remnant kidneys, Am. J. Physiol. 252:F1088-F1094 (1987).

105. B.R. Don, S. Blake, G.A. Kaysen, M. Schambelan, Dietary protein modu-lates glomerular eicosanoid production in rats with experimental renal disease and in normal animals, Kidney Int. 31:267 (1987).

106. L.M. Ruilope, J. Rodicio, R.G. Robles, J. Sancho, B. Miranda, J.P. Granger, J.C. Romero, Influence of a low sodium diet on the renal response to amino acid infusion in humans, Kidney Int. 31:992-999 (1987).

107. M.E. Rosenberg, B.L. Thomas, J.E. Swanson, T.H. Hostetter, Hormonal and glomerular responses to dietary protein intake in human renal disease, Kidney Int. 31:215 (1987).

108. R. Hirschberg and J.D. Kopple, Indomethacin blocks the arginine

induced rise of RPF and GFR in man, Kidney Int. 31:201 (1987).

109. M. Lawlor, W. Lieberthal, R. Perrone, The increase in GFR after a meat meal is mediated by prostaglandins, Kidney Int. 31:208 (1987).

110. M.S. Paller and T.H. Hostetter, Dietary protein increases plasma renin and reduces pressor reactivity to angiotensin II, Am. J. Physiol. 252:F34-F39 (1986).

111. M.M. Levine, M.A. Kirschenbaum, A. Chaudhari, M. Wong, N.S. Bricker, Effect of protein on glomerular filtration rate and prostanoid synthesis in normal and uremic rats, Am. J. Physiol. 251:F635-F641 (1986).

112. A. Benigni, C. Zoja, A. Remuzzi, S. Orisio, A. Piccinelli, G. Remuzzi, Role of renal prostaglandins in normal and nephrotic rats with diet-induced hyperfiltration, J. Lab. Clin. Med. 108:230-240 (1986).

113. S. Anderson, T.W. Meyer, H.G. Rennke, B.M. Brenner, Control of glomerular hypertension limits glomerular injury in rats with reduced renal mass, J. Clin. Invest. 76:612-619 (1985).

114. F.N. Hutchison, M. Schambelan, G.A. Kaysen, Modulation of albuminuria by dietary protein and converting enzyme inhibition, Am. J. Physiol, in press.

115. Y. Taguma, Y. Kitamoto, G. Futake, H. Ueda, H. Momma, M. Ishizaki, H. Takahashi, H. Sakino, Y. Sasaki, Effect of captopril on heavy proteinuria in azotemic diabetics, N. Engl. J. Med. 313:1617-1620 (1985).

116. S. Bjorck, G. Nyberg, H. Mulec, G. Granerus, H. Herlitz, M. Aurell, Beneficial effects of angiotensin converting enzyme inhibition on renal function in patients with diabetic nephropathy, Br. Med. J. 293:471-474 (1986).

117. S. Anderson, H.G. Rennke, B.M. Brenner, Therapeutic advantage of converting enzyme inhibitors in arresting progressive renal disease associated with systemic hypertension in the rat, J. Clin. Invest. 77:1993-2000 (1986).

118. I. Ichikawa and B.M. Brenner, Glomerular action of angiotensin II, Am. J. Med. 76:43-49 (1984).

119. A.Y.M. Chan, M.L. Cheng, and B.D. Myers, Functional response of diseased glomeruli to a large protein meal, Kidney Int. 31:381 (1987).

120. R.C. Blantz, K.S. Konnen, B.J. Tucker, Angiotensin II effect upon the glomerular microcirculation and ultrafiltration, J. Clin. Invest. 57:419-434 (1976).

121. W.L. Henrich, Role of prostaglandins in renin secretion, Kidney Int.

19:822-830 (1981).

122. J.G. Gerber, R.D. Olson, A.S. Nies, Interrelationship between prostaglandins and renin release, Kidney Int. 19:816-821 (1981).

DIABETIC NEPHROPATHY: PATHOPHYSIOLOGY,

CLINICAL COURSE AND SUSCEPTIBILITY

GianCarlo Viberti, M.D., MRCP,
and James D. Walker, BSc, MRCP

Unit for Metabolic Medicine
UMDS (Guy's Hospital Campus),
London SE1 9RT, U.K.

CLINICAL DIABETIC NEPHROPATHY

The clinical diagnosis of diabetic nephropathy in an insulin-dependent patient relies on the detection of persistent proteinuria (i.e., urinary total protein excretion rate greater than 0.5 g/24h) after 10 years or more of the disease, with concomitant diabetic retinopathy and rising arterial pressure, but without signs of other renal disease, urinary tract infection or heart failure. Diabetic glomerulosclerosis is found at renal biopsy in over 90% of the cases, confirming the diagnosis (1,2). Epidemiological studies show that, cumulatively, nephropathy develops in approximately 35% of insulin-dependent patients, and it is mostly in this kind of patient that the natural history of this disorder has been outlined (1,3). Noninsulin-dependent diabetics of European origin seem to develop nephropathy less frequently than insulin-dependent patients, but because of their larger number, the actual number of patients developing end-stage renal failure is approximately the same (4).

Persistent proteinuria is the forerunner of a progressive deterioration in renal function. Glomerular filtration rate (GFR) declines linearly with time at rates ranging between 0.6 and 2.4 ml/min/month (5). The reason for the different rates of decline are largely unknown, but varying degrees of blood pressure control may play a role (6). Indeed, arterial pressure starts to climb at a very early stage of renal involvement in diabetes (7,8). With the fall in GFR, proteinuria and the clearance of major plasma proteins such as albumin and IgG increase. There is a change from a proteinuria characterized by high selectivity for albumin in the early stages, when

the GFR is still normal or only moderately reduced, to a low selectivity proteinuria with proportionally more IgG being cleared when the GFR is markedly
reduced (Fig. 1) (9). In 50% of affected patients, end-stage renal failure
occurs within 7 years of the onset of persistent proteinuria (1).

Three different lines of treatment have been employed in an attempt to
arrest this apparently inexorable evolution to end-stage renal failure. An
obvious approach was the optimization of glycemic control. Controlled
trials of intervention with continuous subcutaneous insulin infusion have
been disappointing, however, in that correction of hyperglycemia has failed
to influence significantly the downhill course of established nephropathy.
Our own experience, spanning a time period from 2 to 6 years in patients
with either persistent or intermittent proteinuria, shows that the average
rate of decline of GFR and the increase in the fractional clearance of albumin and IgG are, by and large, unchanged by improved metabolic control
(Table 1) (10,11). In this series of studies, systemic blood pressure was
maintained stable throughout and diet was unchanged to prevent possible confounding effects of these two factors.

Studies with hypotensive treatment, started early in the course of diabetic renal failure, have had more success in slowing the deterioration of
renal function. In a small but carefully followed series, the rate of
decline of GFR was lowered from 0.9 ml/min./month before treatment to approximately 0.2 ml/min./month during treatment (6). This lower rate of fall
could be maintained for several years and, concomitantly, the albuminuria
was significantly reduced, though not entirely normalized. These results
were obtained in the absence of changes of blood glucose control or dietary
habits. The antihypertensive regimens used in these early studies included
selective beta-blockers, vasodilators and diuretics. More recently, a
number of studies have demonstrated that the administration of angiotensin-
converting enzyme inhibitors to diabetic renal failure patients may have
additional beneficial effects on the progression of renal disease (12-14).
Some workers have claimed that these effects are independent of systemic
blood pressure reduction, and animal studies suggest that they may be related to the lowering of intraglomerular pressure (12,14,15).

The third line of intervention in the treatment of progressive diabetic
nephropathy consists of reduction of dietary protein intake. Surprisingly,
this maneuver, which is known to retard progression in other renal diseases,
has been inadequately explored in diabetic renal failure (16-18). We have
preliminary data in 19 persistently proteinuric, insulin-dependent diabetics

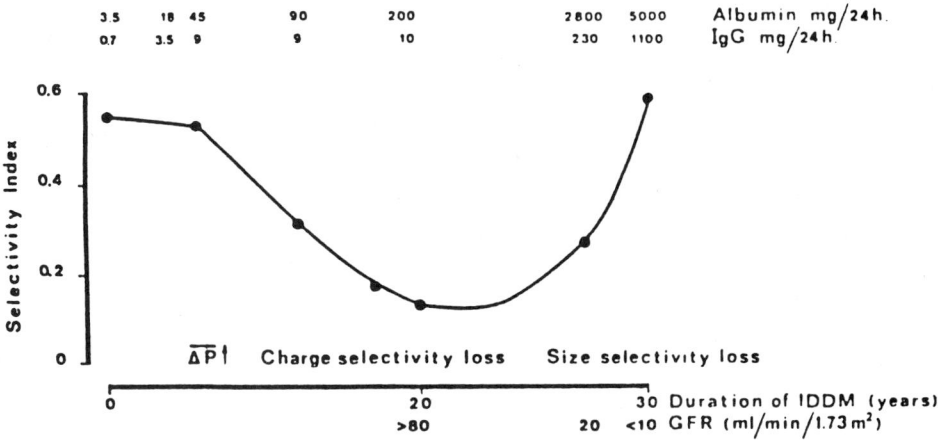

3.5	18	45	90	200	2800	5000	Albumin mg/24h.
0.7	3.5	9	9	10	230	1100	IgG mg/24h.

$\overline{\Delta P} \uparrow$ Charge selectivity loss   Size selectivity loss

0	20	30 Duration of IDDM (years)
	>80	20 <10 GFR (ml/min/1.73m²)

Fig. 1. Diagrammatic representation of the evolution of proteinuria in
diabetic kidney disease. The selectivity index (SI = clearance
of IgG/clearance of albumin) is shown on the vertical axis and
extent of proteinuria with declining GFR and duration of diabetes are shown on the horizontal axis. In early microalbuminuria, the SI remains normal, indicating proportionate increases
in IgG and albumin clearances. As AER increases, within the
microalbuminuria range, SI falls, producing a "high selectivity"
proteinuria. As clinical proteinuria develops and the GFR
declines to a level below 20 ml/min./1.73m², a heavy proteinuria with low selectivity appears. These changes may be due to
a combination of disturbances of intraglomerular pressure and
charge and size selectivity defects. (From ref. 9, with permission.)

Table 1. Effect of Blood Glucose (BG) Control on Rate of Decline of
GFR in Patients with Intermittent and Persistent Clinical
Proteinuria

	Rate of GFR decline (ml/min/month)		
	Conventional BG Control	Strict BG Control	Significance level
Intermittent proteinurics (n=6)	0.9±0.2	0.8±0.3	ns
Persistent proteinurics (n=6)	0.9+0.3	0.7±0.1	ns

Data are mean ± SEM

(15m, 4f), aged between 31 and 67 years (mean 46), studied over a period of
at least 4 years, the last 2 of which were on a restricted protein intake
(19). The low protein diet reduced protein intake to 45 ± 7g/daily, and
this was accompanied by a significant fall in total protein excretion and in
the rate of decline of GFR (Table 2). No significant changes occurred in
average levels of arterial pressure or glycosylated hemoglobin concentrations. Even though the response to low protein diet was heterogeneous, with
some individuals slowing markedly their rate of progression and others showing little change, these findings support the view that low protein diet has
a beneficial and independent effect on the evolution of diabetic renal failure. Further controlled studies are clearly needed to substantiate the prom-
ising results obtained with reduction of blood pressure and dietary protein
intake. However, the overall impression is that present therapeutic means
can at best delay the development of end-stage renal failure, without arresting the progression of the disease.

Table 2. Effect of Low Protein Intake on the Rate of GFR Decline
and Total Urinary Protein Excretion in 19 Clinically
Proteinuric, Insulin-dependent Diabetic Patients

	Normal protein diet (77 ± 5 g/day)	Low protein diet (45 ± 1.6 g/day)	Significance Level (p value)
Total urinary protein (g/24h)	2.9 ± 0.1	1.5 ± 0.1	<0.01
Rate of GFR decline ml/min/month	0.7 ± 0.1	0.14 ± 0.2	<0.001
Glycosylated Hb (%)	9.6 ± 0.3	8.9 ± 0.4	ns
Mean blood pressure (mmHg)	107 ± 2	102 ± 2	ns

Data are mean $\pm$ SEM

Mean blood presure = Diastolic BP + $1/3$ pulse pressure

EARLY MARKERS OF DIABETIC KIDNEY DISEASE

Microalbuminuria

The progressive nature of diabetic nephropathy and our relative therapeutic impotence, together with the elevated mortality associated with this
disorder, emphasize the need for early identification of patients prone to
develop the clinical stage of the disease (20-21). Until some years ago, it
was thought that the increase in albumin excretion above the detection
threshold of current clinical methods (e.g., the Albustix test) was a sudden
event that was preceded by years of normal protein excretion (22). In 1963,
a sensitive and specific radioimmunoassay for measuring urinary albumin in
low concentration was developed (23). Using this method, a number of
authors consistently found that urinary albumin excretion rates (AER) may be
subclinically elevated in nonproteinuric diabetic patients (7,24-26). Approximately 20% of nonproteinuric, insulin-dependent diabetics have albumin
excretion rates above the upper limit of the normal range (2.0-26 mg/24h)
but still well below the excretion rates that are conventionally used to
diagnose clinical albuminuria (i.e., AER greater than 250 mg/24h). This subclinical increase in AER is defined as microalbuminuria (27). The biological variability of AER may be as high as 40-50% (7). Excretion rates may
vary due to the combined effects of posture, exercise, urine flow and the inherent biological variability of albumin excretion. This variability is
similar in diabetic and nondiabetic subjects and is slightly lower during
the night (28-30). Patients with AER near the cutoff level for different
categories of albuminuria are more likely to move into and out of different
classes. Therefore, the extremes of AER ranges, and especially the limit
between normo- and microalbuminuria, are not to be considered as rigid cut-
off values, and common sense should be applied and multiple collections performed before assigning a given AER value to one category or another.

The importance of subclinical elevations of AER remained unknown until
it was shown that some degrees of microalbuminuria could be predictive of
later development of clinical proteinuria. Insulin-dependent diabetic
patients with overnight AER above 30 ug/min. were found to have a 20-fold
risk of developing nephropathy compared with those with AER below 30 ug/min.
after a 14-year follow-up (31). These results were confirmed in two other
independent studies (7,25). The different AER levels that were shown in different studies to predict nephropathy may be explained, at least in part, by
the different duration of follow-up and method of urine collection. The
microalbuminuria of early diabetes appears to be glomerular in origin (9)

and increases with time at an average rate estimated to be about 20% per year (32). Recent findings indicate that abnormal albumin excretion rates, identifying at-risk patients, are absent during the first 5 years of diabetes in conventionally treated, insulin-dependent diabetic patients; this suggests that microalbuminuria is a sign of early disease rather than a marker of susceptibility (33).

Different studies have shown that microalbuminuria is associated with poorer glycemic control and independently and more strongly with raised levels of arterial pressure (7,8,25). Blood pressure in patients with microalbuminuria, although within the so-called normal range, is significantly higher than that of a matched normoalbuminuric group (Table 3). The elevation of blood pressure in these patients is of great interest since it occurs in the absence of any renal hypofunction (the glomerular filtration rate may be either normal or elevated) and raises the question as to whether rises in arterial pressure should be considered a consequence of the renal disease, as generally thought, or rather a phenomenon that may play a casual role in the pathogenesis of diabetic renal failure.

Increased Glomerular Filtration Rate

The occurrence of glomerular hyperfiltration in diabetes mellitus has long been suspected (34). More recently, several studies using accurate techniques for estimating GFR in insulin-dependent diabetic adults and children have confirmed that the GFR is elevated by 20-40% (35-40).

Table 3. Mean Arterial Pressure in Insulin-dependent Diabetic Patients Matched for Age, Sex and Duration, with Albumin Excretion Rates (AER) Above 30 µg/min. (High Risk Group), Below 12 µg/min. (Upper Limit of Normal Range) and Between 12 and 30 µg/min

	Mean blood pressure (mmHg)
AER >30 µg/min (n=12)	103 ± 12
AER 12-30 µg/min (n=16)	92 ± 8*
AER <12 µg/min (n=12)	87 ± 7**

* = p<0.02 versus AER >30 µg/min

** = p<0.001 versus AER >30 µg/min

Data are mean $\pm$ SEM

Approximately 25% of patients have a GFR exceeding the upper limit of the normal range (38). Whether diabetics with GFR within the normal range (84-135 ml/min./1.73m^2) actually have a higher GFR than their putative "nondiabetic" GFR remains uncertain. The proportion of patients with a supranormal GFR thus represents a minimal estimate of the hyperfiltration phenomenon.

The determinants of glomerular hyperfiltration have been directly investigated using micropuncture techniques in diabetic rats and have been shown to consist mainly of renal vasodilation, more accentuated at the afferent than efferent glomerular arteriole, thus resulting in both increased glomerular plasma flow rate and raised mean transglomerular hydraulic pressure gradient (41,42). Glycemic, metabolic and hormonal alterations at least partially mediate these changes in renal hemodynamics. These are summarized in Table 4.

The Role of Eicosanoids

In established diabetic nephropathy, the increased renal production of the vasodilatory eicosanoid PGE$_2$ has been surmised to be important in the maintenance of the GFR since indomethacin acutely reduced GFR (43). Eicosanoids have also been implicated in the genesis of glomerular hyperfiltration although in vitro and animal model studies have failed to provide a uniform picture. Glomeruli isolated from rats with streptozotocin-induced diabetes (STZ) have been reported to produce more prostaglandins than controls (44,45). Species differences seem to be important in this respect, and the BB rat with genetic diabetes has a glomerular prostaglandin production comparable with that of a nondiabetic control rat (45). This genetic model of diabetes in the rat, however, has a different renal complication profile compared with the STZ diabetic rat model in that it develops only mild proteinuria and basement membrane thickening but no mesangial expansion. Moreover, glomerular hyperfiltration is questionable (46-48). A finding of increased synthesis of 6-keto PGF$_{1\alpha}$ by cultured "diabetic" mesangial cells (49) contrasts with other work that reported no difference in the mesangial cell production of PGE$_2$ after development of diabetes (45). Acute administration of indomethacin to short-term STZ diabetic rats caused a reduction in GFR which, from micropuncture studies, was found to be mediated by an increase in afferent arteriolar resistance and a significant decline in ultrafiltration coefficient (50,51). Aspirin administration has been shown to prevent early glomerular hyperfiltration and subsequent decline of GFR and glomerular basement membrane thickening in STZ diabetic rats (52). However, a recent report was unable to show a reduction in GFR following indomethacin

Table 4. Mediators of High GFR in Diabetic Man

MEDIATOR	MAGNITUDE OF EFFECT	COMMENT
Glucose	5-13%	Only with moderate hyperglycaemia. More marked in hyperfilterers.
Ketone bodies	Up to 33%	Pharmacological doses and high physiological doses required.
Insulin	none	No effect per se at levels tested.
Glucagon	6%	Supraphysiological levels required. Greater effect in animals when infused intraportally.
Growth Hormone	7%	Requires several days of administration for effect.
PG System	?	Controversial. Thought to increase GFR if PG predominates over RA system.
RA system	?	Controversial. Possible blunting of RA associated vasoconstrictor effect caused by hyperglycaemia, diabetes and a reduction of glomerular AII receptor number may increase GFR (cf. PG system).
ANP	?	Controversial. May be elevated in diabetes. Uncertain relationship with GFR.

GFR = glomerular filtration rate.
PG = prostaglandin.
RA = renin-angiotensin.
AII = angiotensin II.
ANP = Atrial natriuretic peptide.

infusion even though this markedly reduced urinary PGE_2 (53). These findings are similar to those in long-standing STZ diabetic rats with established glomerular hyperfiltration in which prostaglandin inhibition fails to cause a reduction in the elevated GFR despite a reduction in renal PGE_2 production (51).

In diabetic humans, the findings are no clearer than in the rat model at present. Urinary excretion of PGE_2 and 6-keto $PGF_{1\alpha}$ have been found to be normal by some but elevated by other workers (54-56). Acute intravenous infusion of lysine acetylsalicylic acid and chronic oral administration of piroxicam have been found to reduce GFR and RPF in some insulin-dependent patients (54,55); yet 3 days treatment with oral indomethacin had no effect on GFR in 9 newly diagnosed IDDs, some of whom were hyperfilterers (57).

In both the rat model and in the human situation, there are a number of possible explanations for the disparate findings. In the rat model, different strains of rats at different ages have been used; the determinations of urinary prostaglandins have been made at different intervals after the induction of diabetes; varying levels of glycemia and insulinemia have been present with concomitant volume effects and changes in urine flow and pH. All these may contribute to cause some of the differences. Similarly, in the human studies, the diabetics under study have been far from a homogenous group. Mixed groups of normo- and hyperfilterers have been included at varying levels of glycemia. Diabetics with glomerular hyperfiltration further increase their GFR in response to hyperglycemia, and this effect may be mediated via renal prostaglandins (58,59). Short-term reductions of blood glucose concentrations are accompanied by a reduction in the urinary excretion of 6-keto $PGF_{1\alpha}$ and PGE_2 and a concomitant fall in glomerular filtration (60,61). When these confounding variables are corrected for, as in a recent study, only the urinary excretion of 6-keto $PGF_{1\alpha}$, a compound of endothelial, possibly glomerular, origin, is elevated in diabetics with glomerular

Table 5. Urinary Prostaglandins in Normofiltering and Hyperfiltering Insulin-dependent Diabetics (IDDs) and Controls. Data are Median (Range)

	Normal controls	Normofiltering IDDs	Hyperfiltering IDDs
	N=15	N=10	N=9
GFR (ml/min/ 1.73m^2)	108 (87-128)	121 (105-129)	154 (135-200)
TxB_2 (ng/hr)	4.6 (2.5-10.2)	4.0 (1.8-10.0)	4.4 (2.5-18.9)
PGE_2 (ng/hr)	6.9 (2.7-17.4)	6.5 (1.1-8.9)	6.3 (4.2-32.5)
6-Keto $PGF1_{\alpha}$ (ng/hr)	9.6*(5.2-15.5)	8.8*(1.5-13.8)	17.1 (4.5-33.6)

* p <0.05 compared to hyperfilterers.

hyperfiltration (56) (Table 5). Urinary excretion of PGE_2 and thromboxane B_2 was found to be similar in diabetics with and without hyperfiltration and in normal controls, confirming a previous report (55).

From these conflicting reports, no conclusion can be drawn, but it is possible to postulate that an imbalance between the vasoconstricting and vasodilating forces that control glomerular hemodynamics may contribute to the glomerular hyperfiltration of diabetes. Certainly, later in the course of diabetic nephropathy, the renal prostaglandins appear critically important in maintaining renal plasma flow and GFR (43).

Associations of the Elevated GFR

The elevated GFR in humans is strongly associated with increased kidney volume. Interestingly, even though a large kidney can be associated with a normal GFR, a high GFR is extremely unlikely to occur in normal-size kidneys. Large kidneys reflect hyperplasia and hypertrophy of renal tubules but only hypertrophy of the glomeruli. The latter phenomenon would increase the surface area available for filtration, and a strong correlation has been demonstrated between GFR and glomerular filtering surface in diabetic patients (for review, see ref. 62).

The exact description of the relationship between GFR and albumin excretion rate in nonclinically proteinuric, diabetic patients remains unresolved. Cross-sectionally, an elevated GFR can be associated with either normal or subclinically elevated levels of albumin excretion and, vice versa, microalbuminuria may be accompanied by high or normal GFR (63). Whether there is a chronological relationship between increase in GFR and in albumin excretion rates, with the former preceding the latter, remains to be established in man.

In experimental diabetic animals, the elevated GFR has been implicated in the initiation and progression of diabetic renal disease, but this has not as yet been proven in humans. In one study a small selected group of hyperfiltering diabetic patients (in whom, admittedly, the high GFR was coupled with microalbuminuria) progressed to persistent proteinuria and lost glomerular function, over a 10-year period, at a significantly higher rate than a control group with lower GFR and normal albumin excretion rates (25). This study cannot dissect out the independent contribution of glomerular hyperfiltration to progression of renal damage and loss of glomerular

314

function but is compatible with the view that an elevation of the GFR may play a role in the sequence of events leading to diabetic renal failure.

In a 5-year prospective study of two cohorts of normoalbuminuric, diabetic patients with and without hyperfiltration matched for age, sex and diabetes duration, there was no evidence of progression to persistent proteinuria in those with glomerular hyperfiltration, even though the GFR fell more in this group (64). The follow-up time may, however, be too short to see an effect. Whether the nephromegaly, which invariably accompanies the high GFR, has an independent predictive significance or simply represents an epiphenomenon is unsolved, as is the temporal relationship with glomerular hyperfiltration.

Correction of Early Renal and Systemic Pressure Changes

In sharp contrast to the relative insensitivity to treatment of established diabetic nephropathy, the early-phase abnormalities respond to a number of therapeutic maneuvers. Intensified insulin treatment and strict metabolic control can correct hyperfiltration and microalbuminuria (Table 6) or, at worst, prevent the progressive rise of albumin excretion rates in at-risk individuals (32,65-67). Correction of the marginal elevations in arterial pressure by diuretics or selective beta-blockers also reduces microalbuminuria, and a recent controlled trial has shown a beneficial effect of

Table 6. Effect of Strict Glycaemic Control on Glomerular Filtration Rate (GFR) and Albumin Excretion Rate (AER) in Hyperfiltering and Microalbuminuric Insulin-dependent Patients Respectively

	Ordinary BG control	Strict BG control	Level of significance (p value)
GFR ($ml/min/1.73m^2$) (n=6)	151 ± 6	129 ± 4	$\leqslant 0.001$
AER ($\mu g/min$) (n=10)	30 ± 9	10 ± 3	<0.01

Data are mean $\pm$ SEM. For AER geometric means are used

enalapril on urinary albumin excretion in diabetic patients (68,69). In the diabetic rat, lowering of blood pressure by an angiotensin-converting enzyme inhibitor seems to afford special benefits to the kidney by preventing, through a reduction of intraglomerular pressure, the development of albuminuria and renal histological damage (15). Moreover, short-term studies in microalbuminuric, diabetic patients show that a diet containing approximately 45 grams of protein per day is capable of reducing both albumin excretion rate and the GFR independently of blood glucoseor arterial pressure changes (70). This is consistent with findings in the diabetic rat in whom, at similar levels of hyperglycemia, a protein-restricted diet protects the kidney from hyperfiltration, albuminuria and the consequent histological lesions (71). Recent observations in humans with different protein intakes indicate, interestingly, that vegans, who eat less protein, all of which is vegetable in origin, have significantly lower GFR, albumin excretion rates and arterial blood pressure values than omnivorous subjects (72). A vegan diet, therefore, seems to be associated with renal and systemic changes that should confer protection against nephropathy in diabetes. Other kinds of pharmacological intervention may well be possible in the future, and preliminary reports in the diabetic animal model suggest that administration of aldose reductase inhibitors, which block the sorbitol pathway, may have beneficial effects in preventing the occurrence of, or reducing, proteinuria (73).

Thus, effective measures for correction of early renal abnormalities in diabetes mellitus are available, making overt diabetic nephropathy a potentially preventable condition. However, the correction of an early indicator of disease does not necessarily imply that the disease itself will be abolished, and the results of long-term trials are now eagerly awaited. It remains, therefore, important to recognize different subsets of patients at different stages of the disease, and raises the question as to whether patients susceptible to diabetic nephropathy could be identified at diagnosis of diabetes, by some marker, probably genetic, present well before microalbuminuria or other risk indicators appear.

Perspective for Primary Prevention: Detection of Susceptible Patients

It is clear that only a subset of patients develop nephropathy. The curve distribution of the annual incidence of persistent proteinuria displays a peak at 16-20 years of diabetes which is followed by a considerable reduction in risk thereafter, with patients who survive 30 or more years with diabetes almost free of risk (20). This results in a curve distribution of the cumulative incidence of persistent proteinuria which levels off, after 25 years of diabetes, at around 35% (3). The finding that arterial pressure is elevated in microalbuminuric diabetics without renal hypofunction and that long-term uncomplicated diabetic survivors have low arterial pressures (7,8,25,74) has led us to formulate the hypothesis that an inherited predisposition to raised arterial pressure may be a factor contributing to the susceptibility to nephropathy if diabetes is present. In a recent family study, we have shown that nondiabetic, nonproteinuric parents of proteinuric diabetics have significantly higher blood pressure levels than age and body mass index-matched parents of nonproteinuric diabetics (75) (Fig. 2). A strong correlation was found between arterial pressures in the diabetic patients and those in their respective parents. These results have been complemented by the finding that red blood cell Na+/Li+ counter-transport, a system the activity of which is thought to be genetically determined and is known to be associated with the risk of essential hypertension, is significantly higher in the proteinuric than in nonproteinuric, insulin-dependent diabetics (76,77). We have also found that the exchanger activity was not increased in matched, nondiabetic renal patients with renal function and blood pressure levels comparable with those of diabetic patients with clinical proteinuria, suggesting that high Na+/Li+ counter-transport activity is not a feature of nephrogenic hypertension. These data would further support the view that elevations of blood pressure may play a causal role in the development of diabetic kidney disease rather than being a mere conse-

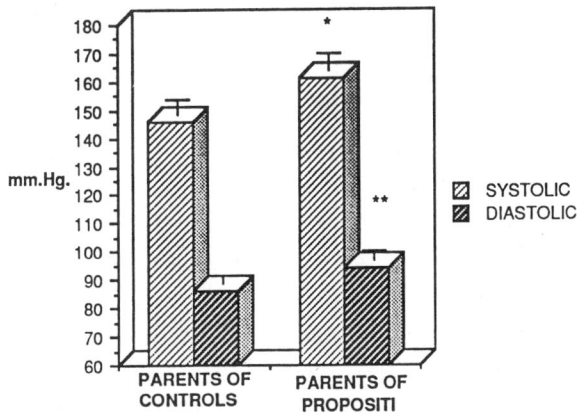

Fig. 2. Systolic and diastolic blood pressure
in the parents of 17 proteinuric (pro-
positi) and 17 nonproteinuric (controls) insulin-dependent patients.
Values are means ± standard errors.
* = p<0.02, ** = p<0.05.

quence of deteriorating renal function. If these preliminary findings are supported by further studies, it may be possible in the future to identify diabetic patients at risk of nephropathy by a positive family history for arterial hypertension or by a cell marker associated with it well before microalbuminuria, indicating early diabetic renal disease, develops. At this stage, primary prevention would then be possible.

REFERENCES

1. A.R. Andersen, J. S. Christiansen, J.K. Andersen, S. Kreiner, T. Deckert, Diabetic nephropathy in Type I (insulin-dependent) diabetes: an epidemiological study, Diabetologia 25:496-501 (1983).

2. A.C. Thomsen, The kidney in diabetes mellitus, Thesis, Copenhagen, Munksgaard (1965).

3. A.S. Krolewski, J.H. Warram, A.R. Christlieb, E.J. Busick, C.R. Kahn, The changing natural history of nephropathy in Type I diabetes, Am. J. Med. 78:785-794 (1985).

4. B. Rettig and S.M. Teusch, The incidence of end-stage renal failure in Type I and Type II diabetes mellitus, Diab. Nephropathy 3:26-27 (1984).

5. G.C. Viberti, R.W. Bilous, D. Mackintosh, H. Keen, Monitoring glomerular function in diabetic nephropathy. A prospective study, Am. J. Med. 74:256-264 (1983).

6. H.H. Parving, A.R. Andersen, U.M. Smidt, E. Hommel, E.R. Mathiesen, P.A. Svendsen, Effect of antihypertensive treatment on kidney function in diabetic nephropathy, Brit. Med. J. 294:1443-1447 (1987).

7. E.R. Mathiesen, B. Oxenboll, K. Johansen, P. Aa Svendsen, T. Deckert, Incipient nephropathy in Type I (insulin-dependent) diabetes, Diabetologia 26:406-410 (1984).

8. M.J. Wiseman, G.C. Viberti, D. Mackintosh, R.J. Jarrett, H. Keen, Glycemia, arterial pressure and microalbuminuria in Type I (insulin-dependent) diabetes mellitus, Diabetologia 26:401-405 (1984).

9. G.C. Viberti and H. Keen, The patterns of proteinuria in diabetes mellitus, Diabetes 33:686-692 (1984).

10. G.C. Viberti, R.W. Bilous, D. Mackintosh, J.J. Bending, H. Keen, Long-term correction of hyperglycemia and progression of renal failure in insulin-dependent diabetes, Br. Med. J. 286:598-602 (1983).

11. J.J. Bending, G.C. Viberti, P.J. Watkins, H. Keen, Intermittent clinical proteinuria and renal function in diabetes: evolution and effect of glycemic control, Br. Med. J. 292:83-86 (1986).

12. S. Bjorck, G. Nyberg, H. Mulec, G. Granerus, H. Herlitz, M. Aurell,

Beneficial effects of angiotensin-converting enzyme inhibition on renal function in patients with diabetic nephropathy, Br. Med. J. 293:471-474 (1986).

13. E. Hommel, H.H. Parving, E. Mathiensen, B. Edsberg, M.D. Nielsen, J. Giese, Effect of captopril on kidney function in insulin-dependent diabetic patients with nephropathy, Br. Med. J. 293:467-470 (1986).

14. A.C. Taguma, Y. Kitamoto, G. Futaki, H. Ueda, H. Monma, M. Ishizakli, H. Takalashli, H. Sekino, Y. Sasaki, Effects of captopril on heavy proteinuria in azotemic diabetics, N. Engl. J. Med. 313:1617-1620 (1985).

15. R. Zatz, T. Dunn, S. Meyer, S. Anderson, H.G. Rennke, B.M. Brenner, Prevention of diabetic glomerulopathy by pharmacological amelioration of glomerular capillary hypertension, J. Clin. Invest. 77: 1925-1930 (1986).

16. J. Bergstrom, Discovery and rediscovery of low-protein diet, Clin. Nephrol. 21(1):29-35 (1984).

17. J.B. Rosman, S. Meijer, W.J. Sluilter, P.M.T. Wee, T. Piers-Becht, A.J.M. Dunker, Prospective randomized trial of early dietary protein restriction in chronic renal failure, Lancet II:1291-1295 (1984).

18. P-O. Attman, H. Bucht, O. Larsson, G. Uddebom, Protein-reduced diet in diabetic renal failure, Clin. Nephrology 19:217-220 (1983).

19. J.J. Bending, R. Dodds, J.D. Walker, H. Keen, G.C. Viberti, Dietary protein restriction delays the progression of diabetic renal failure, Diabetologia 30:598A (1987).

20. K. Borch-Johnsen, P.K. Andersen, T. Deckert, The effect of proteinuria on the relative mortality in Type I (insulin-dependent) diabetes mellitus, Diabetologia 28:590-596 (1985).

21. K. Borch-Johnsen and S. Kreiner, Proteinuria: value as predictor of cardiovascular mortality in insulin-dependent diabetes mellitus, Br. Med. J. 294:1651-1654 (1987).

22. C.E. Mogensen, Renal function changes in diabetes, Diabetes 25:872-879 (1976).

23. H. Keen and C. Chlouverakis, An immunoassay for urinary albumin at low concentration, Lancet II:913-916 (1963).

24. G.C. Viberti, J.C. Pickup, R.J. Jarrett, H. Keen, Effect of control of blood glucose on urinary excretion of albumin and B_2 microglobulin in insulin-dependent diabetics, N. Engl. J. Med. 300:638-642 (1979).

25. C.E. Mogensen and C.K. Christensen, Predicting diabetic nephropathy insulin-dependent patients, N. Engl. J. Med. 311:89-93 (1984).

26. H.H. Parving, E. Hommel, E. Mathiesen, P. Skott, B. Edsberg, M. Bahnsen, M. Lauritzen, P. Hougaard, E. Lauritzen, Prevalence of

microalbuminuria, arterial hypertension, retinopathy and neuropathy in patients with insulin-dependent diabetes, Br. Med. J. 296:156-160 (1988).

27. G.C. Viberti, M.J. Wiseman, S. Redmond, Microalbuminuria: Its history and potential for prevention of clinical nephropathy in diabetes mellitus, Diab. Nephropath. 3:70-82 (1984).

28. B. Feldt Rasmussen and E. Mathiesen, Variability of urinary albumin excretion in incipient diabetic nephropathy, Diab. Nephropath. 3:101-103 (1984).

29. D.J.F. Rowe, H. Bagga, P.B. Betts, Normal variations in the rate of albumin excretion and albumin to creatinine ratios in overnight and daytime urine collections in nondiabetic children, Br. Med. J. 291:693-694 (1985).

30. D.L. Cohen, C.L. Close, G.C. Viberti, The variability of overnight urinary albumin excretion in insulin-dependent and normal subjects, Diab. Med. 4:437-440 (1987).

31. G.C. Viberti, R.D. Hill, R.J. Jarrett, U. Mahamud, H. Keen, Microalbuminuria as predictor of clinical nephropathy in insulin-dependent diabetes mellitus, Lancet I:1430-1432 (1982).

32. B. Feldt-Rasmussen, E.R. Mathiesen, T. Deckert, Effect of two years of strict metabolic control of progression of incipient nephropathy in insulin-dependent diabetes, Lancet II:1300-1304 (1986).

33. C.F. Close, On behalf of the Microalbuminuria Collaborative Study. The prevalence of at-risk microalbuminurics in a population of nonproteinuric, insulin-dependent diabetic subjects, Diabetic Med.:328A (1986).

34. P. Cambier, Application de la theorie de Rehberg a l'etude clinique des affections renales et du diabete, Annales de Medicine 35:273-299 (1934).

35. E. Fiaschi, B. Grassi, G. Andres, La funzione renale nel diabete mellito, Rassegna di fisiopatologia clinica e terapeutica 4:373-410 (1952).

36. J. Ditzel and M. Schwartz, Abnormal glomerular filtration rate in short-term insulin-treated diabetic subjects, Diabetes 16:264-267 (1967).

37. C.E. Mogensen, Glomerular filtration rate and renal plasma flow in short-term and long-term juvenile diabetes mellitus, Scan. J. Clin. Lab. Invest. 28:91-100 (1971).

38. J.S. Christiansen, J. Gammelgaard, M. Frandsen, H.H. Parving, Increased kidney size, glomerular filtration rate and renal plasma flow in short-term insulin-dependent diabetics, Diabetologia 20:451-456 (1981).

39. M.J. Wiseman, G.C. Viberti, H. Keen, Threshold effect of plasma glucose

in the glomerular hyperfiltration of diabetes, Nephron 38:257-260
(1984).

40. G. Stalder and R. Schmid, Severe functional disorders of glomerular cap-
    illaries and renal hemodynamics in treated diabetes mellitus during
    childhood, Ann. Paediatri. 193:129-138 (1959).

41. T.H. Hostetter, J.C. Troy, B.M. Brenner, Glomerular hemodynamics in ex-
    perimental diabetes mellitus, Kidney Int. 19:410-415 (1981).

42. P.K. Jensen, J.S. Christiansen, K. Steven, H.H. Paving, Renal function
    in streptozotocin-diabetic rats, Diabetologia 21:409-414 (1981).

43. E. Hommel, E. Mathiesen, S. Arnold-Larsen, B. Edsberg, U.B. Olsen, H.H.
    Parving, Effects of indomethacin on kidney function in Type I
    (insulin-dependent) diabetic patients with nephropathy, Diabetologia
    30:78-81 (1987).

44. M. Schambelan, S. Blake, J. Sraer, M. Bens, M.P. Nivez, F. Wahbe, In-
    creased prostaglandin production by glomeruli isolated from rats with
    streptozotocin-induced diabetes mellitus, J. Clin. Invest. 75:404-412
    (1985).

45. R. Barnett, L. Scharschmidt, K.O. Young-Hyeh, D. Schlondorff, Comparison
    of glomerular and mesangial prostaglandin synthesis and glomerular
    contraction in two rat models of diabetes mellitus, Diabetes 36:1468-
    1475 (1987).

46. D.M. Brown, M.W. Steffes, P. Thibert, S. Azar, S.H. Mauer, Glomerular
    manifestation of diabetes in the BB rat, Metabolism 32:5131-5135
    (1983).

47. A.F. Nakhooda, A.A. Like, C.I. Chappel, F.T. Murray, B.B. Marliss, The
    spontaneously diabetic Wistar rat: metabolic and morphologic studies,
    Diabetes 26:100-112 (1977).

48. A.J. Cohen, D.M. McCarthy, R.R. Rossetti, Renin secretion by the spon-
    taneously diabetic rat, Diabetes 35:341-346 (1986).

49. J.I. Kreisberg and P.Y. Patel, The effects of insulin, glucose and diabe-
    tes on prostaglandin production by rat kidney glomeruli and cultured
    glomerular mesangial cells, Prostaglandins Leukotrienes Med. 11:431-
    442 (1983).

50. P.K. Jensen, K. Steven, H. Blaehr, J.S. Christiansen, H.H. Parving,
    Effects of indomethacin on glomerular hemodynamics in experimental
    diabetes, Kidney Int. 29:490-495 (1986).

51. P.A. Craven, M.A. Caines, F.R. DeRubertis, Sequential alterations in glo-
    merular prostaglandin and thromboxane synthesis in diabetic rats: re-
    lationship to the hyperfiltration of early diabetes, Metabolism 36:
    95-103 (1987).

52. O.I. Moel, R.L. Safirstein, R.C. McEvoy, W. Hsueh, Effect of aspirin on

experimental diabetic nephropathy, <u>J. Lab. Clin. Med</u>. 110:300-307 (1987).

53. N. Bank, M.A.G. Lahorra, H.S. Aynedjian, D. Schlondorff, Vasoregulatory hormones and the hyperfiltration of diabetes, <u>Am. J. Physiol</u>. 254: F202-F209 (1988).

54. E. Esmatjes, M.R. Fernandez, I. Halpern, J. Camps, J. Gaye et al., Renal hemodynamic abnormalities in patients with short-term insulin-dependent diabetes mellitus: role of renal prostaglandins, <u>J. Clin. Endocrinol. Metab</u>. 60:1231-1236 (1985).

55. S. Gambardella, G. Pugliese, A. Napoli, S. Morano, P. Pietravalle, F. Pugliese, G. Stirati, D. Andreani, Urinary excretion of prostaglandins and renal hemodynamics in Type I (insulin-dependent) diabetes at onset: effect of piroxicam, <u>Diabetologia</u> 29:53A (1986).

56. G.C. Viberti, A. Benigni, E. Bognetti, G. Remuzzi, M.J. Wiseman, Glomerular hyperfiltration and urinary prostaglandins in insulin-dependent diabetes mellitus, In preparation.

57. J.S. Christiansen, B. Feldt-Rasmussen, H.H. Parving, Short-term inhibition of prostaglandin in synthesis has no effect on the elevated glomerular filtration rate of early insulin-dependent diabetes, <u>Diabetic Med</u>. 2:17-20 (1985).

58. M.J. Wiseman, R. Mangili, H. Alberetto, H. Keen, G.C. Viberti, Glomerular response mechanisms to glycemic changes in insulin-dependent diabetics, <u>Kidney Int</u>. 31:1012-1018 (1987).

59. B.L. Kasiske, M.P. O'Donnell, W.F. Keane, Glucose-induced increases in renal hemodynamic function: possible modulation by renal prostaglandins, <u>Diabetes</u> 34:360-364 (1985).

60. D.A. Collier, D.M. Matthews, G. Beel, M.L. Watson, B.F. Clarke, Increased urinary excretion of 6-keto $PGF_{1\alpha}$ and PGE_2 in male insulin-dependent diabetics, <u>Diabetic Med</u>. 3:358A (1986).

61. E. Esmatjes, I. Levy, J. Gaya, F. Rivera, Renal excretion of prostaglandin E_2 and plasma renin activity in Type I diabetes mellitus: relationship to normoglycemia achieved with artificial pancreas, <u>Diabetes Care</u> 10:428-431 (1987).

62. G.C. Viberti and M.J. Wiseman, The kidney in diabetes: significance of early abnormalities, <u>Clinics in Endocrin. Metab</u>. 15:753-783 (1986).

63. G.C. Viberti, R.J. Jarrett, M.J. Wiseman, Predicting diabetic nephropathy (letter), <u>New Engl. J. Med</u>. 311:1256-1257 (1984).

64. S.L. Jones, M.J. Wiseman, G.C. Viberti, H. Keen, Glomerular hyperfiltration and albuminuria - A 5-year prospective study in Type I (insulin-dependent) diabetes mellitus, <u>Diabetologia</u> 30:536A (1987).

65. M.J. Wiseman, A.J. Saunders, H. Keen, G.C. Viberti, Effect of blood

glucose control on increased glomerular filtration rate and kidney size in insulin-dependent diabetes, <u>N. Engl. J. Med</u>. 312:617-621 (1985).

66. J.J. Bending, G.C. Viberti, R.W. Bilous, H. Keen, For the Kroc Collaborative Study Group. Eight-month correction of hyperglycemia in insulin-dependent diabetes mellitus is associated with a significant and sustained reduction of urinary albumin excretion rates in patients with microalbuminuria, <u>Diabetes</u> 34 (Suppl.3):69-73 (1985).

67. K. Dahl-Jorgensen, O. Brinchmann-Hansen, K. Hanssen et al., Effect of near normoglycemia for two years on progression of early diabetic retinopathy, nephropathy and neuropathy: The Oslo Study, <u>Br. Med. J</u>. 293:1195-1199 (1986).

68. C.K. Christensen and C.K. Mogensen, Antihypertensive treatment: long-term reversal of progression of albuminuria in incipient diabetic nephropathy. A longitudinal study of renal function, <u>J. Diab. Complic</u>. 1:45-53 (1987).

69. M. Marre, H. Leblanc, L. Suarez, T.T. Guyenne, J. Menard, P. Pasa, Converting enzyme inhibition and kidney function in normotensive diabetic patients with persistent microalbuminuria, <u>Br. Med. J</u>. 294:1448-1452 (1987).

70. D.L. Cohen, R. Dodds, G.C. Viberti, Effects of protein restriction in insulin-dependent diabetics at risk of nephropathy, <u>Br. Med. J</u>. 294:795-798 (1987).

71. R. Zatz, T.W. Meyer, H. Rennke, B.M. Brenner, Predominance of hemodynamic rather than metabolic factors in the pathogenesis of diabetic glomerulopathy, <u>Proc. Natl. Acad. Sci. USA</u> 82:5963-5967 (1985).

72. M.J. Wiseman, R. Hunt, A. Goodwin, J.L. Gross, H. Keen, G.C. Viberti, Dietary composition and renal function in healthy subjects, <u>Nephron</u> 46:37-42 (1987).

73. A. Beyer-Mears, The polyol pathway, sorbinil and renal dysfunction, <u>Metabolism</u> 35 (Suppl.1):46-54 (1986).

74. W.G. Oakley, D.A. Pyke, R.B. Tattersall, P.J. Watkin, Long-term diabetes. A clinical study of 92 patients after forty years, <u>Q. J. Med</u>. 43:145-156 (1974).

75. G.C. Viberti, H. Keen, M.J. Wiseman, Raised arterial pressure in parents of proteinuric insulin-dependent diabetics, <u>Brit. Med. J</u> 295:575-577 (1987).

76. R. Mangili, J.J. Bending, G.S. Scott, L.K. Li, A. Gupta, G.C. Viberti, Increased sodium-lithium countertransport activity in red cells of patients with insulin-dependent diabetes and nephropathy, <u>N. Engl. J. Med</u>. 318:146-149 (1988).

77. A.S. Krolewski, M. Canessa, J.H. Warram, L.M.B. Laffel, R. Christlieb, W.C. Knowles, L.I. Rand, Predisposition to hypertension and susceptibility to renal disease in insulin-dependent diabetes mellitus, N. Engl. J. Med. 318:140-145 (1988).

THE MOLECULAR, BIOCHEMICAL AND HUMAN PHARMACOLOGY

OF THROMBOXANE A_2 IN RENAL DISEASE

Garret A. FitzGerald, Rosemary Murray, Patricia Price,
and Francesca Catella

Division of Clinical Pharmacology
Vanderbilt University
Nashville, Tennessee 37232

INTRODUCTION

Arachidonic acid is a constituent of the phospholipid domain of biological membranes. Activation of phospholipases results in its release into the intracellular milieu where it is subject to metabolism to biologically active compounds. In most cell types, including the platelet (1), the predominant enzyme involved in catalyzing arachidonate release is phospholipase A_2 rather than phospholipase C (2). It has recently been shown that arachidonate may be subject to direct oxygenation within the cell membrane (3); the biological role of this process remains speculative.

The principle pathways of arachidonate metabolism involve reactions catalyzed by cyclooxygenase, lipoxygenase and epoxygenase enzymes. These pathways give rise respectively to prostaglandins and thromboxane A_2; leukotrienes and lipoxins; and epoxyeicosatrienoic acids (EETs) (4-7). These eicosanoids are formed de novo, in response to a stimulus, and are not stored within cells. While many studies have demonstrated the biological actions of products of these last two pathways *in vitro* and the capacity of intact cells to form many of these compounds, definitive evidence for their functional importance in human disease processes depends upon the development of analytical methodology to define their formation *in vivo* and the availability of specific pharmacologic probes which can be administered safely to man. In this respect, the products of the cyclooxygenase pathway have been subject to more prolonged scrutiny; and although our understanding of their role in human disease is still evolving, recent advances in analytical

325

methodology, the recognition of aspirin as a cyclooxygenase inhibitor for the past fifteen years (8-10), the development of more specific probes and the results of clinical trial (11,12) have served to clarify the importance of these compounds.

MECHANISM OF ACTION OF THROMBOXANE A_2

While the stimuli that lead to arachidonate release from cell membranes are recognized to be very nonspecific, the products formed subsequent to its cyclooxygenation are remarkably cell-specific (Fig. 1). For example, PGD_2 is the major product of mast cells (13), PGI_2 of macrovascular endothelium (14) and PGE_2 of microvascular endothelium (15,16). Thromboxane A_2 is the predominant product formed in platelets and macrophages (17) and is a potent platelet agonist and constrictor of smooth muscle *in vitro*. It is highly evanescent, having a half-life of approximately 30 seconds at physiological pH and is presumed to act as a local mediator (18) rather than

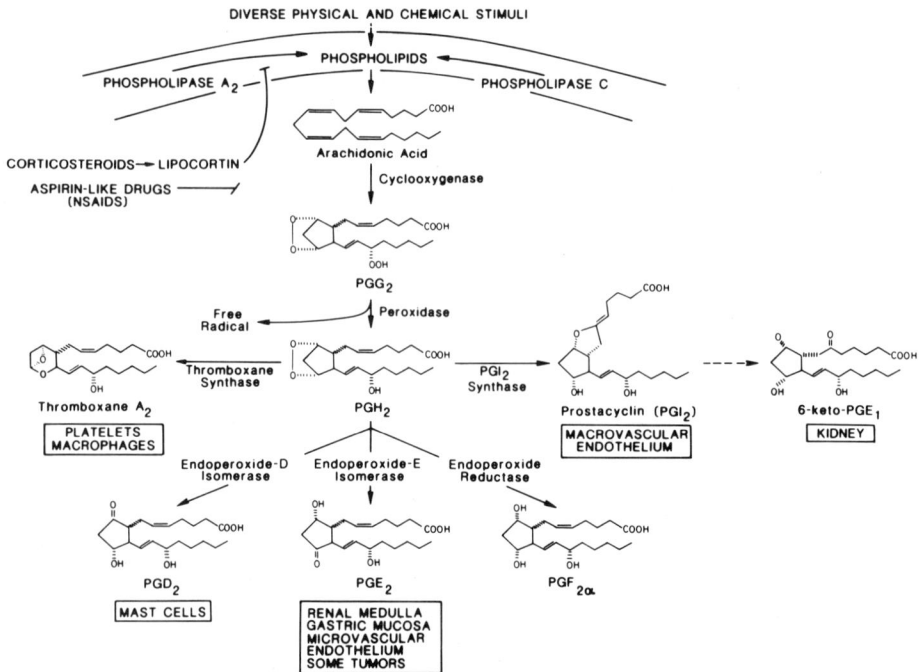

Fig. 1. The cyclooxygenase pathway of arachidonic acid metabolism. Non-specific stimuli lead to arachidonic acid release. Subsequent products of metabolism are relatively cell-specific. (Reproduced with permission from G.A. FitzGerald, Prostaglandins and related compounds, Cecil Loeb Textbook of Medicine, in press, 1988.)

a systemically active autacoid *in vivo*. Thromboxane A_2 is recognized
to act via receptors which have been defined pharmacologically on platelet
and vascular cell membranes (19,20). The development of radiolabeled analogues of chemically synthesized compounds that act as pharmacological agonists (21) and antagonists (22,23) at this receptor has permitted its biochemical definition in human platelets (Fig. 2). Despite the implied presence of thromboxane receptors on other cell types, such as vascular smooth
muscle cell and mesangial cells, the biochemical characterization of such
receptors in human tissues using radioligand binding studies has yet to be
reported. However, this is likely to represent a limitation of the available ligands, as thromboxane analogues can elicit a physiological response
in these cells (contraction) which is accompanied by a biochemical event (an
increase in intracellular calcium), both of which are blocked by prior exposure to receptor antagonists (24,25). Hanaski et al. (26) have identified
thromboxane-ligand binding sites in rat vascular endothelial and smooth
muscle cells (26). Recently, we have obtained evidence for biochemical
heterogeneity among thromboxane receptors in human tissues; thus, the receptor identified in homogenates of human placenta differs from that in human
platelets. Evidence of biochemical heterogeneity among PGE_2 receptors has
previously been reported in the renal tubule (27), and the pharmacological
studies of Mais and colleagues (28) suggest that the vascular thromboxane

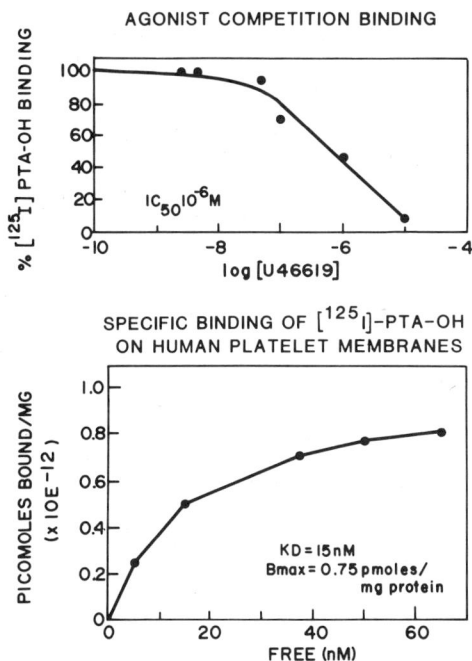

Fig. 2. Biochemical characterization of the thrombox-
ane/prostaglandin re-
ceptor in human plate-
lets. Binding studies
employ the radiola-
beled antagonist
^{125}I-PTA-OH and the
agonist U46619 (see
text). (Reproduced
with permission from
G.A. FitzGerald, D.J.
Fitzgerald, J.A.
Lawson and R. Murray,
Thromboxane biosynthe-
sis and antagonism in
humans, in: Adv.
Prostagland. Thrombox.
Leukot. Res., Vol. 17,
B. Samuelsson, R.
Paoletti and R.W.
Ramwell, eds., Raven
Press, New York, pp.
199-203 [1987].)

327

receptor may differ from that in platelets. At the time of writing, the degree of purification of a thromboxane receptor necessary for sequence information and cloning has not been reported. The disparity among compounds that possess the pharmacological capacity to block the thromboxane agonist-induced responses has resulted in very limited information on structure-activity relationships of a classical nature (Fig. 3). However, the platelet receptor has been solubilized in the detergent CHAPS (29), photoaffinity

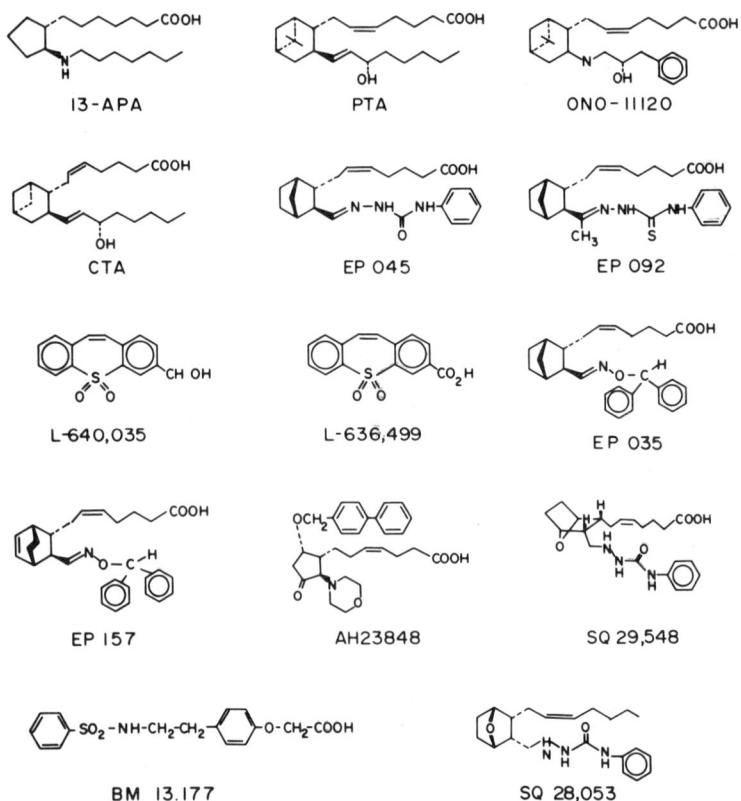

Fig. 3. Structures of compounds identified as pharmacological antagonists of the thromboxane/prostaglandin endoperoxide receptor in human platelets. (Reproduced with permission from G.A. FitzGerald, D.J. Fitzgerald, J.A. Lawson and R. Murray, Thromboxane biosynthesis and antagonism in humans, in: Adv. Prostagland. Thrombox. Leukot. Res., Vol. 17, B. Samuelsson, R. Paoletti and R.W. Ramwell, eds., Raven Press, New York, pp. 199-203 [1987].)

labels have been reported (30), and radioligand binding studies have recently identified the receptor in the HL-60 monocyte-macrophage-derived cell line (31).

The biochemical events that result from the interaction of thromboxane A_2 with its receptor(s) have been most clearly defined in the platelet. What has been described in renal tissues is described elsewhere in this publication by Dunn and Mené (32). Both the pharmacological and biochemical responses to thromboxane A_2 have been inferred from the use of more stable analogues of thromboxane A_2 or its prostaglandin endoperoxide precursor, PGH_2 (Fig. 1). Comparisons of the functional effects of these compounds and biologically generated thromboxane A_2 or thromboxane A_2 derived from a recently synthesized macrolactone (33) imply that they act via a shared or highly analogous receptor (34). Indeed, recent evidence *in vivo* suggests that the endoperoxide might function as a relatively potent agonist at this receptor on platelets and vascular smooth muscle (35), a property of potential consequence during the therapeutic administration of thromboxane synthase inhibitors (vide infra). Prior to describing the biochemical response to such agonists, however, it is appropriate to take note of certain observations that potentially constrain their use as probes of thromboxane receptor(s). Thus, carbocyclic thromboxane A_2 (CTA_2) causes coronary and bronchial smooth muscle constriction *in vivo* and *in vitro* (35) but has been reported to inhibit arachidonate-induced platelet aggregation (35) and to cause only weak, reversible aggregation itself (36); this has been interpreted to reflect its ability to stimulate platelet adenylate cyclase (37), a property that would presumably counteract the effects of thromboxane receptor activation (38). Similarly, both 15-hydroxy-9, 11-epoxymethano-PGF_2 (U46619) and 9,11-epoxymethano-PGF_2 (U44069), compounds commonly employed as receptor agonists, stimulate platelet adenylate cyclase at high concentrations (39), and pinane thromboxane A_2 (used as an antagonist) acts as partial agonist in human platelets (40). Structural analogues which both antagonize the platelet thromboxane receptor and stimulate prostacyclin receptor-linked activation of platelet adenylate cyclase have been described (41). Such limitations of the available probes may, for example, contribute to the reported discrepancy between the rank order for agonists for platelet aggregation and for the displacement of a radiolabeled antagonist from what was assumed to be the same receptor on platelet membranes (29).

Given these constraints, the following information has been acquired with respect to activation of a signal transduction system activated by the

329

THROMBIN

Ca^{++} ①

RECEPTOR

CK ← DAG ← PIP$_2$
② PLC

clP$_3$

Pr$_b$ ⇄ Pr$_b$-P   IP$_3$
(40 kd)

Ca$_c^{++}$

CaCmMLCK

MYOSIN·P ← MYOSIN
(20 kd)

+ACTIN

RESPONSE

Fig. 4. Thrombin interacts with its receptor facilitating (1) calcium entry from extracellular sources, (2) diacylglycerol (DAG) formation by phospholipase C (PLC) catalyzed cleavage of inositol bisphosphate (PLP$_2$) with consequent protein kinase C (CK) activation. This results in phosphorylation of the 40 kD protein (Prh). The increase in inositol trisphosphate (IP$_3$) and its cyclic derivative (clP$_3$) stimulates the release of calcium (Cac) from intracellular stores with consequent activation of calcium calmodulin myosin light chain kinase (Ca Cm myosin). Phosphorylated myosin interacts with actin to trigger the aggregation response (see text).

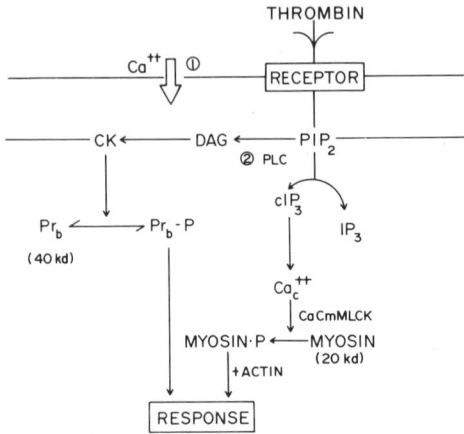

interaction of thromboxane A$_2$ with its receptor in the platelet. Both U46619 and 9,11-epithio-11,12-methano-thromboxane A$_2$ (STA$_2$), a structurally distinct endoperoxide analog, stimulate a transient increase in intracellular free calcium (Ca^{++}i), which is inhibited by the thromboxane receptor antagonist, 13-aza-prostanoic acid (42,43). By analogy with activation of the thrombin receptor (44,45; Fig. 4), U46619 stimulated diacylglycerol (DAG) and the phosphatidic acid (PA formation; 46) presumably reflect calcium-catalyzed activation of phospholipase C and resultant hydrolysis of phosphatidyl inositol-4,5-bisphosphate (PIP$_2$) to form DAG and inositol trisphosphates (IP$_3$; 46). U46619-stimulated DAG and PA formation occurs at concentrations sufficient to initiate platelet shape change but no aggregation and is blocked by the thromboxane receptor antagonist EP045 (46).

Also, analogous to the thrombin-receptor interaction in platelets (Fig. 4), STA$_2$ activates protein kinase C (PKC) and myosin light chain kinase (MLCK), although it caused much less phosphoinositol turnover and PKC activation than MLCK activation (47). Finally, it has recently been described that U46619 and U44069 inhibit the increase in intraplatelet cyclic AMP caused by PGI$_2$ and PGE$_2$ (48,49). A direct inhibitory effect of these analogues in low concentrations on platelet adenylate cyclase has been

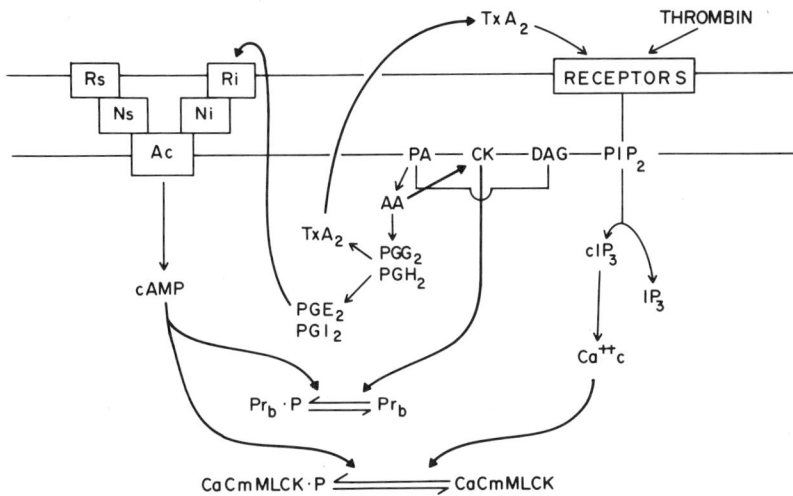

Fig. 5. Thromboxane A_2 may be generated de novo or following phospholipase A_2 catalyzed cleavage of phosphatidic acid (PA). Subsequent to thrombin interaction with its receptor, activation of the thromboxane receptor results in phosphoinositide turnover, a rise in Cac and phosphorylation of the 40 kD protein. By contrast, prostacyclin and PGE_2 activate an adenylate cyclase coupled receptor system (see text).

reported (39). Its correlation with stimulation of a high affinity GTPase suggests an action via receptors (Ri) coupled with a GTP-binding inhibitory protein (Ni). Preliminary evidence suggests the activation of a similar transduction system by thromboxane A_2 in vascular smooth muscle (24). These concepts are outlined in Figure 5.

Thromboxane A_2 formation is likely to act as a recruitment mechanism in human platelets. *In vitro*, several platelet agonists, including epinephrine, platelet activating factor and ADP, stimulate a "primary" aggregation response followed by a thromboxane-dependent "secondary" wave which is abolished by pretreating the platelets with aspirin. Recent experiments, which we have performed using microquantitative analyses by gas chromatography-mass spectrometry (GC/MS) in successive 40 ul aliquots of bleeding time blood, are consistent with this concept operating *in vivo* (50). Thus, while eicosanoids such as PGI_2 and PGE_2 are formed early in the course of the bleeding time, the formation of thromboxane B_2 is gradual and delayed (Fig. 6). This would be consistent with the hypothesis that other agonists serve to initiate the platelet response to vessel injury and that thromboxane A_2 comes into play predominantly as a recruitment mechanism during the phase of platelet-platelet interaction.

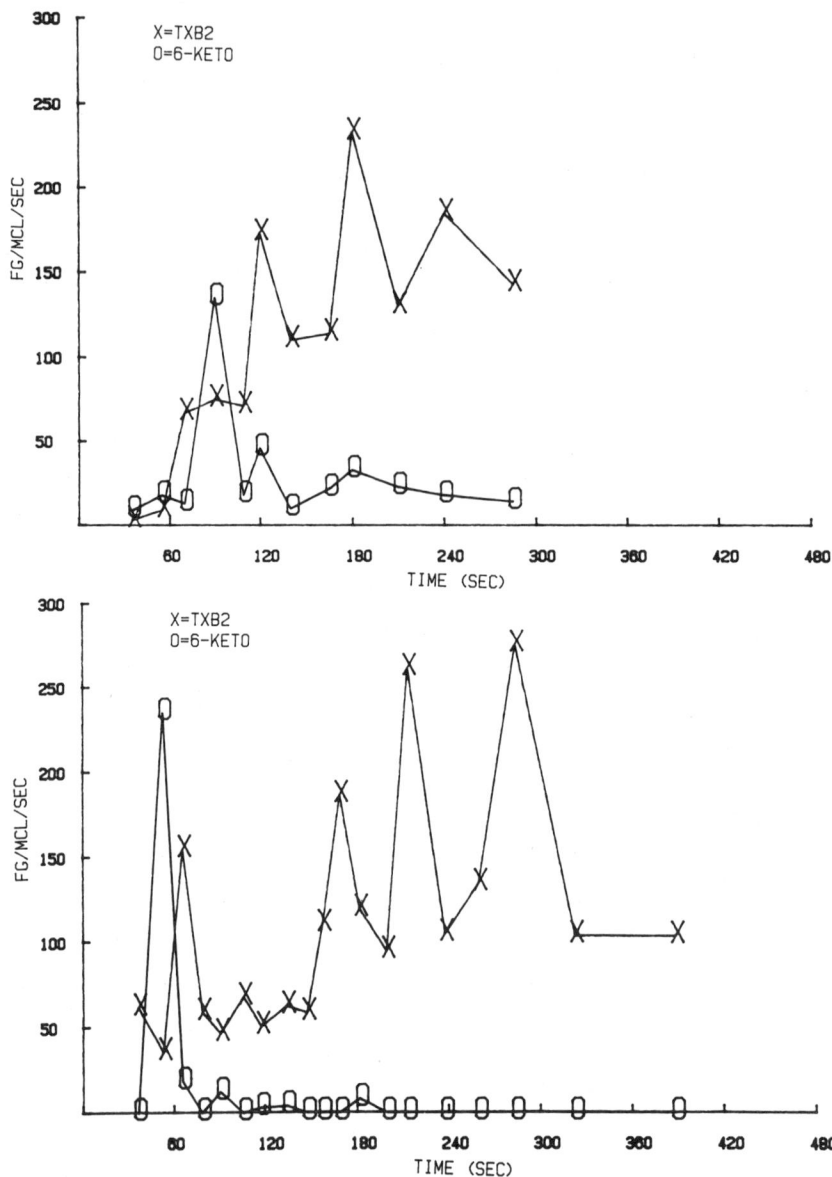

Fig. 6. Temporal formation of the hydration products of prostacyclin (6-keto-PGF$_{1\alpha}$) and thromboxane A$_2$ (TxB$_2$)
in successive 40 µl aliquots of bleeding time blood in
the same individual on separate days (1,2). Prostacyclin formation peaks briefly and earlier than the more
sustained rise in TxB$_2$. (Reproduced with permission
from J. Nowak and G.A. FitzGerald, Prostaglandin endoperoxide reorientation at the platelet vascular interface in man, J. Clin. Invest., January 1989.)

Such a schema would imply the need for sites of regulation to prevent an unrestricted, positive feedback during platelet activation. One logical site for such regulation would be at the level of thromboxane receptor desensitization. The aggregation response of human platelets to U46619 is diminished by prior incubation with this compound (51). We have confirmed these observations, demonstrated that they hold for structurally distinct thromboxane analogues and are homologous with respect to thrombin (52). Desensitization is rapid, with a half-time of about 2-3 minutes, and is receptor-mediated. Accompanying the diminution of the aggregation response is a desensitization of the agonist-induced rise in $Ca^{++}i$ and PKC activity as reflected by phosphorylation of the 40 kD protein (53). The sequence of events appears highly analogous to the desensitization process in the beta adrenergic system (54), initial uncoupling of the thromboxane receptor from its G protein followed by receptor internalization and ultimate degradation (53). The precise nature of the G protein linked to the thromboxane receptor remains undefined, although it appears to be pertussis toxin-insensitive (55).

METABOLISM OF THROMBOXANE A_2 AND ASSESSMENT OF BIOSYNTHESIS *IN VIVO*

Due to its evanescent nature, the disposition of thromboxane A_2 itself has not yet been characterized. However, the reported synthesis of the macrolactone now provides the opportunity to approach this question. The metabolic disposition of the biologically inactive hydrolysis product, thromboxane B_2, has been characterized by Roberts et al. using GC/MS (56). Quantitative assays employing this specific methodology for major products of the two major metabolic pathways (Fig. 7), 2,3-dinor-TxB_2 and 11-dehydro-TxB_2, have now been reported (57,58). Experiments with low-dose aspirin and the recovery time for metabolite excretion following cyclooxygenase inhibition suggest that 85-90% of the quantities of both metabolites excreted in human urine derives from platelets under physiological conditions (59). Analysis of such metabolites in plasma would be expected to minimize the problems of platelet activation *ex vivo* which confounds the use of TxB_2 itself as an index of thromboxane formation *in vivo* (60). Thus, we have demonstrated that plasma 11-dehydro-TxB_2 concentrations do not increase significantly in successive samples drawn via an indwelling catheter despite marked increases of magnitude and variance in the concentration of TxB_2 (61). These observations would be consistent with our previous reports of a marked discrepancy between the capacity of platelets to

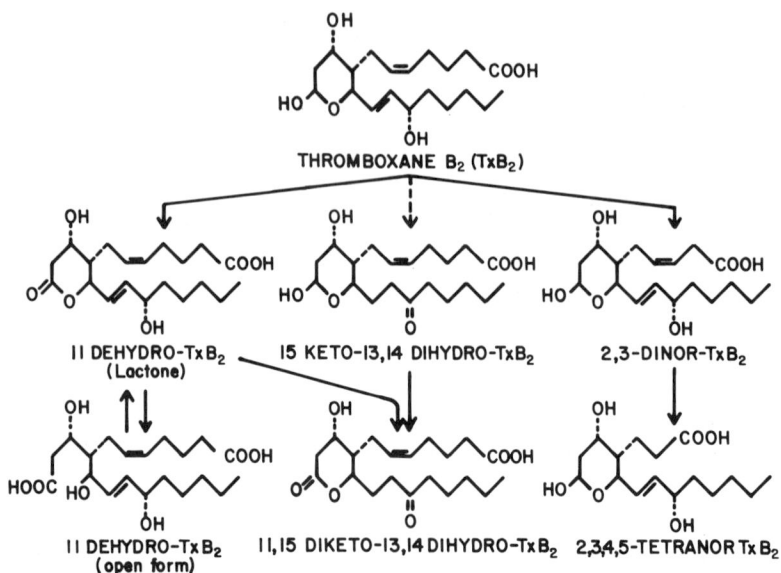

Fig. 7. The major products identified as metabolites of thromboxane B in human urine by L.J. Roberts, II, B.J. Sweetman, J.A. Oates, J. Biol. Chem. 256:8384-8393, 1981. The formation of the 15-keto-13,14-dihydro derivative remains putative. Both the open ring and lactone form of 11-dehydro-TxB$_2$ are measured by the method described in the text.

generate thromboxane A$_2$ and the estimated formation rates under physiological conditions in vivo (62) and the predominant formation of the enzymatic metabolites in tissues such as the liver and lung, rather than in whole blood (61,63). Although attempts to develop radioimmunoassays for 11-dehydro-TxB$_2$ with requisite sensitivity to detect this compound under physiological conditions in plasma have yet to be successful (64,65), qualitatively similar information is usually provided by metabolite analysis in urine, where it is more abundant.

We have found that alteration in excretion of these metabolites reflects platelet activation in a variety of human diseases. For example, patients with independent evidence of platelet activation in vivo and either severe peripheral vascular disease or systemic sclerosis and Raynaud's phenomenon have elevated metabolite excretion (66,67). Metabolite levels in plasma and urine are increased phasically coincident with myocardial ischemia in patients with unstable angina (68); more recently, we have demonstrated that generation of metabolites of thromboxane is markedly increased during coronary thrombolysis with streptokinase and tissue plasminogen activator (69,70). That this reflects platelet activation coincident with the lytic process was supported by our replication of the biochemical events coincident with functional evidence for the importance of platelet activation in a canine model of occlusion-lysis and by the recent evidence of synergy between streptokinase and aspirin in the ISIS-2 trial (71,72).

While excretion of the enzymatic metabolites reflects extrarenal platelet activation, the extent to which they are formed by the kidney *in vivo* is unclear. Preliminary evidence in the perfused rat kidney suggests that renal tissue possesses the enzymatic machinery for generating these compounds, but urinary TxB_2 is most commonly used as a reflection of renal TxA_2 formation. The reasons for supporting the interpretation that urinary TxB_2 largely derives from the kidney have been summarized elsewhere (73) and are mostly indirect. Certainly, increased excretion of TxB_2 has been described following a variety of renal insults in animals and man; these include experimental and human renal lupus (74,75), cyclosporine nephrotoxicity (76), experimental hydronephrosis (77), glycerol-induced renal failure (78), nephrotoxic serum nephritis (79), renal vein constriction (80), and partial renal ablation (81). However, only one study has thus far reported coincidental measurement of TxB_2 and an enzymatic metabolite in urine (74,82). In that case, severe human lupus, the elevation in TxB_2 formation was more pronounced. However, it would be unsurprising in future studies if increases in metabolite excretion, albeit less impressive than in urinary TxB_2, could also be documented in renal disease. Conversely, increased urinary TxB_2, together with 2,3-dinor-TxB_2, has been documented in patients with extrarenal platelet activation and normal renal function (66). Finally, while the renal contribution to urinary concentrations of thromboxane B_2 or its metabolites cannot be precisely defined, the cellular origin of increased renal formation of this eicosanoid *in vivo* must be even more speculative. *In vitro*, PGE_2 is the predominant eicosanoid produced by glomerular cells (83). However, in the presence of excess substrate or following nonspecific stimulation of phospholipases, both glomerular epithelial and mesangial cells synthesize thromboxane A_2, although the vasodilator eicosanoids continue to predominate (84). Similarly, infiltrating platelets and macrophages might substantially augment "renal" thromboxane biosynthesis *in vivo*. This latter mechanism has been speculated to account for the increase in urinary TxB_2 reported in experimental hydronephrosis and in both cardiac and renal allograft rejection (85-87). Despite these observations, it is unlikely that we will ever be able to tie excretion of a particular metabolite of thromboxane to a particular cell within the kidney, except by inference. An example of such an approach would be the study of the turnover time of recovery of elevated urinary TxB$_2$ following aspirin to exclude a role of infiltrating platelets. Less convincingly, a "low" dose of aspirin can be used for a similar purpose; such an approach has been used in experimental lupus nephritis (75). However, we do not know the effects of such doses on other cellular sources of thromboxane *in vivo*; studies of aspirin inhibition of cyclooxygenase from varied cellular sources *in vitro* have provided conflicting information as to excessive sensitivity in the platelet (88,89).

INHIBITION OF THROMBOXANE SYNTHESIS OR ACTION

Theoretically, thromboxane A$_2$ synthesis or action might be modulated by alteration of substrate availability, inhibition of phospholipase action, inhibition of the cyclooxygenase or thromboxane synthase enzymes or blockade of thromboxane receptors. Definitive evidence of altering thromboxane biosynthesis by modulation of phospholipases is currently unavailable; for example, administration of corticosteroids, which have been hypothesized to act via this mechanism, does not alter thromboxane metabolite excretion in

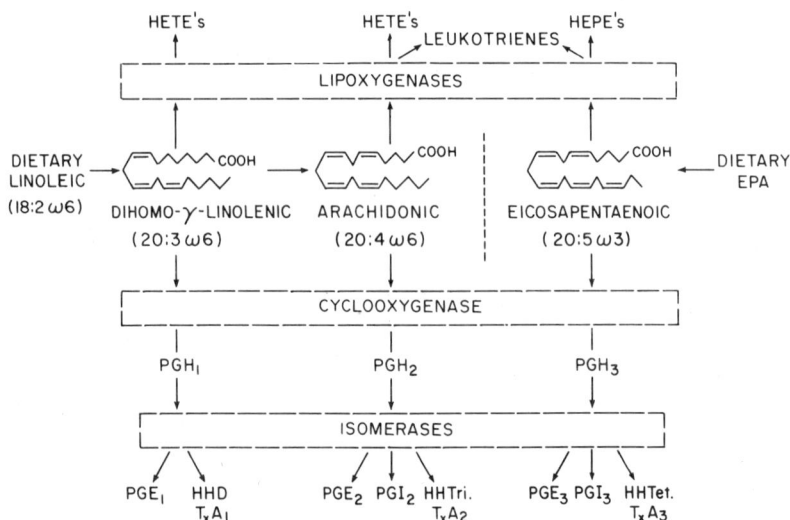

Fig. 8. Formation of mono, bis and trienoic eicosanoid from dihomo-y-linolenic, arachidonic and eicosapentaenoic acids respectively. Analogous compounds may differ in their biological activities dependent upon their degree of unsaturation. (Reproduced with permission from H.R. Knapp and G.A. FitzGerald, Dietary eicosapentaenoic acid and human atherosclerosis, in: Atherosclerosis Reviews, Vol. 13, R.J. Heygeli, ed., Raven Press, New York, pp. 127-143 [1985].)

volunteers, although thromboxane generation by human alveolar macrophages, stimulated *ex vivo*, is inhibited (90,91). No evidence in favor of an effect of steroids on renal thromboxane formation has been provided.

Modulation of Substrate Availability

Certain fish oils contain relatively high quantities of the fatty acids, eicosapentaenoic acid (EPA) and docosahexaenoic acid (DHA), which can be retroconverted to EPA *in vivo* (92). EPA differs from arachidonic acid by the presence of an additional double bond (Fig. 8) and is subject to cyclooxygenation with the resultant formation of trienoic prostaglandins and thromboxane (93). However, by contrast to thromboxane A_2, thromboxane A_3 is virtually biologically inert *in vitro* (94). It has consequently been hypothesized (95) that enriching the diet with EPA at the expense of arachidonate might be of benefit in diseases characterized by platelet activation and excessive smooth muscle constriction. Although preliminary reports of the potential efficacy of fish oil administration in the modulation of experimental cyclosporine nephrotoxicity and on the blood pressure response to salt loading in spontaneously hypertensive rats are encouraging, evidence of benefit from this type of intervention in human renal disease has yet to be reported. Indeed, it seems likely that substrate modification is an inefficient approach to inhibition of thromboxane A_2 biosynthesis. Very high doses (about 10 gm EPA per day) are necessary to reduce platelet formation of this eicosanoid by 85% in man and depressed thromboxane biosynthesis recovers rapidly if the dose is reduced to 2-3 gms per day (93). Due to the nonlinearity of the relationship between inhibition of the platelet capacity to generate thromboxane A_2 and inhibition of actual biosynthesis, it is thought that inhibition of >95% capacity is necessary to modulate significantly thromboxane-dependent platelet activation *in vivo* (96). Indeed, we have recently demonstrated that electrically-induced thrombosis and subsequent therapeutic thrombolysis are associated with a marked increase in both thromboxane A_2 and A_3 formation in dogs fed fish oil at doses that reduce the capacity to form thromboxane A_2 by a mean 90% (97). These observations highlight the importance of documenting the biochemical effects of this type of intervention in the setting of renal disease. To date, no information has been published that addresses the dose response relationship between fish oil administration and renal thromboxane formation in animal models or man. It is, of course, quite possible that fish oil administration might be efficacious in renal disease by mechanisms other than modulation of thromboxane synthesis.

Inhibition of Cyclooxygenase

The renal side effects of nonsteroidal anti-inflammatory drugs have recently been comprehensively reviewed (98). The majority of these derive from inhibition of the biosynthesis of vasodilatory eicosanoids, such as PGI_2 and PGE_2, under circumstances of enhanced vasoconstrictor tone. Under physiological circumstances, such compounds appear to have little influence on renal blood flow (99). Aspirin differs from other cyclooxygenase inhibitors in that it irreversibly inhibits the enzyme (100). Thus, in the platelet, which is incapable of de novo protein synthesis, the effects of repeated doses of aspirin are cumulative until cyclooxygenase inhibition is complete (101), and the synthesis of new platelets is necessary to restore thromboxane biosynthetic capacity.

An objective in the administration of aspirin as a platelet inhibitor has been to achieve inhibition of thromboxane formation without an accompanying diminution in vasodilator eicosanoid formation, so-called biochemical selectivity. This property would be particularly desirable in a setting where the maintenance of renal blood flow had become dependent upon continued synthesis of such compounds. Thus, in patients with renal lupus nephritis, in whom urinary 6-keto-$PGF_{1\alpha}$ and PGE_2 excretion is diminished, ibuprofen, a nonselective cyclooxygenase inhibitor, diminished synthesis of these eicosanoids coincident with thromboxane formation and caused a deterioration of renal function (102). By contrast, administration of sulindac, a compound that selectively inhibited renal thromboxane formation in these patients, did not increase serum creatinine or cause a fall in glomerular filtration rate. The analogous comparative experiments have not been performed with aspirin regimens which have been documented as being respectively selective and nonselective for thromboxane inhibition. Indeed, while it appears that doses of aspirin <100 mg per day are relatively more selective for thromboxane inhibition, this property is not absolute; doses as low as 40 mg per day influence PGI_2 synthesis (103). Absolute biochemical selectivity of aspirin action, with respect to thromboxane formation by platelets, may be achievable by varying rate as well as dose to confine drug action to the presystemic circulation (104,105); but regimens employed in both clinical trials and experimental studies thus far are likely to have had a nonspecific action on eicosanoid formation.

The renal-sparing effect of sulindac reported by Ciabattoni et al. (102) has been ascribed to its acting as a cyclooxygenase inhibitor following metabolism to sulindac sulfoxide. The kidney possesses the capability to back-transform this compound to the prodrug form, thereby diminishing the possibility of intrarenal eicosanoid inhibition. Although this view has been disputed (106,107), studies that integrate both biochemical measurements and indices of renal function in animal models and man lend support for this interpretation, at least at doses of sulindac less than 400 mg per day (102,108). Finally, such strategies are directed at eliminating platelet thromboxane formation selectively. There is no pharmacological strategy available that theoretically could selectively inhibit thromboxane formation by other cellular sources within the kidney.

Thromboxane Synthesis Inhibition

This approach has two theoretical advantages over cyclooxygenase blockade; inhibition of thromboxane formation is selective, and generation of other eicosanoids may be enhanced via rediversion of platelet endoperoxides (109,110; Fig. 9). We have recently demonstrated that endoperoxide rediver-

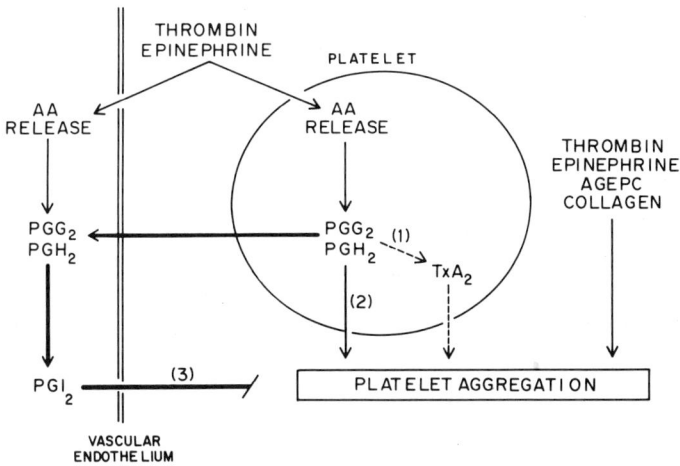

Fig. 9. Following inhibition of thromboxane synthase (1), prostaglandin endoperoxide substrate (PGH$_2$) may accumulate and activate the shared thromboxane PGH$_2$ receptor on platelets (2). Alternatively, platelet PGH$_2$ may be taken up by vascular endothelium and utilized to form prostacyclin (PGI$_2$) by PGI$_2$ synthase. PGI$_2$ inhibits platelet aggregation by all recognized agonists, not just by TxA$_2$ (3). (Reproduced with permission from G.A. FitzGerald, J. Doran and D.M. Fisher, Oxygenase metabolites of arachidonic acid and platelet-vascular interactions, in: Prostaglandins and Cardiovascular Disease, K. Schroer, ed., Springer Verlag, pp. 269-272 [1985].)

sion actually occurs at the platelet-vascular interface in man (111), and second-generation inhibitors are now becoming available that sustain maximal inhibition of thromboxane formation throughout a reasonable dosing interval. There has been evidence of differential sensitivity of platelet and glomerular thromboxane synthase to inhibition by such compounds in the rat (112), but the implications of these observations for the use of thromboxane synthase inhibitors in man is unclear. A remaining theoretical limitation to the action of these drugs has been that accumulated platelet endoperoxides might substitute for the action of the inhibited thromboxane A_2 at their shared receptor (113). Data consistent with this observation have been the synergistic effects observed between synthase inhibitors and receptor antagonists in both the prevention of experimentally induced coronary artery thrombosis (114) and the prolongation of the bleeding time in human volunteers (115). In view of these observations, recent attention has focused upon the development of compounds that possess both pharmacological properties (116).

THROMBOXANE RECEPTOR ANTAGONISTS

A variety of compounds have been synthesized that act as competitive inhibitors of thromboxane agonist-induced responses in platelets and smooth muscle (117-125). Considerable structural disparity exists between these compounds (Fig. 3), several of which have now been administered to man. Some of these drugs have additional properties in addition to thromboxane receptor blockade at high concentrations *in vitro*; these include antagonism of constrictor responses induced by other eicosanoids or serotonin, inhibition of platelet adhesion to foreign surfaces in the presence of cyclooxygenase inhibition and thromboxane synthase inhibition. No compounds have been identified that selectively inhibit thromboxane agonist-induced responses in a smooth muscle preparation but not in the platelet or vice versa.

In animal models, thromboxane antagonists have been shown to prolong dose-dependently the time to electrically induced coronary artery thrombosis (126), to abolish cyclical flow variations in partially obstructed coronary and renal arteries (127,128), to diminish the pulmonary pressor response to endotoxin injection (129), to modulate gastric acid secretion (130), and to synergise with both thromboxane synthase inhibitors and serotonin antagonists in the setting of platelet activation (131,132). Several compounds have been administered to man; the most extensive experience has been with BM13,177, now known as Sulotroban. In general, these compounds have been

well tolerated; occasionally menstrual spotting, excessive prolongation of
the bleeding time and mild reversible elevations of liver enzymes have been
reported. Their safety in pregnant women remains to be established. The
most potent of the compounds presently in man is GR32191. This compound
results in a greater than 2 log-fold shift of the U46619-induced platelet
aggregation dose response curve which persists for up to 6 hours after a
single orally tolerated dose. By contrast, Sulotroban shifts the dose response curve by 1 log unit for approximately 1 hour post dosing in man.
Although clinical investigations employing these compounds are underway in a
variety of settings, few have been reported. However, in view of the marked
increase in thromboxane metabolite excretion following streptokinase administration (133), it is of interest to note the striking diminution in the time
to lysis when Sulotroban was combined with intracoronary streptokinase under
placebo-controlled double blind conditions in man (134). A difficulty with
these drugs is knowing how much of a shift in the platelet dose response
curve ex vivo will be necessary to result in meaningful receptor blockade
in vivo. However, they do offer an approach to the biochemically selective inhibition of thromboxane action and, at the very least, will serve as
extremely useful pharmacological probes for the action of this eicosanoid in
human disease.

THROMBOXANE BIOSYNTHESIS IN RENAL DISEASE

A) Results in Animal Models

Thromboxane A_2 may be formed as a major product of infiltrating cells
within the kidney, such as platelets or macrophages; alternatively, as has
been mentioned, it is formed by isolated renal glomeruli in vitro
(135). Both glomerular epithelial cells and mesangial cells can form thromboxane A_2, although other products such as PGE_2 and PGI_2 tend to predominate (136). Thromboxane agonists are potent renal vasoconstrictors in
vitro (137); and infusion of stable endoperoxide analogues into the canine
renal artery causes a reduction in renal blood flow, glomerular filtration
rate, and electrolyte and water excretion (138). It was thought that the
alterations in salt and water excretion reflected the alterations in renal
blood flow and glomerular filtration rate; however, a direct effect in the
tubule was not excluded. Vasopressin stimulates thromboxane formation by
bladder epithelial cells (139); and the thromboxane agonist, U46619, stimulates water permeability in the urinary bladder of the toad, Bufo
marinus (140).

Evidence for the functional importance of thromboxane biosynthesis in renal pathophysiology was initially provided from animal models. A marked increase in urinary thromboxane B_2 excretion has been noted following renal vein constriction (80) in response to bradykinin stimulation in the rabbit. This coincided with the appearance of macrophages and lymphocytes, together with fibroblast cells in the renal cortex (85). Endotoxin stimulation of macrophage thromboxane formation was demonstrable by renal cortices following 6 hours of renal vein constriction, but not by cortices obtained from unconstricted, contralateral kidneys (85). Other models of renal inflammation associated with exaggerated urinary thromboxane B_2 excretion show a similar histological pattern; these include the ureter-obstructed hydronephrotic kidney, glycerol-induced renal failure and renal venous obstruction (77,78,80). Administration of a thromboxane synthase inhibitor (imidazole) prevents the changes in renal blood flow and GFR induced by ureteric obstruction (141).

Immunologically mediated renal injury has also been associated with altered thromboxane biosynthesis. Nephrotoxic serum nephritis (NSN), produced by administration of heterologous antiglomerular basement membrane (GBM) antibody, is a well-established model of immune nephritis in the rat (142,143). The disease has two stages: an initial heterologous phase, during which complement-dependent polymorphonuclear cells infiltrate during the first three days, followed by an autologous phase production of antibodies to the anti-GBM antibody by infiltrating mononuclear cells. Lianos and Dunn (79) demonstrated that the acute changes (within 3 hours) during the heterologous phase were accompanied by a marked augmentation of glomerular thromboxane formation and were attenuated by the thromboxane synthase inhibitor, OKY-1581. However, during the autologous phase, neither OKY-1581 nor a structurally distinct inhibitor (UK38,485) nor a receptor antagonist (EP092) altered either renal plasma flow or GFR at 14 days (144). By contrast, cyclooxygenase inhibitors markedly depressed both parameters, suggesting a predominant influence of vasodilator eicosanoids, such as PGE_2, at those time points in the model. In a model of immune nephritis produced by injection of bovine serum albumin into New Zealand white rabbits, Saito et al. (145) demonstrated that administration of a thromboxane synthase inhibitor (1-benzyl imidazole) significantly attenuated the proteinuria that develops in this model; GFR was unaffected.

There has been considerable speculation that thromboxane formation within the kidney might influence blood pressure control. The thromboxane agonist, U46619, has been shown to increase norepinephrine efflux dose-dependently from electrically stimulated isolated vas deferens (146), while PGE_2, PGD_2 and, to a lesser extent, PGI_2 reduce norepinephrine release (147,148). It has been postulated that this mechanism may contribute to the attenuation of the rise in blood pressure observed in some, but not all, models of genetic hypertension in rats (149). This effect is delayed beyond when inhibition of platelet thromboxane generation is maximal and has generally been demonstrable at doses that are supramaximal with respect to this effect. Intrarenal thromboxane formation may also interact with the renin-angiotensin system. Thus, the increment in renin secretion following supra-renal aortic clamping in the rat is attenuated by a thromboxane synthase inhibitor (UK38,485), and the diminution of renal blood flow caused by angiotensin II infusion at low perfusion pressures is attenuated by thromboxane synthase inhibition with the same compound (150). Angiotensin II infusion of kidneys obtained from spontaneously hypertensive rats increased renal thromboxane produced (151), and thromboxane formation is also reportedly increased by the contralateral kidney in the 1 kidney, 1 clip, Goldblatt model of hypertension (152). Finally, ablation of >70% of the renal mass in the rat results in hypertension, proteinuria and glomerular sclerosis in the remnant kidney (81). Acute administration of the thromboxane synthase inhibitor increased renal plasma flow and GFR in such animals but not in normal rats or in rats with a remnant kidney that had been pretreated with aspirin. Chronic administration of the same compound increased renal blood flow and GFR, decreased proteinuria and thromboxane excretion, lowered blood pressure and improved renal histology. These phenomena occurred despite persistent evidence of hyperfiltration. The authors speculated that platelet activation and subsequent intraglomerular thrombosis might account for the development of glomerulosclerosis and its attendant hypertension and left ventricular hypertrophy in this model.

Augmented thromboxane production has also been implicated in renal allograft rejection. Thus, rejection was histologically evident three days after transplantation of kidneys from Lewis rats to Brown-Norway recipients (87). The deterioration in *in vivo* inulin clearance significantly correlated with an increment in thromboxane formation by the perfused grafts *ex vivo*, and administration of a thromboxane synthase inhibitor (UK 38485) significantly improved GFR and renal blood flow while reducing urinary thromboxane B_2. Finally, cyclosporine, an agent commonly used in transplant recipients, has been associated with platelet hyperactivity in man (153,154) and exaggerated thromboxane formation *in vivo* in the rat (76). Thromboxane synthase inhibition significantly attenuates the decline in renal function and proteinuria accompanying cyclosporine administration in the rat

(76). Interestingly, the degree of proteinuria in albumin-fed nephrotic rats correlates with augmented glomerular thromboxane production, raising the possibility that the eicosanoid might act relatively nonspecifically as a mediator of protein loss after renal insults (155).

Finally, increased thromboxane formation has been reported in models of murine lupus nephritis. For example, in Kelley et al. (74), intrarenal synthesis of thromboxane B_2 increased as renal function progressively deteriorated in both MRL-1pr and NZBxNZW FI hybrid strains. Inhibition of platelet thromboxane formation by 90% did not retard renal damage; however, for reasons previously mentioned (vide supra), this degree of inhibition does not exclude intrarenal platelets as a source of eicosanoid formation.

B) Results of Clinical Investigations

Less definitive evidence of the functional role of augmented thromboxane formation in man is currently available. With the notable exception of lupus nephritis, urinary thromboxane B_2 excretion is apparently normal in glomerulonephritis (102). It has been hypothesized that platelet activation within glomeruli might contribute to mesangial cell proliferation in a manner analogous to their putative role in atherogenesis (82). In this regard, the report of a prospective, randomized, double blind trial of aspirin (325 mg tid) and dipyridamole (75 mg tid) in patients with Type II membranoproliferative glomerulonephritis is of interest. The rate of decline of renal function and the development of end-stage renal failure was significantly reduced in the active treatment group. However, proteinuria was not influenced by treatment, and data from one-fifth of the patients entering the trial were not subjected to the final analysis (156).

Thromboxane B_2 excretion is markedly increased in patients with lupus nephritis, especially those with histologically severe glomerular disease. By contrast, excretion of the dinor metabolite of thromboxane B_2 is not significantly altered, implying a predominantly renal origin for the augmented formation of this eicosanoid (74). Preliminary evidence suggests that administration of the thromboxane receptor antagonist, sulotorban, to such patients results in a partial, reversible improvement in renal function (157). A recently published (158), uncontrolled, prospective study suggests that aspirin and dipyridamole might retard the deterioration in renal function in patients with diabetic renal disease (nephritis+nephrosis). This hypothesis remains to be rigorously tested. An earlier report of the efficacy of a thromboxane synthase inhibitor in reversing diabetic microalbuminuria (159) could not be replicated (160). We have failed to document abnormal platelet activation in Type I diabetes mellitus *in vivo* (161); whether glomerular thromboxane formation is abnormal in these patients is unknown.

Elevated thromboxane B_2 excretion has also been reported during renal allograft rejection, and in patients with both the hepatorenal syndrome and the hemolytic-uremic syndromes (162,163). The functional importance of these observations is presently unclear. In the case of allograft rejection, the first reports concerned increases of thromboxane B_2 measured by radioimmunoassay in unextracted urine. We have confirmed increments in excretion of this product during or following (but not preceding) some (but not all) cases of clinically apparent rejection, using GC/MS (H.R. Knapp and G.A. FitzGerald, unpublished). Zipser and his colleagues evaluated the functional importance of thromboxane formation in patients with the hepatorenal syndrome using the thromboxane synthase inhibitor, UK38,485 (163). Although the drug had no effect, inhibition of thromboxane B_2 excretion was incomplete; consequently, the results were difficult to interpret. Anecdotal reports have suggested the efficacy of platelet inhibition in the hemolytic uremic syndrome and the related thrombotic thrombocytopenic purpura (164, 165), but a controlled evaluation of this type of intervention has not been performed. Finally, deficient platelet thromboxane formation has been postulated as a contributing factor to the hemostatic defect in chronic renal failure (166,167).

CONCLUSION

Thromboxane A_2 is an evanescent, biologically potent compound that is generated by renal tissues in response to humoral stimuli. Its biological profile suggests that thromboxane may play a role in the regulation of intraglomerular pressure, although other actions, including interactions with the sympathetic and renin-angiotensin systems, are possible. Urinary thromboxane B_2 is a relatively specific marker of renal thromboxane synthesis under physiological conditions *in vivo*, although enzymatic derivatives of this compound may also be formed within the kidney and excreted in urine. In experimental preparations associated with augmented thromboxane synthesis by the kidney, this is often attributable to infiltrating cells, particularly macrophages and platelets. The functional importance of renal thromboxane formation in man is currently the subject of intense investigation.

ACKNOWLEDGEMENTS

Dr. FitzGerald is an Established Investigator of the American Heart Association. Dr. Murray is the recipient of a fellowship from the American Heart Association (Tennessee Affiliate). Dr. Catella is the recipient of a Faculty Department Award from the Pharmaceutical Manufacturers' Association Foundation. This work was supported by a grant (HL30400) from the National Institutes of Health.

REFERENCES

1. P.W. Majerus, S.M. Prescott, S.L. Hofmann, E.J. Neufeld, D.B. Wilson, Uptake and release of arachidonate by platelets, in: "Advances in Prostaglandin, Thromboxane, and Leukotriene Research," Vol. 11, B. Samuelsson, R. Paoletti, and P. Ramwell, eds., Raven Press, New York, pp. 45-52 (1983).

2. J. Bryan Smith, C. Dangelmaier, G. Mauco, Quantitation of arachidonate released during the platelet phosphatidylinositol response to throm-bin, in: "Prostaglandins, Leukotrienes, and Lipoxins. Biochemistry, Mechanism of Action, and Clinical Applications," J.M. Bailey, ed., Plenum Press, New York, pp. 205-211 (1985).

3. A.R. Brash, A.T. Porter, R.L. Maas, Investigation of the selectivity of hydrogen abstraction in the nonenzymatic formation of hydroxyeicosate-traenoic acids and leukotrienes by autoxidation, J. Biol. Chem. 7: 4210-4216 (1986).

4. G.A. FitzGerald, Prostaglandins and related compounds, in: "Cecil Loeb Textbook of Medicine," L.H. Smith, Jr., ed., W.B. Saunders, Philadelphia, (in press) (1988).

5. B. Samuelsson, Leukotrienes: mediators of immediate hypersensitivity and inflammation, Science 220:568-574 (1983).

6. C.N. Serhan, M. Hamberg, B. Samuelsson, Lipoxins: a novel series of bio-logically active compounds, in: "Prostaglandins, Leukotrienes and Lipoxins. Biochemistry, Mechanism of Action, and Clinical Applications," J.M. Bailey, ed., Plenum Press, New York, pp. 3-16 (1985).

7. P. Needleman, J. Turk, B.A. Jakshik et al., Arachidonic acid metabolism, Ann. Rev. Biochem. 55:69-82 (1986).

8. J.R. Vane, Inhibition of prostaglandin synthesis as a mechanism of action for aspirin-like drugs, Nature (New Biology) 231:232-235 (1971).

9. J.B. Smith, A.L. Willis, Aspirin selectivity inhibits prostaglandin

production in human platelets, Nature (New Biology) 231:235-237
(1971).

10. J.W. Burch, N.L. Baenziger, N. Stanford, P.W. Majerus, Sensitivity of
fatty acid cyclooxygenase from human aorta to acetylation by aspirin,
Proc. Natl. Acad. Sci. USA 75:5181-5184 (1978).

11. H.D. Lewis, J.W. Davis, D.G. Archibald et al., Protective effects of aspirin against acute myocardial infarction and death in men with unstable angina, N. Engl. J. Med. 309:396-403 (1983).

12. J.A. Cairns, M. Gent, J. Singer et al., Aspirin, sulfinpyrazone, or both
in unstable angina: Results of a Canadian multicentre trial, N. Engl.
J. Med. 313:1369-1374 (1985).

13. L.J. Roberts, II, R.A. Lewis, K.F. Austen, J.A. Oates, Prostaglandin,
thromboxane, and 12-hydroxy-5,8,10,14-eicosatetraenoic acid production by ionophore-stimulated rat serosal mast cells, Biochim.
Biophys. Acta. 575:185-192 (1979).

14. S.R. Bunting, R.J. Gryglewski, S. Moncada et al., Arterial walls generate from prostaglandin endoperoxides a substance (prostaglandin X)
which relaxes strips of mesenteric and coeliac arteries and inhibits
platelet aggregation, Prostaglandins 12:897-904 (1976).

15. M.E. Gerritsen and C.D. Cheli, Arachidonic acid and prostaglandin endoperoxide metabolism in isolated rabbit and coronary microvessels and
isolated and cultivated coronary microvessel endothelial cells,
J. Clin. Invest. 72:1658-1671 (1984).

16. I.F. Charo, S. Shak, M.A. Karasek, P.M. Davison, I.M. Goldstein, Prostaglandin I_2 is not a major metabolite of arachidonic acid in cultured endothelial cells from human foreskin microvessels, J. Clin.
Invest. 74:914-919 (1984).

17. M. Hamberg, J. Svensson, B. Samuelsson, Thromboxanes - a new group of
biologically active compounds derived from prostaglandin endoperoxides, Proc. Natl. Acad. Sci. USA 72:2994-2998 (1975).

18. G.A. FitzGerald, C. Healy, J. Daugherty, Thromboxane A_2 biosynthesis
in human disease, Fed. Proc. 46:154-158 (1987).

19. J.R. Coleman, P. Humphrey, I. Kennedy, P. Lumley, Prostanoid receptors
- the development of a working classification, Trends Pharmacol. Sci.
59:303-306 (1984).

20. D.E. Mais, R.M. Burch, D.L. Saussy, P.J. Kochel, P.V. Halushka, Binding
of a thromboxane A_2/prostaglandin H_2 receptor antagonist to
washed human platelets, J. Pharm. Exp. Ther. 235:729-734 (1985).

21. N. Liel, D.E. Mais, P.V. Halushka, Binding of a thromboxane A_2/prosta-
glandin H_2 agonist [^{3}H]U46619 to washed human platelets,
Prostaglandins 33:789-797 (1987).

22. P.V. Halushka, J. MacDermot, D. Knapp, T. Eller, D. Saussy, D. Mais, I. Blair, C. Dollery, A novel approach for the study of thromboxane A_2 and prostaglandin H_2 receptors using an ^{125}I-labelled ligand, Biochem. Pharmacol. 34:1165-1170 (1985).

23. D. Mais, D. Knapp, K. Ballard, P. Halushka, N. Hamanaka, Synthesis of thromboxane receptor antagonists with the potential to radiolabel with ^{125}I, Tetrahedron Lett. 25:4207-4210 (1984).

24. G.W. Dorn, II, D. Sens, A. Chaikhouni, D. Mais, P.V. Halushka, Cultured human vascular smooth muscle cells with functional thromboxane A_2 receptors: Measurement of U46619-induced $^{45}Calcium$ efflux, Circulation Res. 60:952-956 (1988).

25. P. Mené and M.J. Dunn, Contractile effects of TxA_2 and endoperoxide analogues on cultured rat glomerular mesangial cells, Am. J. Physiol. 251:F1029-F1035 (1986).

26. K. Hanasaki, K. Nakano, H. Kasai, H. Kurihara, H. Arita, Identification of thromboxane A_2 receptors in cultured vascular endothelial cells of rat aorta, Biochem. Biophys. Res. Comm. 151:1352-1357 (1988).

27. W.L. Smith and A.G. Garcia-Perez, A two-receptor model for the mechanism of action of prostaglandins in the renal collecting tubule, in: "Prostaglandins, Leukotrienes, and Lipoxins. Biochemistry, Mechanism of Action, and Clinical Applications, J.M. Bailey, ed., Plenum Press, New York, pp. 35-45 (1985).

28. D.E. Mais, D.L. Saussy, Jr., A. Chaikhouni, P.J. Kochel, D.R. Knapp, N. Hamanaka, P.V. Halushka, Pharmacologic characterization of human and canine thromboxane A_2/prostaglandin H_2 receptors in platelets and blood vessels: Evidence for different receptors, J. Pharm. Exp. Ther. 233:418-424 (1985).

29. D.L. Saussy, Jr., D.E. Mais, R.M. Burch, P.V. Halushka, Identification of a putative thromboxane A_2/prostaglandin H_2 receptor in human platelet membranes, J. Biol. Chem. 261:3025-3029 (1986).

30. D.E. Mais and P.V. Halushka, Synthesis of an $[^{125}I]$-Azido photoaffinity probe for the human platelet thromboxane A_2/prostaglandin H_2 receptor, J. Labelled Compounds and Radiopharmaceuticals, (in press) (1988).

31. P. Mayeux and P. Halushka, Discovery of a functional thromboxane (Tx) A_2/prostaglandin (PG) H_2 receptor in the human erythroleukemia (HEL) cells, Fed. Proc. 2:A1050 (1988).

32. P. Mené, M. Simonson and M.J. Dunn, Prostaglandins, thromboxane and leukotrienes in the control of mesangial function, in: "Renal Eicosanoids", M.J. Dunn, ed., Plenum Press, New York (1989).

33. S.S. Bhagwat, P.R. Hamann, W.E. Still, S. Bunting, F.A. Fitzpatrick,

Synthesis and structure of the platelet aggregation factor, thromboxane A_2, Nature 315:511-513 (1985).

34. G.A. FitzGerald, I.A.G. Reilly, A.K. Pedersen, The biochemical pharmacology of thromboxane synthase inhibition in man, Circulation 72: 1194-1201 (1985).

35. J.B. Smyth, A. Yanagisawa, R. Zipkin, A.M. Lefer, Constriction of cat coronary arteries by synthetic thromboxane A_2 and its antagonism, Prostaglandins 33:777-782 (1987).

36. A.M. Lefer, E.F. Smith, III, H. Araki, J.R. Smith, D. Aharony, D.A. Claremon, R.I. Magolda, K.C. Nicolaou, Dissociation of vasoconstrictor and platelet aggregatory activities of thromboxane by carbocyclic thromboxane A_2, a stable analog of thromboxane A_2, Proc. Natl. Acad. Sci. USA 77:1706-1710 (1980).

37. R.A. Armstrong, R.L. Jones, V. Peesapati, S.G. Will, N.H. Wilson, Competitive antagonism at thromboxane receptors in human platelets, Br. J. Pharmac. 84:595-607 (1985).

38. F.A. Fitzpatrick and R.R. Gorman, Regulatory role of cyclic adenosin 3',5'-monophosphate on the platelet cyclooxygenase and platelet function, Biochim. Biophys. Acta. 582:44-58 (1979).

39. P.V. Avdonin, I.V. Svitina-Ulitina, V.L. Leytin, V.A. Tkachuk, Interaction of stable prostaglandin endoperoxide analogs U46619 and U44069 with human platelet membranes: Coupling of receptors with high-affinity GTPase and adenylate cyclase, Thromb. Res. 40:101-112 (1985).

40. K.C. Nicolaou, R.L. Magolda, J.B. Smith, J.B. Aharony, E.F. Smith, A.M. Lefer, Synthesis and biological properties of pinane-thromboxane A_2, a selective inhibitor of coronary artery constriction, platelet aggregation, and thromboxane formation, Proc. Natl. Acad. Sci. USA 76:2566-2570 (1979).

41. R.A. Armstrong, R.L. Jones, J. MacDermot, N.H. Wilson, Prostaglandin endoperoxide analogues which are both thromboxane receptor antagonists and prostacyclin mimetics, Br. J. Pharmac. 87:543-551 (1986).

42. J.P. Rybicki, D.L. Venton, G.C. Breton, The thromboxane antagonist, 13-azaprostanoic acid, inhibits arachidonic acid-induced Ca^{2+} release from isolated platelet membrane vesicles, Biochim. Biophys. Acta. 751:66-73 (1983).

43. K. Yasuhiro, J. Yamanish, Y. Furuta, K. Kaibuchi, Y. Takai, H. Fukuzaki, Elevation of cytoplasmic free calcium concentration by stable thromboxane A_2 analogue in human platelets, Biochem. Biophys. Res. Comm. 117:663-669 (1983).

44. R.M. Lyons, N. Stanfrod, P.W. Majerus, Thrombin-induced protein phosphorylation in human platelets, J. Clin. Invest. 56:924-936 (1975).

45. K. Sano, Y. Takai, J. Yamanishi, Y. Nishizuka, A role of calcium-activated phospholipid-dependent protein kinase in human platelet activation. Comparison of thrombin and collagen actions, J. Biol. Chem. 258:2010-2013 (1983).

46. W. Siess, B. Boehlig, P.C. Weber, E.G. Lapetina, Prostaglandin endoperoxide analogues stimulate phospholipase C and protein phosphorylation during platelet shape change, Blood 65:1141-1148 (1985).

47. Y. Kawahara, H. Fukuzaki, K. Kaibuchi, T. Tsuda, M. Hoshijima, A.Y. Takai, Activation of protein kinase C by the action of 9,11-epithio-11,12-methano-thromboxane A_2 (STA$_2$), a stable analogue of thromboxane A_2, in human platelets, Throm. Res. 41:811-818 (1986).

48. O.V. Miller, R.A. Johnson, R.R. Gorman, Inhibition of PGE$_1$-stimulated cAMP accumulation in human platelets by thromboxane A_2, Prostaglandins 13:599-609 (1977).

49. O.V. Miller and R.R. Gorman, Modulation of platelet cyclic nucleotide content by PGE$_1$ and the prostaglandin endoperoxide PGG$_2$, J. Cyclic Nucleotide Res. 2:79-87 (1976).

50. J. Nowak and G.A. FitzGerald, Reorientation of prostaglandin endoperoxide metabolism at the platelet-vascular interface in man, Clin. Res. 35:580A (1987).

51. L.G. Carmo, M. Hatmi, D. Rotilio, B.B. Vargaftig, Platelet desensitization induced by arachidonic acid is not due to cyclo-oxygenase inactivation and involves the endoperoxide receptor, Br. J. Pharmac. 85:849-859 (1985).

52. R. Murray, F. Catella, G.A. FitzGerald, Desensitization of TxA$_2$/PGH$_2$ receptor in human platelets, Circulation 76:iv-483 (1987).

53. R. Murray and G.A. FitzGerald, Regulation of thromboxane receptor activation in human platelets, Proc. Natl. Acad. Sci. (USA) 86:124-128 (1989).

54. D.R. Sibley, R.H. Strasser, M.G. Caron, R.J. Lefkowitz, Homologous desensitization of adenylate cyclase is associated with phosphorylation of the ß-adrenergic receptor, J. Biol. Chem. 260:3883-3886 (1985).

55. L.F. Brass, C.C. Shaller, E.J. Belmonte, Inositol 1,4,5-triphosphate induced granule secretion in platelets. Evidence that the activation of phospholipase C mediated by platelet thromboxane receptors involves a guanine nucleotide binding protein-dependent mechanism distinct from that of thrombin, J. Clin. Invest. 79:1269-1275 (1987).

56. L.J. Roberts, II, B.J. Sweetman, J.A. Oates, Metabolism of thromboxane B$_2$ in man: Identification of twenty urinary metabolites, J. Biol. Chem. 256:8384-8393 (1981).

57. J.L. Lawson, A.R. Brash, J. Doran, G.A. FitzGerald, Measurement of uri-

nary 2,3-dinor-thromboxane B_2 and thromboxane B_2 using bonded-phase phenylboronic acid columns and capillary gas chromatography-negative-ion chemical ionization mass spectrometry, Analyt. Biochem. 150:463-470 (1985).

58. J. Lawson, C. Patrono, G. Ciabattoni, G.A. FitzGerald, Long-lived enzymatic metabolites of thromboxane B_2 in human circulation, Analytical Biochem. 155:198-205 (1986).

59. F. Catella and G.A. FitzGerald, Paired analysis of urinary thromboxane B_2 metabolites in humans, Throm. Res. 47:647-656 (1987).

60. G.A. FitzGerald, A.K. Pedersen, C. Patrono, Analysis of prostacyclin and thromboxane A_2 biosynthesis in cardiovascular disease, (Editorial), Circulation 67:1174-1177 (1983).

61. F. Catella, D. Healy, J. Lawson, G.A. FitzGerald, 11-dehydro-thromboxane B_2: An index of thromboxane formation in the human circulation, Proc. Natl. Acad. Sci. 83:5861-5865 (1986).

62. C. Patrono, G. Ciabattoni, F. Pugliese, A. Pierucci, I.A. Blair, G.A. FitzGerald, Estimated rate of thromboxane secretion into the circulation of normal man, J. Clin. Invest. 77:590-594 (1986).

63. E. Granstrom, P. Westlund, M. Kumlin, A. Nordenstrom, Measurement of thromboxane production in vivo: Metabolic and analytical aspects, Adv. Prostaglandin. Thromboxane and Leukotriene Res. 15:67-70 (1985).

64. M. Kumlin and E. Granstrom, Radioimmunoassay for 11-dehydro-TxB$_2$: A method for monitoring thromboxane production in vivo, Prostaglandins 32:741-767 (1986).

65. G. Ciabattoni, J. Maclouf, F. Catella, G.A. FitzGerald, C. Patrono, Radioimmunoassay of 11-dehydro-thromboxane B_2 in human plasma and urine, Biochem. Biophys. Acta. 918:293-297 (1987).

66. I.A.G. Reilly, J. Doran, B. Smith, G.A. FitzGerald, Increased thromboxane biosynthesis in a human model of platelet activation: Biochemical and functional consequences of selective inhibition of thromboxane synthase, Circulation 73:1300-1309 (1986).

67. I.A.G. Reilly, L. Roy, G.A. FitzGerald, Thromboxane biosynthesis is increased in systemic sclerosis with Raynaud's phenomenon, Br. Med. J. 292:1087-1089 (1986).

68. D.J. Fitzgerald, L. Roy, F. Catella, G.A. FitzGerald, Platelet activation in unstable coronary disease, N. Engl. J. Med. 315:983-989 (1986).

69. D.J. Fitzgerald, F. Catella, L. Roy, G.A. FitzGerald, Marked platelet activation in vivo following intravenous streptokinase in acute myocardial infarction, Circulation 77:142-150 (1988).

70. D.M. Kerins, L. Roy, G.A. FitzGerald, D.J. Fitzgerald, Evidence of

platelet activation during coronary thrombolysis with tissue plasminogen activator in man, <u>Clin. Res</u>. 36:4A (1988).

71. ISIS Collaborative Group, Results of a large randomized trial of intravenous streptokinase and oral aspirin in acute myocardial infarction,
<u>JACC</u> 11:332A (1988).

72. G.A. FitzGerald and D.J. Fitzgerald, Modulation of thromboxane A_2 formation in coronary thrombosis and thrombolysis, <u>JACC</u> (in press)
(1988).

73. C. Patrono, G. Ciabattoni, P. Patrignani, P. Filabozzi, E. Pinca, M.A.
Satta, D. Van Dorne, G.A. Cinotti, F. Pugliese, A. Pierucci, B.M.
Simonetti, Evidence for a renal origin of urinary thromboxane B_2 in
health and disease, <u>in</u>: "Advances in Prostaglandin, Thromboxane, and
Leukotriene Research," Vol. II, B. Samuelsson, R. Paoletti and P.
Ramwell, eds., Raven Press, New York, pp. 493-498 (1983).

74. V.E. Kelley, S. Sneve, S. Musinski, Increased renal thromboxane production in murine lupus nephritis, <u>J. Clin. Invest</u>. 77:252-259 (1986).

75. C. Patrono, G. Ciabattoni, G. Remuzzi et al., Functional significance of
renal prostacyclin and thromboxane A_2 production in patients with
systemic lupus erythematosus, <u>J. Clin. Invest</u>. 76:1011-1018 (1985).

76. T.M. Coffman, D.R. Carr, W.E. Varger, P.E. Klotman, Evidence that renal
prostaglandin and thromboxane production is stimulated in chronic cyclosporine nephrotoxicity, <u>Transplantation</u> 43:282-285 (1987).

77. A.R. Morrison, K. Nishikawa, P. Needleman, Thromboxane A_2 biosynthesis
in the ureter obstructed isolated perfused kidney of the rabbit, <u>J.
Pharmacol. Exp. Ther</u>. 205:1-8 (1978).

78. J. Benabe, S. Klahr, M.D. Hoffman et al., Production of thromboxane
A_2 by the kidney in glycerol-induced acute renal failure,
<u>Prostaglandins</u> 19:333-367 (1980).

79. E.A. Lianos, G.A. Andres, M.J. Dunn, Glomerular prostaglandin and thromboxane synthesis in rat nephrotoxic serum nephritis, <u>J. Clin. Invest</u>.
72:1439-1448 (1983).

80. R. Zipser, S. Myers, P. Needleman, Exaggerated prostaglandin and thromboxane synthesis in the renal vein-constricted rabbit, <u>Circ. Res</u>. 47:
231-237 (1980).

81. M.L. Purkerson, J.H. Joist, J. Yates, A. Valdes, A. Morrison, S. Klahr,
Inhibition of thromboxane synthesis ameliorates the progressive kid-
ney disease of rats with subtotal renal ablation, <u>Proc. Natl. Acad.
Sci. USA</u> 82:193-197 (1985).

82. C. Patrono and A. Pierucci, Renal effects of nonsteroidal anti-inflammatory drugs in chronic glomerular disease, <u>Am. J. Med</u>. 81 (Suppl.2B):
71-83 (1986).

83. D. Schlondorff, Renal prostaglandin synthesis. Sites of production and specific actions of prostaglandins, Am. J. Med. 81:1-11 (1986).

84. L. Scharschmidt, M. Simonson, M.J. Dunn, Glomerular prostaglandins, angiotensin II, and nonsteroidal anti-inflammatory drugs, Am. J. Med. 81:31-42 (1986).

85. D. Schwartz, K. DeSchryver-Kecskemeti, P. Needleman, Renal arachidonic acid metabolism and cellular changes in the rabbit renal vein-constricted kidney: Inflammation as a common process in renal injury models, Prostaglandins 27:605-613 (1984).

86. M.L. Foegh, B. Khirabadi, P.W. Ramwell, Prolongation of experimental cardiac allograft survival with thromboxane-related drugs, Transplantation 40:124-129 (1985).

87. T.M. Coffman, W.E. Yarger, P.E. Klotman, Functional role of thromboxane production by acutely rejecting renal allografts in rats, J. Clin. Invest. 75:1242-1248 (1985).

88. N.L. Baenziger, M.J. Dillender, P.W. Majerus, Cultured human skin fibroblasts and arterial cells produce a labile platelet inhibitory prostaglandin, Biochem. Biophys. Res. Comm. 78:294-301 (1977).

89. J.W. Burch, N.L. Baenziger, N. Stanford, P.W. Majerus, Sensitivity of fatty acid cyclooxygenase from human aorta to acetylation by aspirin, Proc. Natl. Acad. Sci. USA 75:5181-5184 (1978).

90. R.J. Sebaldt, J.A. Oates, G.A. FitzGerald, Eicosanoid biosynthesis and leukotriene B_4 release during steroid administration in man, Clin. Res. 35:387A (1987).

91. R.J. Sebaldt, J.R. Sheller, J.A. Oates, G.A. FitzGerald, Effects of high-dose glucocorticosteroid administration on eicosanoid biosynthesis by human alveolar macrophages, Clin. Res. 569A (1988).

92. S. Fischer, A. Vischer, V. Preac-Mursic, P.C. Weber, Dietary docosahexaenoic acid is retroconverted in man to eicosapentaenoic acid, which can be quickly transformed to prostaglandin I_3, Prostaglandins 34: 367-375 (1987).

93. H.R. Knapp, I.A.G. Reilly, P. Alessandrini, G.A. FitzGerald, In vivo indexes of platelet and vascular function during fish-oil administration in patients with atherosclerosis, N. Engl. J. Med. 314:937-942 (1986).

94. P. Needleman, A. Raz, M.S. Minkes, J.A. Ferrendelli, H. Sprecher, Triene prostaglandins: prostacyclin and thromboxane biosynthesis and unique biological properties, Proc. Natl. Acad. Sci. USA 176:944-948 (1979).

95. A. Leaf and P.C. Weber, Cardiovascular effects of n-3 fatty acids, N. Engl. J. Med. 318:549-556 (1988).

96. I.A.G. Reilly and G.A. FitzGerald, Inhibition of thromboxane formation

in vivo and *ex vivo*: Implications for therapy with platelet inhibitory drugs, <u>Blood</u> 69:180-186 (1987).

97. G.A. Braden, D.J. Fitzgerald, H.R. Knapp, G.A. FitzGerald, Increased thromboxane $(Tx)A_2$ biosynthesis during coronary thrombosis and thrombolysis with n-3 fatty acid (FA) supplementation, <u>Circulation</u> 78:11-120 (1988).

98. D.M. Clive and J.S. Stoff, Renal syndromes associated with nonsteroidal anti-inflammatory drugs, <u>N. Engl. J. Med.</u> 310:563-572 (1984).

99. J.R. Sedor, E.W. Davidson, M.J. Dunn, Effects of nonsteroidal anti-inflammatory drugs in healthy subjects, <u>Am. J. Med.</u> 81(Suppl.2B): 59-70 (1986).

100. J.W. Burch, N. Stanford, P.W. Majerus, Inhibition of platelet prostaglandin synthetase by oral aspirin, <u>J. Clin. Invest.</u> 61:314-319 (1978).

101. P. Patrignani, P. Filabozzi, C. Patrono, Selective cumulative inhibition of platelet thromboxane production by low-dose aspirin in healthy subjects, <u>J. Clin. Invest.</u> 69:1366-1372 (1982).

102. G. Ciabattoni, G.A. Cinotti, A. Pierucci et al., Effects of sulindac and ibuprofen in patients with chronic glomerular disease. Evidence for the dependence of renal function on prostacyclin, <u>N. Engl. J. Med.</u> 310:279-283 (1984).

103. G.A. FitzGerald, J.A. Oates, J. Hawiger, R.L. Maas, L.J. Roberts, A.R. Brash, Endogenous synthesis of prostacyclin and thromboxane and platelet function during chronic aspirin administration in man, <u>J. Clin. Invest.</u> 71:676-688 (1983).

104. A.K. Pedersen and G.A. FitzGerald, Dose-related pharmacokinetics of aspirin: Presystemic acetylation of platelet cyclooxygenase in man, <u>N. Engl. J. Med.</u> 311:1206-1211 (1984).

105. A.K. Pedersen, J. Nowak, G.A. FitzGerald, Slow administration of low-dose aspirin: Enhanced inhibition of platelet cyclooxygenase, <u>Circulation</u> 72:772A (1985).

106. G. Laffi, G. Daskalopoulos, I. Kronborg, W. Hsueh, P. Gentalini, R.D. Zipser, Effects of sulindac and ibuprofen in patients with cirrhosis and ascites. An explanation for the renal-sparing effect of sulindac, <u>Gastroenterology</u> 90:182-187 (1986).

107. D.C. Brater, S. Anderson, B. Baird, W.B. Campell, Effects of ibuprofen, naproxen, and sulindac on prostaglandins in men, <u>Kidney Int.</u> 27:66-73 (1985).

108. L-O. Eriksson, B. Beermann, M. Kallner, Renal function and tubular transport effects of sulindac and naproxen in chronic heart failure, <u>Clin. Pharmacol. Ther.</u> 42:646-654 (1987).

109. A.J. Marcus, B.B. Weksler, E.A. Jaffe, M.J. Broekman, Synthesis of pros-
     tacyclin from platelet-derived endoperoxides by cultured human endo-
     thelial cells, J. Clin. Invest. 66:979-985 (1980).

110. A.I. Schafer, D.D. Crawford, M.A. Gimbrone, Unidirectional transfer of
     prostaglandin endoperoxides between platelets and endothelial cells,
     J. Clin. Invest. 73:1105-1111 (1984).

111. J. Nowak and G.A. FitzGerald, Prostaglandin endoperoxide reorientation
     at platelet vascular interface in man, J. Clin. Invest., (in press)
     (1988).

112. P. Patrignani, P. Filabozzi, F. Catella, F. Pugliese, C. Patrono, Dif-
     ferential effects of dazoxiben, a selective thromboxane synthase in-
     hibitor, on synthase inhibitor, on platelet and renal prostaglandin-
     endoperoxide metabolism, J. Pharm. Exp. Ther. 228:472-477 (1984).

113. V. Bertele and G. De Gaetano, Potentiation by dazoxiben, a thromboxane
     synthetase inhibitor, of platelet aggregation inhibitory activity of
     a thromboxane receptor antagonist and of prostacyclin, Eur. J.
     Pharmacol. 85:331 (1982).

114. G.A. FitzGerald, J. Fragetta, D.J. Fitzgerald, Thromboxane (Tx) syn-
     thase inhibition and TxA_2/endoperoxide receptor antagonism in a
     chronic canine model of coronary thrombosis: Effects on TxA_2 and
     prostacyclin (PGI_2) biosynthesis, Adv. Prostaglandin. Thromboxane
     and Leukotriene Res. 17:496-500 (1987).

115. P. Gresele, J. Arnout, H. Deckmyn, J. Vermylen, Endogenous anti-aggre-
     gatory prostaglandins can contribute to inhibition of hemostasis: A
     pharmacological study in vivo in humans, in: "Advances in
     Prostaglandin, Thromboxane, and Leukotriene Research," Vol. 17, B.
     Samuelsson, R. Paoletti and P.W. Ramwell, eds., Raven Press, New
     York, pp. 248-253 (1987).

116. J. Van Reempts, B. Van Deuren, M. Borgers, F. De Clerck, R 68 070, a
     combined TxA_2-synthetase/TxA_2-prostaglandin endoperoxide recep-
     tor inhibitor, reduces cerebral infarct size after photochemically
     initiated thrombosis in spontaneously hypertensive rats, Thromb.
     Haemost. 58(1):182, 671A (1987).

117. R.A. Armstrong, R.L. Jones, V. Peesapati, S.G. Will, N.H. Wilson, Compe-
     titive antagonism at thromboxane receptors in human platelet, Br. J.
     Pharmac. 84:595-607 (1985).

118. R.A. Coleman, I. Kennedy, R.L.G. Sheldrick, AH 6809, a prostanoid EP_1
     receptor-blocking drug, Br. J. Pharmac. Proc. Suppl. 85:273P (1985).

119. D.N. Harris, R. Greenberg, M.B. Phillips, I.M. Michel, H.J. Goldenberg,
     M.F. Haslanger, T.E. Steinbacher, Effects of SQ 27,427, a thrombox-
     ane A_2 receptor antagonist in the human platelet and isolated

smooth muscle, _Eur. J. Pharmacol_. 103:9-18 (1984).

120. D.N. Harris, M.B. Phillips, I.M. Michel, H.J. Goldenberg, J.E. Heikes, P.W. Sprague, M.J. Antonaccio, 9-Homo-9,11-epoxy, 5,13-prostadienoic acid analogues: Specific stable agonist (SQ 26,538) and antagonist (SQ 26,536) of the human platelet thromboxane receptor, _Prostaglandins_ 22:295-307 (1981).

121. M.L. Ogletree, D.N. Harris, R. Greenberg, M.F. Haslanger, M. Nakane, Pharmacological actions of SQ 29,548, a novel selective thromboxane antagonist, _J. Pharmacol. Exp. Ther_. 234:435-441 (1985).

122. Huzoor-Akbar, A. Mukhopadhyay, K.S. Anderson, S.S. Navran, K. Romstedt, D.D. Miller, D.R. Feller, Antagonism of prostaglandin-mediated responses in platelets and vascular smooth muscle by 13-azaprostanoic acid analogs, _Biomed. Pharmacol_. 34:641-647 (1985).

123. G.C. Le Breton, J.P. Lipowski, H. Feinberg, D.L. Venton, T. Ho, K.K. Wu, Antagonism of thromboxane A_2/prostaglandin H_2 by 13-azaprostanoic acid prevents platelet deposition to the de-endothelialized rabbit aorta _in vivo_, _J. Pharmacol. Exp. Ther_. 229:80-84 (1984).

124. M. Fujioka, T. Nagao, H. Kuriyama, Actions of the novel thromboxane A_2 antagonists, ONO-1270 and ONO-3708, on smooth muscle cells of the guinea-pig basilar artery, _Archives of Pharmacol_. 334:468-474 (1986).

125. M. Katsura, T. Miyamoto, N. Hamanaka, K. Kondo, T. Terada, Y. Ohgaki, A. Kawasaki, M. Tsuboshima, _In vitro_ and _in vivo_ effects of new powerful thromboxane antagonists (3-alkylamino pinane derivatives), _in_: "Advances in Prostaglandin, Thromboxane and Leukotriene Research," Vol. 11, B. Samuelsson, R. Paoletti and P. Ramwell, eds., Raven Press, New York, pp. 351-357 (1983).

126. D.J. Fitzgerald, J. Doran, E.K. Jackson, G.A. FitzGerald, Coronary vascular occlusion mediated through thromboxane A_2-prostaglandin endoperoxide receptor activation _in vivo_, _J. Clin. Invest_. 77:496-502 (1986).

127. W.A. Schumacher, H.J. Goldenberg, D.N. Harris, M.L. Ogletree, Effect of thromboxane receptor antagonists on renal artery thrombosis in the cynomolgus monkey, _J. Pharmacol. Exp. Ther_. 242:460-466 (1987).

128. J.H. Ashton, M.L. Ogletree, I.M. Michel, P. Golino, J.M. McNatt, A.L. Taylor, S. Raheja, J. Schmitz, L.M. Buja, W.B. Campbell, J.T. Willerson, Cooperative mediation by serotonin S_2 and thromboxane A_2/prostaglandin H_2 receptor activation of cyclic flow variations in dogs with severe coronary artery stenoses, _Circulation_ 76: 952-959 (1987).

129. P.G. Kuehl, J.M. Bolds, J.E. Lloyd, J. Snapper, G.A. FitzGerald, Throm-

boxane A_2/prostaglandin endoperoxide activation mediates bronchial and hemodynamic responses to endotoxemia in the conscious sheep, Am. J. Physiol. 254 (Regulatory Integrative Comp. Physiol. 23):R310-R319 (1988).

130. J.J. Reeves and R. Stables, Thromboxane receptors can modulate gastric acid secretion in the rat, Prostaglandins 34:829-840 (1987).

131. P. Golino, J.H. Ashton, P. Glas-Greenwalt, J. McNatt, L.M. Buja, J.T. Willerson, Mediation of reocclusion by thromboxane A_2 and serotonin after thrombolysis with tissue-type plasminogen activator in a canine preparation of coronary thrombosis, Circulation 77:676-684 (1988).

132. D.J. Fitzgerald, J. Fragetta, G.A. FitzGerald, Prostaglandin endoperoxides modulate the response to thromboxane synthase inhibition during coronary thrombosis, J. Clin. Invest., (submitted) (1988).

133. D.J. Fitzgerald, F. Catella, L. Roy, G.A. FitzGerald, Marked platelet activation in vivo following intravenous streptokinase in acute myocardial infarction, Circulation 77:142-150 (1988).

134. Ettie et al, (submitted) (1988).

135. D. Schlondorff, The glomerular mesangial cell: An expanding role for a specialized pericyte, FASEB J. 1:272-281 (1987).

136. J. Sraer, W. Siess, L. Moulonguet-Dolerias, J-P. Oudinet, F. Dray, R. Ardaillou, In vitro prostaglandin synthesis by various rat renal preparations, Biochim. Biophys. Acta. 710:45-52 (1982).

137. J.G. Gerber, E. Ellis, J. Hollifield, A.S. Nies, Effect of prostaglandin endoperoxide analogue on canine renal function, hemodynamics and renin release, Eur. J. Pharmacol. 53:239-246 (1979).

138. R. Loutzenhiser, M. Epstein, C. Horton, P. Sonke, Reversal of renal and smooth muscle actions of the .....Am. J. Physiol. 250:F619-F626 (1986).

139. R.M. Burch and P.V. Halushka, Thromboxane and stable prostaglandin endoperoxide analogs stimulate water permeability in the toad urinary bladder, J. Clin. Invest. 66:1251-1257 (1980).

140. R.M. Burch and P.V. Halushka, Alterations in extracellular calcium concentration differentially influence prostaglandin and thromboxane synthesis in epithelial cells from the toad urinary bladder, Archives Biochem. Biophys. 229:90-97 (1984).

141. F. Goto, E.K. Jackson, A. Ohnishi, W. Herzer, R.A. Branch, Effect of cyclooxygenase and thromboxane synthase inhibition on the response to angiotensin II in the hypoperfused canine kidney, J. Pharmacol. Exp. Ther. 242:799-803 (1987).

142. N.M. Thomson, S.R. Holdsworth, E.F. Glasgow, D.K. Peters, R.C. Atkins,

Mechanisms of injury in experimental glomerulonephritis, in: "Progress in Glomerulonephritis," P. Kincaid-Smith, Aprice and R.C. Atkins, eds., John Wiley and Sons, New York, pp. 51-72 (1979).

143. E.R. Unanue and F.J. Dixon, Experimental glomerulonephritis, VI. The autologous phase of nephrotoxic nephritis, J. Exp. Med. 121:715-725 (1965).

144. J.E. Stork and M.J. Dunn, Hemodynamic roles of thromboxane A_2 and prostaglandin E_2 in glomerulonephritis, J. Pharmacol. Exp. Ther. 233:672-678 (1985).

145. H. Saito, T. Ideura, J. Takeuchi, Effects of a selective thromboxane A_2 synthetase inhibitor on immune complex glomerulonephritis, Nephron 36:38-45 (1984).

146. G.J. Trachte, Thromboxane agonist (U46619) potentiates norepinephrine efflux from adrenergic nerves, J. Pharmacol. Exp. Ther. 237:473-477 (1986).

147. P. Hedqvist, Control by prostaglandin E_2 of sympathetic neurotransmission in the spleen, Life Sci. 9:269-278 (1970).

148. P. Hedqvist, Basic mechanisms of prostaglandin action on autonomic neurotransmission, Ann. Rev. Pharmacol. Toxicol. 17:259-279 (1977).

149. E.K. Jackson, H.D. Uderman, W.A. Herzer, R.A. Branch, Attenuation of noradrenergic neurotransmission by the thromboxane synthetase inhibitor, UK38,485, Life Sci. 35:221-228 (1984).

150. E.K. Jackson, F. Goto, H.D. Uderman, R.J. Workman, R.A. Branch, Inhibition of thromboxane A_2 (TxA_2) synthetase attenuates the renin secretion response to suprarenal aortic clamping, but not to sodium arachidonate, Proceedings of the SCOR-Hypertension Symposium, Sept. 23-24, Cleveland, Ohio (1983).

151. R.A. Branch, F. Goto, W. Ohnishi, W. Herzer, E.K. Jackson, Renal thromboxane A_2 release stimulated by angiotensin II in the hypoperfused canine kidney, in: "Advances in Prostaglandin, Thromboxane, and Leukotriene Research," Vol. 17, B. Samuelsson, R. Paoletti and P.W. Ramwell, eds., Raven Press, New York, pp. 749-756 (1987).

152. S.I. Himmelstein, W.E. Yarger, P.E. Klotman, Altered eicosanoid production by the contralateral kidney in the two-kidney, one clip Goldblatt hypertensive rat, J. Hyperten. 4(suppl.5):S165-S167 (1986).

153. Y. Vanrenterghem, T. Lerut, L. Roels, J. Gruwetz, P. Michielsen, Thromboembolic complications and haemostatic changes in cyclosporin-treated cadaveric kidney allograft recipients, Lancet v:8436-1002 (1985).

154. H. Cohen, G.H. Nield, R. Patel, I.J. Mackie, Machin and S.J. Machin,

Evidence for chronic platelet hyperaggregability and *in vivo* activation in cyclosporin-treated renal allograft recipients, Thromb. Res. 49:91-101 (1988).

155. G. Remuzzi, A. Benigni, N. Perico, Renal prostaglandins and hyperfiltration, in: "Advances in Prostaglandin, Thromboxane, and Leukotriene Research," Vol. 17, B. Samuelsson, R. Paoletti and P.W. Ramwell, eds., Raven Press, New York, pp. 719-724 (1987).

156. J.V. Donadio, C.F. Anderson, J.C. Mitchell et al., Membrano-proliferative glomerulonephritis. A prospective clinical trial of platelet-inhibitor therapy, N. Engl. J. Med. 310:1421-1426 (1984).

157. A. Pierucci, B.M. Simonetti, G. Pecci, G. Mavrikakis, S. Feriozzi, G.A. Cinotti, G. Ciabattoni, C. Patrono, Acute effects of a thromboxane receptor antagonist on renal function in patients with lupus nephritis, Kidney Int. 31:283A (1987).

158. J.V. Donadio, Jr., D.M. Ilstrup, K.E. Holley, J.C. Romero, Platelet-inhibitor treatment of diabetic nephropathy: a 10-year prospective study, Mayo Clin. Proc. 63:3-15 (1988).

159. A.H. Barnett, K. Wakelin, B.A. Leatherdale, J.R. Britton, A. Polak, J. Bennett, M. Toop, D. Rowe, K. Dallinger, Specific thromboxane synthetase inhibition and albumin excretion rate in insulin-dependent diabetes, Lancet 1:1322 (1984).

160. H. Tyler, Pfizer, Inc., Personal communication (1985).

161. P. Alessandrini, J. McRae, S.S. Feman, G.A. FitzGerald, Thromboxane biosynthesis and platelet function in endogenous biosynthesis of Type I diabetes mellitus, N. Engl. J. Med., (in press) (1988).

162. R.P. Zipser, S. Hoef, P.F. Speckart, P.K. Zia, R. Horton, Prostaglandins: Modulators of renal function and pressor resistance in chronic liver disease, J. Clin. End. Metab. 48:895-900 (1979).

163. R.D. Zipser, G.H. Radvan, I.J. Kronberg, R. Duke, T.E. Little, Urinary thromboxane B_2 and prostaglandin E_2 in the hepatorenal syndrome: Evidence for increased vasoconstrictor and decreased vasodilator factors, Gastroent. 84:697-703 (1983).

164. G.A. FitzGerald, R.L. Maas, R. Stein, J.A. Oates, L.J. Roberts, Intravenous prostacyclin in thrombotic thrombocytopenic purpura, Annals Intern. Med. 95:319-322 (1981).

165. G. Remuzzi and N. Perico, Prostacyclin in thrombotic thrombocytopenic purpura and hemolytic uremic syndrome, in: "Prostaglandins in Clinical Medicine. Cardiovascular and Thrombotic Disorders," K. Wu and E.C. Rossi, eds., Year Book Medical Publishers, Chicago, pp. 319-324 (1981).

166. G. Remuzzi, M. Livio, A.E. Cavenaghi, D. Marchesi, G. Mecca, M.B.

Donati, G. De Gaetano, Unbalanced prostaglandin synthesis and plasma factors in uremic bleeding. A hypothesis, Thromb. Res. 13:531-536 (1978).

167. M.C. Smith and M.J. Dunn, Impaired platelet thromboxane production in renal failure, Nephron 29:133-137 (1981).

STUDIES OF RENAL EICOSANOID SYNTHESIS IN VIVO AND IN VITRO

Giovanni Ciabattoni and Francesco Pugliese

Department of Pharmacology
Catholic University School of Medicine, Rome and
II Clinica Medica, Cattedra di Nefrologia
Rome University "La Sapienza," Rome, Italy

INTRODUCTION

The kidney is a rich source of cyclooxygenase and lipoxygenase enzymes
which catalyze the conversion of arachidonic acid to prostaglandins (PGs),
thromboxane (Tx) and leukotrienes. These eicosanoids are autacoids which
have been shown to have biologic actions in vitro and in vivo. Although the
biochemical analysis of renal eicosanoids in vitro has been easily accom-
plished, their identification and quantification in vivo have been shown to
be more complex. The present chapter describes some analytic methodologies
for the study of the renal eicosanoid system, both in vitro and in vivo. In
vitro studies of eicosanoid synthesis by glomerular cells in response to
injury will be addressed. In vivo measurement of primary (unmetabolized)
PGs and Tx in humans and clinical conditions and pharmacological evidence in
favor of a renal origin of the urinary PGI_2 breakdown product, i.e.,
6-keto-$PGF_{1\alpha}$ and TxB_2, will be discussed.

IN VITRO MEASUREMENT OF RENAL EICOSANOIDS

The kidney is an heterogeneous organ composed of functionally distinct
anatomic structures. Each structure, in turn, may be composed by multiple
cell types with different and specific functions within the structure
itself. Thus, different parts of the kidney may have different eicosanoid
synthetic capacity, quantitatively and qualitatively. Detailed information
regarding eicosanoid biosynthesis by different parts of the kidney can be
provided by in vitro techniques involving both tissue separation and cell

361

cultures (1). Preparations of intact tissues present experimental limitations because of the artifactual stimulation of eicosanoid synthesis due to
isolation procedures as well as to ischemia. Additionally, the cellular
heterogeneity of intact tissues complicates the analysis of eicosanoid biosynthesis at cellular level. Preparations of renal cells in culture provide
a practical alternative to evaluate the specific products of arachidonate
metabolism of each cell type. Moreover, cell cultures can be easily manipulated without the interfering effects of tissue extraction and purification
on phospholipase, resulting in an artificial stimulation of eicosanoid synthesis.

GLOMERULAR EICOSANOID SYNTHESIS

Eicosanoid synthesis by isolated glomeruli has been intensively investigated. Human glomeruli synthesize mainly PGI_2 and some amounts of
TxA_2, PGE_2 and $PGF_{2\alpha}$ (2,3). Rat glomeruli produce predominantly
$PGF_{2\alpha}$ and PGE_2 and less PGI_2 and TxA_2 (4,5). The glomerulus is a
complex microvascular structure composed of at least three major cell types,
i.e., endothelial, epithelial and mesangial cells. The precise cellular
sites of glomerular eicosanoid production have been identified using glomerular cell cultures.

According to Petrulis et al. (6) and Sraer et al. (7), PGE_2 is the
major prostaglandin synthesized by primary cultures of rat glomerular epithelial cells, whereas $PGF_{2\alpha}$, PGI_2 and TxB_2 are produced in smaller
amounts. However, when these cells are cloned, the principal prostaglandin
synthesized is PGI_2 (8). Thus, there seems to exist a quantitative difference in the PG biosynthesis between these two different populations of
cells in vitro. Whether this difference reflects different cell populations, or rather a change in the cellular PG biosynthetic capacity depending
upon the number of passages the cells undergo, remains to be seen. Both
primary cultures and cloned populations of rat mesangial cells synthesize
large amounts of PGE_2 and less $PGF_{2\alpha}$ and PGI_2 (7,9). Human mesangial
cells synthesize predominantly PGI_2 and no TxA_2 (10), whereas human
glomerular epithelial cells seem to produce only negligible amounts of PGs
and TX (11).

CELL CULTURE APPROACHES TO ANALYSIS OF GLOMERULAR EICOSANOID SYNTHESIS IN
RESPONSE TO INJURY

Changes in biosynthetic activity represent a prominent feature of the

362

glomerular cellular response to injury. Production of eicosanoids by glomerular cells can possibly be activated by a number of inflammatory factors, i.e., complement factors (12,13), oxygen radicals (14), opsonized zymosan (15), IL-1 (16), or platelet activating factor (15). Because of both pro- and antiinflammatory effects, as well as vasoactivity (see ref. 17 for a review), eicosanoids are believed to play a major pathophysiologic role in the evolution of glomerulonephritis. A pathogenic aspect of glomerular damage is the alteration of the anionic properties of the cellular and structural components of the glomerular capillary loops. Insults that alter the composition of negatively charged molecules of the glomerulus have been associated with alteration in the permselectivity characteristics of the filtration membrane and consequent proteinuria (18). Fixed anionic groups on glomerular epithelial cells are known to be a major component of the charge barrier to anionic circulating macromolecules (19-21). Alterations of these structures have been described in some models of glomerular injury (22,23). Negatively charged proteins have been identified also on the mesangial cell surface (24). Platelet cationic proteins have been shown to bind strongly to the anionic sites of the capillary wall and the mesangium after intravenous injection into rats (25). Induced platelet aggregation in the renal microvasculature of rabbits is followed by a marked proliferation of mesangial cells (26). Furthermore, beneficial effects of heparin have been described in the rats with remnant kidney (27), a model of glomerular injury characterized by mesangial cell proliferation and glomerulosclerosis. In the light of these data, we have selected cloned rat glomerular epithelial cell and primary cultures of rat mesangial cells as models in vitro for the study of a possible relationship between insults that may neutralize the negative charges of glomerular polyanion and changes in eicosanoid synthesis.

EICOSANOID SYNTHESIS BY CLONED RAT GLOMERULAR EPITHELIAL CELLS

Cloned rat glomerular epithelial cells were kindly provided by Dr. J. Kreisberg (San Antonio, Texas). We grew these cells in the medium described by Kreisberg et al. (28) at $37^{\circ}C$ and in an atmosphere of 95% air and 5% CO2. The profile of PG synthesis by cloned glomerular epithelial cells was in general agreement with that reported previously by these investigators (8). As shown in Table 1, cloned glomerular epithelial cells synthesized principally PGI_2, measured as the stable hydration product 6-keto-$PGF_{1\alpha}$, and less PGE_2, under both basal and A23187 (1 ug/ml) stimulated conditions. TxB_2, the stable hydration product of TxA_2, was undetectable either under basal conditions or after A23187 stimulation. Baseline

Table 1. Glomerular Epithelial PG Synthesis

Addition	PGE$_2$	6-keto-PGF$_{1\alpha}$	TXB$_2$
		ng / mg protein / h	
none	0.10±0.07 (8)	0.44±0.16 (20)	ND (4)
Sulindac sulfide (1 μg/ml)	ND (4)	ND (4)	
A23187 (1.5 μg/ml)	0.65±0.18 (8)	3.10±1.50 (8)	ND (4)
A23187 + Sulindac	ND (4)	0.14±0.30 (4)	

Data are represented as mean±SD. The numbers in parentheses represent the number of observations. ND = not detectable.

6-keto-PGF$_{1\alpha}$ and PGE$_2$ production was completely suppressed by sulindac sulfide (1 ug/ml), an inhibitor of cyclooxygenase; when sulindac was added in the presence of A23187, PGE$_2$ and 6-keto-PGF$_{1\alpha}$ synthesis was suppressed. Since 6-keto-PGF$_{1\alpha}$ was the major product of the endoperoxide metabolism in cloned glomerular epithelial cells, we measured only this product in most of the experiments carried out on these cells.

EICOSANOID SYNTHESIS BY RAT MESANGIAL CELLS

Mesangial cells from rat glomeruli were isolated and cultured as reported (7). The profile of PG synthesis by rat mesangial cells did not substantially differ from that described by other investigators in primary cultures (7,9). PGE$_2$ was the major product of arachidonate cyclooxygenation, while 6-keto-PGF$_{1\alpha}$ was produced in smaller quantities. TxB$_2$ was undetectable either in the basal state or in the presence of A23187. Fenoprofen (10 ug/ml), a known inhibitor of cyclooxygenase, completely suppressed either basal or A23187-stimulated PG synthesis (Table 2). Due to its predominance, we measured only PGE$_2$ production in most of experiments on glomerular mesangial cells.

RAT GLOMERULAR EPITHELIAL CELL EICOSANOID SYNTHESIS IN RESPONSE TO CELL SURFACE CHARGE NEUTRALIZATION BY DIFFERENT POLYCATIONS

For these experiments cells were preincubated for 1 hour in veronal buffer (3 mM, pH 7.4) containing either poly-L-lysine (MW 3,500), cytochrome C (MW 12,00), lysozyme (MW 14,000), or protamine sulfate (MW 7,000). The veronal buffer was formulated as a low-ionic-strength buffer to promote

Table 2. Mesangial PG Synthesis

Addition	PGE_2	6-keto-$PGF_{1\alpha}$	TXB_2
		ng / mg protein / h	
none	2.03±0.7 (21)	0.20±0.05 (4)	ND (4)
Fenoprofen (10 µg/ml)	ND (8)	ND (4)	ND (4)
A23187 (1.5 µg/ml)	12.85±1.83 (6)	0.92±0.30	ND (4)
A23187 + Fenoprofen	ND (6)	ND (4)	ND (4)

Data are represented as mean±SD. The numbers in parentheses represent the number of observations. ND = not detectable.

interaction between the cell surface anionic charges and the experimental polycations. After preincubating, the cells were washed three times with RPMI medium, and 1 ml of fresh 37°C RPMI was added to the cells followed by a second period of incubation for the indicated time periods. Pretreatment of rat glomerular epithelial cells with poly-L-lysine evoked a rapid accumulation of 6-keto-$PGF_{1\alpha}$ in the supernatant media. At the earliest time point of incubation (5 min.), 6-keto-$PGF_{1\alpha}$ concentration was signifi-

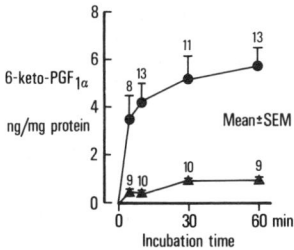

Fig. 1

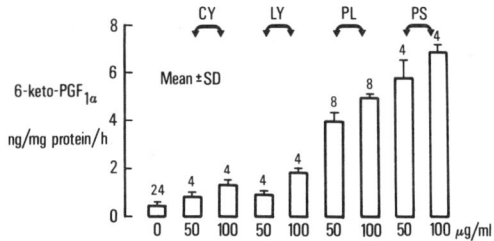

Fig. 2

Fig. 1. Time dependence for accumulating 6-keto-$PGF_{1\alpha}$ in the medium by either Poly-L-lysine (circles) pretreated rat glomerular epithelial cells or controls (triangles). The number over each point represents the number of observations.

Fig. 2. 6-keto-$PGF_{1\alpha}$ production by rat glomerular epithelial cells after treatment with Cytochrome C (CY), Lysozyme (LY), Poly-L-lysine (PL), or Protamine Sulfate (PS). Each bar represents the mean ± SD. The numbers over each bar represent the number of observations. All values were significantly above control values at p<0.01.

cantly elevated (p<0.001) as compared with that of control. Maximal 6-keto-PGF$_{1\alpha}$ concentration was reached after 30-60 minutes of incubation (Fig. 1).

A stimulatory effect on 6-keto-PGF$_{1\alpha}$ production by glomerular epithelial cells was obtained in a dose-dependent fashion with a variety of other polycations (Fig. 2). The effectiveness of these different positively-charged macromolecules to stimulate glomerular epithelial cell 6-keto-PGF$_{1\alpha}$ production strongly suggests that stimulatory activity is dependent upon polycationic character, rather than some nonspecific action of these macromolecules. A known polyanion such as heparin did not stimulate eicosanoid synthesis; on the contrary, heparin inhibited the effect of polycations (data not shown) suggesting that this was likely due to polycation binding to the anionic sites on the cell surface. This conclusion is also strengthened by the recent findings of Singh et al. (29) that the same cationic macromolecules as we used in our experiments are capable of binding and neutralizing the glomerular epithelial cell surface charge.

Whether these changes in eicosanoid synthetic activity, induced experimentally in cultured glomerular epithelial cells, can be extrapolated to in vivo situations remains speculative. However, loss of epithelial polyanion associated with changes in eicosanoid synthesis has recently been reported in adriamycin-induced nephrosis in rat model (30). In this model the pathological changes in the glomerulus were confined to glomerular visceral epithelial cells. Such changes include the replacement of podocytes with flattened expanses of epithelial cell cytoplasm (that is, "fusion of foot processes") and loss of epithelial anionic sites. At least two modes of glomerular polyanion loss during the inflammatory process in glomerulonephritis can be proposed: (a) removal or degradation of anionic sites by some destructive process (31) and (b) their neutralization by a polycation (32). We can speculate that adriamycin can activate a renal and/or an extrarenal cell to synthesize a positively charged macromolecule which may be responsible for histological and biochemical changes seen in this model of glomerular injury.

RAT MESANGIAL CELL EICOSANOID SYNTHESIS IN RESPONSE TO CELL SURFACE CHARGE NEUTRALIZATION BY POLY-L-LYSINE

For these experiments, rat mesangial cells were rinsed three times with 1 ml of 37°C PBS and then incubated for the indicated time periods at 37°C in PBS containing poly-L-lysine with molecular weight 47,000. Consistent with the observation of Shier, Dubourdieu and Durkin (33) in cultures

of 3T3 mouse fibroblasts, we found that 47,000 was a minimum molecular
weight for poly-L-lysine to interact with the mesangial cell surface anionic
sites, when incubated in balanced salt solution. The addition of poly-L-
lysine to rat mesangial cells resulted in a time-dependent increase in
PGE_2 production. At the earliest time point of incubation (5 minutes)
PGE_2 concentration in the media was significantly higher than that of control (p<0.001). Maximal PGE_2 levels were reached after 30-60 minutes of
incubation (Fig. 3). As shown in Fig. 4, the biosynthetic response of rat
mesangial cells to poly-L-lysine was dose-dependent, with the minimum effective dose at 0.5 ug/ml and the maximal response observed at 3 to 5 ug/ml of
poly-L-lysine. To examine an ionic interaction between poly-L-lysine and the
mesangial cell surface, mesangial cells were incubated with poly-L-lysine in
the presence of two polyanions, heparin and native bovine serum albumin
(BSA). Addition of increasing doses of heparin produced a dose-dependent
inhibition of poly-L-lysine (5 ug/ml) evoked PGE_2 synthesis by mesangial
cells. A maximal inhibition of 95% was observed with 2.5 U/ml of heparin
(Fig. 5). BSA (2 mg/ml) completely abolished the stimulatory effect of
poly-L-lysine (5 ug/ml) on PGE_2 production by mesangial cells (data not
shown). Thus, two different polyanions were capable of inhibiting the
ability of poly-L-lysine to stimulate PG production, indicating that the
most probable type of interaction was an electrostatic interaction of

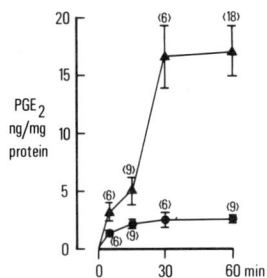

Fig. 3

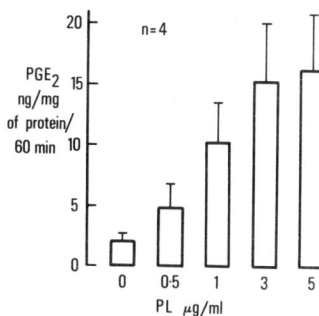

Fig. 4

Fig. 3. Time-dependent increase in PGE_2 production by rat mesangial
cells in presence (triangles) or in absence (circles) of Poly-
L-lysine, 5 ug. At the earliest experimental time point (5 min.),
production by PL-treated cells was significantly higher than that of
controls (p<0.001). The number over each point represents the
number of observations. Values are mean SEM.

Fig. 4. Poly-L-lysine (PL) dose-dependent increase in PGE_2 production
by rat mesangial cells. Each bar represents the mean ± SD of four
observations. Value at each PL concentration was compared with
value at zero PL. P was <0.001 at 0.5 ug/ml and at all higher
concentrations.

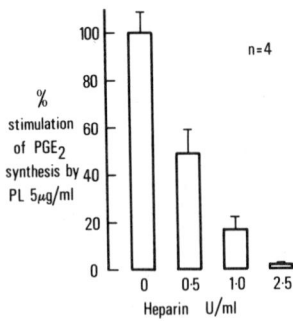

Fig. 5. Effect of increasing concentrations of heparin on PGE_2 production by PL-treated rat mesangial cells. Each bar represents the mean $\pm$ SD of four experiments. At all concentrations of heparin, comparison with control (0 heparin) was statistically significant at $p < 0.001$.

poly-L-lysine with negatively charged structures on the mesangial cell surface.

In vivo, cationic proteins released from platelet and leukocytes might also bind to glomerular polyanionic sites by electrostatic interaction during glomerular disease. Indeed, it has been reported that platelet factor 4, a cationic protein, binds strongly to anionic structures of capillary wall and mesangial cells in sections of human and rat renal cortex (25). Furthermore, platelet-derived growth factor, another cationic platelet secretory protein, has recently been shown to stimulate PGE_2 synthesis by rat mesangial cells in culture (H. E. Abboud, personal communication). Changes in glomerular synthesis of eicosanoids, as well as functional and structural alterations of the glomerulus including increased glomerular blood flow, mesangial cell proliferation and glomerulosclerosis, are prominent features of some experimental and clinical glomerular diseases, i.e., severe renal ablation in the rat (34,35) and diabetic nephropathy (36,37). In these situations, the mechanism underlying the increase in mesangial proliferation and the renal cellular source of increased eicosanoid synthesis is not clear.

GLOMERULAR INFLAMMATORY DISEASE

Changes in eicosanoid synthesis represent a feature of the biosynthetic reaction of the cell to injury. Despite the increasing evidence for enhanced intrarenal production of eicosanoids in a number of experimental as well as clinical forms of glomerulonephritis, the cause and the cellular source of these products has remained controversial. These cells could be infiltrating blood elements such as leukocytes, monocytes and platelets, recruited and activated to synthesize eicosanoids at the site of glomerular injury (38-40), as well as glomerular epithelial and mesangial cells (4-11). The latter intrinsic glomerular cells could possibly be activated by a number of inflammatory factors (12-16). Alterations in anionic struc-

tures of the glomerulus as well as increased renal eicosanoid synthesis have
been described in immune and nonimmune-mediated glomerular disease, such as
serum nephrotoxic nephritis (41,42), lupus glomerulonephritis (43,44), and
diabetic nephropathy (36,45,46). Furthermore, increased glomerular production of eicosanoids has been observed in glomeruli isolated from rats with
remnant kidney (34). In this model, progressive glomerular damage can be
prevented partially by heparin through a mechanism that does not involve
only anticoagulation (27). Our results provide an example in vitro of glomerular cellular biosynthetic response to nonimmune injury and suggest that
a similar mechanism may be, at least partially, operating also in some forms
of glomerular disease.

IN VIVO MEASUREMENT OF RENAL EICOSANOIDS

Despite the ease with which renal eicosanoid synthesis has been deline-
ated in vitro, study of eicosanoid turnover in vivo, especially in humans,
has proven to be very complex. Unlike other mediators, eicosanoids formed
within the kidney are not stored following synthesis, but are either locally
metabolized or removed from the kidney in the lymphatic and venous drainage,
or excreted into the urine (see ref. 47 for a review). Attempts to estimate
in vivo renal eicosanoid synthesis have relied upon assessment of renal secretory and excretory rates of primary (unmetabolized) compounds. Measurement of unmetabolized prostaglandins in renal venous plasma has been reported by several authors in different animal species, including humans (see
ref. 48 for a review). This attempt to estimate renal prostaglandin synthesis relies upon the assumption that renal, venous, unmetabolized prostaglandins are presumably derived from direct drainage of cellular sites of
synthesis and also from venous removal of prostaglandins reabsorbed by
tubules. However, data reported by direct assay in renal venous plasma and
differences described between venous and arterial plasma prostaglandin content (assuming that such a difference may reflect the intrarenal synthesis)
appear to be unreliable or impractical for the following reasons: (a) prostaglandins are not circulating substances in peripheral blood, and the reported levels in renal arterial and venous plasma may be due to artifactual synthesis during sampling; (b) renal venous plasma levels of unmetabolized prostaglandins seem to be very low (49); (c) active platelet prostaglandin synthesis is present, during and after sampling, and this contribution is only
partially prevented by cyclooxygenase inhibitors; (d) this approach to in
vivo study of renal eicosanoid synthesis relies on an invasive procedure,
which hampers the possibility of an extensive clinical investigation.

Samuelsson and coworkers first identified the major urinary metabolites of infused PGE_2 in humans (50,51) by using quantitative assays employing gas chromatography/mass spectrometry (GC/MS). A similar approach was followed by the same authors to identify the major urinary metabolites of infused $PGF_{2\alpha}$ in healthy subjects (52,53). These studies demonstrated that the infused amounts of primary prostaglandins disappear from the circulation with half-lives of a few minutes only, followed by the excretion of degradation products into the urine up to 8 hours after infusion. The profile of metabolites recovered in urine does not include the parent compound (i.e., the infused prostaglandin), except minimal traces, thus suggesting that primary (unmetabolized) prostaglandins are not circulating substances that may be filtered by the kidney and excreted unchanged into the urine. Subsequently, measurement of unmetabolized PGE_2 and $PGF_{2\alpha}$ in human urine, as a reflection of intrarenal synthesis, was reported by Frolich et al. (54). These two compounds were detected in urine from female subjects by gas chromatography/mass spectrometry. Their origin was determined by stimulating renal PG synthesis by arachidonic acid or angiotensin infusion. Arachidonic acid, when infused into one renal artery of dog, led to a significant increase in the excretion rate of PGE_2. Angiotensin II infusion led to a significant ipsilateral increase of the excretion rate of PGE_2 and $PGF_{2\alpha}$, and IV infusion of angiotensin in man led to an increased urinary elimination of PGE_2 (54). More recently, Zipser and Martin infused tritium labeled PGE_2, 6-keto-$PGF_{1\alpha}$ and TXB_2 into the renal artery or brachial vein of volunteers (55). Following renal arterial infusion, 32.2% of 6-keto-$PGF_{1\alpha}$, 13.5% of TXB_2 and 3.9% of PGE_2 were excreted unmetabolized. If tritiated PGI_2 was infused intravenously into a brachial vein, 2.7% of the infusate was excreted as 6-keto-$PGF_{1\alpha}$ in urine (55). The authors concluded that urinary levels of TXB_2 and 6-keto-$PGF_{1\alpha}$ could reflect systemic production of TxA_2 and PGI_2, respectively, while systemic PGE_2 synthesis would not be expected to significantly contribute to the urinary levels of this compound. However, it is reasonable to believe that exogenously administered prostaglandins may saturate transport systems as well as metabolizing enzymes; therefore, the PG excretory rates are not truly representative of the normal situation in vivo. Moreover, it should be stressed that TxA_2 and PGI_2 are not circulating substances (see ref. 56 for a review).

The advantages of urinary measurement of renal prostaglandins are (a) concentrations two orders of magnitude higher than in renal venous plasma, (b) absence of artifactual platelet synthesis, (c) noninvasive technique allowing extensive clinical investigations in healthy subjects as well as in hospitalized or ambulatory patients, (d) continuous rather than episodic measurement of renal prostaglandin synthesis, (e) evaluation of renal as well as extrarenal (systemic metabolites) cyclooxygenase activity. However, several methodological problems as well as biologic variables can affect urinary measurement in humans (see ref. 57 for a review). These include (a) sampling and storage of urine before extraction, (b) purification procedures, (c) assay methods, (d) possible contribution from extrarenal sources and (e) biologic variables. Storage of urine before extraction may variably affect urinary PGE_2 and $PGF_{2\alpha}$ concentrations as a function of temperature and time. PGE_2 seems to be very susceptible to this kind of degradation. Up to 60% of the initial PGE_2 can be lost during storage of urine at $4^\circ C$ for 24 hr. (58). $PGF_{2\alpha}$ (58), TXB_2 and 6-keto-$PGF_{1\alpha}$ are more stable than PGE_2 (unpublished data). However, in order to avoid variable losses, urine should be frozen immediately after voiding and kept at $-20^\circ C$ until extracted. Considering the methodological problems arising from assay procedures--i.e., purification of urine by chromatographic techniques and quantitative analysis by RIA or GC/MS--the presence in urine of a large array of structurally related eicosanoids and the low concentration of primary PGs may represent a serious problem, if RIA methods are employed. In particular, it must be stressed that RIA simply measures the inhibition of the binding of labeled antigen to antibody, and this inhibition can be caused by a multitude of completely unrelated factors. If the system is not examined critically, this nonimmunologic inhibition will be interpreted as being caused by large amounts of the measured compounds. Therefore, given a particular antibody, the specificity of the reaction depends on a constant (i.e., the affinity and conformation of binding sites) and a variable (i.e., the composition of the biologic fluid to be examined and its degree of purification). In considering the methodological variables inherent in RIA techniques, it should be emphasized that the use of classical criteria of specificity, such as ascertainment of the identical immunochemical behavior of the unknown immunoreactivity with authentic PGs (dilution and recovery studies), is inadequate to validate RIA measurements of urinary prostanoids. Different anti-PGE_2 (57,59), anti-6-keto-$PGF_{1\alpha}$ (60) and anti-TxB_2 (44) sera that satisfied such requirements yielded completely different estimates of the measured eicosanoid. Characterization of the measured immunoreactivity on TLC revealed the presence of two or more peaks of immunoreactivity, only one co-migrating with the substance which should have been specifically recognized by antisera (44,57,59,60). Few antisera tested recognized a single peak of immunoreactive material migrating in a fashion identical to that of the reference standard. These antisera gave an estimate of urinary prostanoids in agreement with an independent technique, i.e., GC/MS. Thus validation criteria for RIA measurements of urinary prostanoids should include, besides the binding characteristic and cross reactivity of antisera, also the simultaneous use of multiple antisera, the identical chromatographic behavior of standards and unknown, and comparison with an independent method of analysis (see ref. 61 for a review). A possible origin of urinary eicosanoids from extrarenal sources should be taken into consideration in males, since trace amounts of seminal fluid may contribute a highly variable fraction of the measured urinary prostaglandins in men (49). Finally, it should be considered that urinary unmetabolized PG excretion rate may be highly variable in female subjects during the menstrual cycle, with no apparent pattern (59). A satisfactory degree of reproducibility can be found when overnight specimens are collected under standard conditions and immediately frozen (57-59).

EVIDENCE FOR RENAL ORIGIN OF URINARY 6-KETO-PGF$_{1\alpha}$ AND TXB$_2$ IN MAN

Although no evidence has been provided for a specific origin of urinary prostaglandins and thromboxane from a well-defined nephron segment, the intrarenal compartmentalization of the eicosanoid system suggests that the urinary excretion rate of PGE$_2$ and PGF$_{2\alpha}$ mainly reflects a medullary synthesis, while 6-keto-PGF$_{1\alpha}$ and TxB$_2$ might reflect a prevalent cortical cyclooxygenase activity. Since cortical functions, such as blood flow and glomerular filtration, appear to be dependent upon eicosanoid synthesis in many pathophysiological conditions, attention has been focused in the past years on the urinary excretion rate of 6-keto-PGF$_{1\alpha}$ and TxB$_2$ as a reflection of the renal synthesis of the unstable parent compounds.

An essential feature in showing that a given urinary eicosanoid is of renal origin is to demonstrate that it is not filtered and excreted by the kidney. The secretion rate of endogenous PGI$_2$ into the circulation is extremely low (62), and 6-keto-PGF$_{1\alpha}$ in human plasma is in the low picogram range (see ref. 56 for a review). Similarly, TxB$_2$ is undetectable in human plasma when care is taken to avoid artifacts induced by sampling (56). Experiments performed to estimate the endogenous secretion rate of TxB$_2$ in the human circulation indicate that the secretion rate is extremely low, corresponding to maximal concentration of endogenous TxB$_2$ in plasma in the range of 1-2 pg/ml (63,64). The reported levels of 6-keto-PGF$_{1\alpha}$ (60) and TxB$_2$ (44) in human urine could hardly be derived from the renal

clearance of circulating compounds if we consider the tight binding to plasma proteins. A renal origin of urinary 6-keto-PGF$_{1\alpha}$ is also supported by data obtained by PGI$_2$ infusion at low constant rate, very close to the total body PGI$_2$ production into healthy volunteers after the endogenous synthesis had been blocked by indomethacin (57). Under physiologic conditions, the estimated endogenous PGI$_2$ rate of entry in circulation in humans is below 100 pg/kg/min. (62). Based on these data, PGI$_2$ was infused at a constant rate of 50 pg/kg/min. over a 2 hr. period into 5 healthy volunteers after endogenous PGI$_2$ synthesis had been blocked by indomethacin (200 mg/day for two days) (57). As shown in Table 3, urinary excretion rate of 6-keto-PGF$_{1\alpha}$ and PGE$_2$ were unchanged under these experimental conditions, thus suggesting that under physiological circumstances it is unlikely that urinary 6-keto-PGF$_{1\alpha}$ is derived from circulating PGI$_2$.
Similar evidence, suggesting an intrarenal origin of urinary TxB$_2$, may be sought by infusion of exogenous TxB$_2$ into healthy volunteers. The rate of thromboxane secretion into the circulation of normal humans was estimated as 0.11 ng/kg/min. by quantification of a major urinary metabolite, 2,3-dinor-TxB$_2$ (63). In four healthy, male volunteers, we infused vehicle alone and TxB$_2$ at 0.1, 1.0 and 5.0 ng/kg/min. for 6 hr. in a first series of experiments (63), and for 4 hr. in a second series (65). Volunteers were pretreated with aspirin at a dose of 325 mg/day in order to suppress endogenous platelet TxB$_2$ production. Urinary TxB$_2$ and 2,3-dinor-TxB$_2$ were measured in the first series of experiments before, during and up to 24 hr. after the infusion by means of GC/MS-validated RIA techniques (44,63). The fractional elimination of 2,3-dinor-TxB$_2$ in urine was independent of the rate of TxB$_2$ infusion (Table 4) and averaged 5.3 $\pm$ 0.8% (mean $\pm$ SD) (63). In contrast, urinary TxB$_2$ fractional excretion rate was increased by the

Table 3. Urinary Excretory Rates of 6-Keto-PGF$_{1\alpha}$ and PGE$_2$ Before and After PGI$_2$ Infusion (50 pg/kg/min.)

	U-6-keto-PGF$_{1\alpha}$ ng/h	U-PGE$_2$ ng/h	U-6-keto-PGF$_{1\alpha}$/ U-PGE$_2$ ratio
Before PGI$_2$	4.53±3.35	10.12±4.96	0.450±0.18
During PGI$_2$	4.49±2.54	10.76±7.20	0.511±0.27

Numbers are mean±SD; data from ref. 58

Table 4. Fractional Excretion of TxB_2 and TxB_2
Metabolites After TxB_2 Infusion

Rate of TXB_2 infusion ng/kg·min^{-1}	TXB_2 %	2,3-dinor-TXB_2[*] %	11-dehydro-TXB_2[**] %
0.1	0	6.2±1.6	7.5±1.5
1.0	0.7±0.4	4.7±1.7	6.8±0.4
5.0	1.4±0.5	4.9±1.2	6.2±1.5

Numbers are mean±SD; the fractional excretion rates of TXB_2 and
its metabolites are expressed for the 24 h period commencing with
each infusion; n = 4 healthy volunteers; *data from ref. 63

** data from ref. 65

infusion rates of 1.0 and 5.0 ng/kg/min. but not by the lowest infusion rate
(Table 4). In the second series of experiments, the same volunteers were
infused with the same doses of TxB_2 for 4 hr., and 11-dehydro-TxB_2 was
measured in urine by GC/MS-validated RIA techniques (66). Like 2,3-dinor-
TxB_2, the fractional elimination of 11-dehydro-TxB_2 was independent of
the rate of TxB_2 infusion (Table 4) and averaged 6.8 ± 0.7% (mean ± SD)
(65). Interpolation of metabolite values, obtained in aspirin-free periods
onto the linear relationship between the quantities of infused TxB_2 and
the amount of 11-dehydro-TxB_2 excreted in excess of control values, permitted calculation of the mean rate of entry of TxA_2 into the circulation as
0.12 ng/kg/min. (65), a value that is nearly identical to that calculated by
quantification of 2,3-dinor-TxB_2 (63). When exogenous TxB_2 is infused
at a rate very close to the rate of entry of endogenous TxA_2 into the
bloodstream, urinary levels of TxB_2 are unmodified (Table 5), thus strongly suggesting that the measured levels of TxB_2 in human urine under normal
conditions reflect an intrarenal synthesis, rather than a systemic production. Additional evidence in favor of renal origin of urinary TxB_2 and
6-keto-$PGF_{1\alpha}$ can be derived from experiments in healthy volunteers with
loop diuretics, which induce a parallel increase of urinary PGE_2, $PGF_{2\alpha}$,
TxB_2 (67) and 6-keto-$PGF_{1\alpha}$ (60), thus suggesting a common renal origin
of these compounds.

Indirect evidence suggesting a renal origin of 6-keto-$PGF_{1\alpha}$ is also
provided by Bartter's syndrome. Increased urinary PG excretion rate in this
clinical condition has been reported (68). Also urinary 6-keto-$PGF_{1\alpha}$ excretion rate is increased as compared with normal subjects (69). However,
urinary excretion rate of 2,3-dinor-6-keto-$PGF_{1\alpha}$, a systemic metabolite of

Table 5. Excretion of TxB_2 (ng/hr.): The Effects of TxB_2 Infusion

	TXB_2 infusion rate ng/kg·min^{-1}			
	0	0.1	1.0	5.0
Pre-infusion	3.1±0.6	4.0±2.8	2.9±1.4	3.2±0.6
Infusion	3.2±1.0	3.7±1.7	*25.6±14.3	*216.5±93.8
Post-infusion I	2.8±1.3	3.5±1.2	3.3±1.0	11.1±14.3
Post-infusion II	3.0±1.0	3.7±1.7	2.8±1.1	4.0±1.4

Numbers are mean±SD; *p < 0.01 vs vehicle infusion.

endogenous PGI_2 synthesis (62), was comparable with that of normal subjects, thus suggesting that systemic synthesis of PGI_2 in these patients was normal (70). Similarly, systemic lupus erythematosus (SLE) has recently been identified as a clinical condition associated with increased urinary excretion rate of TxB_2, especially in patients with histologically documented active renal lesions (44). However, the urinary excretion rate of 2,3-dinor-TxB_2, indicative of nonrenal TxA_2 production (71), was normal in these patients (44).

STUDIES WITH RENAL SPARING CYCLOOXYGENASE INHIBITORS

Selective renal cyclooxygenase inhibitors include sulindac (72,73), low-dose aspirin (74), sulfinpyrazone (75) and TxA_2-synthase inhibitors (76-78). Nonsteroidal antiinflammatory drugs (NSAIDs) compete in a dose-dependent manner with substrate (arachidonic acid) to bind to cyclooxygenase. A similar action is shared by sulfinpyrazone (a uricosuric agent); while, in contrast to other NSAIDs, aspirin irreversibly acetylates the hydroxyl group of a serine residue at the active site of cyclooxygenase, thereby inactivating the enzyme permanently (79,80). Acetylation of platelet cyclooxygenase by oral aspirin is dose-dependent, cumulative with repeated administration and selective (74). Daily administration of low-dose aspirin does not reduce the urinary excretion of PGE_2, $PGF_{2\alpha}$, 6-keto- $PGF_{2\alpha}$ (74) and TxB_2 (81), despite a nearly complete inhibition of platelet TxB_2 synthesis (74). Moreover, it should be noted that aspirin treatment at doses of 0.5 and 1 g daily is required to reduce significantly, by 40% and 60%, the urinary excretion rate of 6-keto-$PGF_{1\alpha}$ (74), while lower doses of aspirin

largely suppress vascular PGI_2 production (82). The effects of 4 weeks
dosing with aspirin (25 mg b.i.d.) plus dipyridamole (150 mg b.i.d.) in 8
healthy volunteers (4 men and 4 women, age range 28-40 years) on renal and
extrarenal PGI_2 and thromboxane production are shown in Figures 6 and 7.
Urinary excretion rate of 6-keto-$PGF_{1\alpha}$ and TxB_2 were measured by previously described RIA techniques (44,60). The major urinary metabolites of
PGI_2 (62) and TxA_2 (71), 2,3-dinor-6-keto-$PGF_{1\alpha}$ and 2,3-dinor-TxB_2,
respectively, were measured by newly developed RIA techniques (83). Both
systemic (extrarenal) metabolites of PGI_2 and thromboxane are decreased by
60% and 80%, respectively (Fig. 6 and 7). Serum TxB_2, an ex vivo index of
platelet cyclooxygenase activity (84) was inhibited by 98% (Fig. 7). Despite this systemic effect, the co-administration of aspirin and dipyridamole failed to modify urinary TxB_2 excretion rate (Fig. 7). A similar
low-dose aspirin (0.45 mg/kg per day) had been previously reported not to
change urinary 6-keto-$PGF_{1\alpha}$ excretion rate (74).

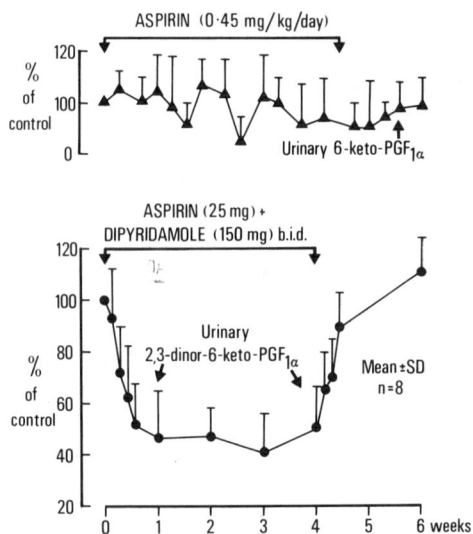

Fig. 6. Lower panel: urinary excretion of 2,3-dinor-6-keto-
$PGF_{1\alpha}$ expressed as percentage of control values during
and after chronic dosing
(4 weeks) with aspirin plus
dipyridamole in 8 healthy
volunteers. Upper panel:
urinary excretion of 6-keto-
$PGF_{1\alpha}$ expressed as percentage of control values in
similar experiments (74).

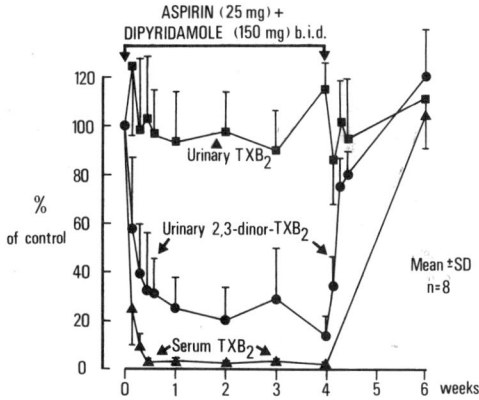

Fig. 7. Urinary excretion of TxB$_2$ (squares) and 2,3-dinor-TxB$_2$ (circles) and serum TxB$_2$ (triangles) expressed as percentage of control values during and after chronic dosing (4 weeks) with aspirin plus dipyridamole in 8 healthy volunteers (74).

Sulindac is a NSAID that differs from other NSAIDs in that it spares renal cyclooxygenase activity. In 1980, we reported that renal PG excretion in healthy women, and in patients with Bartter's syndrome, was not decreased by sulindac (73), possibly because of its redox pro-drug nature confining formation and/or access of its reactive sulfide metabolite to extrarenal sites of cyclooxygenase activity (85). We extended our first observations to female patients with chronic glomerular disease (72). In this condition, sulindac treatment was not associated with any statistically significant change of renal function, in contrast to the consistent reduction of glomerular filtration rate and renal blood flow caused by ibuprofen (72). Other investigators extended and confirmed our observations (86-94). However, other studies reported conflicting evidence regarding the biochemical selectivity of sulindac (95-100). It should be noted that in the 6 studies (95-100) describing some degree of reduction of urinary PG excretion after sulindac administration, the majority of patients and healthy volunteers were men (see ref. 101 for a review). It is possible that an extrarenal source, i.e., seminal fluid (even in trace amounts) or other secretions of the low genito-urinary tract, might contribute to the urinary PG excretion rate in men, and hence to the sulindac-induced reductions, especially if PGE$_2$ is used as index of renal PG synthesis (49). It is noteworthy that when 6-keto-PGF$_{1\alpha}$ was used as index of renal cyclooxygenase activity, only one of the published studies, assessing the effect of sulindac treatment, described a significant reduction of urinary 6-keto-PGF$_{1\alpha}$ (100). Interestingly, this occurred in five patients with cirrhosis and ascites with plasma concentrations of the active sulfide metabolite of sulindac fourfold higher than drug levels measured in healthy controls (see ref. 101 for a full discussion of this argument). A possible explanation of the apparent reduction

377

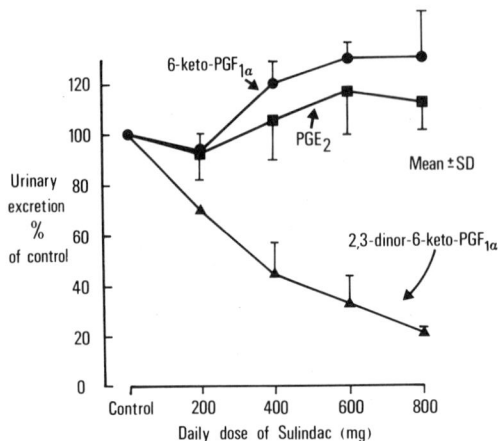

Fig. 8. Urinary excretion of 6-keto-PGF$_{1\alpha}$ (circles), PGE$_2$ (squares) and 2,3-dinor-6-keto-PGF$_{1\alpha}$ (triangles) expressed as percentage of control values during chronic dosing with sulindac in 7 healthy volunteers. Each dose was given for 7 days in successive weeks. Urinary measurements were performed during the last 3 days of each week (102).

of urinary PGE$_2$-like immunoreactivity associated with sulindac may be related to reduced urinary excretion of systemic (extrarenal) metabolites cross-reacting with anti-PG sera. Indirect evidence of such an occurrence is presented in Figures 8 and 9, which show the effects of sulindac at four different doses on in vivo indexes of PGI$_2$ and thromboxane production by renal as well as extrarenal sites. Seven healthy subjects (5 men and 2 women between ages of 28 and 40 yrs) received 200, 400, 600 and 800 mg sulindac per day, each dose given for 7 days. Indexes of renal cyclooxygenase activity were urinary PGE$_2$, 6-keto-PGF$_{1\alpha}$ (Fig. 8) and TxB$_2$ (Fig. 9), while extrarenal PGI$_2$ and TxA$_2$ synthesis was monitored by urinary excretion of their 2,3-dinor metabolites. Serum TxB$_2$ was also measured as an index of platelet cyclooxygenase activity. Each subject served as his own control and was subjected to 3 daily examinations under control conditions and during each week of dosing with sulindac. It is clear that sulindac dose-dependently decreases urinary excretion rates of 2,3-dinor-6-keto-PGF$_{1\alpha}$ (Fig. 8) and 2,3-dinor-TxB$_2$ (Fig. 9), as well as serum TxB$_2$ concentrations (Fig. 9), without affecting urinary levels of PGE$_2$ and 6-keto-PGF$_{1\alpha}$ (Fig. 8) and TxB$_2$ (Fig. 9). These results extend our previous observations of a selective sparing of renal cyclooxygenase activity by sulindac in man (72,73) and strongly suggest an intrarenal origin of urinary 6-keto-PGF$_{1\alpha}$ and TxB$_2$. Moreover, the sulindac-induced suppression of extrarenal cyclooxygenase metabolite excretion could explain an apparent reduction in urinary PG-like immunoreactivity reported in some studies employing RIA techniques to quantitate urinary excretion. Although RIA has the advantage of a great sensitivity and a potential specificity for a quick

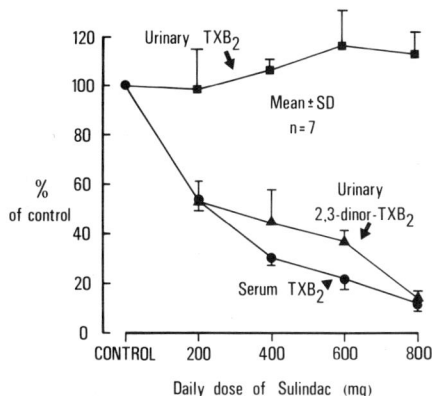

Fig. 9. Urinary excretion of TxB$_2$ (squares) and 2,3-dinor-TxB$_2$ (triangles) and serum TxB$_2$ (circles) expressed as percentage of control values during chronic dosing with sulindac in 7 healthy volunteers. Each dose was given for 7 days in successive weeks. Urinary measurements were performed during the last 3 days of each week (102).

and reliable measurement of eicosanoids in urine, it is liable to all forms of artifacts possibly affecting an antigen antibody reaction, i.e., nonspecifically interfering factors and compounds, chemically related to the substance to be measured, which may cross-react with antisera. If we suppress extrarenal sites of eicosanoid synthesis by any given NSAID, the consequence will be a simultaneous reduction in the urinary excretion of both unmetabolized (renal) eicosanoids and potentially cross-reacting systemic (extrarenal) metabolites. Sulindac, although selectively sparing renal eicosanoid synthesis, reduces up to 80% extrarenal cyclooxygenase activity.

As with sulindac, sulfinpyrazone when given to healthy volunteers reduced platelet thromboxane production, without affecting renal cyclooxygenase activity (75). A possible common mechanism for this selectivity could be proposed since, as with sulindac (85), the effectiveness of sulfinpyrazone in inhibiting cyclooxygenase activity is dependent on reduction to an active sulfide metabolite (103). Finally, further evidence for renal origin of urinary thromboxane is given by the use of selective TxA$_2$ synthase inhibitors in man. We reported the effects of three orally active Tx-synthase inhibitors, i.e., dazoxiben (76,77), OKY-046 (78) and UK-38,485 (77) on renal PG endoperoxide metabolism in healthy subjects. Despite variable biological half life of platelet Tx-synthase inhibition (4 to > 8 h), the three drugs caused only a marginal reduction of urinary TxB$_2$ excretion of approximately 30% at doses maximally inhibiting platelet TxB$_2$ production (76-78). Redirection of PG endoperoxide metabolism to prostaglandins was demonstrated in platelets, but not in the kidney. These results, demonstrating quantitatively and qualitatively diverse effects on Tx-synthase inhibitors in different TxA_2 producing cells, support an origin or urinary TxB_2 other than circulating blood elements.

CONCLUSIONS

We have described methods that allow a reliable assessment of renal eicosanoids in vitro as well as in vivo. In vitro measurement of these compounds is limited to evaluate the synthetic capacity of the isolated structures in response to a given stimulus. Measurement of unmetabolized PG in renal venous plasma is unreliable because of artifactual platelet synthesis during sampling; whereas measurement of unmetabolized prostaglandins in urine, although raising technical problems, provides a time-integrated measure of renal PG synthesis in vivo. Urinary TxB_2 and 6-keto-$PGF_{1\alpha}$ excretion rate reflects, primarily, intrarenal TxA_2 and PGI_2 synthesis in humans. Their origin is presumably from the cortex, although no direct proof can be provided for such a hypothesis. The use of appropriate analytic techniques for measurement of renal eicosanoids may offer the opportunity to clarify the metabolism of arachidonic acid in the modulation of renal hemodynamics and in the development of glomerular disease.

REFERENCES

1. D. Schlondorff and R. Ardaillou, Prostaglandin and other arachidonic acid metabolites in the kidney, Kidney Int. 29:108-119 (1986).

2. R.A. Stahl, M. Paravicini, P. Schollmeyer, Angiotensin II stimulation of PGE_2 and 6-keto-$PGF_{1\alpha}$ formation by isolated human glomeruli, Kidney Int. 26:30-34 (1984).

3. J. Sraer, N. Ardaillou, J.D. Sraer, R. Ardaillou, In vitro prostaglandin synthesis by human glomeruli and papillae, Prostaglandins 23:855-864 (1983).

4. A. Hassid, M. Konieczkowski, M.J. Dunn, Prostaglandin synthesis in isolated rat kidney glomeruli, Proc. Natl. Acad. Sci. USA 76:1155-1159 (1979).

5. J. Sraer, J.D. Sraer, D. Chansel, F. Russo-Marie, B. Kouznetzova, R. Ardaillou, Prostaglandin synthesis by isolated rat renal glomeruli, Mol. Cell Endocrinol. 16:29-37 (1979).

6. A.S. Petrulis, M. Aikawa, M.J. Dunn, Prostaglandin and thromboxane synthesis by rat glomerular epithelial cells, Kidney Int. 20:469-474 (1981).

7. J. Sraer, J. Foidart, D. Chansel, P. Mahieu, B. Kouznetzova, R.

Ardaillou, Prostaglandin synthesis by mesangial and epithelial glomerular cultured cells, FEBS Letters 15:420-424 (1979).

8. J.I. Kreisberg, M.J. Karnowsky, L. Levine, Prostaglandin production by homogeneous cultures of rat glomerular epithelial and mesangial cells, Kidney Int. 22:355-359 (1982).

9. L.A. Scharschmidt and M.J. Dunn, Prostaglandin synthesis by rat glomerular mesangial cells in culture. Effects of angiotensin II and arginine vasopressin, J. Clin. Invest. 71:1756-1764 (1983).

10. N. Ardaillou, Y. Hagege, M.P. Nivez, R. Ardaillou, D. Schlondorff, Vasoconstrictor-evoked prostaglandin synthesis in cultured human mesangial cells, Am. J. Physiol. 248:F240-F246 (1985).

11. N. Ardaillou, M.P. Nivez, G. Striker, R. Ardaillou, Prostaglandin synthesis by human glomerular cells in culture, Prostaglandins 26:778-784 (1983).

12. D. Lovett, G.M. Hänsch, K. Resch, D. Gemsa, Activation of glomerular mesangial cells by terminal complement components: stimulation of prostanoid and interleukin 1-like factor release, (abstract), Immunobiol. 168:34-35 (1984).

13. G.M. Hänsch, K. Rother, J. Gunther, S. Filsinger, R.B. Sterzel, Terminal complement components stimulate prostanoid production of cultured glomerular epithelial cells, (abstract), Kidney Int. 29:277 (1986).

14. L. Baud, M.P. Nivez, D. Chansel, R. Ardaillou, Stimulation by oxygen free radicals of prostaglandin production by rat renal glomeruli, Kidney Int. 20:332-345 (1981).

15. D. Schlondorff, J.A. Satriano, J. Hagege, J. Perez, L. Baud, Effect of platelet activating factor and serum-treated zymosan on prostaglandin E_2 synthesis, arachidonic acid release, and contraction of cultured rat mesangial cells, J. Clin. Invest. 73:1227-1231 (1984).

16. D. Lovett, D. Gemsa, K. Resch, Stimulation of mesangial cell prostanoid secretion by interleukin 1, (abstract), Kidney Int. 19:282 (1986).

17. J.E. Stork, M.A. Rahman, M.J. Dunn, Eicosanoids in experimental and human renal disease, Am. J. Med. 80 (suppl.1A):34-35 (1986).

18. M.W. Seiler, M.A. Venkatachalam, R.S. Contran, Glomerular epithelium: structural alterations induced by polycations, Science 189:390-393 (1975).

19. W.M. Deen, B.D. Myers, B.M. Brenner, The glomerular barrier to macromolecules: theoretical and experimental considerations, in: "Contemporary Issues In Nephrology," B.M. Brenner and J.H. Stein, eds., Churchill Livingstone, New York, Vol. 9, pp. 1-29 (1982).

20. B.M. Brenner, T.H. Hostetter, D. Humes, Molecular basis of proteinuria of glomerular origin, N. Engl. J. Med. 298:826-833 (1978).

21. E.B. Blau and J.E. Haas, Glomerular sialic acid and proteinuria in human renal disease, Lab. Invest. 28:477-481 (1973).

22. D. Kerjaschki, A.T. Vernillo, M.G. Farquhar, Reduced sialylation of podocalixin - the major sialoprotein of the rat kidney glomerulus - in aminonucleoside nephrosis, Am. J. Pathol. 118:343-349 (1985).

23. R.L. Vernier, D.J. Klein, S.P. Sisson, J.D. Mahan, T.R. Oegema, D.M. Brown, Heparan sulfate-rich anionic sites in the human glomerular basement membrane. Decreased concentration in congenital nephrotic syndrome, N. Engl. J. Med. 309:1001-1009 (1983).

24. J.B. Foidart, Y.S. Picard, R.J. Winaud, P.R. Mahieu, Tissue culture of normal rat glomeruli: glycosaminoglycan biosynthesis by homogenous epithelial and mesangial cell populations, Renal Physiol. 3:169-173 (1980).

25. J.L. Barnes, S.P. Levine, M.A. Venkatachalam, Binding of platelet factor four (PF4) to glomerular polyanions, Kidney Int. 25:759-765 (1984).

26. L. Jorgensen, M.F. Glynn, T. Hovig, E.A. Murphy, M.R. Buchanan, J.F. Mustard, Renal lesions and rise in blood pressure caused by adenosine diphosphate-induced platelet aggregation in rabbits, Lab. Invest. 23: 347-357 (1970).

27. J.L. Olson, Role of heparin as a protective agent following reduction of renal mass, Kidney Int. 25:376-382 (1984).

28. J.I. Kreisberg, R.L. Hoover, M.J. Karnowsky, Isolation and characterization of rat glomerular epithelial cells in vitro, Kidney Int. 14:21-30 (1978).

29. A.K. Singh, B.S. Kasinath, E.J. Lewis, Discrete anionic domains on the surface of glomerular epithelial cells (GEC) (abstract), Clin. Res. 34:609A (1986).

30. G. Remuzzi, L. Imberti, M. Rossini, C. Morelli, C. Carminati, M.G. Cattaneo, T. Bertani, Increased glomerular thromboxane synthesis as a possible cause of proteinuria in experimental nephrosis, J. Clin. Invest. 75:94-101 (1985).

31. L.J. Rosenzweig and Y.S. Kanvar, Removal of sulphated (heparan sulphate) or nonsulphated (hyaluronic acid) glycosaminoglycans results in increased permeability of the glomerular basement membrane to ^{125}I-bovine serum albumin, Lab. Invest. 47:177-184 (1982).

32. V.M. Vehaskari, E.R. Root, F.G. Germuth, A.M. Robson, Glomerular charge and urinary protein excretion: effects of systemic and intrarenal polycation infusion in the rat, Kidney Int. 22:127-135 (1982).

33. W.T. Shier, D. Dubourdieu, J.P. Durkin, Polycations as prostaglandin synthesis inducers. Stimulation of arachidonic acid release and prostaglandin synthesis in cultured fibroblasts by poly(L-lysine) and other

synthetic polycations, <u>Biochim. Biophys. Acta</u> 793:238-250 (1984).

34. R.A. Stahl, M. Kudelka, M. Paravicini, P. Schollmeyer, Prostaglandin and thromboxane formation in glomeruli from rats with reduced renal mass, <u>Nephron</u> 42:252-257 (1986).

35. T.H. Hostetter, J.L. Olson, H.G. Rennke, M.A. Venkatachalam, B.M. Brenner, Hyperfiltration in remnant nephrons: a potentially adverse response to renal ablation, <u>Am. J. Physiol</u>. 241:F85-F93 (1981).

36. M. Schambelan, S. Blake, J. Sraer, M. Beres, M.P. Nivez, F. Wahbe, Increased prostaglandin production by glomeruli isolated from rats with streptozotocin-induced diabetes mellitus, <u>J. Clin. Invest</u>. 75:404-412 (1985).

37. T.H. Hostetter, J.L. Troy, B.M. Brenner, Glomerular hemodynamics in experimental diabetes mellitus, <u>Kidney Int</u>. 19:410-415 (1981).

38. I.M. Goldstein, C.L. Malmsten, H. Kindhal, H.B. Kaplan, D. Radmark, B. Samuelsson, G. Weisman, Thromboxane generation by human peripheral blood polymorphonuclear leukocytes, <u>J. Exp. Med</u>. 148:787-792 (1979).

39. J. Morley, M.A. Bray, R.W. Jones, D.H. Nugteren, D.A. Van Dorp, Prostaglandin and thromboxane production by human and guinea-pig macrophages and leukocytes, <u>Prostaglandins</u> 17:730-736 (1979).

40. P.S. Needleman, S. Moncada, S. Bunting, J.R. Vane, M. Hamberg, B. Samuelsson, Identification of an enzyme in platelet microsomes which generates thromboxane A_2 from prostaglandin endoperoxides, <u>Nature</u> 261:558-560 (1976).

41. J.I. Kreisberg, D.B. Wayne, M.J. Karnowsky, Rapid and focal loss of negative charge associated with mononuclear cell infiltration early in nephrotoxic serum nephritis, <u>Kidney Int</u>. 16:290-300 (1979).

42. E.A. Lianos, G.A. Andres, M.J. Dunn, Glomerular prostaglandin and thromboxane synthesis in rat nephrotoxic serum nephritis, <u>J. Clin. Invest</u>. 72:1439-1448 (1983).

43. G. Camussi, C. Tetta, G. Segoloni, R. Coda, A. Vercellone, Localization of neutrophil cationic proteins and loww of anionic charges in glomeruli of patients with systemic lupus erythematosus, <u>Clin. Immun. Immunopathol</u>. 24:299-314 (1982).

44. C. Patrono, G. Ciabattoni, G. Remuzzi, E. Gotti, S. Bombardieri, O. Di Munno, G. Tartarelli, G.A. Cinotti, B.M. Simonetti, A. Pierucci: Functional significance of renal prostacyclin and thromboxane A_2 production in patients with systemic lupus erythematosus, <u>J. Clin. Invest</u>. 76:1011-1018 (1985).

45. D.M. Brown, D.J. Klein, A.F. Michael, T.R. Oegema, ^{35}S-glycosaminoglycan and ^{35}S-glycopeptide metabolism by diabetic glomeruli and aorta, <u>Diabetes</u> 31:418-425 (1982).

46. M.P. Cohen, M.L. Surma, Effects of diabetes on in vivo metabolism of ^{35}S-labeled glomerular basement membrane, <u>Diabetes</u> 33:8-12 (1984).

47. M.J. Dunn, Renal prostaglandins, <u>in</u>: "Renal Endocrinology," M.J. Dunn, ed., Williams and Wilkins, Baltimore, pp. 1-74 (1983).

48. C. Patrono and F. Pugliese, The involvement of arachidonic acid metabolism in the control of renin release, <u>J. Endocrinol. Invest</u>. 3:193-201 (1980).

49. C. Patrono, Å. Wennmalm, G. Ciabattoni, J. Nowak, F. Pugliese, G.A. Cinotti, Evidence for an extrarenal origin of urinary prostaglandin E_2 in healthy men, <u>Prostaglandins</u> 18:623-629 (1979).

50. M. Hamberg and B. Samuelsson, The structure of the major urinary metabolite of prostaglandin E_2 in man, <u>J. Am. Chem. Soc</u>. 91:2177-2178 (1969).

51. M. Hamberg and B. Samuelsson, On the metabolism of prostaglandin E_1 and E_2 in man, <u>J. Biol. Chem</u>. 246:6713-6721 (1971).

52. E. Granström and B. Samuelsson, On the metabolism of $PGF_{2\alpha}$ in female subjects, <u>J. Biol. Chem</u>. 246:5254-5263 (1971).

53. E. Granström and B. Samuelsson, On the metabolism of prostaglandin $F_{2\alpha}$ in female subjects. II. Structures of six metabolites, <u>J. Biol. Chem</u>. 246:7470-7485 (1971).

54. J.C. Frölich, T.W. Wilson, B.J. Sweetman, M. Smigel, A.S. Nies, K. Carr, J.T. Watson, J.A. Oates, Urinary prostaglandins: identification and origin, <u>J. Clin. Invest</u>. 55:763-770 (1975).

55. R.D. Zipser and K. Martin, Urinary excretion of arterial blood prostaglandins and thromboxane in man, <u>Am. J. Physiol</u>. 242:E171-177 (1982).

56. G.A. FitzGerald, A.K. Pedersen, C. Patrono, Analysis of prostacyclin and thromboxane biosynthesis in cardiovascular disease, <u>Circulation</u> 67:1174-1177 (1983).

57. F. Pugliese and G. Ciabattoni, Investigations of renal arachidonic acid metabolites by radioimmunoassay, <u>in</u>: "Prostaglandins and The Kidney," M.J. Dunn, C. Patrono, G.A. Cinotti, eds., Plenum Press, New York, pp. 83-98 (1983).

58. G. Ciabattoni, F. Pugliese, E. Pinca, G.A. Cinotti, A. De Salvo, M.A. Satta, C. Patrono, Biologic and methodologic variables affecting urinary prostaglandin measurement, <u>in</u>: "Advances in Prostaglandin, Thromboxane and Leukotriene Research," B. Samuelsson, P.W. Ramwell, R. Paoletti, eds., Raven Press, New York, Vol. 6, pp. 207-212 (1980).

59. G. Ciabattoni, F. Pugliese, G.A. Cinotti, C. Patrono, Methodologic problems in the radioimmunoassay of prostaglandin E_2 and $F_{2\alpha}$ in human urine, <u>in</u>: "Radioimmunoassay of Drugs and Hormones in Cardiovascular Medicine," A. Albertini, M. Da Prada, B.A. Peskar, eds.,

Elsevier/North Holland, Amsterdam, pp. 265-280 (1979).

60. C. Patrono, F. Pugliese, G. Ciabattoni, P. Patrignani, A. Maseri, S. Chierchia, B.A. Peskar, G.A. Cinotti, B.M. Simonetti, A. Pierucci, Evidence for a direct stimulatory effect of prostacyclin on renin release in man, J. Clin. Invest. 69:231-239 (1982).

61. C. Patrono, Validation criteria for radioimmunoassay, in: "Radioimmunoassay in Basic and Clinical Pharmacology," C. Patrono and B.A. Peskar, eds., Handb. Exp. Pharm., Springer-Verlag, Berlin, Vol. 82, pp. 213-225 (1987).

62. G.A. FitzGerald, A.R. Brash, P.F. Falardeau, J.A. Oates, Estimated rate of prostacyclin secretion into the circulation of normal man, J. Clin. Invest. 68:1272-1276 (1981).

63. C. Patrono, G. Ciabattoni, F. Pugliese, A. Pierucci, I.A. Blair, G.A. FitzGerald, Estimated rate of thromboxane secretion into the circulation, J. Clin. Invest. 77:590-594 (1986).

64. J.A. Lawson, C. Patrono, G. Ciabattoni, G.A. FitzGerald, Long-lived enzymatic metabolites of TxB_2 in the human circulation, Anal. Biochem. 155:198-205 (1986).

65. G. Ciabattoni, F. Pugliese, A. Pierucci, B.M. Simonetti, C. Patrono, Fractional conversion of thromboxane B_2 to urinary 11-dehydro-thromboxane B_2 in man, J. Clin. Invest., submitted for publication.

66. G. Ciabattoni, J. Maclouf, F. Catella, G.A. FitzGerald, C. Patrono, Radioimmunoassay of 11-dehydro-thromboxane B_2 in human plasma and urine, Biochim. Biophys. Acta 918:293-297 (1987).

67. G. Ciabattoni, F. Pugliese, G.A. Cinotti, G. Stirati, R. Ronci, G. Castrucci, A. Pierucci, C. Patrono, Characterization of furosemide-induced activation of the renal prostaglandin system, Eur. J. Pharmacol. 60:181-187 (1979).

68. J.R. Gill, Jr., J.C. Frölich, R.E. Bowden, A.A. Taylor, H.R. Keiser, W.S. Hannsjorg, J.A. Oates, F.C. Bartter, Bartter's syndrome: a disorder characterized by high urinary prostaglandins and a dependence of hyperreninemia on prostaglandin synthesis, Am. J. Med. 61:43-51 (1976).

69. H.G. Güllner, C. Cerletti, F.C. Bartter, J.B. Smith, J.R. Gill, Jr., Prostacyclin overproduction in Bartter's syndrome, Lancet II:767-768 (1976).

70. M.L. Watson, J.R. Gill, R.A. Branch, J.A. Oates, A.R. Brash, Systemic prostaglandin I_2 synthesis is normal in patients with Bartter's syndrome, Lancet II:368-369 (1983).

71. L.J. Roberts, B.J. Sweetman, J.A. Oates, Metabolism of thromboxane B_2 in man, J. Biol. Chem. 256:8384-8393 (1981).

72. G. Ciabattoni, G.A. Cinotti, A. Pierucci, B.M. Simonetti, M. Manzi, F. Pugliese, P. Barsotti, G. Pecci, F. Taggi, C. Patrono, Effects of sulindac and ibuprofen in patients with chronic glomerular disease, N. Engl. J. Med. 310:279-283 (1984).

73. G. Ciabattoni, F. Pugliese, G.A. Cinotti, C. Patrono, Renal effects of anti-inflammatory drugs, Eur. J. Rheumatol. Inflam. 3:210-221 (1980).

74. P. Patrignani, P. Filabozzi, C. Patrono, Selective cumulative inhibition of platelet thromboxane production by low-dose aspirin in healthy subjects, J. Clin. Invest. 69:1366-1372 (1982).

75. F. Catella, F. Pugliese, P. Patrignani, P. Filabozzi, G. Ciabattoni, C. Patrono, Differential platelet and renal effects of sulfinpyrazone in man, (Abstract), Clin. Res. 32:239A (1984).

76. P. Patrignani, P. Filabozzi, F. Catella, F. Pugliese, C. Patrono, Differential effects of dazoxiben, a selective thromboxane-synthase inhibitor, on platelet and renal prostaglandin-endoperoxide metabolism, J. Pharmacol. Exp. Ther. 228:472-477 (1984).

77. C. Patrono, G. Ciabattoni, P. Filabozzi, F. Catella, L. Forni, M. Segni, P. Patrignani, F. Pugliese, B.M. Simonetti, A. Pierucci, Drugs, Prostaglandin and Renal Function, in: "Advances in Prostaglandin, Thromboxane and Leukotriene Research," G.G. Neri Serneri, J.C. McGiff, R. Paoletti, G.N.R. Born, eds., Raven Press, New York, Vol. 3, pp. 131-139 (1985).

78. P. Patrignani, F. Catella, P. Filabozzi, F. Pugliese, A. Pierucci, B.M. Simonetti, L. Forni, M. Segni, C. Patrono, Differential effects of OKY-046, a selective thromboxane-synthase inhibitor, on platelet and renal prostaglandin endoperoxide metabolism, (Abstract), Clin. Res. 32:246A (1984).

79. G.J. Roth, N.S. Stanford, P.W. Majerus, Acetylation of prostaglandin synthase by aspirin, Proc. Natl. Acad. Sci. USA 72:3073-3077 (1975).

80. F.J. Van der Ouderaa, M. Buytenhek, D.H. Nugteren, D.A. Van Dorp, Acetylation of prostaglandin endoperoxide synthase with acetylsalicylic acid, Eur. J. Biochem. 109:1-8 (1980).

81. C. Patrono, G. Ciabattoni, P. Patrignani, P. Filabozzi, E. Pinca, M.A. Satta, D. Van Dorne, G.A. Cinotti, F. Pugliese, A. Pierucci, B.M. Simonetti, Evidence for a renal origin of urinary TxB_2 in health and disease, in: "Advances in Prostaglandin, Thromboxane and Leukotriene Research," B. Samuelsson, R. Paoletti, P. Ramwell, eds., Raven Press, New York, Vol. 11, pp. 493-498 (1983).

82. G.A. FitzGerald, J.A. Oates, J. Hawiger, R.L. Maas, L.J. Roberts II, J.A. Lawson, A.R. Brash, Endogenous biosynthesis of prostacyclin and thromboxane and platelet function during chronic administration of

aspirin in man, J. Clin. Invest. 71:676-688 (1983).

83. G. Ciabattoni, A.H. Boss, L. Daffonchio, J. Daugherty, G.A. FitzGerald, F. Catella, F. Dray, C. Patrono, Radioimmunoassay measurement of 2,3-dinor metabolites of prostacyclin and thromboxane in human urine, in: "Advances in Prostaglandin, Thromboxane and Leukotriene Research," B. Samuelsson, R. Paoletti, P.W. Ramwell, eds., Raven Press, New York, Vol. 17, pp. 598-602 (1987).

84. C. Patrono, G. Ciabattoni, E. Pinca, F. Pugliese, G. Castrucci, A. De Salvo, M.A. Satta, B.A. Peskar, Low-dose aspirin and inhibition of thromboxane B_2 production in healthy subjects, Thromb. Res. 17:317-327 (1980).

85. D.E. Duggan, L.E. Hare, C.A. Ditzler, B.W. Lei, C. Kwan, The disposition of sulindac, Clin. Pharmacol. Ther. 21:326-335 (1977).

86. J.R. Sedor, S.L. Williams, A.N. Chremos, C.L. Johnson, M.J. Dunn, Effects of sulindac and indomethacin on renal prostaglandin synthesis, Clin. Pharmacol. Ther. 36:85-91 (1984).

87. J.L. Nadler, M. McKay, V. Campese, J. Vrbanac, R. Horton, Evidence that prostacyclin modulates the vascular actions of calcium in man, J. Clin. Invest. 77:1278-1284 (1986).

88. R. Vriesendorp, D. De Zeeuw, P.E. De Jong, A.J.M. Donker, J.J. Pratt, G.K. van der Hem, Reduction of urinary protein and prostaglandin E_2 excretion in the nephrotic syndrome by nonsteroidal anti-inflammatory drugs, Clin. Nephrol. 25:105-110 (1986).

89. A. Salvetti, R. Pedrinelli, P. Alberici, A. Magagna, B. Abdel-Haq, The influence of indomethacin and sulindac on some pharmacological actions of atenolol in hypertensive patients, Br. J. Clin. Pharmacol. 17:108S-111S (1984).

90. A. Salvetti, R. Pedrinelli, A. Magagna, P. Ugenti, Differential effects of selective and non-selective prostaglandin-synthesis inhibition on the pharmacological responses to captopril in patients with essential hypertension, Clin. Sci. 63:261S-263S (1982).

91. I.B. Puddey, L.J. Beilin, R. Vandongen, R. Banks, I. Rouse, Differential effects of sulindac and indomethacin on blood pressure in treated essential hypertensive subjects, Clin. Sci. 69:327-336 (1985).

92. C.D. Mistry, C.J. Lote, R. Gokal, W.J. Currie, M. Vandenburg, M.P. Mallick, Effects of sulindac on renal function and prostaglandin synthesis in patients with moderate chronic renal insufficiency, Clin. Sci. 70:501-505 (1986).

93. F. Guarner, C. Guarner, J. Prieto, I. Colina, J. Quiroga, J. Casas, R. Freixa, J. Posello, E. Gelpi, J. Balanzo, Increased synthesis of systemic prostacyclin in cirrhotic patients, Gastroenterology 90:687-694

(1986).

94. G. Laffi, G. La Villa, M. Pinzani, G. Ciabattoni, P. Patrignani, M. Mannelli, F. Cominelli, P. Gentilini, Altered renal and platelet arachidonic acid metabolism in cirrhosis, Gastroenterology 90:274-282 (1986).

95. U.G. Svendsen, J. Gerstoft, T.M. Hansen, P. Christensen, I.B. Lorenzen, The renal excretion of prostaglandins and changes in plasma renin during treatment with either sulindac or naproxen in patients with rheumatoid arthritis and thiazide-treated heart failure, J. Rheumatol. 11:779-782 (1984).

96. D.C. Brater, S. Anderson, B. Baird, W.B. Campbell, Effects of ibuprofen, naproxen and sulindac on prostaglandins in men, Kidney Int. 27:66-73 (1985).

97. D.G. Roberts, J.G. Gerber, J.S. Barnes, G.O. Zerbe, A.S. Nies, Sulindac is not renal-sparing in man, Clin. Pharmacol. Ther. 38:258-265 (1985).

98. K.J. Berg and T. Talseth, Acute effects of sulindac and indomethacin in chronic renal failure, Clin. Pharmacol. Ther. 37:447-452 (1985).

99. G. Laffi, G. Daskalopoulos, I. Kronborg, W. Hsueh, P. Gentilini, R.D. Zipser, Effects of sulindac and ibuprofen in patients with cirrhosis and ascites. An explanation for the renal-sparing effect of sulindac, Gastroenterology 90:182-187 (1986).

100. E. Quintero, P. Gines, V. Arroyo, A. Rimola, J. Camps, J. Gaya, A. Guevara, M. Rodamilans, J. Rodes, Sulindac reduces the urinary excretion of prostaglandins and impairs renal function in cirrhosis with ascites, Nephron 42:298-303 (1986).

101. C. Patrono and M.J. Dunn, The clinical significance of inhibition of renal prostaglandin synthesis, Kidney Int. 32:1-12 (1987).

102. G. Ciabattoni, A.H. Boss, P. Patrignani, F. Catella, B.M. Simonetti, A. Pierucci, F. Pugliese, P. Filabozzi, C. Patrono, Effects of sulindac on renal and extrarenal eicosanoid synthesis, Clin. Pharmacol. Ther. 41:380-383 (1987).

103. A. Kirstein Pedersen and P. Jakobsen, Sulfinpyrazone metabolism during long-term therapy, Br. J. Clin. Pharmacol. 11:597-603 (1981).

THE ROLE OF EICOSANOIDS IN HUMAN GLOMERULAR DISEASE

Alessandro Pierucci and Giulio Alberto Cinotti

Department of Internal Medicine II,
Division of Nephrology,
University of Rome "La Sapienza"
Rome, Italy

INTRODUCTION

A large body of evidence suggests a complex role of renal prostaglandins (PGs) in the control of cortical and medullary functions, i.e., renal blood flow (RBF), glomerular filtration rate (GFR), renin release, urinary concentrating mechanism, and electrolyte and water excretion. Several reviews of renal arachidonic acid metabolism can be recommended (1-4). PGs are considered to be ubiquitous autacoids, that is, substances that act predominantly at their production site. Since different nephron segments have highly specialized structures and functions, the synthesis and functions of the various PGs should be considered separately for each segment, rather than globally for the whole kidney (3). Although no direct evidence exists for a differential origin of any given urinary eicosanoid from a well-defined nephron segment in human kidney, a working hypothesis can be put forward relating cellular sites of biosynthesis and potential actions of eicosanoids to urinary excretion (Table 1). Prostacyclin (PGI_2) and prostaglandin ($PG)E_2$ are powerful vasodilators and appear to serve an important role in maintaining renal vasodilation. However, in the human kidney, PGI_2 is primarily involved in the modulation of cortical events, i.e., RBF, GFR and renin release, while PGE_2 is largely confined to controlling medullary functions, i.e., medullary blood flow, sodium and chloride reabsorption in the loop of Henle, and collecting-tubule response to vasopressin. It is important to point out that PGs do not appear to play a major role in the maintenance of renal hemodynamics under normal conditions (11). The importance of renal PGs for the maintenance of RBF and GFR can be demonstrated only under conditions in which the vasoconstrictor system (angiotensin II,

389

catecholamines, vasopressin) is activated (12). Situations in which the
activity of the various neurohumoral vasoactive mechanisms is increased
include sodium depletion, anesthetic or surgical stress, renal ischemia, and
clinical conditions such as congestive heart failure and cirrhosis with
ascites. All the above-mentioned states are characterized by an ineffective
circulatory volume with decreased renal perfusion. Although PGs were first
identified in 1930, it was not until 1971 that nonsteroidal antiinflammatory
drugs (NSAIDs) or aspirin-like drugs were shown to work by inhibiting PG synthesis (13). In 1976 Donker and coworkers demonstrated that indomethacin
reduced effective renal plasma flow (RPF) and GFR in chronic glomerular
disease (14). Since then, attention has been focused on the role of PGs in
the regulation of renal hemodynamics in renal disease. NSAIDs were introduced for treatment of patients with nephrotic syndrome by Fieschi and
Bianchi (15). They observed that phenylbutazone induced a reduction in proteinuria (15). De Vries and coworkers extended those observations by demonstrating the antiproteinuric effect of phenylbutazone and aminophenazone
(16). Moreover, they observed that the decrease in proteinuria induced by
these drugs was often associated with a mild but reversible decrease in creatinine clearance (16). NSAIDs are effective antiproteinuric agents, and
indomethacin has been the most extensively studied in the treatment of proteinuria in the nephrotic syndrome (17). However, whether NSAIDs therapy
favorably influences the long-term prognosis of nephrotic syndrome is still
a matter of debate, and prospective, placebo-controlled studies have not

Table 1. Renal Eicosanoids: Sites of Synthesis and Physiologic Actions

BIOLOGICALLY ACTIVE EICOSANOIDS	STABLE URINARY DERIVATIVES	SITES OF SYNTHESIS	ACTIONS
PGI_2	6-keto-PGF_{1alpha}	-Glomerulus -Arterioles	-Mesangial relaxation -Vasodilation -Renin release
TXA_2	TXB_2	-Glomerulus	-Mesangial contraction -Vasoconstriction
PGE_2	PGE_2	-Medullary interstitium -Tubules	-Vasodilation -Natriuresis -Diuresis
PGF_{2alpha}	PGF_{2alpha}	-Medullary interstitium -Tubules	- ?

PG = prostaglandin; TX = thromboxane
Data are based largely on information derived from: (1,5,6,7,8,9,10)

been performed as yet. Our specific aims are to describe (a) the pattern of urinary PGs, as a reflection of renal cyclooxygenase activity in human chronic glomerular disease, and (b) the renal functional effects of cyclooxygenase inhibitors and those of a specific thromboxane A_2 receptor antagonist. Furthermore, we will focus on human studies and refer to animal work only to the extent that it has direct relevance to the clinical situation or provides otherwise unavailable information.

MEASUREMENT OF RENAL ARACHIDONATE METABOLISM IN MAN

Over the last ten years, increasing evidence has suggested that the urinary excretion of primary unmetabolized PGs is largely a result of renal synthesis, as originally suggested by Frolich et al. in healthy women (5). Although measurements of primary PGs have been attempted in the venous effluent of the kidney, the major limitations of this approach are linked to the invasive procedure and the mechanical activation of platelets during and after sampling (18). Urine, besides having a PG concentration that is two orders of magnitude greater than that of renal venous plasma, has the additional advantages of being virtually free of platelet eicosanoids and of providing a noninvasive and continuous, rather than episodic, measure of renal PG synthesis. However, it should be pointed out that the potential contamination of male urine with seminal fluid (19) reduces the accuracy of urinary PG measurements in males, and consequently, human studies should use female subjects whenever possible. In addition to PGE_2 and $PGF_{2\alpha}$, human urine contains measurable amounts of 6-keto-$PGF_{1\alpha}$ (8), chemically stable hydration product of PGI_2 and thromboxane (TX)B_2, chemically stable hydration product of TXA_2 (7). Measuring the biosynthesis of renal PGs and TX through assays of urine is based on the assumption that several eicosanoid breakdown products and primary PGs in urine are derived from the kidney (20). It has been shown that the respective hydration products of TXA_2 and PGI_2, i.e. TXB_2 and 6-keto-$PGF_{1\alpha}$, are more likely to reflect renal biosynthesis than dinor metabolites, i.e. 2-3 dinor TXB_2 and 2-3 dinor 6-keto-$PGF_{1\alpha}$ (18). For example, urinary TXB_2 excretion is increased in patients with systemic lupus erythematosus (SLE), but the increment in urinary TXB_2 excretion is unaccompanied by a corresponding increase in 2-3 dinor TXB_2 excretion (9).

Gas chromatography-mass spectrometry (GC/MS) as well as radioimmunoassay (RIA) have been employed to quantitate urinary PG and TXB_2 excretion (20). The following validation criteria for RIA, which should be used to assess the specificity of urinary PGs and TX measurements are (a) characterization of the chromatographic pattern of distribution of the extracted PG-like immunoreactivity on thin-layer or high-pressure liquid chromatography, (b) use of multiple antisera, and (c) comparison with GC/MS determinations (21). This approach can provide meaningful information on renal PGs and TXA$_2$ synthesis in a number of pathophysiological conditions such as chronic glomerular disease.

PATTERN OF URINARY PROSTAGLANDINS IN HUMAN CHRONIC RENAL DISEASE

Urinary PGE$_2$

A number of studies have been carried out to quantitate the excretion of PGE$_2$ in chronic renal parenchymal disease. However, the results in patients with renal failure are conflicting. In fact, the inclusion of male patients in some studies is a confounding factor (Table 2). With the exception of patients with severe renal failure in whom the excretion rate of PGE$_2$ was profoundly diminished (24,27) and patients with SLE in whom urinary PGE$_2$ excretion was moderately increased (22), most patients with mild to moderate renal failure have been reported to have normal excretory rates (23-25,28). The impaired renal PGE$_2$ synthesis in severe renal failure (24,27) may be related to loss of renal mass; however, the decreased PGE$_2$ excretion which appears to parallel the progression of renal failure does not exclude an accelerated production in the remaining intact PGE$_2$-synthesizing cells (27). Blum et al. (23) and Ruilope et al. (24) reported that patients with moderate renal insufficiency had higher PGE$_2$ excretion than a control group. However, the inclusion of male patients in both studies might have contributed to a highly variable fraction of the measured urinary PGE$_2$. In fact, when the urinary PGE$_2$ excretion of female patients with mild-moderate renal failure was compared with age- and sex-matched control subjects, it appeared slightly elevated (not significant) (27,28). In most of the above-mentioned studies, no statistically significant correlation has been demonstrated between the urinary excretion of PGE$_2$ and any of the measured parameters of cortical or medullary function in either patients or controls.

In a comparative study on 16 female patients with proven chronic glomerular disease and normal renal function or mild renal impairment (C_{creat} 91 $\pm$ 19 ml/min./1.73 m^2; mean $\pm$ SD), and 20 healthy women (C_{creat} 110 $\pm$ 5 ml/min./1.73 m^2; mean $\pm$ SD), we found that both groups excreted PGE$_2$ at almost identical rates, i.e. 7.6 $\pm$ 2.7 vs. 7.4 $\pm$ 2.4 ng per hour (29) (Fig. 1). Moreover, although urinary PGE$_2$ excretion was significantly increased

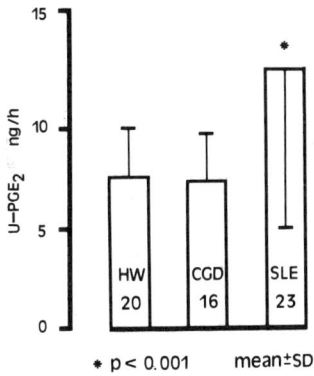

Fig. 1

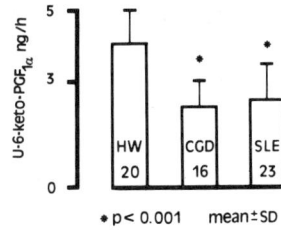

Fig. 2

Fig. 1. Urinary PGE$_2$ excretion in healthy women (HW), chronic glomerular disease (CGD) and systemic lupus erythematosus (SLE).

Fig. 2. Urinary 6-keto-PGF$_{1\alpha}$ excretion in healthy women (HW), chronic glomerular disease (CGD) and systemic lupus erythematosus (SLE).

in a group of 23 female patients with lupus nephritis (13 patients with active and 10 patients with inactive renal lesions), only 8 patients (6 with active renal lesions) had urinary PGE$_2$ excretion in excess of 2 standard deviations of the normal mean (9) (Fig. 1). Since interstitial mononuclear cells are frequently found in focal and diffuse proliferative forms of lupus nephritis (30), the increased urinary PGE$_2$ excretion in 35% of SLE patients could represent an inflammatory response involving mononuclear cells in the interstitial spaces. It should be pointed out that urinary PGE$_2$ did not correlate with either GFR or RPF to any statistically significant extent (9).

Urinary 6-keto-PGF$_{1\alpha}$

We have observed a reduced excretion of 6-keto-PGF$_{1\alpha}$ in female patients with proven chronic glomerular disease (29), including lupus nephritis (9) (Fig. 2). In these patients the urinary excretion of 6-keto-PGF$_{1\alpha}$ showed a statistically significant positive correlation with both GFR and RPF (9,29). Nadler et al. (31) have recently reported that in 7 patients (diabetes mellitus, essential hypertension, SLE) with moderate chronic renal failure (C$_{creat}$ 46 ± 5 ml/min.; mean ± SEM), the 24-hour excretion of 6-keto-PGF$_{1\alpha}$ was similar to that of 12 normal subjects (164 ±

Table 2. PGE$_2$ Excretion In Human Chronic Renal Disease

AUTHORS	NUMBER OF PATIENTS, SEX	DIAGNOSIS	RENAL FAILURE	URINARY PGE$_2$ (mean±SEM)	URINARY PGE$_2$ IN CONTROLS (mean±SEM)
Kimberly R.P. et al, (22) 1978	n = 7 F	Systemic Lupus Erythematosus (Clinical criteria and renal biopsy in 4)	*Mild* (C_{In}: 85±9.2 ml/min)	42.7±6.4 ng/h	29.0±1.9 ng/h n = 25 F
Blum M. et al, (23) 1981	n = 20 13 M 7 F	Chronic GN and Chronic IN (Clinical criteria and renal biopsy in most cases)	*Mild* (C_{Cr}: 106±14 ml/min, n = 7)	801±83 ng/24h	765±65 ng/24h n = 29, Sex: NR
			Moderate (C_{Cr}: 36±4.28 ml/min, n = 7)	2560±381 ng/24h	
			Severe (C_{Cr}: 11.8±0.42 ml/min, n = 6)	555±104 ng/24h	
Ruilope L. et al, (24) 1981	n = 46 22 M 24 F	Chronic GN (n = 30) Chronic IN (n = 16) (N.S in 9)	*Moderate* (C_{Cr}: 20-60 ml/min, n = 15)	1.18±0.20 ug/24h	1.36±0.13 ug/24h n = 27 (16M,11F)
			Mild-Moderate (NS) (C_{Cr}: 55-90 ml/min, n = 9)	2.69±1.17 ug/24h	
			Severe (C_{Cr}: <20 ml/min, n = 22)	0.24±0.25 ug/24h	
Ruilope L. et al, (25) 1982	n = 37 21 M 16 F	Chronic GN (n = 17) Chronic IN (n = 10) Adult polycystic kidney disease (n = 10) (Hypertension in 21: 12 M, 9 F)	*Moderate-Severe* (C_{Cr}: above 25 ml/min/1.73 m^2) *Normotensive* (n = 16) *Hypertensive* (n = 21)	1.1±0.2 ug/24h 1.1±0.1 ug/24h	1.3±0.1 ug/24h n = 27 (16 M, 11 F)

Reference	n	Diagnosis	Condition	Value		NR
Mordechai S. et al, (26) 1985	n = 6 M	Chronic Renal Failure (Clinical criteria and renal biopsy in 3)	Moderate-Severe (C_Cr: 16.1-43.8 ml/min/1.73 m²)	2.114±634 ng/24h		NR
Lebel M., Grose J.H., (27) 1986	n = 41 25 M 16 F	*Cortical renal lesions* (n = 15: Chronic GN in 12, Acute GN in 1, Polycystic kidney in 2) *Interstitial lesions* (n = 9: Chronic IN in 6, Hereditary tubulo-interstitial nephritis in 2, Medullary cystic disease in 1) (Clinical criteria and renal biopsy in 16)	*Without Renal Failure* (C_in: >80 ml/min, n = 7)	367±73 ng/24h (M)	107±16 ng/24h (F)	277±41 ng/24h (M = 11)
			Moderate (C_in: >25 ml/min, n = 10)			130±13 ng/24h (F = 11)
			Severe (C_in: <25 ml/min, n = 8)	81±17 ng/24h (M)	68±17 ng/24h (F)	
			End Stage Renal Failure (on hemodialysis n = 16)	(n=17) 65±21 ng/24h (M = 8)	(n=8) 29±16 ng/24h (F = 8)	
Mistry C.D. et al, (28) 1986	n = 9 4 M 5 F	Chronic Renal Failure	Moderate (C_Cr: 24.7-54.6 ml/min/1.73 m²)	47.2±13.6 ng/h n = 5 F		30.85±6.7 ng/h n = 4 F

M = Males, F = Females, GN = Glomerulonephritis, IN = Interstitial Nephritis, NS = Nephrotic Syndrome, NR = Not Reported
ALL VALUES ARE REPORTED AS IN ORIGINAL PAPERS (means±SEM)

20 ng/g of creatinine vs. 185 $\pm$ 37 ng/g of creatinine; mean $\pm$ SEM). However, 6-keto-PGF$_{1\alpha}$, but not PGE$_2$, was markedly reduced in 7 other
patients with hyporeninemic hypoaldosteronism (diabetes mellitus, essential
hypertension, SLE) (41.8 $\pm$ 7 ng/g of creatinine) with comparable moderate
renal failure (C_{creat} 36 $\pm$ 5 ml/min.). In agreement with our findings,
Mistry et al. (28) have shown that the urinary excretion of 6-keto-PGF$_{1\alpha}$
was significantly lower in 9 patients with established moderate renal failure (C_{creat} 24.7-54.6 ml/min./1.73 m^2) than in healthy women (19.6 $\pm$ 8.7
ng/h vs. 45.86 $\pm$ 5.86 ng/h; mean $\pm$ SEM). The reduced excretion of PGI$_2$
hydration product in chronic glomerular disease may be the consequence of a
reduced glomerular, as well as arteriolar, mass available for PGI$_2$ synthesis. Other hypotheses that could explain the selective reduction in the
synthesis of PGI$_2$ are (a) inhibition of PGI$_2$-synthase by hydroperoxy-
derivatives of arachidonate (6) or (b) feedback regulation of PGI$_2$ synthesis due to the hemodynamic changes associated with progressive glomerular
sclerosis (33). To the extent that urinary excretion of 6-keto-PGF$_{1\alpha}$ is
primarily a reflection of the renal synthesis of the unstable parent compound, the reduction of renal PGI$_2$ production seems to indicate that the
suggested adaptive increases in glomerular capillary pressure and flow
(reviewed in ref. 34), characterizing progressive glomerular sclerosis, are
not associated with increased PGI$_2$ production.

Urinary Thromboxane B$_2$

We also described enhancement of urinary TXB$_2$, as a reflection of
intrarenal synthesis of vasoconstrictor TXA$_2$, in patients with lupus nephritis (9) but not in other forms of chronic glomerular disease (9,29). Moreover, SLE patients with active renal lesions differ from those with inactive
renal lesions by having a twofold higher TXB$_2$ excretion rate (9) (Fig. 3).
Furthermore, a significant inverse correlation was found between urinary
TXB$_2$ and creatinine clearance (9). In contrast to urinary TXB$_2$ excretion, no increase was seen in 2,3-dinor- TXB$_2$, a major metabolite that primarily reflects systemic thromboxane production (Table 3), thus suggesting
that the increased urinary TXB$_2$ excretion reflects intrarenal production
(9). TXA$_2$ is a potent vasoconstrictor, in addition to its well-known role
in platelet aggregation, and may have an immunoregulatory action. Several
reports indicate involvement of TXA$_2$ in the pathogenesis of heterogeneous
experimental models of kidney disease such as ureteral obstruction (35), subtotal nephrectomy (36), murine lupus nephritis (37) and in some models of
immune and nonimmune glomerular injury (38-41). In addition to vasoconstrictive activity, TxA$_2$ could regulate glomerular ultrafiltration coefficient

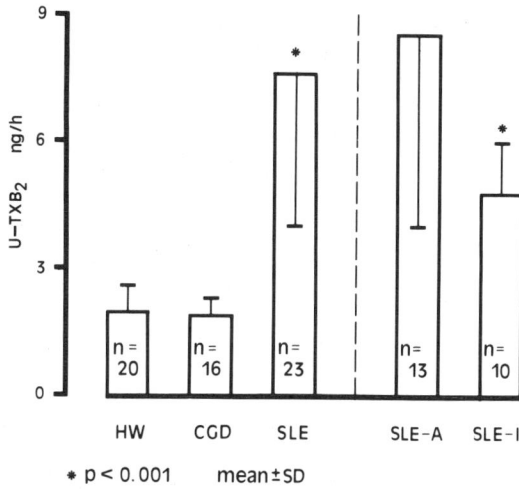

Fig. 3. Urinary TXB_2 excretion in healthy women (HW), chronic glomerular disease (CGD) and systemic lupus erythematosus (SLE). On the right of the dashed line, values of TXB_2 excretion in SLE with active (A) and inactive (I) renal lesions are reported.

(Kf) through an effect on the glomerular mesangium, causing contraction of mesangial cells and a decrease in the area available for filtration (42).

TxA_2 is synthesized, in vitro, by glomerular epithelial and mesangial cells (43,44) as well as by a variety of circulating inflammatory cells, including neutrophils (45), macrophages (46) and platelets (47).

Table 3. Excretion of TXA_2 Metabolites In Healthy Women (HW) and SLE, Active (A) and Inactive (I)

Subjects	Urinary TXB_2 ng/h	2,3-dinor-TXB_2 ng/h
HW (N=20)	2.2±0.3	13.4±4.5
SLE-A (N=13)	9.6±5.6	14.7±4.7
SLE-I (N=10)	4.7±1.9	13.1±4.4

Data are from (9)

Recently, it has been demonstrated in two animal models of membranous nephropathy that induction of immunologic glomerular injury is associated with an increase in TXB_2 (and PGE_2) synthesis by isolated glomeruli (48,49). These studies prove that the increased glomerular synthesis of TXA_2 (and PGE_2) must result from enhanced glomerular production of eicosanoids, rather than from contributions by inflammatory cells, since these models of glomerular injury develop in the absence of infiltrating, inflammatory cells.

Interstitial, mononuclear, inflammatory infiltrates are found in many different types of human primary glomerular disease (50). In lupus nephritis the prevalence and severity of interstitial inflammation correlates with the pattern of glomerular injury, being greater in the active, endocapillary proliferative forms (30). By light microscopical analyses, these populations of mononuclear inflammatory cells infiltrating the renal interstitium are composed of lymphocytes, monocytes, plasma cells and polymorphonuclear leukocytes (30). The potential cellular sources of intrarenal TXA_2 in lupus nephritis (9) include parenchymal renal cells, blood-derived mononuclear cells and platelets, or biochemical cooperation among these different cell types. Macrophages have been shown to be prominent participants in immunologically induced glomerular damage in human disease (51-53), and evidence that they are important mediators of glomerular injury has also been accumulating (54). Interstitial monocyte-macrophage cells are frequently found in renal biopsies of patients with SLE, and these infiltrates are correlated with the presence of active lesions (50). On the other hand, macrophages have a key role in mediating inflammatory reactions, and they release large quantities of inflammatory mediators, including TXA_2 (55). Platelets are a major source of TXA_2 at sites of platelet aggregation. They are also a potential source of a number of mediators possibly affecting immune-complex deposition, capillary permeability, mesangial function and glomerular hemodynamics (56). Moreover, platelet-derived growth factors can stimulate the proliferation of endothelial (57) and mesangial (58) cells. Interestingly, glomerular thrombi occur frequently in proliferative lupus nephritis and may represent an important prognostic factor in predicting whether glomerular sclerosis will subsequently develop (59). In summary, the cellular sources of intrarenal TXA_2 production and the mechanisms responsible for its enhanced synthesis in lupus nephritis remain entirely speculative (9,37). Although increments of TXA_2 may contribute to the deterioration of renal function in lupus nephritis, the precise role of enhanced intrarenal TXA_2 production can be best assessed by studying TXA_2 synthase inhibitors or specific receptor antagonists (see below).

FUNCTIONAL EFFECTS OF SELECTIVE AND NONSELECTIVE CYCLOOXYGENASE INHIBITORS
AND A SPECIFIC THROMBOXANE A_2-RECEPTOR ANTAGONIST

Selective and Nonselective Cyclooxygenase Inhibitors

Numerous NSAIDs are presently available for clinical use. These drugs
have become increasingly popular because of their effectiveness in a broad
range of clinical disorders, such as arthritis and dysmenorrhea. Since
several clinically important effects depend on their capacity for inhibition
of eicosanoid synthesis, most widely used NSAIDs have been tested for their
short-term and/or long-term effects on renal PG synthesis in healthy subjects as well as in patients with renal, hepatic, rheumatic, and cardiovascular disease. These include indomethacin (60-66), ibuprofen (29,60,61,65,67,
68), naproxen (67-71), fenoprofen (67), diclofenac (70,72), piroxicam (70,
71,73), flurbiprofen (74) and sulindac (28,29,62-65,69,73,75-80).

These NSAIDs have been reported to inhibit 50 to 80% of renal as well as
extrarenal cyclooxygenase activity in health and disease (21). On the other
hand, sulindac (75) and low-dose aspirin selectively spare renal cyclooxygenase activity (81). The biochemical selectivity of low-dose aspirin is probably related to a different rate of resynthesis of the enzyme (82,83) which
occurs within hours in the glomeruli, while several days are needed to produce new non-acetylated platelets. In fact, irreversible acetylation of
cyclooxygenase by aspirin requires de novo synthesis of the enzyme in order
to restore normal PGs and TXA_2 formation. Such a process can easily occur
within hours in most nucleated cells but cannot take place in platelets that
lack the biochemical capacity for protein synthesis (84). In the case of
sulindac, the selective sparing of renal cyclooxygenase activity in health
(11) and disease (28,29,62,75,78) probably is related to renal enzymes capable of oxidizing the active agent to the inactive metabolites, sulindac sulfide or sulindac sulfone. Several studies (28,62,63) have confirmed our
original observations that sulindac spares renal cyclooxygenase activity
(29,75). However, a number of other studies in health (68,77) as well as in
disease (69,76,78,79) contain conflicting data regarding the biochemical selectivity of sulindac (reviewed in refs. 21 and 85).

Recently, Ciabattoni et al. (80) have reported that during chronic administration of sulindac (200, 400, 600 and 800 mg/day, each dose given for 7
days in successive weeks) in healthy subjects, both 2,3-dinor-TXB_2 and
2,3-dinor-6-keto-$PGF_{1\alpha}$, reflecting systemic synthesis of TXA_2 and PGI_2
respectively, showed a dose-dependent reduction ranging between 45% and

80%. In contrast, the urinary excretion of 6-keto-PGF$_{1\alpha}$ and TXB$_2$ did not change significantly throughout the study. These results thus extend previous observations of selective sparing of renal cyclooxygenase activity by sulindac in humans and demonstrate that this selectivity is not simply related to an overall weaker enzyme inhibition.

Nonsteroidal Antiinflammatory Drugs in Chronic Glomerular Disease

The effects of cyclooxygenase inhibitors have been studied in several animal models of immune and nonimmune renal disease (reviewed in ref. 10). Stork and Dunn (86) have studied the effects of meclofenamate and indomethacin in a rat model of nephrotoxic serum nephritis (NSN). Short-term treatment with either cyclooxygenase inhibitor on day 14 of the autologous phase resulted in a 50% decrease in both RPF and GFR, suggesting that vasodilator PGE$_2$ (rat glomeruli produce mainly PGE$_2$ and less PGI$_2$ [87]) is hemodynamically important to preserve renal function in the rat model of NSN. Kaizu et al. (88) have studied the effects of indomethacin and saralasin (a competitive inhibitor of angiotensin II action) in a rat model of autologous NSN. Indomethacin infusion resulted in a reduction in RBF and GFR and a rise in renal vascular resistance. Saralasin reversed these abnormalities when infused with indomethacin, and an increase in GFR was noted when saralasin was infused alone. These hemodynamic changes may be consistent with increased action of vasodilator PGs opposing the action of an activated renin-angiotensin system (88). Recently, it has also been demonstrated that short-term administration of indomethacin significantly reduced RPF and GFR in the animal model of membranous nephropathy (49). The long-term consequences of selective versus nonselective cyclooxygenase inhibition have been studied in a rabbit model of NSN (90). Aspirin, at a dose (100 mg/kg per day during the autologous phase) that almost completely inhibits cyclooxygenase activity in both circulating cells and the kidney, worsened the morphologic expression of NSN and negatively influenced the clinical course of the disease (90). In contrast, sulindac, at a dose (60 mg/kg per day) that suppressed circulating platelet cyclooxygenase activity by 90% but substantially spared renal PG synthesis, prevented extracapillary proliferation and reduced proteinuria without negative influence on glomerular hemodynamics (90). The beneficial effect of sulindac was interpreted as being related to suppression of eicosanoid synthesis in blood-derived infiltrating cells, a potential source of inflammatory mediators contributing to glomerular damage.

Kelley et al. (37) administered ibuprofen (8-9 mg/kg per day) for one

week to two different strains of mice with autoimmune lupus nephritis. The
NSAID did not alter TXB_2 or PGE_2 in the renal cortex or medulla, although it suppressed platelet TXB_2 production by 90%. Long-term treatment
with ibuprofen, begun at two months of age, did not modify development of
renal disease or 12-month survival in NZBxW mice with spontaneous lupus nephritis (91). However, ibuprofen did change the site of immune complex localization from the subepithelial to the subendothelial aspect of the glomerular
basement membrane, a shift not associated with alterations in renal function
(91). The first report demonstrating that the maintenance of compensated
kidney function in chronic glomerular disease depended on endogenous production of PGs is that of Donker et al. (14). They reported a significant reduction of renal function after administration of indomethacin in patients
with chronic glomerular disease maintained on normal sodium intake. These
results were subsequently confirmed by others in diverse forms of renal parenchymal diseases (Table 4).

In a study of 10 patients with various forms of proven chronic glomerular disease (29), we found that the administration of ibuprofen for one week
(1200 mg per day--approximately half of therapeutic dose) significantly reduced creatinine and sodium para-aminohyppurate clearance by 27% and 32%
respectively, regardless of initial renal function. All patients were kept
on a controlled diet (sodium 100 mmol per day, potassium 80 mmol per day)
for at least one week before and during the study. Moreover, there was an
inverse relation between the reduction of RPF and GFR and the basal values
for urinary excretion of 6-keto-$PGF_{1\alpha}$ (r=-0.67 and -0.55, respectively;
p<0.01). Thus, reduction of urinary 6-keto-$PGF_{1\alpha}$ by ibuprofen suggests
that the renal function of these patients is critically dependent on PGI_2
production. Interestingly, all patients responded in a qualitatively similar fashion, which suggests that chronic glomerular damage and the accompanying hemodynamic and biochemical changes, independent of the nature of the
initial insult, represent a major determinant of the PG-dependence of renal
function (29).

Recently, Mistry et al. (28) reported a modest reduction in creatinine
and ^{99m}Tc-DTPA clearance during sulindac treatment (400 mg/day for 9 days)
in 9 patients with established chronic renal failure. In contrast, [131]I-
hyppuran clearance, plasma renin activity and electrolyte excretion were not
altered to any statistically significant extent. It is worth noting that
both urinary excretion of 6-keto-$PGF_{1\alpha}$ and PGE_2, measured only in female
patients, remained unchanged during sulindac administration. On the other
hand, Berg and Talseth (76) reported a 47% reduction in urinary PGE_2

Table 4. Renal Functional Effects Of Nonselective Cyclooxygenase
Inhibitors

AUTHORS	NUMBER OF PATIENTS	DIAGNOSIS	NSAIDs	DOSE AND DURATION OF TREATMENT	GFR Δ%	RPF Δ%
Donker A.J.M. et al, (14) 1976	n = 10	Renal parenchymal disease including transplant kidney	INDOMETHACIN	150 mg/day; 3 days	-12%	-19%
Arisz L. et al, (92) 1976	n = 19	Chronic glomerulo-nephritis	INDOMETHACIN (+ Na restriction)	150 mg/day; 7 days	-35%	-23% in 16 = in 3
Berg K.J. (93) 1977	n = 6	"Non-oedematous renal failure"	ACETYLSALI-CYLIC ACID	750 mg i.v.(acutely)	-54%	-66%
Kimberly R.P., Plotz P.H. (94) 1977	n = 23	Systemic Lupus Erythematosus	ACETYLSALI-CYLIC ACID	4.8 g/day; at least 7 days	>-58% in 11	NR
Donker A.J.M. et al, (95) 1978	n = 25	Nephrotic syndrome of varying origin	INDOMETHACIN	150 mg/day; 3 days	-25%	-17%
				150 mg/day; 1-3 years	-26%	-19%
Kimberly R.P. et al, (22) 1978	n = 7	Systemic Lupus Erythematosus	ASPIRIN	3.3 to 6.6 g/day; 7-8 days	-14%	-29%
Kimberly R.P. et al, (67) 1978	n = 3	Systemic Lupus Erythematosus	IBUPROFEN NAPROXEN	2.4 g/day; 3 days 750 mg/day; 10 days	-60% -60%	NR NR
			FENOPROFEN-CALCIUM	NR ; 6 days	-40%	NR
Tiggeler R.G.W.L., (96) 1979	n = 33	Varying chronic glo-merular diseases	INDOMETHACIN	150 mg/day; 5 days	-11%	NR
Ciabattoni G. et al, (29) 1984	n = 10	Varying chronic glomerular disease including lupus nephritis	IBUPROFEN	1200 mg/day; 7 days	-28%	-35%
Vriesendorp R. et al, (97) 1985	n = 10	Nephrotic syndrome of varying origin	INDOMETHACIN	150 mg/day; 7 days (n = 10)	-34%	-17%
			NAPROXEN (+low sodium intake and hydrochloro-thiazide 50 mg/day)	750 mg/day; 7 days (n = 5) 1500 mg/day; 7 days (n = 5)	-17% -17%	-10% -7%
Vriesendorp R. et al, (74) 1986	n = 7	Nephrotic syndrome of varying origin	INDOMETHACIN	150 mg/day; 7-10 days	-19%	NR
			DICLOFENAC	200 mg/day; 7-10 days	-19%	NR
			FLURBIPROFEN (+low sodium intake and hydrochloro-thiazide 50 mg/day)	200 mg/day; 7-10 days	-19%	NR
Toto R.D. et al, (98) 1986	n = 12	Chronic renal insuf-ficiency of varying etiologies	INDOMETHACIN KETOPROFEN INDOMETHACIN	50mg; acute study 50mg; acute study 150mg/day; 5 days	-18% -45% =	-44% -24% =
			KETOPROFEN	200mg/day; 5 days	=	=

NSAIDs : Non steroidal anti-inflammatory drugs, NR : Not Reported = : Unmodified

excretion during sulindac administration (400 mg/day) in 8 patients with
chronic renal failure, despite no changes in creatinine clearance.

In summary, studies on the renal functional effects of nonselective cyclooxygenase inhibitors suggest that vasodilatory PGs are important in preserving RPF and GFR in renal parenchymal disease. Most NSAIDs are antiproteinuric agents, especially if the patient is sodium-depleted. However,
long-term inhibition of renal cyclooxygenase might result in hemodynamically
mediated reductions in RPF and GFR and also in structural changes that may
not be reversible. Surprisingly, a possible contributory role of inhibition
of PG synthesis to the etiology of analgesic nephropathy has not been investigated extensively. Thus, the nephrotoxic potential of nonselective cyclooxygenase inhibitors, both in terms of consistent initial reduction in renal
function and the unknown risk of permanent renal impairment, argues against
the long-term use of these agents in renal disease (21). In this regard, a
remote and unproven potential long-term use for these drugs might be the prevention of progressive loss of renal function in the face of renal failure
and significant loss of nephron population (100).

Nonsteroidal Antiinflammatory Drugs in Reduced Renal Mass for Kidney Transplant Donation

The adaptive response to the reduction of renal mass in the rat is characterized by hyperfiltration in the remnant nephrons, which results from
elevation of the glomerular capillary plasma flow rate and hydraulic pressure (101). In addition, the reduction of renal mass leads to hypertrophy
and hyperplasia of remaining kidney tissue (reviewed in ref. 102). In the
remnant kidney of rats subjected to extensive renal ablation, the elevated
glomerular capillary pressure and flow are associated with progressive proteinuria and eventual glomerular sclerosis (101). However, the factors responsible for progressive renal failure in this model are incompletely understood (103). It is possible that vasodilatory PGs play a role in this hyperfiltration, since renal PG synthesis is stimulated by reduction in renal
mass and since the adaptive increase in renal function appears to require
cyclooxygenase activity (104-106).

Nath et al. (106) studied the effect of a short-term indomethacin infusion (5 mg/kg intravenously) on glomerular hemodynamics in euvolemic, subtotally nephrectomized rats. Indomethacin reduced both the remnant kidney
GFR and the single-nephron GFR (20% and 23% respectively) by decreasing glomerular plasma flow rate and K_f. These investigators also found that

absolute urinary excretion of PGE_2, 6-keto-$PGF_{1\alpha}$ and TXB_2 was striking-
ly increased in nephrectomized rats compared with controls (106). These
data were interpreted to suggest that adaptive increments in vasodilatory PG
synthesis sustain, at least in part, hyperfiltration in the remnant nephron
due to their effects on intrarenal vascular resistances and K_f (106). A
role of the renal PG system in the maintenance of the adaptive changes to a
diminished nephron population has also been suggested by Kirscherbaum and
Serros (107) in a rabbit model. The administration of two cyclooxygenase inhibitors resulted in a marked and significant reduction in endogenous creatinine clearance (107). More recently, Rubinger et al. (108) studied the
effects of indomethacin (33.3 mg/kg in 48 hours) and sulindac (30 mg/kg in
48 hours) in 5/6 nephrectomized rats. Indomethacin reduced the 24-hour urinary sodium excretion and the fractional excretion of sodium, independently
of changes in GFR, possibly as the result of direct tubular transport mechanisms. In contrast, sulindac had no effect on renal function or electrolyte
excretion in rats with reduced renal mass. Indomethacin suppressed urinary
PGE_2 and TXB_2 excretion in these animals, whereas sulindac had no effect
(108).

Evidence against the hypothesis that hyperperfusion and hyperfiltration
cause glomerular injury and sclerosis comes from work by Purkerson et al.
(36) who demonstrated increased urinary excretion of vasoconstrictor TXB_2
in rats with ablation of more than 70% of the renal mass. Long-term oral administration of OKY-1581, a selective TXA_2-synthase inhibitor, increased
RPF and GFR, decreased proteinuria, lowered blood pressure and improved
renal histology (36). OKY-1581 had no effect in normal rats or in ablated
rats pretreated with aspirin, suggesting that the effects of the drug were
due to inhibition of TXA_2 synthesis. Of particular interest was the finding that despite the increase in hyperfiltration and hyperperfusion seen
with the administration of OKY-1581, there was amelioration of the renal
disease in this model (36). As suggested by Klahr, platelet aggregation and
intraglomerular thrombosis may play a key role in the development of glomerulosclerosis in this experimental model of renal disease (reviewed in ref.
109). Interestingly, eicosapentaenoic acid, a potential substrate for cyclooxygenase, accelerates the rate of glomerular sclerosis in renoprival nephropathy (110), although it delays progression of immune-mediated renal disease
(111). However, the relevance of intrarenal changes in eicosanoid production (induced by eicosapentaenoic acid) to these different patterns of response remains to be established (reviewed in ref. 112).

The above-mentioned studies provide evidence for a functionally impor-

404

tant effect of vasodilator PG synthesis on renal function in renoprival states. Moreover, it has been reported that renal PGs may also participate in the mediation of compensatory renal growth in the uninephrectomized rat (113). The first report in uninephrectomized human beings demonstrating that the maintenance of RPF and GFR depends on endogenous production of PGs is that of Donker (114). Indomethacin (50 mg tid for three consecutive days) reduced RPF and GFR in 5 healthy uninephrectomized subjects both during a normal sodium intake (20% and 20%, respectively) and during sodium restriction (15% and 20%, respectively) (114). As living kidney donors for renal transplantation provide a useful study model to evaluate prospectively the compensatory hyperfiltration developing in remnant nephrons following renal ablation in human beings, Simonetti et al. (115) recently evaluated the influence of unilateral nephrectomy on GFR and PGI_2 synthesis and the dependence of GFR on intact renal cyclooxygenase activity in kidney donors. Four living kidney donors--one man and three women--40 to 49 years old, on normal fixed daily sodium and potassium were studied under control conditions and after one week of ibuprofen treatment (1.2 g/day) on three separate occasions; that is, (a) at one month prior to nephrectomy, (b) at one month postnephrectomy, and (c) at one year postnephrectomy. The obtained results demonstrated that GFR was not modified either prior to nephrectomy or at one month postunilateral nephrectomy. In contrast, one year after ablation, a comparable inhibition of PGI_2 synthesis was associated with a statistically significant reduction in GFR (Fig. 4) (115). Unilateral nephrectomy did not change the urinary 6-keto-$PGF_{1\alpha}$ excretion rate. However, considering the hypothetical contribution of each kidney (approximately 50%) to total renal function, a rise occurred in 6-keto-$PGF_{1\alpha}$ and GFR (86 ± 2.1% and 45 ± 2.3% at 1 month postuninephrectomy, respectively, and 49 ± 1.5% and 68 ± 1.8% at 1 year postuninephrectomy, respectively) (115). It should be pointed out that the increase of urinary 6-keto-$PGF_{1\alpha}$ excretion was not accompanied by any change of urinary 2,3-dinor-6-keto-$PGF_{1\alpha}$ (a major metabolite that reflects systemic synthesis of PGI_2) (116), thus suggesting that the increased urinary 6-keto-$PGF_{1\alpha}$ reflects an increased intrarenal production of PGI_2.

Thromboxane A_2-synthase Inhibitors and Thromboxane A_2-receptor Antagonists in Renal Disease

Besides being explored as potential antithrombotic agents, TX-synthase inhibitors have been used to investigate the role of TXA_2-dependent deterioration of renal function in experimental models of renal disease. Lianos et al. (38) showed increased production of TXB_2 in glomeruli isolated from

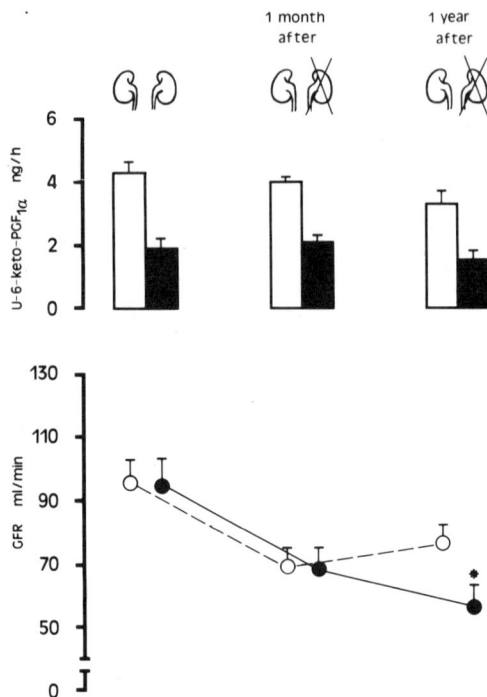

Fig. 4. Mean ± SD of urinary (U)-6-keto-PGF$_{1\alpha}$ excretion rate and GFR before and after nephrectomy under basal conditions (□ ○) and after 1 week of ibuprofen treatment (1.2 g/day) (■ ●). *p<0.05 (reproduced from reference 115).

NSN rats two hours after nephrotoxic serum administration (heterologous phase). The increments in TXB$_2$ were associated with decrements in GFR and filtration fraction over a 3-hour period following induction of NSN (38). Selective inhibition of TXA$_2$ synthesis by pretreatment with OKY-1581 (a pyridine derivative) or UK-38485 (an imidazole compound) partially prevented the acute decrements in RPF and GFR (38). In NSN, however, vasoconstrictor TXA$_2$ appears to mediate only the early decrements of both RPF and GFR (38), whereas at later stages (autologous phase), renal hemodynamics seem to be dependent upon vasodilator PGE$_2$ synthesis (86). In fact, long-term therapy with TXA$_2$-synthase inhibitor OKY-1581 had no effect on RPF and GFR (86). On the other hand, Rahman et al. (49) reported that acute administration of the TXA$_2$-synthesis inhibitor, UK-38485, and a TXA$_2$-receptor antagonist, EP-092, to a rat model of membranous nephropathy did not affect GFR or RPF. Thus, increments of glomerular TXA$_2$ are hemodynamically

unimportant in both the autologous phase of rat NSN (86) and in a noninfil-
trative model of immune-mediated glomerular disease (49).

Unilateral ureteric obstruction causes vasoconstriction resulting in a
decrease in RPF and GFR. Hydronephrotic kidneys produce increased amounts
of TXA_2 when studied in vitro (35,117). Using OKY-1581, Kawasaki and
Needleman (118) showed a dose-related inhibition of TXA_2 in the perfused
hydronephrotic rabbit kidney. Moreover, Yarger et al. (119) demonstrated
that intrarenal infusion of the TXA_2-synthase inhibitor, imidazole, into
rats with obstructive nephropathy improved renal function in vivo. Strand
et al. (120), however, were unable to confirm these observations using the
same inhibitor. More recently, it has been reported that administration of
imidazole into the renal artery of a rat model of unilateral hydronephrosis
improves renal function slightly, whereas administration of UK-37248 or
UK-38485 (two imidazole-substituted derivatives) doubles RBF and excretory
function (121). Thus, it seems that the hydronephrotic kidney produces
greater amounts of TXA_2 both in vitro and in vivo, and this eicosanoid is
an important mediator of vasoconstriction in experimental models of ureteral
obstruction.

TXA_2 also appears to play a role in renal allograft rejection.
Coffman et al. (122) investigated the role of TXA_2 in mediating the reduction in renal function of acute renal allograft rejection in the rat. In
addition to showing increased TXB_2 production by ex vivo perfused renal
allograft, they demonstrated that acute administration of UK-37248 was effective in decreasing urinary TXB_2 excretion as well as in increasing RBF and
GFR.

The role of TXA_2 (and PGs) in acute renal failure remains uncertain
(10). However, Benabe et al. (123) showed increased TXB_2 production by
isolated perfused rabbit kidney, previously injured by glycerol administration. Moreover, Sraer et al. (124) showed increased synthesis of TXB_2
(and PGE_2) in isolated glomeruli from glycerol-injured rats. Interestingly, Stegmeier et al. (125) reported the effects of a new selective TXA_2-
receptor antagonist, BM-13505, in glycerol-induced acute renal failure in
rabbits. Glycerol injection stimulated urinary TXB_2 excretion. Concomitant administration of glycerol and BM-13505 prevented the increase of serum
creatinine, and only minimal morphological changes were observed (125).
Thus, BM-13505 appears to protect against the development of glycerol-
induced acute renal failure.

<u>Nonsteroidal Antiinflammatory Drugs and a Thromboxane A_2-receptor</u>
<u>Antagonist in Systemic Lupus Erythematosus</u>

We have evaluated the effects of ibuprofen treatment for one week in patients with renal disease secondary to SLE and compared these responses with those of patients with other types of chronic glomerular disease (Fig. 5). Our results suggest a functional prevalence of vasodilator PGs over vasoconstrictor TXA_2 since ibuprofen reduced GFR and RBF in both groups. However, although patients with chronic glomerular disease and patients with lupus nephritis showed a similar percentage reduction of both creatinine (25% vs. 26%) and para-aminohyppurate (35% vs. 34%) clearances, a significant difference became apparent when individual changes of GFR and RPF were calculated per unit of change in urinary 6-keto-$PGF_{1\alpha}$ and PGE_2 excretion (9). Thus, the average decrease in both clearances was approximately 50% lower in SLE patients than in patients with chronic glomerular disease, as a function of the reduction of urinary 6-keto-$PGF_{1\alpha}$ excretion and urinary PGE_2 excretion (9). We believe that in SLE patients the simultaneous suppression of enhanced intrarenal TXA_2 synthesis as well as inhibition of PGE_2 and PGI_2 partially attenuates the functional consequences of cyclo-oxygenase inhibition, as compared with other forms of chronic glomerular disease in which TXA_2 is produced normally (9).

Recently the effects of three orally active TXA_2-synthase inhibitors, (dazoxiben [126], OKY-046 [127] and UK-38485 [128]) have been examined in healthy subjects. Despite the variable biological half-lives of platelet

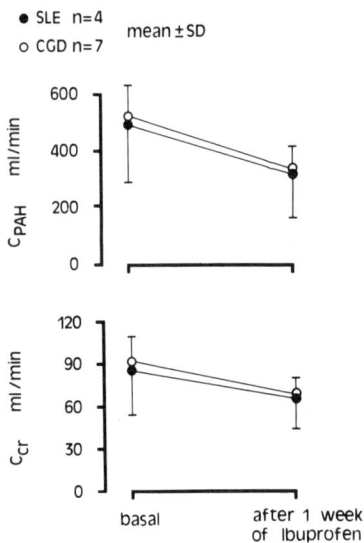

Fig. 5. Short-term effects of ibuprofen (1200 mg/day) on para-aminohyppurate (PAH) and creatinine (Cr) clearances in chronic glomerular disease (CGD) and systemic lupus erythematosus (SLE).

TXA$_2$-synthase inhibitors (4 to over 8 hours), the drugs caused only a marginal reduction of urinary TXB$_2$ excretion, approximately 30%, at doses maximally inhibiting platelet TXB$_2$ production. Moreover, oral dosing with OKY-046 (600 mg/day for 3 days) was not associated with any statistically significant changes of GFR and RPF in 4 healthy subjects (unpublished observations). Similar results were also obtained in 8 patients with the hepatorenal syndrome, a condition characterized by enhanced renal TXA$_2$ production (129). No consistent improvement of renal function occurred in these patients despite a 54% reduction of urinary TXB$_2$ excretion after dazoxiben administration (129). On the basis of these results, we decided to examine the acute effects of a selective and competitive antagonist of TXA$_2$ receptors (130) to assess the pathophysiologic role of enhanced intrarenal TXA$_2$ production in lupus nephritis.

BM-13177, a sulphonamide derivative (Fig. 6), has recently been shown to antagonize TXA$_2$-induced platelet aggregation and vascular contraction in vitro (131-134) and has been tested in healthy volunteers (135,136). This drug is eliminated predominantly by the kidney and, as its renal clearance is greater than that of creatinine, it must be eliminated through tubular secretion (136). After oral administration, only 52% of BM-13177 was found unchanged in the urine, whereas after intravenous dosing, the fraction of unchanged drug in the urine was approximately 90% (136). Because of its short half-life in human circulation (136), we decided to test its acute effects when given by continuous IV infusion. We studied 7 patients with proven lupus nephritis in a double-blind, placebo-controlled, cross-over study with the TXA$_2$-antagonist BM-13177. All patients satisfied the

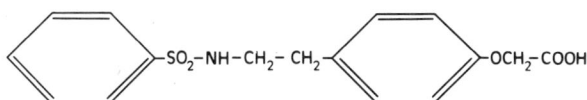

CHEMICAL NAME

4-[2-(benzene-sulfonylamino)ethyl]-phenoxyacetic acid

MOLECULAR WEIGHT

335.38

Fig. 6. BM 13177 Structural Formula.

following criteria for inclusion: (a) histologic evidence of diffuse proliferative lupus nephritis, (b) a GFR of 40 to 100 ml/min., (c) age over 18
years, (d) informed consent. The investigation was subdivided into four
phases. Following a basal (phase 0) period, the randomized study began with
a continuous infusion of the drug at a dose of 187.5 mg/h or placebo for two
days (phase 1). This was followed by a three-day wash-out (phase 2). In
phase 3 placebo or BM-13177 was infused again for 2 days. Inulin and para-
aminohippurate clearances and bleeding time were assessed on days 1, 3 and 8
of the study. Urinary sodium excretion and blood pressure were measured
every day. Inulin clearance increased from 71 ± 15 to 89 ± 15 ml/min. during TxA_2 blockade and returned to 69 ± 16 ml/min. during placebo infusion
($p < 0.001$ for both changes). PAH clearance increased from 335 ± 49 to 427
± 52 after BM-13177 and returned to 330 ± 50 ml/min. after placebo ($p <$
0.05). Bleeding time values were significantly increased from 40% to 100%
during BM-13177 infusion. By contrast, no statistically significant changes
were observed in blood pressure and body weight. These results suggest that
the increased TXA_2 synthesis, due either to resident renal cells or to
cells infiltrating the kidney, has pathophysiological effects on RPF and GFR
in patients with diffuse proliferative lupus nephritis. It is likely that
when the effects of TXA_2 are inhibited by the TXA_2-receptor antagonist,
arteriolar dilation and mesangial relaxation both occur (42,137). Consequently, RBF and GFR are hemodynamically controlled by a balance between
constrictor and dilator factors (42).

CONCLUSIONS

We have reviewed the pattern of urinary PGs and TXB_2 as a reflection
of intrarenal synthesis in human renal disease. There is abundant evidence
that vasodilatory prostaglandins play an important regulatory role in these
settings. Moreover, we have discussed our contention that intrarenal PGI_2
is a major determinant of glomerular hemodynamics not only in human glomerulonephritis but also after unilateral nephrectomy. In contrast to other
forms of chronic glomerular disease, lupus nephritis is characterized by enhanced intrarenal TXA_2 production and reduced intrarenal synthesis of
PGI_2. Inhibition of PG synthesis after unilateral nephrectomy and chronic
glomerular disease--including lupus nephritis--by the administration of nonselective cyclooxygenase inhibitors, removes the PGI_2-mediated vasodilatory effect and thereby accentuates vasoconstriction. The nephrotoxic potential of nonselective cyclooxygenase inhibitors, in terms of both consistent
inital reduction in renal function and the unknown risk of permanent renal
impairment, argues against an indiscriminate long-term use of these agents

for the prevention of nonimmunologic mechanisms of progressive glomerular damage in various chronic glomerular diseases. The use of a selective and competitive TXA$_2$-receptor antagonist allowed us to evaluate the pathophysiologic role of enhanced intrarenal TXA$_2$ production by unidentified cellular sources in lupus nephritis. However, the potential benefits of TXA$_2$-receptor antagonist should be explored in prospective controlled trials to clarify the contribution of intrarenal TXA$_2$ production in the development of renal damage.

BIBLIOGRAPHY

1. M.J. Dunn, Renal prostaglandins, in: "Renal Endocrinology," Williams and Wilkins, eds., Baltimore., 1-74 (1983).

2. D. Schlondorff and R. Ardaillou, Prostaglandins and other arachidonic acid metabolites in the kidney, Kidney Int. 29:108-119 (1986).

3. D. Schlondorff, Renal prostaglandin synthesis, Am. J. Med. 81 (suppl. 2B):1-11 (1986).

4. M.J. Dunn, Renal prostaglandins, in: "Contemporary Nephrol., Vol. 4," S. Klahr and S.G. Massry, eds., Plenum, New York, pp. 133-194 (1987).

5. J.C. Frolich, F.W. Wilson, B.J. Sweetman, M. Smigel, A.S. Nies, K. Carr, J.T. Waison, J.A. Oates, Urinary prostaglandins: identification and origin, J. Clin. Invest. 55:763-770 (1975).

6. R.J. Gryglewski, S. Bunting, S. Moncada, R.J. Flower, J.R. Vane, Arterial walls are protected against deposition of platelet thrombi by a substance (prostaglandin X) which they make from prostaglandin endoperoxides, Prostaglandins 12:685-713 (1976).

7. G. Ciabattoni, F. Pugliese, G.A. Cinotti, G. Stirati, R. Ronci, G. Castrucci, A. Pierucci, C. Patrono, Characterization of furosemide-induced activation of the renal prostaglandin system, Eur. J. Pharmacol. 60:181-187 (1979).

8. C. Patrono, F. Pugliese, G. Ciabattoni, P. Patrignani, A. Maseri, S. Chierchia, B.A. Peskar, G. A. Cinotti, B.M. Simonetti, A. Pierucci, Evidence for a direct stimulatory effect of prostacyclin on renin release in man, J. Clin. Invest. 69:231-239 (1982).

9. C. Patrono, G. Ciabattoni, G. Remuzzi, E. Gotti, S. Bombardieri, O. Di Munno, G. Tartarelli, G.A. Cinotti, B.M. Simonetti, A. Pierucci, Functional significance of renal prostacyclin and thromboxane A$_2$ production in patients with systemic lupus erythematosus, J. Clin. Invest. 76:1011-1018 (1985).

10. J.E. Stork, M.A. Rahman, M.J. Dunn, Eicosanoids in experimental and human renal disease, Am. J. Med. 80 (suppl.1A):34-45 (1986).

11. J.R. Sedor, E.W. Davidson, M.J. Dunn, Effects of nonsteroidal antiinflam-
    matory drugs in healthy subjects, Am. J. Med. 81 (suppl.2B): 58-70
    (1986).

12. M.J. Dunn, Nonsteroidal antiinflammatory drugs and renal function, Ann.
    Rev. Med. 35:411-428 (1984).

13. S.H. Ferreira, S. Moncada, J.R. Vane, Indomethacin and aspirin abolish
    prostaglandin release from the spleen, Nature 231:237-239 (1971).

14. A.J.M. Donker, L. Arisz, J.R.H. Brentjens, G.K. van der Hem, H.J.G.
    Hollemans, The effect of indomethacin on kidney function and plasma
    renin activity in man, Nephron 17:288-296 (1976).

15. A. Fieschi, B. Bianchi, II fenilbutazone. Un farmaco attivo sulla albu-
    minuria, Progr. Med. 9:257-263 (1955).

16. L.A. de Vries, S.P. ten Holt, J.J. van Daatselaar, A. Mulder, J.G.G.
    Borst, Characteristic renal excretion patterns in response to physio-
    logical, pathological and pharmacological stimuli, Clin. Chim. Acta
    5:915-937 (1960).

17. R. Vriesendorp, A.J.M. Donker, D. de Zeeuw, P.E. de Jong, G.K. van der
    Hem, Effects of nonsteroidal anti-inflammatory drugs on proteinuria,
    Am. J. Med. 81 (suppl.2B):84-94 (1986).

18. G.A. FitzGerald, A.K. Pedersen, C. Patrono, Analysis of prostacyclin
    and thromboxane biosynthesis in cardiovascular disease, Circulation
    67:1174-1177 (1983).

19. C. Patrono, A. Wennmaln, G. Ciabattoni, J. Nowak, F. Pugliese, G.A.
    Cinotti, Evidence for an extra-renal origin of urinary prostaglandin
    E_2 in healthy men, Prostaglandins 18:623-629 (1979).

20. F. Catella, J. Novak, G.A. FitzGerald, Measurement of renal and non-
    renal eicosanoid synthesis, Am. J. Med. 81 (Suppl.2B):23-29 (1986).

21. C. Patrono and A. Pierucci, Renal effects of nonsteroidal anti-inflamma-
    tory drugs in chronic glomerular disease, Am. J. Med. 81 (suppl.2B):
    71-83 (1986).

22. R.P. Kimberly, J.R. Gill, Jr., R.E. Browden, H.R. Keiser, P.H. Plotz,
    Elevated urinary prostaglandins and the effects of aspirin on renal
    function in lupus erythematosus, Ann. Intern. Med. 89:336-341 (1978).

23. M. Blum, S. Bauminger, A. Algueti, E. Kisch, D. Ayalon, A. Aviram, Uri-
    nary prostaglandin-E_2 in chronic renal disease, Clin. Nephrol. 15:
    87-89 (1981).

24. L. Ruilope, C. Bernis, R. Garcia-Robles, J. Alcazar, A. Barrientos,
    J.A.F. Tresguerres, J.L. Rodicio, Urinary prostaglandin E_2 in
    chronic renal disease, Clin. Nephrol. 16:215 (letter) (1981).

25. L. Ruilope, R.G. Robles, C. Bernis, A. Barrientos, J. Alcazar, J.A.F.
    Tresguerres, J. Sancho, J.L. Rodicio, Role of renal prostaglandin

E_2 in chronic renal disease and hypertension, <u>Nephron</u> 32:202-206 (1982).

26. S. Mordechai, M. Rathaus, J. Shapira, J. Bernheim, Urinary prostaglandins E_2 and $F_{2\alpha}$ in chronic renal failure, <u>Nephron</u> 40:152-154 (1985).

27. M. Lebel and J.H. Grose, Abnormal renal prostaglandin production during the evolution of chronic nephropathy, <u>Am. J. Nephrol.</u> 6:96-100 (1986).

28. C.D. Mistry, C.J. Lote, R. Gokal, W.J.C. Currie, M. Vandenburg, N.P. Mallick, Effects of sulindac on renal function and prostaglandin synthesis in patients with moderate chronic renal insufficiency, <u>Clin. Sci.</u> 70:501-505 (1986).

29. G. Ciabattoni, G.A. Cinotti, A. Pierucci, B.M. Simonetti, F. Pugliese, P. Barsotti, G. Pecci, F. Taggi, C. Patrono, Effects of sulindac and ibuprofen in patients with chronic glomerular disease, <u>N. Engl. J. Med.</u> 310:279-283 (1984).

30. V.D. D'Agati, G.B. Appek, D. Estes, D.M. Knowles II, C.L. Pirani, Monoclonal antibody identification of infiltrating mononuclear leukocytes in lupus nephritis, <u>Kidney Int.</u> 30:573-581 (1986).

31. J.L. Nadler, O.L. Frederick, W. Hsueh, R. Horton, Evidence of prostacyclin deficiency in the syndrome of hyporeninemic hypoaldosteronism, <u>N. Engl. J. Med.</u> 314:1015-1020 (1986).

32. J. Sraer, N. Ardaillou, J.D. Sraer, R. Ardaillou, In vitro prostaglandin synthesis by human glomeruli and papillae, <u>Prostaglandins</u> 23:855-864 (1982).

33. B.M. Brenner, T.W. Meyer, T.H. Hostetter, Dietary protein intake and the progressive nature of kidney disease: the role of hemodynamically mediated glomerular injury in the pathogenesis of progressive glomerular sclerosis in aging, renal ablation, and intrinsic renal disease, <u>N. Engl. J. Med.</u> 307:652-659 (1982).

34. T.W. Meyer and B.M. Brenner, The contribution of glomerular hemodynamic alterations to progressive renal disease, <u>in</u>: "The progressive nature of renal disease, Contemporary Issues in Nephrology," Vol. 14., B.M. Brenner and J.H. Stein, eds., Churchill Livingstone, New York, 1-16 (1986).

35. T. Okegawa, P.E. Jonas, K. De Schryver, A. Kawasaki, P. Needleman, Metabolic and cellular alterations underlying the exaggerated renal prostaglandin and thromboxane synthesis in ureter obstruction in rabbits, <u>J. Clin. Invest.</u> 71:81-90 (1983).

36. M.L. Purkerson, J.H. Joist, J. Yates, A. Valdes, A. Morrison, S. Klahr, Inhibition of thromboxane synthesis ameliorates the progressive

kidney disease of rats with subtotal renal ablation, <u>Proc. Natl. Acad. Sci. USA</u> 82:193-197 (1985).

37. V.E. Kelley, S. Sneve, S. Musinski, Increased renal thromboxane production in murine lupus nephritis, <u>J. Clin. Invest.</u> 77:252-259 (1986).

38. E.A. Lianos, C.A. Andres, M.J. Dunn, Glomerular prostaglandin and thromboxane synthesis in rat nephrotic serum nephritis, <u>J. Clin. Invest.</u> 72:1439-1448 (1983).

39. G. Remuzzi, L. Imberti, M. Rossini, C. Morelli, C. Carminati, G.M. Cattaneo, T. Betani, Increased glomerular thromboxane synthesis as a possible cause of proteinuria in experimental nephrosis, <u>J. Clin. Invest.</u> 75:94-101 (1985).

40. K.F. Badr, V.E. Kelley, H.G. Rennke, B.M. Brenner, Roles for thromboxane A_2 and leukotrienes in endotoxin-induced acute renal failure, <u>Kidney Int.</u> 30:474-480 (1986).

41. C. Zoja, A. Benigni, P. Verroust, P. Ronco, T. Bertani, G. Remuzzi, Indomethacin reduces proteinuria in passive Heymann nephritis in rats, <u>Kidney Int.</u> 31:1335-1343 (1987).

42. L.A. Scharschmidt, E. Lianos, M.J. Dunn, Arachidonate metabolites and the control of glomerular function, <u>Fed. Proc.</u> 42:3058-3063 (1983).

43. A.S. Petrulis, G. Aikawa, M.J. Dunn, Prostaglandin and thromboxane synthesis by rat glomerular epithelial cells, <u>Kidney Int.</u> 20:469-474 (1981).

44. J. Sraer, J. Foidart, D. Chansel, P. Mahieu, R. Ardaillou, Prostaglandin synthesis by rat isolated glomeruli and glomerular cultured cells, <u>Int. J. Biochem.</u> 12:203-207 (1980).

45. J.M. Goldstein, C.L. Malmsten, H. Kindahl, H.P. Kaplan, D. Radmark, B. Samuelsson, G. Weissman, Thromboxane generation by human peripheral blood polymorphonuclear leukocytes, <u>J. Exp. Med.</u> 148:787-792 (1978).

46. K. Brune, M. Galah, H. Kahn, Pharmacological control of prostaglandin and thromboxane release from macrophages, <u>Nature</u> 274:261-263 (1978).

47. P. Needleman, S. Moncada, S. Bunting, J.R. Vane, M. Hamberg, B. Sammuelsson, Identification of an enzyme in platelet microsomes which generates thromboxane A_2 from prostaglandin endoperoxides, <u>Nature</u> 261:558-560 (1976).

48. R.A.K. Stahl, S. Adler, P.J. Baker, Y.P. Chen, P.M. Pritzl, W.G. Couser, Enhanced glomerular prostaglandin formation in experimental membranous nephropathy, <u>Kidney Int.</u> 31:1126-1131 (1987).

49. M.A. Rahman, S.N. Emancipator, M.J. Dunn, Immune complex effects on glomerular eicosanoid production and renal hemodynamics, <u>Kidney Int.</u> 31: 1317-1326 (1987).

50. F. Ferrario, A. Castiglione, G. Coalsanti, G. Barbiano di Belgioioso, S. Bertoli, G. D'Amico, The detection of monocytes in human glomerulonephritis, Kidney Int. 28:513-519 (1985).

51. R.C. Atkins, S.R. Holdworth, E.F. Glasgow, F.C. Matthews, The macrophage in human progressive glomerulonephritis, Lancet i:830-832 (1976).

52. G. Monga, G. Mazzucco, G. Barbiano di Belgioioso, G. Busnack, The presence and possible role of monocyte infiltration in human chronic proliferative glomerulonephritis, Am. J. Pathol. 94:271-284 (1979).

53. A.B. Magil, L.D. Wadsworth, M. Loewen, Monocytes and human renal glomerular disease. A quantitative evaluation, Lab. Invest. 44:27-33 (1981).

54. R.T. McCluskey and A.K. Bhan, Cell-mediated immunity in renal disease, Human Pathol. 17:146-153 (1986).

55. J.B. Lefkowith, T. Okegawa, K. De Schryver-Kecskemeti, P. Needleman, Macrophage-dependent arachidonate metabolism in hydronephrosis, Kidney Int. 26:10-17 (1984).

56. J.S. Cameron, Platelets in glomerular disease, Ann. Rev. Med. 35:175-180 (1984).

57. P.E. Di Corleto and D.F. Bowen-Pope, Cultural endothelial cells produce a platelet-derived growth factor-like protein, Proc. Natl. Acad. Sci. USA 80:1919-1923 (1983).

58. H.E. Abboud, E. Poptic, P. Di Corleto, Production of platelet-derived growth factor-like protein by rat mesangial cells in culture, J. Clin. Invest. 80:675-683 (1987).

59. K.S. Kant, V.E. Pollak, M.A. Weiss, H.I. Glueck, M.A. Miller, E.V. Hess, Glomerular thrombosis in systemic lupus erythematosus: prevalence and significance, Medicine (Baltimore) 60:71-86 (1981).

60. J.R. Gill, J.C. Frolich, R.E. Bowden, A.A. Taylor, H.R. Keiser, H.W. Seyberth, J.A. Oates, F.C. Bartter, Bartter's syndrome. A disorder characterized by high urinary prostaglandins and a dependence of hyperreninemia on prostaglandin synthesis, Am. J. Med. 61:43-51 (1976).

61. R.D. Zipser, J.C. Hoefs, P.F. Speckart, P.K. Zia, R. Horton, Prostaglandins: modulators of renal function and pressor resistance in chronic liver disease, J. Clin. Endocr. Metab. 48:895-900 (1979).

62. A. Salvetti, R. Pedrinelli, A. Magagna, P. Ugenti, Differential effects of selective and non-selective prostaglandin-synthesis inhibition of the pharmacological responses to captopril in patients with essential hypertension, Clin. Sci. 63:261s-263s (1982).

63. J.R. Sedor, S.L. Williams, A.N. Chremos, C.L. Johnson, M.J. Dunn, Effects of sulindac and indomethacin on renal prostaglandin synthesis, Clin. Pharmacol. Ther. 36:85-91 (1984).

64. I.B. Puddey, L.J. Beilin, R. Vandongen, R. Banks, I. Rouse, Differential

effects of sulindac and indomethacin on blood pressure in treated essential hypertensive subjects, Clin. Sci. 69:327-336 (1985).

65. B. Trimarco, A. DeSimone, A. Cuocolo, B. Ricciarelli, M. Volpe, P. Patrignani, L. Sacca, M. Condorelli, Role of prostaglandins in the renal handling of a salt load in essential hypertension, Am. J. Cardiol. 55:116-121 (1985).

66. L. Ruilope, R.G. Robles, C. Paya, J.M. Alcazar, E. Miravalles, J. Sancho-Rof, J. Rodicio, F.G. Knox, J. C. Romero, Effects of long-term treatment with indomethacin on renal function, Hypertension 8: 677-684 (1986).

67. R.P. Kimberly, R.E. Bowden, H.R. Keiser, P.H. Plotz, Reduction of renal function by newer non-steroidal anti-inflammatory drugs, Am. J. Med. 64:804-807 (1978).

68. D.C. Brater, S. Anderson, B. Baird, W.B. Campbell, Effects of ibuprofen, naproxen, and sulindac on prostaglandin in man, Kidney Int. 27:66-73 (1985).

69. U.G. Svendsen, J. Gerstoff, T.M. Hansen, P. Christensen, I.B. Lorenzen, The renal excretion of prostaglandins and changes in plasma renin during treatment with either sulindac or naproxen in patients with rheumatoid arthritis and thiazide-treated heart failure, J. Rheumatol. 11:779-782 (1984).

70. G. Ciabattoni, G. Bianchi-Porro, I. Caruso, M. Furmagalli, F. Pugliese, C. Patrono, Differential inhibition of prostaglandin and thromboxane synthesis in human tissues by nonsteroidal anti-inflammatory drugs, Clin. Res. 32:462A (Abs.) (1984).

71. D.G. Wong, J.D. Spence, L. Lamki, D. Freeman, J.W.D. McDonald, Effect of nonsteroidal anti-inflammatory drugs on control of hypertension by beta-blockers and diuretics, Lancet i:997-1001 (1986).

72. J. Laurent, D. Belghiti, C. Bruneau, G. Lagrue, Diclofenac, a nonsteroidal anti-inflammatory drug, decreases proteinuria in some glomerular diseases: a controlled study, Am. J. Nephrol. 7:198-202 (1987).

73. F. Pugliese, B.M. Simonetti, G.A. Cinotti, G. Ciabattoni, F. Catella, S. Vastano, A. Ghidini Ottonelli, A. Pierucci, Differential interaction of piroxicam and sulindac with the antihypertensive effect of propranolol, Eur. J. Clin. Invest. 14(2):54 (Abs.) (1984).

74. R. Vriesendorp, D. De Zeeuw, P.E. De Jong, A.J.M. Donker, J.J. Pratt, G.K. van der Hem, Reduction of urinary protein and prostaglandin E_2 excretion in the nephrotic syndrome by nonsteroidal anti-inflammatory drugs, Clin. Nephrol. 25:105-110 (1986).

75. G. Ciabbattoni, F. Pugliese, G.A. Cinotti, C. Patrono, Renal effects of antiinflammatory drugs, Eur. J. Rheumatol. Inflamm. 3:210-221 (1980).

76. K.J. Berg and T. Talseth, Acute renal effects of sulindac and indometha-
    cin in chronic renal failure, Clin. Pharmacol. Ther. 37:447-452
    (1985).

77. D.G. Roberts, J.C. Gerber, J.S. Barnes, G.O. Zerbe, A.S. Nies, Sulindac
    is not renal-sparing in man, Clin. Pharmacol. Ther. 38:258-265
    (1985).

78. G. Laffi, G. Daskalopoulos, I. Kromborg, W. Hsueh, P. Gentilini, R.D.
    Zipser, Effects of sulindac and ibuprofen in patients with cirrhosis
    and ascites, Gastroenterology 90:182-187 (1986).

79. E. Quintero, P. Gine's, V. Arroya, A. Rimola, J. Camps, J. Gaya, A.
    Guevara, M. Rodamilans, J. Rode's, Sulindac reduces the urinary excre-
    tion of prostaglandins and impairs renal function in cirrhosis with
    ascites, Nephron 42:298-303 (1986).

80. G. Ciabattoni, A.H. Boss, P. Patrignani, F. Catella, B.M. Simonetti, A.
    Pierucci, F. Pugliese, P. Filabozzi, C. Patrono, Effects of sulindac
    on renal and extrarenal eicosanoid synthesis, Clin. Pharmacol. Ther.
    41:380-383 (1987).

81. P. Patrignani, P. Filabozzi, C. Patrono, Selective cumulative inhibition
    of platelet thromboxane production by low-dose aspirin in healthy sub-
    jects, J. Clin. Invest. 69:1366-1372 (1982).

82. A.K. Pedersen and G.A. FitzGerald, Dose-related kinetics of aspirin, N.
    Engl. J. Med. 311:1206-1211 (1984).

83. C. Patrono, G. Ciabattoni, P. Patrignani, F. Pugliese, P. Filabozzi, F.
    Catella, G. Davi, L. Forni, Clinical pharmacology of platelet cyclo-
    oxygenase inhibition, Circulation 72:1177-1184 (1985).

84. C. Patrono and P. Patrignani, Clinical pharmacology of acetylsalicylic
    acid as an antiplatelet agent, Atherosclerosis Reviews, R.J. Hegyeli,
    ed., Raven Press, New York, 12:51-61 (1984).

85. C. Patrono and M.J. Dunn, The clinical significance of inhibition of
    renal prostaglandin synthesis, Kidney Int. 32:1-12 (1987).

86. J.E. Stork and M.J. Dunn, Hemodynamic roles of thromboxane A_2 and pros-
    taglandin E_2 in glomerulonephritis, J. Pharmacol. Exp. Ther. 233:
    672-678 (1984).

87. J. Sraer, J.D. Sraer, D. Chansel, F. Russo-Marie, B. Kouznetzova, R.
    Ardaillou, Prostaglandin synthesis by isolated rat renal glomeruli,
    Mol. Cell. Endocrinol. 16:29-37 (1979).

88. K. Kaizu, D. Marsh, R. Zipser, R.J. Glassock, Role of prostaglandins and
    angiotensin II in experimental glomerulonephritis, Kidney Int. 28:
    629-635 (1985).

89. M.J. Dunn and L.A. Scharschmidt, Prostaglandins modulate the glomerular
    actions of angiotensin II, Kidney Int. 31(Suppl. 20):S95-S101 (1987).

90. T. Bertani, A. Benigni, F. Cutillo, G. Rocchi, C. Morelli, C. Carminati, P. Verroust, G. Remuzzi, Effect of aspirin and sulindac in rabbit nephrotoxic nephritis, J. Lab. Clin. Med. 107:261-268 (1986).

91. V.E. Kelley, S. Izvi, P.V. Halushka, Effect of ibuprofen, a fatty acid cyclooxygenase inhibitor, on murine lupus, Clin. Immunol. Immunopathol. 25:223-231 (1982).

92. L. Arisz, A.J.M. Donker, J.R.H. Brenjens, G.K. van der Hem, The effect of indomethacin on proteinuria and kidney function in the nephrotic syndrome, Acta. Med. Scand. 199:121-125 (1976).

93. K.J. Berg, Acute effects of acetylsalicylic acid in patients with chronic renal insufficiency, Europ. J. Clin. Pharmacol. 11:111-116 (1977).

94. R.P. Kimberly and P.H. Plotz, Aspirin-induced depression of renal function, N. Engl. J. Med. 296:418-424 (1977).

95. A.J.M. Donker, J.R.H. Brentjens, L. Arisz, G.K. van der Hem, Treatment of the nephrotic syndrome with indomethacin, Nephron 22:374-381 (1978).

96. R.G.W.L. Tiggeler, B. Hulme, P.G.A.B. Wijdeveld, Effect of indomethacin on glomerular permeability in the nephrotic syndrome, Kidney Int. 16: 312-321 (1979).

97. R. Vriesendorp, A.J.M. Donker, D. De Zeeuw, P.E. De Jong, G.K. van der Hem, Antiproteinuric effect of naproxen and indomethacin, Am. J. Nephrol. 5:236-242 (1985).

98. R.D. Toto, S.A. Anderson, D. Brown-Cartwright, J.P. Kokko, D.C. Brater, Effects of acute and chronic dosing of NSAIDs in patients with renal insufficiency, Kidney int. 30:760-768 (1986).

99. L. Scharschmidt, M. Simonson and M.J. Dunn, Glomerular prostaglandins, angiotensin II, and nonsteroidal anti-inflammatory drugs, Am. J. Med. 81 (Suppl.2B):30-42 (1986).

100. V.E. Torres, Present and future of the nonsteroidal anti-inflammatory drugs in nephrology, Mayo Clin. Proc. 57:389-393 (1982).

101. T.H. Hostetter, J.L. Olson, H.G. Rennke, M.A. Venkatachalam, B.M. Brenner, Hyperfiltration in remnant nephrons: a potentially adverse response to renal ablation, Am. J. Physiol. 241:F85-F93 (1981).

102. L. Fine, The biology of renal hypertrophy, Kidney Int. 29:619-634 (1986).

103. M.L. Purkerson, P.E. Hoffsten, S. Klahr, Pathogenesis of the glomerulopathy associated with renal infarction in rats, Kidney int. 9:407-417 (1976).

104. B. Hahne, G. Selen, A. Erik, G. Persson, Indomethacin inhibits renal functional adaptation to nephron loss, Renal Physio., Basel, 7:13-21 (1984).

105. R.A.K. Stahl, S. Kudelka, M. Paravincini, P. Schollmeyer, Prostaglandin and thromboxane formation in glomeruli from rats with reduced renal mass, Nephron 42:252-257 (1986).

106. K.A. Nath, D.H. Chmielewski, T.H. Hostetter, Regulatory role of prosta-noids in glomerular microcirculation of remnant nephrons, Am. J. Physiol. 252:829-837 (1987).

107. M.A. Kirschenbaun and E.R. Serros, Effect of prostaglandin inhibition on glomerular filtration rate in normal and uremic rabbits, Prostaglandins 22:245-254 (1981).

108. D. Rubinger, Y. Frishberg, A. Eldor, M.M. Popovtzer, The effect of sup-pression of prostaglandin synthesis on renal function in rats with intact and reduced renal mass, Prostaglandins 30:651-668 (1985).

109. S. Klahr, M. Heifets, M.B. Purkerson, The influence of anticoagulation on the progression of experimental renal disease, in: "The progres-sive nature of renal disease. Contemporary Issues in Nephrology, Vol. 14, B.M. Brenner and J.H. Stein, eds., Churchill Livingstone, New York, pp. 45-64 (1986).

110. L.A. Scharschmidt, N.B. Gibbons, L. McGarry, P. Berger, M. Axelrod, R. Janis, Y.H. Ko, Effects of dietary fish oil on renal insufficiency in rats with subtotal nephrectomy, Kidney Int. 32:700-709 (1987).

111. V.E. Kelley, A. Ferretti, S. Izvi, T.B. Strom, A fish oil diet rich in eicosapentaenoic acid reduces cyclooxygenase metabolites, and suppres-ses lupus in MRL-lpr mice, J. Immunol. 134:1914-1919 (1985).

112. U.O. Barcelli and V.E. Pollak, Prostaglandins and progressive renal insufficiency, in: "The progressive nature of renal disease, Contemporary Issues in Nephrology," Vol. 14, B.M. Brenner and J.H. Stein, eds., Churchill Livingstone, New York, pp. 65-80 (1986).

113. J.L. Logan, S.M. Lee, B. Benson, U.F. Michael, Inhibition of compensa-tory renal growth by indomethacin, Prostaglandins 31:253-261 (1986).

114. A.J. Donker, The effect of indomethacin on renal function and glomeru-lar protein loss, in: "Prostaglandins and the Kidney," M.J. Dunn, C. Patrono, G.A. Cinotti, eds., Plenum, New York, pp. 251-262 (1983).

115. B.M. Simonetti, R. Tersigni, G. Ciabattoni, I. Cardamone, A. Messina, G. Stirati, A. Pierucci, Renal hyperfiltration in kidney donors: its dependence on renal prostacyclin, in: "Adv. Prostaglandin, Thromboxane, Leukotriene Res.," Vol. 17, B. Samuelsonn, R. Paoletti, P.W. Ramwell, eds., Raven Press, New York, pp. 757-760 (1987).

116. A. Pierucci, R. Tersigni, I. Cardamone, G. Stirati, G. Ciabattoni, B.M. Simonetti, Renal hyperfiltration in kidney donors: its depen-dence on renal prostacyclin, Xth Int. Congr. Nephrol., London, Abstracts Book, p. 513 (1987).

117. A.R. Morrison, A. Nishikawa, P. Needleman, Thromboxane A_2 biosynthesis in the ureter-obstructed isolated perfused kidney of the rabbit, J. Pharmacol. Ther. 205:1-8 (1978).

118. A. Kawasaki and P. Needleman, Contribution of thromboxane to renal resistance changes in the isolated perfused hydronephrotic rabbit kidney, Circ. Res. 50:486-490 (1982).

119. W.E. Yarger, D.D. Schocken, R.H. Harris, Obstructive nephropathy in the rat: possible roles for the renin-angiotensin system, prostaglandins and thromboxane in postobstructive renal function, J. Clin. Invest. 65:400-412 (1980).

120. J.C. Strand, B.S. Edwards, M.E. Anderson, J.C. Romero, F.G. Knox, Effect of imidazole on renal function in unilateral ureteral-obstructed rat kidneys, Am. J. Physiol. 240:F508-F514 (1981).

121. P.E. Klotman, S.R. Smith, B.D. Volpp, T.M. Coffman, W.E. Yarger, Thromboxane synthetase inhibition improves function of hydronephrotic rat kidneys, Am. J. Physiol. 250:F282-F287 (1986).

122. T.M. Coffman, W.E. Yarger, P.E. Klotman, Functional role of thromboxane production by acutely rejecting renal allografts in rats, J. Clin. Invest. 75:1242-1248 (1985).

123. J.E. Benabe, S. Klahr, M.K. Hoffman, A.R. Morrison, Production of thromboxane A_2 by the kidney in glycerol-induced acute renal failure in the rabbit, Prostaglandins 19:333-347 (1980)

124. J.D. Sraer, L. Moulonguet-Doleris, F. Delarve, J. Sraer, R. Ardaillou, Prostaglandin synthesis by glomeruli isolated from rats with glycerol-induced acute renal failure, Cir. Res. 49:775-783 (1981).

125. K. Stegmeier, F. Hartig, J. Pill, H. Patscheke, Prevention of glycerol-induced acute renal failure in rabbits by the thromboxane-antagonist BM-13505, Proceedings of the European Dialysis and Transplant Association, A.M. Davison and P.J. Guillou, eds., B. Tindall, 22: 1012-1016 (1985).

126. P. Patrignani, P. Filabozzi, F. Catella, F. Pugliese, C. Patrono, Differential effects of dazoxiben, a selective thromboxane-synthase inhibitor, on platelet and renal prostaglandin-endoperoxide metabolism, J. Pharmacol. Exp. Ther. 288:472-477 (1984).

127. P. Patrignani, F. Catella, P. Filabozzi, F. Pugliese, A. Pierucci, B.M. Simonetti, L. Forni, M. Segni, C. Patrono, Differential effects of OKY-046, a selective thromboxane-synthase inhibitor, on platelet and renal prostaglandin endoperoxide metabolism, Clin. Res. (Abstr.) 32:246A (1984).

128. C. Patrono, G. Ciabattoni, P. Filabozzi, F. Catella, L. Forni, M. Segni, P. Patrignani, F. Pugliese, B.M. Simonetti, A. Pierucci,

Drugs, prostaglandins, and renal function, "Advances in Prostaglandin, Thromboxane and Leukotriene Research," G.G. Neri Serneri et al., eds., Raven Press, New York, 13:131-139 (1985).

129. R.D. Zipser, I. Kronborg, G. Radvan, T. Reynolds, G. Daskalopoulos, Therapeutic trial of thromboxane synthesis inhibition in the hepato-renal syndrome, Gastroenterology 87:1228-1232 (1984).

130. G. Ciabattoni, A. Pierucci, B.M. Simonetti, G. Pecci, G. Mavrikakis, S. Feriozzi, G.A. Cinotti, C. Patrono, Acute effects of a thromboxane receptor antagonist on renal function in patients with lupus nephritis, (Abstr.) Clin. Res. 35:544A (1987).

131. B. Muller-Beckmann, K. Stegmeier, W. Schmitt, H. Patscheke, Evidence for competitive antagonism of BM-13177 to U-46619 (thromboxane-mimetic) induced constriction of rabbit aorta, Ircs Med. Sci. 11: 1100-1101 (1983).

132. H. Patscheke and K. Stegmeier, BM-13177 is a selective antagonist of prostaglandin H$_2$ and thromboxane in human platelets, Ircs Med. Sci. 12:9-10 (1984).

133. H. Patscheke and K. Stegmeier, Investigations on a selective non-prostanoic thromboxane antagonist, BM-13177, in human platelets, Thromb. Res. 33:277-288 (1984).

134. H. Patscheke, K. Stegmeier, B. Muller-Beckmann, G. Sponer, C. Staiger, G. Neugebauer, Inhibitory effects of the selective thromboxane receptor antagonist BM-13177 on platelet aggregation, vasoconstriction and sudden death, Biomed. Biochim. Acta. 43:312-318 (1984).

135. P. Gresele, J. Arnout, W. Janssens, H. Deckmyin, J. Lemmens, J. Vermylen, BM-13177, A selective blocker of platelet and vessel wall thromboxane receptors is active in man, Lancet i (8384):991-994 (1984).

136. H. Patscheke, C. Staiger, G. Neugebauer, B. Kaufmann, K. Strein, R. Endele, K. Stegmeier, The pharmacokinetic and pharmacodynamic profiles of the thromboxane A$_2$ receptor blocker BM-13177, Clin. Pharmacol. Ther. 39:145-150, 1986.

137. P. Mene' and M.J. Dunn, Contractile effects of TXA$_2$ and endoperoxide analogues on cultured rat glomerular mesangial cells, Am. J. Physiol. 251:F1029-F1035 (1986).